T0324837

Negative Binomial Regression

Second Edition

This second edition of *Negative Binomial Regression* provides a comprehensive discussion of count models and the problem of overdispersion, focusing attention on the many varieties of negative binomal regression. A substantial enhancement from the first edition, the text provides the theoretical background as well as fully worked out examples using Stata and R for most every model having commercial and R software support. Examples using SAS and LIMDEP are given as well. This new edition is an ideal handbook for any researcher needing advice on the selection, construction, interpretation, and comparative evaluation of count models in general, and of negative binomial models in particular.

Following an overview of the nature of risk and risk ratio and the nature of the estimating algorithms used in the modeling of count data, the book provides an exhaustive analysis of the basic Poisson model, followed by a thorough analysis of the meanings and scope of overdispersion. Simulations and real data using both Stata and R are provided throughout the text in order to clarify the essentials of the models being discussed. The negative binomial distribution and its various parameterizations and models are then examined with the aim of explaining how each type of model addresses extra-dispersion. New to this edition are chapters on dealing with endogeny and latent class models, finite mixture and quantile count models, and a full chapter on Bayesian negative binomial models. This new edition is clearly the most comprehensive applied text on count models available.

JOSEPH M. HILBE is a Solar System Ambassador with NASA's Jet Propulsion Laboratory at the California Institute of Technology, an Adjunct Professor of statistics at Arizona State University, and an Emeritus Professor at the University of Hawaii. Professor Hilbe is an elected Fellow of the American Statistical Association and elected Member of the International Statistical Institute (ISI), for which he is the founding Chair of the ISI astrostatistics committee and Network. He is the author of *Logistic Regression Models*, a leading text on the subject, co-author of *R for Stata Users* (with R. Muenchen), and of both *Generalized Estimating Equations* and *Generalized Linear Models and Extensions* (with J. Hardin).

Negative Binomial Regression

Second Edition

JOSEPH M. HILBE

Jet Propulsion Laboratory,
California Institute of Technology and
Arizona State University

CAMBRIDGE
UNIVERSITY PRESS

University Printing House, Cambridge CB2 8BS, United Kingdom

One Liberty Plaza, 20th Floor, New York, NY 10006, USA

477 Williamstown Road, Port Melbourne, VIC 3207, Australia

4843/24, 2nd Floor, Ansari Road, Daryaganj, Delhi - 110002, India

79 Anson Road, #06-04/06, Singapore 079906

Cambridge University Press is part of the University of Cambridge.

It furthers the University's mission by disseminating knowledge in the pursuit of education, learning and research at the highest international levels of excellence.

www.cambridge.org
Information on this title: www.cambridge.org/9780521198158

© J. M. Hilbe 2007, 2011

This publication is in copyright. Subject to statutory exception and to the provisions of relevant collective licensing agreements, no reproduction of any part may take place without the written permission of Cambridge University Press.

First published 2007
Reprinted with corrections 2008
Second edition 2011
Reprinted with corrections 2012
10th printing 2016

A catalogue record for this publication is available from the British Library

Library of Congress Cataloging in Publication data
Hilbe, Joseph.
Negative binomial regression / Joseph M. Hilbe. – 2nd ed.
p. cm.
Includes bibliographical references and index.
ISBN 978-0-521-19815-8 (hardback)
1. Negative binomial distribution. 2. Poisson algebras. I. Title.
QA161.B5H55 2011
519.2′4 – dc22 2010051121

ISBN 978-0-521-19815-8 Hardback

Additional resources for this publication at http://works.bepress.com/joseph_hilbe/

Cambridge University Press has no responsibility for the persistence or accuracy of URLs for external or third-party internet websites referred to in this publication, and does not guarantee that any content on such websites is, or will remain, accurate or appropriate.

Contents

Preface to the second edition

The aim of this book is to present a detailed, but thoroughly clear and understandable, analysis of the nature and scope of the varieties of negative binomial model that are currently available for use in research. Modeling count data using the standard negative binomial model, termed NB2, has recently become a foremost method of analyzing count response models, yet relatively few researchers or applied statisticians are familiar with the varieties of available negative binomial models, or how best to incorporate them into a research plan.

Note that the Poisson regression model, traditionally considered as the basic count model, is in fact an instance of NB2 – it is an NB2 with a heterogeneity parameter of value 0. We shall discuss the implications of this in the book, as well as other negative binomial models that differ from the NB2. Since Poisson is a variety of the NB2 negative binomial, we may regard the latter as more general and perhaps as even more representative of the majority of count models used in everyday research.

I began writing this second edition of the text in mid-2009, some two years after the first edition of the text was published. Most of the first edition was authored in 2006. In just this short time – from 2006 to 2009/2010 – a number of advancements have been made to the modeling of count data. The advances, however, have not been as much in terms of new theoretical developments, as in the availability of statistical software related to the modeling of counts. Stata commands have now become available for modeling finite mixture models, quantile count models, and a variety of models to accommodate endogenous predictors, e.g. selection models and generalized method of moments. These commands were all authored by users, but, owing to the nature of Stata, the commands can be regarded as part of the *Stata* repertoire of capabilities.

R has substantially expanded its range of count models since 2006 with many new functions added to its resources; e.g. zero-inflated models, truncated, censored, and hurdle models, finite-mixture models, and bivariate count models, etc. Moreover, R functions now exist that allow non-parametric features to be added to the count models being estimated. These can assist in further adjusting for overdispersion identified in the data.

SAS has also enhanced its count modeling capabilities. SAS now provides the ability of estimating zero-inflated count models as well as the NB1 parameterization of the negative binomial. Several macros exist that provide even more modeling opportunities, but at this time they are still under construction. When the first edition of this book was written, only the Poisson and two links of the negative binomial as found in SAS/STAT **GENMOD** were available in SAS.

SPSS has added the **Genlin** procedure to its functions. **Genlin** is the SPSS equivalent of Stata's **glm** command and SAS's **GENMOD** procedure. **Genlin** provides the now standard GLM count models of Poisson and three parameterizations of negative binomial: log, identity, and canonical linked models. At this writing, SPSS supports no other count models. LIMDEP, perhaps one of the most well-respected econometric applications, has more capabilities for modeling count data than the others mentioned above. However, there are models we discuss in this text that are unavailable in LIMDEP.

This new edition is aimed to update the reader with a presentation of these new advances and to address other issues and methodologies regarding the modeling of negative binomial count data that were not discussed in the first edition. The book has been written for the practicing applied statistician who needs to use one or more of these models in their research. The book seeks to explain the types of count models that are appropriate for given data situations, and to help guide the researcher in constructing, interpreting, and fitting the various models. Understanding model assumptions and how to adjust for their violations is a key theme throughout the text.

In the first edition I gave Stata examples for nearly every model discussed in the text. LIMDEP was used for examples discussing sample selection and mixed-effects negative binomial models. Stata code was displayed to allow readers to replicate the many examples used throughout the text. In this second edition I do the same, but also add R code that can be used to emulate Stata output. Nearly all output is from Stata, in part because Stata output is nicely presented and compact. Stata commands also generally come with a variety of post-estimation commands that can be used to easily assess fit. We shall discuss these commands in considerable detail as we progress through the book. However, R users can generally replicate Stata output insofar as possible by pasting the source code available on the book's website into the R script editor and running it. Code is also available in tables within the appropriate area of discussion. R output is given for modeling situations where Stata does not have the associated command. Together the two programming languages provide the researcher with the ability to model almost every count model discussed in the literature.

I should perhaps mention that, although this text focuses on understanding and using the wide variety of available negative binomial models, we also address several other count models. We do this for the purpose of clarifying a corresponding or associated negative binomial model. For example, the Poisson model is examined in considerable detail because, as previously mentioned, it is in fact a special case of the negative binomial model. Distributional violations of the Poisson model are what has generally motivated the creation and implementation of other count models, and of negative binomial models in particular. Therefore, a solid understanding of the Poisson model is essential to the understanding of negative binomial models.

I believe that this book will demonstrate that negative binomial models are core to the modeling of count data. Unfortunately they are poorly understood by many researchers and members of the statistical community. A central aim of writing this text is to help remedy this situation, and to provide the reader with both a conceptual understanding of these models, and practical guidance on the use of software for the appropriate modeling of count data.

New subjects discussed in the second edition

In this edition I present an added examination of the nature and meaning of risk and risk ratio, and how they differ from odds and odds ratios. Using 2×2 and $2\times k$ tables, we define and interpret risk, risk ratio, and relative risk, as well as related standard errors and confidence intervals. We provide two forms of coefficient interpretation, and some detail about how they are related. Also emphasized is how to test for model dispersion. We consider at length the meaning of extra-dispersion, including both under- and overdispersion. For example, a model may be both Poisson overdispersed and negative binomial under-dispersed. The nature of overdispersion and how it can be identified and accommodated is central to our discussion. Also of prime importance is an understanding of the consequences that follow when these assumptions are violated.

I also give additional space to the NB-C, or canonical negative binomial. NB-C is unlike all other parameterizations of the negative binomial, but is the only one that directly derives from the negative binomial probability mass function, or PMF. It will be discovered that certain types of data are better modeled using NB-C; guidelines are provided on its applicability in research.

In the first edition I provided the reader with Stata code to create Poisson, NB2 and NB1 synthetic count models. We now examine the nature of these types of synthetic models and describe how they can be used to understand the relationship of models to data. In addition, synthetic models are provided to estimate two-part synthetic hurdle models. This code can be useful as a paradigm for the optimization or maximum likelihood estimation of more complex count models. Also discussed in this edition are marginal effects and discrete change, which have a prominent place in econometrics, but which can also be used with considerable value in other disciplines. I provide details on how to construct and interpret both marginal effects and discrete change for the major models discussed in the text.

New models added to the text from the first edition include: finite mixture models, quantile count models, bivariate negative binomial, and various methods used to model endogeneity. We previously examined negative binomial sample selection models and negative binomial models with endogenous stratification, but we now add instrumental variables, generalized method of moments, and methods of dealing with predictors having missing values. Stata now supports these models.

Finally, Stata 11 appeared in late July 2009, offering several capabilities related to our discussion which were largely unavailable in the version used with the first edition of this text. In particular, Stata's **glm** command now allows maximum likelihood estimation of the negative binomial heterogeneity parameter, which it did not in earlier versions. R's **glm** function and SAS's STAT/**GENMOD** procedure also provide the same capability. This option allows easier estimation and comparative evaluation of NB2 models.

I created the code and provided explanations for the development and use of a variety of synthetic regression models used in the text. Most were originally written using pseudo-random number generators I published in 1995, but a few were developed a decade later. When Stata offered their own suite of pseudo-random number generators in 2009, I re-wrote the code for constructing these synthetic models using Stata's code. This was largely done in the first two months of 2009. The synthetic models appearing in the first edition of this book, which are employed in this edition, now use the new code, as do the other synthetic models. In fact, Stata code for most of the synthetic models appearing in this text was published in the *Stata Journal* (Volume 10:1, pages 104–124). Readers are referred to that source for additional explanation on constructing synthetic models, including many binomial models that are not in this book. It should be noted that the synthetic Stata models discussed in the text have corresponding R scripts provided to duplicate model results.

R data files, functions, and scripts written for this text are available in the COUNT package that can be downloaded on CRAN. I recommend Muenchen and Hilbe (2010), *R for Stata Users* (Springer) for Stata users who wish to understand the logic of R functions. For those who wish to learn more about Stata, and how it can be used for data management and programming, I refer you to the books published by Stata Press. See *www.stata-press.com*. The websites which readers can access to download data files and user-authored commands and functions are:

Cambridge University Press: www.cambridge.org/9780521198158

Data sets, software code, and electronic *Extensions* to the text can be downloaded from: http://works.bepress.com/joseph_hilbe/

Errata and a post-publication *Negative Binomial Regression Extensions* can be found on the above sites, as well as at: http://www.statistics.com/hilbe/nbr.php. Additional code, text, graphs and tables for the book are available in this electronic document.

I should mention that practicing researchers rarely read this type of text from beginning straight through to the end. Rather, they tend to view it as a type of handbook where a model they wish to use can be reviewed for modeling details. The text can indeed be used in such a manner. However, most of the chapters are based on information contained in earlier chapters. Material explained in Chapters 2, 5, 6, and 7 is particularly important for mastery of the discussion of negative binomial models presented in Chapters 8 and 9. Knowledge of the material given in Chapters 7 and 8 is fundamental to understanding extended Poisson and negative binomial models. Because of this, I encourage you to make an exception and to read sequentially though the text, at least through Chapter 9. For those who have no interest at all in the derivation of count models, or of the algorithms by which these models are estimated, I suggest that you only skim through Chapter 4. Analogical to driving a car and understanding automotive mechanics, one can employ a sophisticated negative binomial model without understanding exactly how estimation is achieved. But, again as with a car, if you wish to re-parameterize a model or to construct an entirely new count model, understanding the discussion in Chapter 4 is essential. However, I have reiterated certain important concepts and relationships throughout the text in order to provide background for those who have not been able to read earlier chapters, and to emphasize them for those who may find reminding to be helpful.

To those who made this a better text

I wish to acknowledge several individuals who have contributed specifically to the second edition of *Negative Binomial Regression*. Chris Dorger of Intel kindly read through the manuscript, identifying typos and errors, and re-derived the derivations and equations presented in the text in order to make certain that they are expressed correctly. He primarily checked Chapters 2 through 10. Dr. Dorger provided assistance as the book was just forming, and at the end, prior to submission of the manuscript to the publisher. He spent a substantial amount of time working on this project. Dr. Tad Hogg, a physicist with the Institute of Molecular Manufacturing, spent many hours doing the same, and provided excellent advice regarding the clarification of particular subjects. James Hardin, with whom I have co-authored two texts on statistical modeling and whom I consider a very good friend, read over the manuscript close to the end, providing valuable insight and help, particularly with respect to Chapters 13 and 14. Andrew Robinson of the University of Melbourne provided invaluable help in setting up the COUNT package on CRAN, which has all of the R functions and scripts for the book, as well as data in the form of data frames. He also helped in re-designing some of the functions I developed to run more efficiently. We are now writing a book together on the *Methods of Statistical Model Estimation* (Chapman & Hall/CRC). Gordon Johnston of SAS Institute, author of the GENMOD procedure and longtime friend, provided support on SAS capabilities related to GEE and Bayesian models. He is responsible for developing the SAS programs used for the examples in Section 15.3, and read over Sections 15.1 and 15.2, offering insightful advice and making certain that the concepts expressed were correct. Allan Medwick, who recently earned a Ph.D. in statistics at the University of Pennsylvania read carefully through the entire manuscript near the end, checking for readability and typos; he also helped running the SAS models in Chapter 14. The help of these statisticians in fine-tuning the text has been invaluable, and I deeply appreciate their assistance. I am alone responsible for any remaining errors, but readers should be aware that every attempt has been made to assure accuracy.

I also thank Professor James Albert of Bowling Green University for his insightful assistance in helping develop R code for Monte Carlo simulations. Robert Muenchen, co-author with me of *R for Stata Users* (2010), assisted me in various places to streamline R code, and at times to make sure my code did exactly as I had intended. The efforts of these two individuals also contributed much to the book's value.

Again, as in the first edition, I express my gratitude and friendship to the late John Nelder, with whom I have discussed these topics for some 20 years. In fact,

I first came to the idea of including the negative binomial family in standard generalized linear models software in 1992 while John and I hiked down and back up the Grand Canyon Angel Trail together. We had been discussing his newly developed *kk* add-ons to GenStat software, which included rudimentary negative binomial code, when it became clear to me that the negative binomial, with its heterogeneity parameter as a constant, is a full-fledged member of the single parameter exponential family of distributions, and as such warranted inclusion as a GLM family member. It was easy to then parameterize the negative binomial probability distribution into exponential family form, and abstract the requisite GLM link, mean, variance, and inverse link functions – as any other GLM family member. It was also clear that this was not the traditional negative binomial, but that the latter could be developed by simple transformations. His visits to Arizona over the years, and our long discussions, have led to many of the approaches I now take to GLM–related models, even where a few of my findings have run counter to some of the positions he initially maintained.

I must also acknowledge the expertise of Robert Rigby and Mikis Stasinopoulos, authors of R's **gamlss** suite of functions, who upon my request rewrote part of the software in such a manner that it can now be used to estimate censored and truncated count models, which were not previously available using R. These capabilities are now part of **gamlss** on CRAN. Also to be acknowledged is Masakazu Iwasaki, Clinical Statistics Group Manager for Research & Development at Schering-Plough K.K., Tokyo, who re-worked code he had earlier developed into a viable R function for estimating bivariate negative binomial regression. He amended it expressely for this text; it is currently the only bivariate negative binomial function of which I am aware, and is available in the COUNT package. Malcolm Faddy and David Smith provided code and direction regarding the implementation of Faddy's namesake distribution for the development of a new count model for both under- and overdispersion. Robert LaBudde, President of Least Cost Formulations, Ltd., and adjunct professor of statistics at Old Dominion University, provided very useful advice related to R programming, for which I am most grateful. The above statisticians have helped add new capabilities to R, and have made this text a more valuable resource as a result.

The various participants in my **Statistics.com** web course, *Modeling Count Data*, should be recognized. Nearly all of them are professors teaching statistics courses in various disciplines, or researchers working in the corporate world or for a governmental agency. Their many questions and comments have helped shape this second edition. Elizabeth Kelly of the Los Alamos National Laboratories in New Mexico and Kim Dietrich, University of Washington, provided very helpful assistance in reviewing the manuscript at its final stage, catching several items that had escaped others, including myself. They also offered suggestions for clarification at certain points in the discussion, many of which were taken. Denis Shah, Department of Plant Pathology at Kansas State University, in particular, asked many insightful questions whose answers found their way into the text. I also wish to again express my appreciation to Diana Gillooly, statistics editor at Cambridge University Press, for her advice and for her continued confidence in the value of my work. Clare Dennison, assistant editor (Math/Computer science) at Cambridge, is also to be thanked for her fine help with the text, as are Joanna Endell-Cooper (production editor), Mairi Sutherland (freelance editor), Simon Crump (production manager), and Nicola Philps (Production Editor Manufacturing). They very kindly accommodated my many amendments. Ms Sutherland provided excellent editorial advice when required, reviewing every word of the manuscript. I alone am responsible for any remaining errors. In addition, I acknowledge the longtime assistance and friendship of Pat Branton of *Stata* Corp, who has helped me in innumerable ways over the past 20 years. Without her support over this period, this book would very likely not have been written.

Finally, I wish to thank the members of my family, who again had to lose time I would have otherwise spent with them. To my wife, Cheryl, our children, Heather, Michael, and Mitchell, and to my constant companion, Sirr, a white Maltese, I express my deepest appreciation. I dedicate this text to my late parents, Rader John and Nadyne Anderson Hilbe. My father, an engineering supervisor with Douglas Aircraft during the Second World War, and a UCLA mathematics professor during the early 1940s, would have very much appreciated this volume.

1

Introduction

1.1 What is a negative binomial model?

The negative binomial regression model is a truly unusual statistical model. Typically, those in the statistical community refer to the negative binomial as a single model, as we would in referring to Poisson regression, logistic regression, or probit regression. However, there are in fact several distinct negative binomial models, each of which are referred to as being a negative binomial model. Boswell and Patil (1970) identified 13 separate types of derivations for the negative binomial distribution. Other statisticians have argued that there are even more derivations. Generally, those who are using the distribution as the basis for a statistical model of count data have no idea that the parameterization of the negative binomial they are employing may differ from the parameterization being used by another. Most of the time it makes little difference how the distribution is derived, but, as we shall discover, there are times when it does. Perhaps no other model has such a varied pedigree.

I will provide an outline here of the intertwining nature of the negative binomial. Unless you previously have a solid background in this area of statistics, my overview is not likely to be completely clear. But, as we progress through the book, its logic will become evident.

The negative binomial model is, as are most regression models, based on an underlying probability distribution function (PDF). The Poisson model is derived from the Poisson PDF, the logistic regression model is derived from the binomial PDF, and the normal linear regression model (i.e. ordinary least squares), is derived from the Gaussian, or normal, PDF. However, the traditional negative binomial, which is now commonly symbolized as NB2 (Cameron and Trivedi, 1986), is derived from a Poisson–gamma mixture distribution. But such a mixture of distibutions is only one of the ways in which the negative binomial PDF can be defined. Unless otherwise specified, when I refer to a negative binomial model, it is the NB2 parameterization to which I refer.

The nice feature of this parameterization is that it allows us to model Poisson heterogeneity. As we shall discover, the mean and variance of the Poisson PDF are equal. The greater the mean value, the greater is the variability in the data as measured by the variance statistic. This characteristic of the data is termed equidispersion and is a distributional assumption of Poisson data. Inherent in this assumption is the requirement that counts are independent of one another. When they are not, the distributional properties of the Poisson PDF are violated, resulting in extra-dispersion. The mean and variance can no longer be identical. The form of extra-dispersion is nearly always one of overdispersion. That is, the variance is greater in value to that of the mean.

The negative binomial model, as a Poisson–gamma mixture model, is appropriate to use when the overdispersion in an otherwise Poisson model is thought to take the form of a gamma shape or distribution. The same shape value is assumed to hold across all conditioned counts in the model. If different cells of counts have different gamma shapes, then the negative binomial may itself be overdispersed; i.e the data may be both Poisson and negative binomial overdispersed. Random-effects and mixed-effects Poisson and negative binomial models are then reasonable alternatives.

What if the shape of the extra correlation, or overdispersion, is not gamma, but rather another identifiable shape such as inverse Gaussian? It is possible to construct a Poisson-inverse Gaussian distribution, and model. This distribution is formally known as a Holla distribution, but is better known to most statisticians as a PIG function. Unfortunately there is no closed form solution for the PIG model; estimation is therefore typically based on quadrature or simulation. It is not, however, a negative binomial, but can be used when the data takes its form.

What if we find that the shape of overdispersion is neither gamma nor inverse Gaussian? Poisson-lognormal models have been designed as well, but they too have no closed form. If overdispersion in the data takes no identifiable shape, most statisticians employ a negative binomial. There are other alternatives though – for instance, quantile count models. We spend some time later in the text evaluating models that address these data situations.

It may appear that we have gotten off track in discussing non-negative binomial methods of handling Poisson overdispersion. However, these methods were derived specifically because they could better model the data than available negative binomial alternatives. Each one we mentioned is based on the mixture approach to the modeling of counts. Knowledge of the negative binomial regression model therefore entails at least a rudimentary acquaintance with its alternatives.

I should mention here that the form of the mixture of variances that constitute the core of the Poisson–gamma mixture is $\mu + \mu^2/\nu$ where μ is the Poisson variance and μ^2/ν is the gamma variance; ν is the gamma shape parameter, and corresponds to extra dispersion in the mixture model meaning of negative binomial, as described above. Conceived of in this manner, there is an indirect relationship between ν and the degree of overdispersion in the data. A negative binomial model based on this joint definition of the variance becomes Poisson when ν approaches infinity. However, it is perfectly permissible, and more intuitive, if ν is inverted so that there is a direct relationship between the parameter and extra correlation. The standard symbol for the heterogeneity or overdispersion parameter given this parameterization of the negative binomial variance is α. Sometimes you may find r or k symbolizing ν, and confusingly some have used k to represent α. We shall use the symbols r for ν for the indirect relationship and α for the directly related heterogeneity parameter. All current commercial software applications of which I am aware employ the α parameterization. R's **glm** and **glm.nb** functions employ θ, or $1/\alpha$.

The origin of the negative binomial distribution is not as a Poisson–gamma mixture, which is a rather new parameterization. The earliest definitions of the negative binomial are based on the binomial PDF. Specifically, the negative binomial distribution is characterized as the number of failures before the rth success in a series of independent Bernoulli trials. The Bernoulli distribution is, as you may recall, a binomial distribution with the binomial denominator set at one (1). Given r as an integer, this form of the distribution is also known as a *Pascal* distribution, after mathematician Blaise Pascal (1623–1662). However, for negative binomial models, r is taken as a real number greater than 0, although it is rarely above four.

It is important to understand that the traditional negative binomial model can be estimated using a standard maximum likelihood function, or it can be estimated as a member of the family of generalized linear models (GLM). A negative binomial model is a GLM only if its heterogeneity parameter is entered into the generalized linear models algorithm as a constant. We shall observe the consequences of this requirement later in the text.

In generalized linear model theory the link function is the term that linearizes the relationship of the linear predictor, $x'\beta$, and the fitted value, μ or $\hat{y}$. In turn, μ is defined in terms of the inverse link. Given that generalized linear models are themselves members of the single parameter exponential family of distributions, the exponential family log-likelihood for count models can be expressed as

$$\sum y\theta + b(\theta) + c(y) \tag{1.1}$$

with θ as the link, $b(\theta)$ as the cumulant from which the mean and variance functions are derived, and $c(y)$ as the normalization term guaranteeing that the probability sums to 1. For the GLM negative binomial, the link, $\theta = -\ln((1/(\alpha\mu)) + 1)$, and the inverse link, which defines the fitted value, is $b'(\theta) = \mu$, or $1/(\alpha(\exp(x\beta) - 1))$. The GLM inverse link is also symbolized as η.

The traditional NB2 negative binomial amends the canonical link and inverse link values to take a log link, $\ln(\mu)$, and exponential inverse link, $\exp(\mathrm{x}'\beta)$. These are the same values as the canonical Poisson model. When the negative binomial is parameterized in this form, it is directly related to the Poisson model. As a GLM, the traditional NB2 model is a log-linked negative binomial, and is distinguished from the canonical form, symbolized as NB-C.

We shall discover when we display the derivations of the negative binomial as a Poisson–gamma mixture, and then from the canonical form defined as the number of failures before the *r*th success in a series of independent Bernoulli trials, that both result in an identical probability function when the mean is given as μ. When the negative binomial PDF is parameterized in terms of $x'\beta$, the two differ.

There are very good reasons to prefer the NB2 parameterization of the negative binomial, primarily because it is suitable as an adjustment for Poisson overdispersion. The NB-C form is not interpretable as a Poisson type model, even though it is the canonical form derived directly from the PDF. We shall discuss its interpretation later in the text. We shall also show how the negative binomial variance function has been employed to generalize the function.

The characteristic form of the canonical and NB2 variance functions is $\mu + \alpha\mu^2$. This value can be determined as the second derivative of the cumulant, or $b''(\theta)$. A linear negative binomial has been constructed, termed NB1, that parameterizes the variance as $\mu + \alpha\mu$. The NB1 model can also be derived as a form of Poisson–gamma mixture, but with different properties resulting in a linear variance. In addition, a generalized negative binomial has been formulated as $\mu + \alpha\mu^p$, where p is a third parameter to be estimated. For NB1, $p = 1$; for NB2, $p = 2$. The generalized negative binomial provides for any reasonable value of p. Another form of negative binomial, called heterogeneous negative binomial, NB-H, is typically a NB2 model, but with α parameterized. A second table of estimates is presented that displays coefficients for the influence of predictors on the amount of overdispersion in the data.

We have seen that the negative binomial can be understood in a variety of ways. All of the models we have discussed here are negative binomial; both the NB2 and NB1 are commonly used when extending the negative binomial to form models such as mixtures of negative binomial models, or when employed in panel models. Knowing which underlying parameterization of negative binomial is being used in the construction of an extended negative binomial model is essential when we are evaluating and interpreting it, which is our subject matter.

Now that we have provided an overview of the landscape of the basic negative binomial models, we take a diversion and provide a brief history of the negative binomial. Such an historical overview may help provide a sense of how the above varieties came into existence, and inform us as to when and why they are most effectively used.

1.2 A brief history of the negative binomial

If we are to believe Isaac Todhunter's report in his *History of the Mathematical Theory of Probability from the Time of Pascal to that of Laplace* (1865), Pierre de Montmort in 1713 mentioned the negative binomial distribution in the context of its feature as the number of failures, *y*, before the *k*th success in a series of binary trials. As a leading mathematician of his day, Montmort was in constant communication with many familiar figures in the history of statistics, including Nicholas and Jacob Bernoulli, Blaise Pascal, Brook Taylor (Taylor series) and Gottfried Leibniz (credited, along with Newton, with the discovery of the Calculus). He is said to have alluded to the distribution in the second edition of his foremost work, *Essay on the Analysis of Games of Chance* (1708), but did not fully develop it for another five years.

The Poisson distribution upon which Poisson regression is based, originates from the work of Siméon Poisson (1781–1840). He first introduced the distribution as a limiting case of the binomial in his, *Research on the Probability of Judgments in Criminal and Civil Matters* (1838). Later in the text we derive the Poisson from the binomial, demonstrating how the two distributions relate. Poisson regression developed as the foremost method of understanding the distribution of count data, and later became the standard method used to model counts. However, as previously mentioned, the Poisson distribution assumes the equality of its mean and variance – a property that is rarely found in real data. Data that have greater variance than the mean are termed *Poisson overdispersed*, but are more commonly designated as simply *overdispersed*.

Little was done with either the earliest definition of negative binomial as derived by Montmort, or with Poisson's distribution for describing count data, until the early twentieth century. Building on the work originating with Gauss (1823), who developed the normal, or Gaussian, distribution, upon which ordinary least squares (OLS) regression is based, the negative binomial was again derived by William Gosset, under his pen name, Student, in 1907 while working under Karl Pearson at his Biometric Laboratory in London (Student, 1907). In the first paper he wrote while at the laboratory, he derived the negative binomial while investigating the sampling error involved in the counting of yeast cells with a haemocytometer. The paper was published in *Biometrika*, and appeared a year earlier than his well regarded papers on the sampling error of the mean and correlation coefficient (Jain, 1959). However, G. Udny Yule is generally, but arguably, credited with formulating the first negative binomial distribution based on a 1910 article dealing with the distribution of the number of deaths that would occur as a results of being exposed to a disease (i.e. how many deaths occur given a certain number of exposures). This formulation stems from what is called inverse binomial sampling. Later Greenwood and Yule (1920) derived the negative binomial distribution as the probability of observing y failures before the rth success in a series of Bernoulli trials, replicating in a more sophisticated manner the work of Montmort. Three years later the contagion or mixture concept of the negative binomial originated with Eggenberger and Polya (1923). They conceived of the negative binomial as a compound Poisson distribution by holding the Poisson parameter, λ, as a random variable having a gamma distribution. This was the first derivation of the negative binomial as a Poisson–gamma mixture distribution. The article also is the first to demonstrate that the Poisson parameter varies proportionally to the *chi*2 distribution with 2 degrees of freedom.

Much of the early work on the negative binomial during this period related to the *chi*2 distribution, which seems somewhat foreign to the way in which we now understand the distribution. During the 1940s, most of the original work on count models came from George Beall (1942), F. J. Anscombe (1949), and Maurice Bartlett (1947). All three developed measures of transformation to normalize non-normal data, with Bartlett (1947) proposing an analysis of square root transforms on Poisson data by examining variance stabilizing transformations for overdispersed data. Anscombe's work entailed the construction of the first negative binomial regression model, but as an intercept-only non-linear regression. Anscombe (1950), later derived the negative binomial as a series of logarithmic distributions, and discussed alternative derivations as well, for example: (1) inverse binomial sampling; (2) heterogeneous Poisson sampling where λ is considered as proportional to a *chi*2; (3) the negative binomial as a population growth model; and (4) the negative binomial derived from a geometric series. Evans (1953) developed what we now refer to as the NB1, or linear negative binomial, parameterization of the negative binomial.

Leroy Simon (1961), following his seminal work differentiating the Poisson and negative binomial models (1960), was the first to publish a maximum likelihood algorithm for fitting the negative binomial. He was one of the many actuarial scientists at the time who were engaged in fitting the Poisson and negative binomial distributions to insurance data. His work stands out as being the most sophisticated, and he was perhaps cited more often for his efforts in the area than anyone else in the 1960s. Birch (1963) is noted as well for being the first to develop a single predictor maximum likelihood Poisson regression model which he used to analyze tables of counts. It was not until 1981 that Plackett first developed a single predictor maximum likelihood negative binomial while working with categorical data which he could not fit using the Poisson approach.

Until the mid-1970s, parameterizing a non-linear distribution such as logit, Poisson, or negative binomial, so that the distributional response variable was conditioned on the basis of one of more explanatory predictors, was not generally conceived to be as important as understanding the nature of the underlying distribution itself, i.e. determining the relationships that obtain between the various distributions. When considering the negative binomial distribution, for example, the major concern was to determine how it related to other distributions – the *chi*2, geometric, binomial and Bernoulli, Poisson, gamma, beta, incomplete beta, and so forth. Regression model development was primarily thought to be a function of the normal model, and the transformations that could be made to both the least squares response and predictors.

It was not until 1981 that the first IBM personal computer became available to the general public, an event that changed forever the manner in which statistical modeling could be performed. Before that event, most complex statistical analyses were done using mainframe computers, which were usually at a remote site. Interactive analyses simply did not occur. Computer time was both time-consuming and expensive.

The emphasis on distributional properties and relationships between distributions began to change following the development of generalized linear models (GLM) by John Nelder and R. W. M. Wedderburn (1972). The new emphasis was on the construction of non-linear models that incorporated explanatory predictors. In 1974 Nelder headed a team of statisticians, including Wedderburn and members of the statistical computing working party of the Royal Statistical Society, to develop GLIM (Generalized Linear Interactive Modeling), a software application aimed to implement GLM theory. GLIM software allowed users to estimate GLM models for a limited set of exponential family members, including, among others, the binomial, Poisson, and, for a constant value of its heterogeneity parameter, the negative binomial. Although GLIM did not have a specific option for negative binomial models, one could use the *open* option to craft such a model.

Incorporated into GLIM software was the ability to parameterize Poisson models as rates. Nelder had developed the notion of offsets as a side exercise, only to discover that they could be used to model counts as incident rate ratios, which as we shall discover was a considerable advancement in statistical modeling. The traditional negative binomial model can also be parameterized in this manner.

In 1982 Nelder joined with Peter McCullagh to write the first edition of *Generalized Linear Models*, in which the negative binomial regression model was described, albeit briefly. The second edition of the text appeared in 1989 (McCullagh and Nelder, 1989), and is still regarded as the premiere text on the subject. GLM-based negative binomial regression software was only available as a user-defined macro in GLIM until 1992 when Nelder developed what he called the *kk* system for estimating the negative binomial as a GenStat macro. The first implementation of the negative binomial as part of a GLM software algorithm did not occur until 1993 (Hilbe), with the software developed for both Stata and Xplore. The algorithm included links for the estimation of the traditional log-linked negative binomial, NB2, the canonical model, NB-C, and a negative binomial with an identity link. We shall discuss the use of an identity linked negative binomial later in the text.

In 1994 Hilbe (1994) developed a SAS macro for the NB2 model using SAS's GENMOD procedure, SAS's GLM modeling tool. The macro estimated the negative binomial heterogeneity parameter using a damping method adapted from a method first advanced by Breslow (1984) of reducing the Pearson dispersion to a value approximating 1. In late 1994, Venables posted a GLM-based NB2 model to StatLib using S-Plus, and SAS (Johnston) incorporated the negative binomial into its GENMOD procedure in 1998, with the same links offered in Stata and Xplore. SPSS did not offer a GLM procedure until 2006 with the release of version 15. A negative binomial option with all three links was included.

R has offered its users GLM-based negative binomial models through the **glm.nb** and **negative.binomial** functions, which are functions in the **MASS** package that is normally included when installing R from the web. Packages such as **gamlss** and **pscl** also provide negative binomial options, but they are based on full maximum likelihood estimation.

Currently nearly all GLM software includes a negative binomial family, and several of the major statistical applications, like Stata and LIMDEP, offer independent maximum likelihood negative binomial commands.

Full maximum likelihood models were also being developed for extended negative binomial models during this time. Geometric hurdle models were developed by Mullahy (1986), with a later enhancement to negative binomial hurdle models. Prem Consul and Felix Famoye have developed various forms of generalized negative binomial models using generalized maximum likelihood, as well as other mixture models. They have worked singly as well as jointly for some 30 years investigating the properties of such models – but the models have never gained widespread popularity.

William Greene's LIMDEP was the first commercial package to offer maximum likelihood negative binomial regression models to its users (2006a [1987]). Stata was next with a maximum likelihood negative binomial (1994). Called **nbreg**, Stata's negative binomial command was later enhanced to allow modeling of both NB1 and NB2 parameterizations. In 1998, Stata offered a generalized negative binomial, **gnbreg**, in which the heterogeneity parameter itself could be parameterized. It should be emphasized that this command does not address the generalized negative binomial distribution, but rather it allows a generalization of the scalar overdispersion parameter such that parameter estimates can be calculated showing how model predictors comparatively influence overdispersion. Following LIMDEP, I have referred to this model as a heterogeneous negative binomial, NB-H, since the model extends NB2 to permit observed sources of heterogeneity in the overdispersion parameter.

Gauss and MATLAB also provide their users with the ability to estimate maximum likelihood negative binomial models. In Matlab one can use the maximum likelihood functions to rather easily estimate NB2 and NB1 models. Gauss provides modules in their *Count* library for handling NB2, as well as truncated and censored negative binomial models. Only LIMDEP and R provide both truncated and censored negative binomial modeling capability. In the meantime, LIMDEP has continuously added to its initial negative binomial offerings. It currently estimates many of the negative binomial-related models that shall be discussed in this monograph. In 2006 Greene developed a new parameterization of the negative binomial, NB-P, which estimates both the traditional negative binomial ancillary parameter, as well as the exponent of the second term of the variance function.

I should perhaps reiterate that the negative binomial has been derived and presented with different parameterizations. Some authors employ a variance function that clearly reflects a Poisson–gamma mixture; this is the case when the Poisson variance defined as μ and the gamma as μ^2/ν, is used to create the negative binomial variance characterized as $\mu + \mu^2/\nu$. This parameterization is the same as that originally derived by Greenwood and Yule (1920). An inverse relationship between μ and ν was also used to define the negative binomial variance in McCullagh and Nelder (1989), to which some authors refer when continuing this manner of representation.

However, shortly after the publication of that text, Nelder and Lee (1992) developed his *kk* system, a user-defined negative binomial macro written for use with GenStat software. In this system he favored the direct relationship between α and μ^2 – resulting in a negative binomial variance function of $\mu + \alpha\mu^2$. Nelder (1994) continued to prefer the direct relationship in his subsequent writings. Still, referencing the 1989 work, a few authors have continued to use the originally defined relationship, even as recently as Faraway (2006).

The direct parameterization of the negative binomial variance function was first suggested by Bliss and Owen (1958) and was favored by Breslow (1984) and Lawless (1987) in their highly influential seminal articles on the negative binomial. In 1990s, the direct relationship was used in the major software implementations of the negative binomial: Hilbe (1993b, 1994b) for XploRe and Stata, Greene (2006a) for LIMDEP, and Johnston (1997) for SAS. The direct parameterization was also specified in Hilbe (1994b), Long (1997), Cameron and Trivedi (1998), and most articles and books dealing with the subject. Recently Long and Freese (2003, 2006), Hardin and Hilbe (2001, 2007), and a number of other recent authors have employed the direct relationship as the preferred variance function. It is rare now to find current applications using the older inverse parameterization.

The reason for preferring the direct relationship stems from the use of the negative binomial in modeling overdispersed Poisson count data. Considered in this manner, α is directly related to the amount of overdispersion in the data. If the data are not overdispersed, i.e. the data are Poisson, then $\alpha = 0$. Increasing values of α indicate increasing amounts of overdispersion. Since a negative binomial algorithm cannot estimate $\alpha = 0$, owing to division by zero in the estimating algorithm, values for data seen in practice typically range from 0.01 to about 4.

Interestingly, two books have been recently published, Hoffmann (2004) and Faraway (2006), asserting that the negative binomial is not a true generalized linear model. However, the GLM status of the negative binomial depends on whether it is a member of the single-parameter exponential family of distributions. If we assume that the overdispersion parameter, α, is known and is ancillary, resulting in what has been called a LIMQL (limited information maximum quasi-likelihood) model (see Greene, 2003), then the negative binomial is a GLM. On the other hand, if α is considered to be a parameter to be estimated, then the model may be estimated as FIMQL (full information maximum quasi-likelihood), but it is not strictly speaking a GLM.

Finally, it should be reiterated that Stata's **glm** command, R's **glm.nb** function, and SAS's **GENMOD** procedure are IRLS (iteratively reweighted least squares)-based applications in which the negative binomial heterogeneity parameter, α, is estimated using an external maximum likelihood mechanism, which then inserts the resulting value into the GLM algorithm as a constant. This procedure allows the GLM application to produce maximum likelihood estimates of α, even though the actual estimation of α is done outside the IRLS scheme.

1.3 Overview of the book

The introductory chapter provides an overview of the nature and scope of count models, and of negative binomial models in particular. It also provides a brief overview of the history of the distribution and the model, and of the various software applications that include the negative binomial among their offerings.

Chapter 2 provides an extended definition and examples of risk, risk ratio, risk difference, odds, and odds ratio. Count model parameter estimates are typically parameterized in terms of risk ratios, or incidence rate ratios. We demonstrate the logic of a risk ratio with both 2×2 and $2\times k$ tables, as well as showing how to calculate the confidence intervals for risk ratios. We also give a brief overview of the relationship of risk and odds ratio, and of when each is most appropriately used.

Chapter 3 gives a brief overview of count response regression models. Incorporated in this discussion is an outline of the variety of negative binomial models that have been constructed from its basic parameterization. Each extension from the base model is considered as a response to a violation of model assumptions. Enhanced negative binomial models are identified as solutions to the respective violations.

Chapter 4 examines the two major methods of parameter estimation relevant to modeling Poisson and negative binomial data. We begin by illustrating the construction of distribution-based statistical models. That is, starting from a probability distribution, we follow the logic of establishing the estimating equations that serve as the focus of the fitting algorithms. Given that the Poisson and traditional negative binomial (NB2) are members of the exponential family of distributions, we define the exponential family and its constituent terms. In so doing we derive the iteratively reweighted least squares (IRLS) algorithm and the form of the algorithm required to estimate the model parameters. We then define maximum likelihood estimation and show how the modified Newton–Raphson algorithm works in comparison to IRLS. We discuss the reason for differences in output between the two estimation methods, and explain when and why differences occur.

Chapter 5 provides an overview of fit statistics which are commonly used to evaluate count models. The analysis of residuals – in their many forms – can inform us about violations of model assumptions using graphical displays. We also address the most used goodness-of-fit tests, which assess the comparative fit of related models. Finally, we discuss validation samples, which are used to ascertain the extensibility of the model to a greater population.

Chapter 6 is devoted to the derivation of the Poisson log-likelihood and estimating equations. The Poisson traditionally serves as the basis for deriving the traditional parameterization of the negative binomial; e.g. NB2. Nonetheless, Poisson regression remains the fundamental method used to model counts. We identify how overdispersion is indicated in Poisson model output, and provide guidance on the construction of prediction and conditional effects plots, which are valuable in understanding the comparative predictive effect of the levels of a factor variable. We then address marginal effects, which are used primarily for continuous predictors, and discrete change for factor variables. Finally, we discuss what can be called the rate parameterization of count models, in which counts are considered as being taken within given areas or time periods. The subject relates to the topic of offsets. We use both synthetic and real data to explain the logic and application of using offsets.

Chapter 7 details the criteria that can be used to distinguish real from apparent overdispersion. Simulated examples are constructed that show how apparent overdispersion can be eliminated, and we discuss how overdispersion affects count models in general. Finally, scaling of standard errors, application of robust variance estimators, jackknifing, and bootstrapping of standard errors are all evaluated in terms of their effect on inference. An additional section related to negative binomial overdispersion is provided, showing that overdispersion is a problem for all count models, not simply for Poisson models. Finally, the *score* and *Lagrange multiplier* tests for overdispersion are examined. This chapter is vital to the development of the negative binomial model.

In Chapter 8 we define the negative binomial probability distribution function (PDF) and proceed to derive the various statistics required to model the canonical and traditional forms of the distribution. Additionally, we derive the Poisson–gamma mixture parameterization that is used in maximum likelihood algorithms, and provide Stata and R code to simulate the mixture model. Throughout this chapter it becomes clear that the negative binomial is a full member of the exponential family of generalized linear models. We discuss the nature of the canonical form, and the problems that have been claimed to emanate when applying it to real data. We then re-parameterize the canonical form of the model to derive the traditional log-linked form (NB2).

Chapter 9 discusses the development and interpretation of the NB2 model. Examples are provided that demonstrate how the negative binomial is used to accommodate overdispersed Poisson data. Goodness-of-fit statistics are examined with a particular emphasis on methods used to determine whether the negative binomial fit is statistically different from a Poisson. Marginal effects for negative binomial models are discussed, expanding on the previous presentation of Poisson marginal effects.

Chapter 10 addresses alternative parameterizations of the negative binomial. We begin with a discussion of the geometric model, a simplification of the negative binomial where the overdispersion parameter has a value of 1. When the value of the overdispersion parameter is zero, NB2 reduces to a Poisson model. The geometric distribution is the discrete correlate of the negative exponential distribution. We then address the interpretation of the canonical link derived in Chapter 4. We thereupon derive and discuss how the linear negative binomial, or NB1, is best interpreted. Finally, the NB2 model is generalized in the sense that the ancillary or overdispersion parameter itself is parameterized by user-defined predictors for generalization from scalar to observation-specific interpretation. NB2 can also be generalized to construct three-parameter negative binomial models. We look at a few important examples.

Chapter 11 deals with two common problems faced by researchers handling real data. The first deals with count data which structurally exclude zero counts. The other relates to count data having excessive zeros – far more than defined by usual count distributions.

Zero-truncated and zero-inflated Poisson (ZIP) and negative binomial (ZINB) models, as well as hurdle models, have been developed to accommodate these two types of data situations. Hurdle models are typically used when the data have excessive zero counts, much like zero-inflated models, but can also be used to assess model underdispersion. We comparatively examine zero-inflated and hurdle models, providing guidance on when they are optimally applied to data.

Chapter 12 discusses truncated and censored data and how they are modeled using appropriately adjusted Poisson and negative binomial models. Two types of parameterizations are delineated for censored count models: econometric or dataset-based censored data, and survival or observation-based censored data. Stata code is provided for the survival parameteriation; R code is provided for the econometric parameterizations.

Chapter 13 focuses on the problem of handling endogeneity. First, we address finite mixture models where the response is not assumed to have a single distribution, but rather, it is assumed that the response is generated from two or more separate shapes. Finite mixture models may be composed of the same distributions, or of differing distributions, e.g. Poisson-NB2. Typically the same distribution is used. Also discussed in this chapter are methods of dealing with endogenous predictors. Specifically addressed are the use of two-stage instrumental variables and the generalized methods of moments approaches of accommodating endogeneity. We expand the presentation to include negative binomial with endogenous multinomial treatment variables, endogeneity resulting from measurement error, and various methods of sample selection. We conclude with an overview of endogenous switching models and quantile count models.

Chapter 14 addresses the subject of negative binomial panel models. These models are used when the data are either clustered or when they are in the form of longitudinal panels. We first address generalized estimating equations (GEE), the foremost representative of population averaging methods, and then we derive and examine unconditional and conditional fixed-effects and random-effects Poisson and negative binomial regression models. Mixed count models are also addressed, e.g. random-intercept and random-coefficient multilevel negative binomial models.

Chapter 15 provides an overview to Bayesian modeling in general and of the analysis of negative binomial models. These are newly developed models, with limited software support. Much of our discussion will therefore be prospective, describing models being developed.

Appendix A presents an overview of how to construct and interpret interaction terms for count models. Appendix B provides a listing of data files, functions, and scripts found in the text. All data and related functions, commands and scripts may be downloaded from the text's website: http://www.cambridge.org/9780521198158 or, http://www.stata.com/bookstore/nbr2.html

R users should install and load the COUNT package, which contains all of the data in data frames, and book's R functions and scripts. The R user must access the COUNT package directory under scripts, and can paste them into the R script editor to run. They may be amended to work with your own data situations as well. Once COUNT is installed and loaded, you can observe the available function names by typing on the command line "ls("package:COUNT");" for a list of the available dataframes, type "data(package= "COUNT")." To obtain the structure of the function, ml.nb2, type "str(ml.nb2)." or "help(ml.nb2)." The same is the case for any other function. You can also type *help(ml.nb2)*.

It should be noted that many of the Stata commands in the book were published on the Boston School of Economics Statistical Software Center (SSC) website: [http://ideas.repec.org/s/boc/bocode.html].

Regarding notation: File and command/function names and $\mathbf{I} \times \mathbf{J}$ matrix terms ($\mathbf{I}, \mathbf{J} \neq 1$) are in bold; variables are italicized. Hats have not been put over estimated parameters ($\hat{\beta}$) or predicted ($\hat{\mu}$) variables; I assume that they are understood given the context. I did this to keep complex equations easier to read. For consistency, I have employed this method throughout the text.

2

The concept of risk

2.1 Risk and 2×2 tables

The notion of risk lays at the foundation of the modeling of counts. In this chapter we discuss the technical meaning of risk and risk ratio, and how to interpret the estimated incidence rate ratios that are displayed in the model output of Poisson and negative binomial regression. In the process, we also discuss the associated relationship of risk difference as well as odds and odds ratios, which are generally understood with respect to logistic regression models.

Risk is an exposure to the chance or probability of some outcome, typically thought of as a loss or injury. In epidemiology, risk refers to the probability of a person or group becoming diseased given some set of attributes or characteristics. In more general terms, the risk that an individual with a specified condition will experience a given outcome is the probability that the individual actually experiences the outcome. It is the proportion of individuals with the risk factor who experience the outcome. In epidemiological terms, risk is therefore a measure of the probability of the incidence of disease. The attribute or condition upon which the risk is measured is termed a risk factor, or exposure. Using these terms then, risk is a summary measure of the relationship of disease (outcome) to a specified risk factor (condition). The same logic of risk applies in insurance, where the term applies more globally to the probability of any type of loss.

Maintaining the epidemiological interpretation, relative risk is the ratio of the probability of disease for a given risk factor compared with the probability of disease for those not having the risk factor. It is therefore a ratio of two ratios, and is often simply referred to as the *risk ratio*, or, when referencing counts, the *incidence rate ratio* (IRR). Parameter estimates calculated when modeling counts are generally expressed in terms of incidence risk or rate ratios.

Table 2.1 *Partial Titanic data as grouped*

obs	survive	cases	class	sex	age
1	14	31	3	0	0
2	13	13	2	0	0
3	1	1	1	0	0
4	13	48	3	1	0
...					

It is perhaps easier to understand the components of risk and risk ratios by constructing a table of counts. We begin simply with a 2×2 table, with the response on the vertical axis and risk factor on the horizontal. The response, which in regression models is also known as the dependent variable, can represent a disease outcome, or any particular occurrence.

For our example we shall use data from the 1912 *Titanic* survival log. These data are used again for an example of a negative binomial model in Chapter 9, but we shall now simply express the survival statistics in terms of a table. The response term, or outcome, is *survived*, indicating that the passenger survived. Risk factors include *class* (i.e. whether the passenger paid for a first-, second-, or third-class ticket), *sex*, and *age*. Like *survived*, *age* is binary (i.e. a value of 1 indicates that the passenger is an adult, and 0 indicates a child). *Sex* is also binary (i.e. a value of 1 indicates that the passenger is a male, and 0 indicates a female).

Before commencing, however, it may be wise to mention how count data is typically stored for subsequent analysis. First, data may be stored by observations. In this case there is a single record for each *Titanic* passenger. Second, data may be grouped. Using the *Titanic* example, grouped data tell us the number of passengers who survived for a given covariate pattern. A covariate pattern is a particular set of unique values for all explanatory predictors in a model. Suppose the partial table shown in Table 2.1.

The first observation indicates that 14 passengers survived from a total of 31 who were third-class female children passengers. The second observation tells us that all second-class female children passengers survived. So did all first-class female children. It is rather obvious that being a third-class passenger presents a higher risk of dying, in particular for boys.

There are 1,316 observations in the **titanic** data set we are using (I have dropped crew members from the data). In grouped format, the dataset is reduced to 12 observations – with no loss of information.

Table 2.2 R: *Basic tabulation of Titanic data: survived on age*

```
library(COUNT)        # use for remainder of book
data(titanic)         # use for remainder of Chapter
attach(titanic)       # use for remainder of Chapter
library(gmodels)      # must be pre-installed
CrossTable(survived, age, prop.t=FALSE, prop.r=FALSE,
      prop.c=FALSE, prop.chisq=FALSE)
```

Third, we may present the data in terms of tables, but in doing so only two variables are compared at a time. For example, we may have a table of *survived* and *age*, given as follows

```
. tab survived age
          | Age (Child vs Adult) |
 Survived |      child    adults | Total
----------+----------------------+------
       no |         52       765 |   817
      yes |         57       442 |   499
----------+----------------------+------
    Total |        109     1,207 | 1,316
```

This form of table may be given in paradigm form as:

```
              x
          0       1
   ---+--------------+
    0 |  A       B   |  A+B
y     |              |
    1 |  C       D   |  C+D
   ---+--------------+
        A+C     B+D
```

Many epidemiological texts prefer to express the relationship of risk factor and outcome or response by having *x* on the vertical axis and *y* on the horizontal. Moreover, you will also find the 1 listed before the 0, unlike the above table. However, since the majority of software applications display table results as above, we shall employ the same format here. Be aware, though, that the relationships of A, B, C, and D we give are valid only for this format; make the appropriate adjustment when other paradigms are used. It is the logic of the relationships that is of paramount importance.

Using the definition of risk and risk ratio discussed above, we have the following relationships.

The risk of y given $x = 1$:

$$D/(B + D) \tag{2.1}$$

The risk of y given $x = 0$:

$$C/(A + C) \tag{2.2}$$

The risk ratio (relative risk) of y given $x = 1$ compared with $x = 0$:

$$\frac{D/(B + D)}{C/(A + C)} = \frac{D(A + C)}{C(B + D)} = \frac{AD + CD}{BC + CD} \tag{2.3}$$

We use the same paradigm with the values from the **titanic** data of *survived* on *age*. Recall that adults have the value of *age* = 1, children of *age* = 0.

```
                0          1
Survived | child      adults |
---------+-------------------+
0     no |    52         765 |
         |                   |
1    yes |    57         442 |
---------+-------------------+
   Total |   109       1,207
```

The risk of survival given that a passenger is an adult: 442/1207 = 0.36619718
The risk of survival given that a passenger is a child: 57/109 = 0.52293578
The risk ratio (relative risk) of survival for an adult compared with a child:

$$\frac{442/1207}{57/109} = \frac{0.36619718}{0.52293578} = 0.7002718$$

This value may be interpreted as:

The likelihood of survival was 30% less for adults than for children.

Or, since $1/0.70027 = 1.42802$,

The likelihood of survival was 43% greater for children than for adults.

2.2 Risk and 2×*k* tables

The same logic we have used for 2×2 tables applies with respect to 2×*k* tables. Using the same **titanic** data, we look at the relationship of *survived* on the three levels of passenger class. We would hope that the class of ticket paid for by a passenger would have no bearing on survival but, given the year and social conditions, we suspect that first-class passengers had a greater likelihood of survival than second- or third-class passengers.

A cross-tabulation of *survived* on *class* produces Table 2.3 and Stata code.

Table 2.3 R: *Basic tabulation of Titanic data: survived on class*

```
CrossTable(survived, class, prop.t=FALSE, prop.r=FALSE,
  prop.c=FALSE, prop.chisq=FALSE, dnn=c('survived','class'))
```

```
. tab survived class
          |          class (ticket)
 Survived | 1st class  2nd class  3rd class | Total
----------+---------------------------------+------
       no |       122        167        528 |   817
      yes |       203        118        178 |   499
----------+---------------------------------+------
    Total |       325        285        706 | 1,316
```

As may be recalled from elementary regression modeling, when employing dummy or indicator variables for the levels of a categorical variable, one of the levels is regarded as the reference. You may select any level as the reference level; for ease of interpretation, it is generally preferred to select either the highest or lowest level. Some statisticians, however, depending on the data at hand, use the level having the most observations as the referent. Which level to use as the referent should be based on what makes the most interpretative sense for the model, not on some pre-determined criterion.

This time we select third-class passengers as the reference level. Comparing the risk of level 2 with level 3 (reference), we have:

SECOND CLASS

$$\frac{(118/285)}{(178/706)} = 1.6421841$$

and

FIRST CLASS

$$\frac{(203/325)}{(178/706)} = 2.4774071$$

Although we have not yet discussed Poisson regression, which is the basic model for estimating risk, we can test the above calculations by employing the model to this data, as in Table 2.4.

The values are identical.

I used a generalized linear models command to obtain the incidence rate ratios, which is another term for risk ratios in this context. Note that I used

Table 2.4 R: *Poisson model with robust standard errors*

```
titanic$class <- relevel(factor(titanic$class), ref=3)
tit3 <- glm(survived ~ factor(class), family=poisson, data=titanic)
irr <- exp(coef(tit3))                              # vector of IRRs
library("sandwich")
rse <- sqrt(diag(vcovHC(tit3, type="HC0")))        # coef robust SEs
irr*rse                                             # IRR robust SEs
```

```
. tab class, gen(class)
. glm survived class2 class1, fam(poi) eform nohead vce(robust)

-----------------------------------------------------------------
         |              Robust
survived |      IRR   Std. Err.      z    P>|z|  [95% Conf. Interval]
---------+-------------------------------------------------------
  class2 | 1.642184   .1572928    5.18  0.000   1.361105   1.981309
  class1 | 2.477407    .192782   11.66  0.000   2.126965   2.885589
-----------------------------------------------------------------
```

the *noheader* option so that only a table of exponentiated estimates is displayed. As will be discussed at greater length in Chapter 4, Poisson risk ratios are calculated as the exponentiation of the Poisson regression coefficients.

The risk ratios are interpreted as:

Second-class passengers had a 64% greater likelihood of survival than third-class passengers.
First-class passengers had a some two and a half greater likelihood of survival than third-class passengers.

It should be recalled that when the response is binary, as in the above model, the exponentiated coefficients are not incidence rate ratios, but are simply risk ratios. Only when the response is a count is it meaningful to refer to the relationship of response to risk factor as an incidence rate ratio (IRR).

2.3 Risk ratio confidence intervals

Risk ratio confidence intervals require a knowledge of the standard error and level of α. The standard error used, however, is the standard error of the Poisson coefficient, adjusted by a robust or sandwich estimator. We shall discuss such an estimator later in the book.

The coefficient standard errors (SE) for the above calculated risk ratios may be determined by use of the following equation:

$$\ln(\text{SE}) = \sqrt{\frac{1}{D} - \frac{1}{B+D} + \frac{1}{C} - \frac{1}{A+C}} \tag{2.4}$$

For the survival on age relationship, we have

```
                  0           1
Survived | child      adults |
---------+-------------------+
 0    no |     52        765 |
         |                   |
 1   yes |     57        442 |
---------+-------------------+
   Total |    109      1,207
```

Substituting the above values into the formula results in:

$$\ln(\text{SE}) = \sqrt{\frac{1}{442} - \frac{1}{1207} + \frac{1}{57} - \frac{1}{109}} \tag{2.5}$$

```
. di sqrt(1/442 - 1/1207 + 1/57 - 1/109)
.09901258
```

The Stata command, **di**, is an abbreviation for **display**, which indicates calculation of a numeric computation. Next, we take the natural log of the risk ratio to obtain a Poisson coefficient:

```
. di ln(.7002718)
-.35628673
```

We now have the coefficient and its standard error. To calculate the confidence intervals of the coefficient, we have:

$$\text{CONFIDENCE INTERVAL:} \quad \beta_k \pm z^{\alpha/2}\text{SE}(\beta_k) \tag{2.6}$$

where $z^{\alpha/2}$ is a quantile from the normal distribution. A two-sided p-value of 0.05 is a z-score or value of 1.96.

```
. di abs(invnormal(.05/2))
1.959964
```

Traditionally, a p-value of 0.05 is expressed as $\alpha = 0.05$, corresponding to a confidence interval 95% (1 – 0.05); $z^{\alpha/2}$ indicates a z-score with $\alpha/2$. A 95% confidence interval is standard in fields such as biostatistics, epidemiology, econometrics, social statistics, and environmental statistics. β_k is the coefficient on the kth predictor, and se(β_k) is its standard error. Applied to our model,

COEFFICIENT: LOWER 95% CI

```
. di -.35628673 - 1.96*.09901258
-.55035139
```

COEFFICIENT: UPPER 95% CI

```
. di -.35628673 + 1.96*.09901258
-.16222207
```

Exponentiation of each coefficient confidence interval results in the 95% risk ratio confidence intervals

RISK RATIO: LOWER 95% CI

```
. di exp(-.55035139)
.57674711
```

RISK RATIO: UPPER 95% CI

```
. di exp(-.16222207)
.85025237
```

A Poisson risk ratio model demonstrates the accuracy of the hand calculations, even in light of rounding error.

Table 2.5 R: *Poisson model with risk ratio robust standard errors*

```
titanic2 <- glm(survived ~ age, family=poisson, data=titanic)
irr2 <-exp(coef(titanic2))
library("sandwich")
sqrt(diag(vcovHC(titanic2, type="HC0")))
rse2 <- sqrt(diag(vcovHC(titanic2, type="HC0")))
irr2 * rse2
```

```
. glm survived age, fam(poi) vce(robust) eform nohead

-----------------------------------------------------------------------
         |              Robust
survived |       IRR  Std. Err.      z   P>|z|  [95% Conf.Interval]
---------+-------------------------------------------------------------
     age | .7002718   .0693621   -3.60  0.000   .5767066   .8503121
-----------------------------------------------------------------------
```

The interpretation of the confidence interval may be expressed as:

Upon repeated sampling, the true risk ratio will be found within the interval 0.58 to.85 approximately 95% of the time.

A risk ratio of 1 indicates that there is no difference in risk between the levels of the risk factor. It follows then that if the confidence interval of the risk factor contains the value of 1, it does not significantly explain the response, or outcome. That is, a confidence interval not including the value of 1 indicates that there is a significant difference between the two levels of the risk factor. Note that the confidence interval in the regression output above does not include 1. There is, therefore, a significant difference in the survival of adults and children.

A note on the R code for the robust standard errors of risk ratios such as the example from Table 2.4, or incidence rate ratios for models where the response is a count. Unlike abstracting standard errors from the model variance–covariance matrix, robust standard errors for exponentiated coefficients are based on the *delta method*. Simply, such standard errors are calculated as

$$SE_{RR} = \exp(\beta) SE_{robust}(\beta) \tag{2.7}$$

where the final term is the robust standard error of the coefficient. Using the above example from Table 2.4, we display the results of the coefficient and its robust standard error, which can be used to calculate the robust standard error for the risk ratio. The meaning of *robust* with respect to standard errors will be discussed in detail in Section 7.3.3. Robust standard errors are also known as *sandwich* or *empirical* standard errors, in difference to the *model* standard errors typically displayed in regression output. Note that the formula (2.7) for calculating standard errors using the delta method is only applicable for the standard errors of the exponentiated coefficients of discrete response regression models.

Table 2.6 R: *Poisson model with robust standard errors*

```
titanic3 <- glm(survived ~ age, family=poisson, data=titanic)
library("sandwich")
coef(titanic3)
sqrt(diag(vcovHC(titanic3, type="HC0")))
```

```
. glm survived age, nolog fam(poi) robust nohead

------------------------------------------------------------------------
         |               Robust
survived |      Coef.   Std. Err.     z   P>|z|    [95% Conf. Interval]
---------+--------------------------------------------------------------
     age | -.3562867  .0990502   -3.60  0.000   -.5504216   -.1621519
   _cons | -.6482966    .09152   -7.08  0.000   -.8276725   -.4689207
------------------------------------------------------------------------

. di exp(-.3562867) * .0990502
.06936206
```

We find the value of 0.0693621, calculated using the delta method, to be identical with the results displayed in the previous risk ratio model. However, confidence intervals for risk ratios (or for incidence rate ratio count models) are determined by exponentiating the model-based confidence intervals. We shall examine these relationships in more detail in Chapters 6 and 7.

Given the same logic as employed above for the confidence intervals of risk ratios, the confidence interval of a regression coefficient that does not include the value of 0 indicates that the predictor significantly contributes to the model.

2.4 Risk difference

Risk difference is also referred to as *absolute risk reduction*, and can be used as a measure of risk reduction. The calculation of the statistic is quite simple if the components of a relative risk ratio have been computed. An example will make the relationship clear.

Consider the **titanic** data we have used to determine risks and risk ratios. Given the outcome, *survived*, and risk factor, *age*, as displayed in the Table 2.2, the risk of an adult surviving is: 442/1,207 = 0.36619718

The risk of child surviving: $57/109 = 0.52293578$

The risk difference: $0.36619718 - 0.52293578 = -0.1567386$

The survival rate for adults was some 16 percentile points lower than for children.

A model for this relationship is traditionally obtained by modeling a binomial generalized linear model with an identity link. However, the same coefficients and standard errors are obtained by modeling the data with an identity linked Poisson model with robust or sandwich standard errors. R's **glm** function does not permit an identity-link binomial model; we may obtain the identical results, however, using the identity-Poisson approach.

R: *Poisson model with identity link and robust standard errors*

```
idpoi <-glm(survived ~ age, family=poisson(link=identity),
  data=titanic)
summary(idpoi)
library(sandwich)
sqrt(diag(vcovHC(idpoi, type="HC0")))
```

```
. glm survived age, fam(poisson) link(id) nohead robust

------------------------------------------------------------------
         |             Robust
survived |      Coef. Std. Err.      z  P>|z|   [95% Conf. Interval]
---------+--------------------------------------------------------
     age | -.1567384   .049829  -3.15 0.002  -.2544015   -.0590754
   _cons |  .5229356 .0478591  10.93 0.000   .4291336    .6167377
------------------------------------------------------------------
```

Note that the model intercept, _cons, is the same value as the risk of *age* = 0 (i.e. of a child surviving).

Risk difference is provided on most epidemiological tables, but it is rarely used in a regression context (i.e. where the risk factor is adjusted by other explanatory predictors). A researcher must also be keenly aware of exactly what a difference in risk represents. For example, a 2% (0.02) risk difference may not represent a substantial difference when the difference is between 90% and 92% compared with 1% and 3%, where the increase is threefold. This problem does not affect relative risk ratios. We shall not discuss it further, but interesting research could be done on its importance and interpretational relationship to risk ratios.

2.5 The relationship of risk to odds ratios

The odds of an event occurring is simply the probability of an event occurring divided by the probability of the event not occurring.

$$\text{Odds} = \frac{p}{1-p} \tag{2.8}$$

p is the probability of an event occurring ($0 < p < 1$).

When comparing the odds of each level of a risk factor, x, the relationship of the odds of $x = 1$ to $x = 0$ is called the odds ratio, OR:

$$\text{OR} = \frac{p_1/(1-p_1)}{p_0/(1-p_0)} = \frac{p_1(1-p_0)}{p_0(1-p_1)} \tag{2.9}$$

When the probability of $x = 0$ is low (e.g. less than 10%), then the odds ratio and risk ratio are fairly close in value. Lower probabilities result in the two values getting closer. We return to the paradigm table used in 2.1 for examining the risk ratio given a 2×2 table.

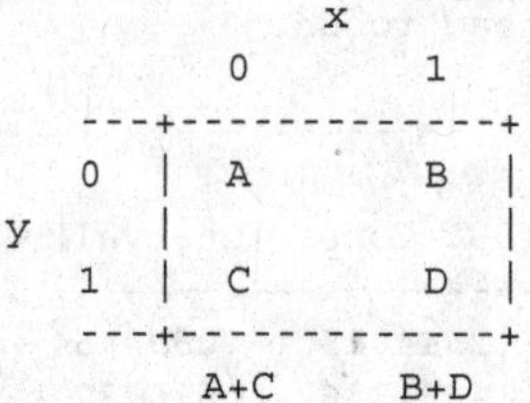

The odds of $x = 1$ is calculated as D/B.
The odds of $x = 0$ is calculated as C/A.
The odds ratio is the ratio of the two above odds:

$$\mathrm{OR} = \frac{\mathrm{D/B}}{\mathrm{C/A}} = \frac{\mathrm{AD}}{\mathrm{BC}} \tag{2.10}$$

The numerator of the rightmost term of equation 2.10 is the product of the number of cases for which $y = 1$, $x = 1$ and the number of cases for which $y = 0$, $x = 0$. This relationship may be expressed as "($y = 1$, $x = 1$) * ($y = 0$, $x = 0$)." The denominator is then the product of the two cross-terms, ($y = 1, x = 0$) * ($y = 0, x = 1$). Note also that the right term of (2.10) is identical to the left side of the right term of (2.3). The risk ratio is the same as (2.10), but with the product CD added to both the numerator and denominator. Therefore, it is plain that if C and D are small compared with A and B respectively, the values of the odds ratio and risk ratio will be close.

For example, using the same **titanic** data from (2.1):

```
           0         1
Survived | child     adults |
---------+------------------+
0     no |   52         765 |
         |                  |
1    yes |   57         442 |
---------+------------------+
   Total |  109       1,207
```

The odds of survival for an adult is $442/765 = 0.57777778$
The odds of survival of a child is $57/52 = 1.0961538$
The odds of an adult surviving compared to a child is

$$\frac{442/765}{57/52} = \frac{0.57777778}{1.0961538} = 0.52709554$$

We may compare the above calculations with a logistic regression model on the data.

R: *Poisson model of survived on age*

```
titanic3 <-glm(survived ~ age, family=binomial, data=titanic)
exp(coef(titanic3))
```

```
. glm survived age, fam(bin) nohead eform

------------------------------------------------------------------------
         |              OIM
survived | Odds Ratio Std. Err.    z  P>|z|  [95% Conf. Interval]
---------+--------------------------------------------------------------
     age |  .5270955 .1058718  -3.19 0.001  .3555642    .7813771
------------------------------------------------------------------------
```

The results are consistent. The odds of a child surviving compared with an adult is:

```
. di 1/.52709554
1.8971893
```

The odds of a child surviving on the Titanic is nearly twice that of adults.

Note that the risk of adults surviving compared to children is some 53%.

Risk ratios cannot be directly calculated from case-control data. In this type of study subjects are selected into the study based on having a condition or disease. They are compared with a control group of subjects who do not have the condition. The problem is that such data do not include the baseline or population values for subjects having the condition. It may be rare, or common. Calculating odds ratios, though, is not affected by this lack of information. However, if risk studies include population offsets, as will be described later in the text, it is possible to model risk given case-control studies, as well as prospective and experimental data. On the other hand, in the absence of offsets, odds ratios can be used in the place of risk when the incidence of a condition is low. Case-control studies are often used for rare conditions (e.g. disease), and may therefore be modeled for risk by means of a logistic model. See Hilbe (2009) for a complete discussion of these relationships.

2.6 Marginal probabilities: joint and conditional

The terms joint and conditional probabilities are frequently alluded to when discussing tables of counts, as well as statistical models. Later we shall discuss what are called marginal models, and will use marginals to define the calculation of probabilities. In this section we shall provide an outline as to what these two terms mean, and how they may be calculated for a 2×3 table.

Consider a table from the *Titanic* data providing the counts of passenger survival status based on their ticket class. A table was displayed in Section 2.2, but will be re-displayed here, with *Y* representing survival status (0/1) and *X* representing class (1, 2, 3)

```
                                   X
                            class (ticket)
           Survived | 1st class  2nd class  3rd class | Total
           ---------+-----------------------------------+------
   Y       0     no |       122        167        528 |   817
           1    yes |       203        118        178 |   499
margin ->  ---------+-----------------------------------+------
              Total |       325        285        706 | 1,316
                                                      ^
                                                    margin
```

The table margins are the lines separating the cell counts and summary values. The lowest horizontal and right vertical lines are the margins for this table. *Joint probabilities* are calculated for each cell as the cell value divided by the total number of counts in the table. Therefore, the joint probability of first class passengers who survived is 0.154. That is 15.4% of all passengers were survivors with first class tickets. We can also use the term joint probability in defining *marginal probabilities*; for example, 38% (0.3792) of the passengers survived, or 25% (0.247) of the passengers are first class. A table of joint probabilities is given below.

Joint probabilities

0.0927	0.1269	0.4012	0.6208
0.1543	0.0897	0.1353	0.3792
0.2470	0.2166	0.5365	1.000

Values outside margins are termed marginal probabilities. *Conditional probabilities*, on the other hand, are calculated with the marginals serving as the denominator. Conditional probabilities are provided for both the *Y* and *X* referenced cells, with each marginal term defining a set of conditional probabilities. The two tables below provide the conditional probabilities of *X* and *Y* respectively.

Conditional probabilities:

```
         Pr(Y|X)                                   Pr(X|Y)
            X                                         Y
    1       2       3                        1       2       3

  +-----------------------------        +-----------------------
0 | .3754  .5860  .7479 | .6208      0 | .1493   .2044   .6463 | 1
  |---------------------|              |-----------------------|
1 | .6246  .4140  .1521 | .3792      1 | .4068   .2365   .3567 | 1
  +-----------------------------        +-----------------------
    1       1       1                    .2470   .2166   .5365
```

For *X* marginals: of first-class passengers, 62% (0.6246) survived, while 15% of the third-class passengers survived. For *Y* marginals: 15% (0.1493) of those who failed to survive were first-class passengers. Of course, with added predictors, or table levels, the number of marginals and associated probabilities increase.

Observed versus predicted probabilities may be calculated for each level of a model, with overall differences indicating model fit. Cells, and groups of cells, in which the differences between observed and predicted vary more than other levels do may indicate relationships which the model poorly understands.

Summary

Nearly all count models are parameterized in terms of incidence rate ratios. As we shall discover, rate ratios are calculated from Poisson and most negative binomial models by exponentiating the respective model coefficients. Rate ratios are also calculated from multinomial logistic models, where the categories or levels of the response are independent of one another. Any order to the levels result in any one of a variety of ordered or partially ordered categorical response models; e.g. proportional odds model. Likewise, counts are considered to be independent of one another. We shall discuss this requirement in more detail later in the text since it directly affects the generation of negative binomial models.

3

Overview of count response models

Count response models are a subset of discrete response regression models. Discrete models address non-negative integer responses. Examples of discrete models include:

Binary:	binary logistic and probit regression
Proportional:	grouped logistic, grouped complementary loglog
Ordered:	ordinal logistic and ordered probit regression
Multinomial:	discrete choice logistic regression
Count:	Poisson and negative binomial regression

A count response consists of any discrete response of counts: for example, the number of hits recorded by a Geiger counter, patient days in the hospital, and goals scored at major contests. All count models aim to explain the number of occurrences, or counts, of an event. The counts themselves are intrinsically heteroskedastic, right skewed, and have a variance that increases with the mean of the distribution.

3.1 Varieties of count response model

Poisson regression is traditionally conceived of as the basic count model upon which a variety of other count models are based.[1] The Poisson distribution may be characterized as

$$f(y;\lambda) = \frac{e^{-\lambda_i}(\lambda_i)^{y_i}}{y_i!}, \quad y_i = 0, 1, 2, \ldots, n_i; \lambda > 0 \tag{3.1}$$

[1] The statistical analysis of frequency tables of counts, and statistical inference based on such tables, appears to have first been developed by Abu al-Kindi (801–873CE), a Persian mathematician who lived in present day Iraq. He was the first to use frequency analysis for cryptoanalysis, and can be regarded as the father of modeling count data, as well as perhaps the father of statistics. He was also primarily responsible for bringing "Arabic" numerals to the attention of scholars in the West.

where the random variable y is the count response and parameter λ is the mean. Often, λ is also called the rate or intensity parameter. Unlike most other distributions, the Poisson does not have a distinct scale parameter. Rather, the scale is assumed equal to 1.

In statistical literature, λ is also expressed as μ when referring to Poisson and traditional negative binomial (NB2) models. Moreover, μ is the standard manner in which the mean parameter is expressed in generalized linear models (GLM). Since we will be using the **glm** command or function for estimating many Poisson and negative binomial examples in this text, we will henceforth employ μ in place of λ for expressing the mean of a GLM model. We will use λ later for certain non-GLM count models.

The Poisson and negative binomial distributions may also include an exposure variable associated with μ. The variable t is considered to be the area in, or length of time during, which events or counts occur. This is typically called the *exposure*. If $t = 1$, then the Poisson probability distribution reduces to the standard form. If t is a constant, or varies between events, then the distribution can be parameterized as

$$f(y; \mu) = \frac{e^{-t_i \mu_i} (t_i \mu_i)^{y_i}}{y_i!} \tag{3.2}$$

When included in the data, the natural log of t is entered as an offset into the model estimation. Playing an important role in estimating both Poisson and negative binomial models, offsets are discussed at greater length in Chapter 6. A unique feature of the Poisson distribution is the relationship of its mean to the variance – they are equal. This relationship is termed *equidispersion*. The fact that it is rarely found in real data has driven the development of more general count models, which do not assume such a relationship.

Poisson counts, y, are examples of a Poisson process. Each event in a given time period or area that is subject to the counting process is independent of one another; they enter into each period or area according to a uniform distribution. The fact that there are y events in period A has no bearing on how many events are counted in period B. The rate at which events occur in a period or area is μ, with a probability of occurrence being μ times the length of the period or size of the area.

The Poisson regression model derives from the Poisson distribution. The relationship between μ, β, and x – the fitted mean of the model, parameters, and model covariates or predictors, respectively – is parameterized such that $\mu = \exp(x\beta)$. Here $x\beta$ is the linear predictor, which is also symbolized as η within the context of GLM. Exponentiating $x\beta$ guarantees that μ is positive for all values of η and for all parameter estimates. By attaching the subscript i

to μ, y, and x, the parameterization can be extended to all observations in the model. The subscript can also be used when modeling non-iid (independent and identically distributed) observations.

It should be explicitly understood that the response, or dependent variable, of a Poisson or negative binomial regression model, y, is a random variable specifying the count of some identified event. The explanatory predictors, or independent variables, X, are given as nonrandom sets of observations. Observations within predictors are assumed independent of one another, and predictors are assumed to have minimal correlation between one another.

As shall be described in greater detail later in this book, the Poisson model carries with it various assumptions in addition to those given above. Violations of Poisson assumptions usually result in overdispersion, where the variance of the model exceeds the value of the mean. Violations of equidispersion indicate correlation in the data, which affects standard errors of the parameter estimates. Model fit is also affected. Chapter 7 is devoted to this discussion.

A simple example of how distributional assumptions may be violated will likely be instructional at this point. We begin with the base count model – the Poisson. The Poisson distribution defines a probability distribution function for non-negative counts or outcomes. For example, given a Poisson distribution having a mean of 2, some 13% of the outcomes are predicted to be zero. If, in fact, we are given an otherwise Poisson distribution having a mean of 2, but with 50% zeros, it is clear that the Poisson distribution may not adequately describe the data at hand. When such a situation arises, modifications are made to the Poisson model to account for discrepancies in the goodness of fit of the underlying distribution. Models such as zero-inflated Poisson and zero-truncated Poisson directly address such problems.

The above discussion regarding distributional assumptions applies equally to the negative binomial. A traditional negative binomial distribution having a mean of 2 and an ancillary parameter of 1.5 yields a probability of approximately 40% for an outcome of zero. When the observed number of zeros substantially differs from the theoretically imposed number of zeros, the base negative binomial model can be adjusted in a manner similar to the adjustments mentioned for the Poisson. It is easy to construct a graph to observe the values of the predicted counts for a specified mean. The code can be given as in Table 3.1.

Note the differences in predicted zero count for Poisson compared with the negative binomial with heterogeneity parameter, α, having a value of 1.5. If α were 0.5 instead of 1.5, the predicted zeros would be 25%. For higher means, both predicted zero counts get progressively smaller. A mean of 5 results in approximately 1% predicted zeros for the Poisson distribution and 24% for the negative binomial with alpha of 1.5.

Table 3.1 R: *Poisson vs. negative binomial PDF*

```
obs <- 11
mu <- 2
y <- 0:10
yp2 <- (exp(-mu)*mu^y)/exp(log(gamma(y+1)))
alpha <- 1.5
amu <- mu*alpha
ynb2 = exp(
      y*log(amu/(1+amu))
    - (1/alpha)*log(1+amu)
    + log(gamma(y +1/alpha))
    - log(gamma(y+1))
    - log(gamma(1/alpha))
)
plot(y, ynb2, col="red", pch=5,
  main="Poisson vs Negative Binomial PDFs")
  lines(y, ynb2, col="red")
  points(y, yp2, col="blue", pch=2)
  lines(y, yp2, col="blue")
  legend(4.3,.40,
    c("Negative Binomial: mean=2, a=1.5",
      "Poisson: mean=2"),
    col=(c("red","blue")),
    pch=(c(5,2)),
    lty=1)
# FOR NICER GRAPHIC
zt <- 0:10                 #zt is Zero to Ten
x <- c(zt,zt)              #two zt's stacked for use with ggplot2
newY <- c(yp2, ynb2)       #Now stacking these two vars
Distribution <- gl(n=2,k=11,length=22,
label=c("Poisson","Negative Binomial")
)
NBPlines <- data.frame(x,newY,Distribution)
library("ggplot2")
ggplot(NBPlines, aes(x,newY,shape=Distribution,
  col=Distribution)) + geom_line() + geom_point()
```

```
. set obs 11
. gen byte mu = 2                 // mean = 2
. gen byte y = _n-1
* POISSON
. gen yp2 = (exp(-mu)*mu^y)/exp(lngamma(y+1))
```

```
* NEGATIVE BINOMIAL
. gen alpha = 1.5                // NB2 alpha=1.5
. gen amu = mu*alpha
. gen ynb2 = exp(y*ln(amu/(1+amu)) - (1/alpha)*ln(1+amu) +
      lngamma(y +1/alpha)- lngamma(y+1) - lngamma(1/alpha))
. label var yp2 "Poisson: mean=2"
. label var ynb2 "Negative Binomial: mean=2, a=1.5"
. graph twoway connected ynb2 yp2 y, ms(T d) ///
    title(Poisson vs Negative Binomial PDFs)
```

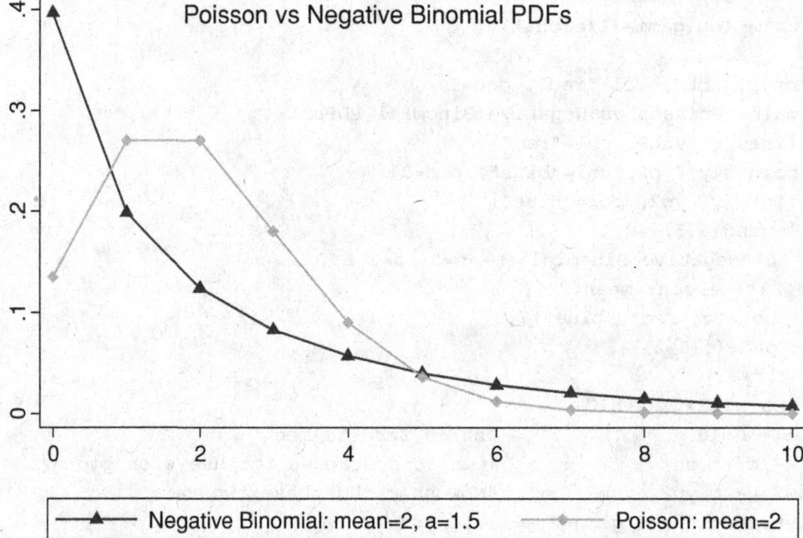

Figure 3.1 Poisson versus negative binomial PDF at mean = 2

Early on, researchers developed enhancements to the Poisson model, which involved adjusting the standard errors in such a manner that the presumed overdispersion would be dampened. Scaling of the standard errors was the first method developed to deal with overdispersion from within the GLM framework.

It is a particularly easy tactic to take when the Poisson model is estimated as a generalized linear model. We shall describe scaling in more detail in Section 7.3.1. Nonetheless, the majority of count models require more sophisticated adjustments than simple scaling.

Again, the negative binomial is normally used to model overdispersed Poisson data, which spawns our notion of the negative binomial as an extension of the Poisson. However, distributional problems affect both models, and negative binomial models themselves may be overdispersed. Both models can be extended in similar manners to accommodate any extra correlation or dispersion in the data that result in a violation of the distributional properties of each respective distribution (Table 3.1). The enhanced or advanced Poisson or negative binomial model can be regarded as a solution to a violation of the distributional assumptions of the primary model.

Table 3.2 enumerates the types of extensions that are made to both Poisson and negative binomial regression. Thereafter, we provide a bit more detail as to the nature of the assumption being violated and how it is addressed by each type of extension. Later chapters are devoted to a more detailed examination of each of these model types.

Earlier in this chapter we described violations of Poisson and negative binomial distributions as related to excessive zero counts. Each distribution has an expected number of counts for each value of the mean parameter; we saw how for a given mean, an excess – or deficiency – of zero counts results in overdispersion. However, it must be understood that the negative binomial has an additional ancillary or heterogeneity parameter, which, in concert with the value of the mean parameter, defines (in a probabilistic sense) specific expected values of counts. Substantial discrepancies in the number of counts, i.e. how many zeros, how many 1s, how many 2s, and so forth, observed in the data from the expected frequencies defined by the given mean and ancillary parameter (NB model), result in correlated data and hence overdispersion. The first two items in Table 3.2 directly address this problem.

Table 3.2 *Violations of distributional assumptions*

1. No zeros in data
2. Excess zeros in data
3. Data separable into two or more distributions
4. Censored observations
5. Truncated data
6. Data structured as panels: clustered and longitudinal data
7. Some responses occur based on the value of another variable
8. Endogenous variables in model

Violation 1: The Poisson and negative binomial distributions assume that zero counts are a possibility. When the data to be modeled originate from a generating mechanism that structurally excludes zero counts, then the Poisson or negative binomial distribution must be adjusted to account for the missing zeros. Such model adjustment is not used when the data can have zero counts, but simply do not. Rather, an adjustment is made only when the data must be such that it is not possible to have zero counts. Hospital length of stay is a good example. When a patient enters the hospital, a count of 1 is given. There are no lengths of stay recorded as zero days. The possible values for data begin with a count of 1. Zero-truncated Poisson and zero-truncated negative binomial models are normally used for such situations.

Violation 2: The Poisson and negative binomial distributions define an expected number of zero counts for a given value of the mean. The greater the mean, the fewer zero counts are expected. Some data, however, come with a high percentage of zero counts – far more than are accounted for by the Poisson or negative binomial distribution. When this occurs statisticians have developed regression models called zero-inflated Poisson (ZIP) and zero-inflated negative binomial (ZINB). The data are assumed to come from a mixture of two distributions where the structural zeros from a binary distribution are mixed with the non-negative integer outcomes (including zeros) from a count distribution. Logistic or probit regression is typically used to model the structural zeros, and Poisson or negative binomial regression is used for the count outcomes. If we were to apply a count model to the data without explicitly addressing the mixture, it would be strongly affected by the presence of the excess zeros. This inflation of the probability of a zero outcome is the genesis of the zero-inflated name.

Violation 3: When the zero counts of a Poisson or negative binomial model do not appear to be generated from their respective distributions, one may separate the model into two parts, somewhat like the ZIP and ZINB models above. However, in the case of hurdle models, the assumption is that a threshold must be crossed from zero counts to actually entering the counting process. For example, when modeling insurance claims, clients may have a year without claims – zero counts. But when one or more accidents occur, counts of claims follow a count distribution (e.g. Poisson or negative binomial). The logic of the severability in hurdle models differs from that of zero-inflated models. Hurdle models are sometimes called zero-altered models, giving us model acronyms of ZAP and ZANB.

Like zero-inflated models, hurdle or zero-altered algorithms separate the data into zero versus positive counts. However, the binary component of a hurdle model is separate from the count component. The binary component consists of two values, 0 for 0 counts and 1 for positive counts; the count component is a zero-truncated count model with integer values greater than 0. For zero-inflated models, the binary component is overlapped with the count component. The binary component model estimates the count of 0, with the count component estimating the full range of counts. The count of 0 is mixed into both components. Care must be taken when interpreting the binary component of zero-inflated compared with hurdle models.

The binary component of zero-inflated models is typically a logit or probit model, but complementatry loglog models are also used. The binary component of hurdle models is usually one of the above binomial models, but can be a censored-at-one Poisson, geometric or negative binomial as well. Zero-inflated likelihood functions therefore differ considerably from the likelihood functions of similar hurdle models. We shall address these differences in more detail in Chapter 11.

Violation 4: At times certain observations are censored from the rest of the model. With respect to count response models, censoring takes two forms. In either case a censored observation is one that contributes to the model, but for which exact information is missing.

The traditional form, which I call the econometric or cut parameterization, revalues censored observations as the value of the lower or upper valued non-censored observation. Left-censored data take the value of the lowest non-censored count; right-censored data take the value of the highest non-censored count. Another parameterization, which can be referred to as the survival parameterization, considers censoring in the same manner as is employed with survival models.That is, an observation is left-censored to when events are known to enter into the data; they are right-censored when events are lost to the data due to withdrawal from the study, loss of information, and so forth. The log-likelihood functions of the two parameterizations differ, but the parameter estimates calculated are usually not too different.

Violation 5: Truncated observations consist of those that are entirely excluded from the model, from either the lower, or left, or higher, or right side of the distribution of counts. Unlike the econometric parameterization of censoring described in Violation 4, truncated data are excluded, not revalued, from the model.

Violation 6: Longitudinal data come in the form of panels. For example, in health studies, patients given a drug may be followed for a period of time to ascertain effects occurring during the duration of taking the drug. Each patient may have one or more follow-up tests. Each set of patient observations is considered to be a panel. The data consist of a number of panels. However, observations within each panel cannot be considered independent – a central assumption of maximum likelihood theory. Within-panel correlation result in overdispersed data. Clustered data result in similar difficulties. In either case, methods have been developed to accommodate extra correlation in the data due to the within-panel correlation of observations. Such models, however, do require that the panels themselves are independent of one another, even though the observations within the panels are not. Generalized estimating equations (GEE), fixed-effects models, random-effects models, and mixed-effects and multilevel models have been widely used for such data.

Violation 7: Data sometimes come to us in such a manner that an event does not begin until a specified value of another variable has reached a certain threshold. One may use a selection model to estimate parameters of this type of data. Greene (1994) summarizes problems related to selection models.

Violation 8: At times there are variables not in the model which affect the values of predictors that are in the model. These influence the model, but only indirectly. They also result in model error, that is, variability in the model that is not accounted for by variables in the model. A standard model incorporates this extra variation into the error term. Models have been designed, however, to define endogenous variables that identify extra variation in the data that is not incorporated into existing predictors. Endogeneity can arise from a variety of sources (e.g. errors in measuring model predictors, variables that have been omitted from the model, and that have not been collected for inclusion in the data, predictors that have been categorized from continuous predictors, and for which information has been lost in the process, sample selection errors in gathering the data). The problem of endogeneity is of considerable importance in econometrics, and should be in other disciplines as well. Endogeneity is central to the discussion of Chapter 13, but is also inherent in the models addressed in Chapters 12 and 14.

Table 3.3 provides a schema of the major types of negative binomial regression models. A similar schema may also be presented characterizing varieties of the Poisson model. Some exceptions exist, however. Little development work has been committed to the exact statistical estimation of negative binomial parameters and standard errors. However, substantial work has been done on Poisson models of this type – particularly by Cytel Corp, manufacturers of LogXact software, and Stata Corporation. Additionally, models such as heterogeneous negative binomial, and NB-P have no comparative Poisson model.

3.2 Estimation

There are two basic approaches to estimating models of count data. The first is by full maximum likelihood estimation (MLE or FMLE), and the second is by an iteratively re-weighted least squares (IRLS) algorithm, which is based

Table 3.3 *Varieties of negative binomial model*

Model	
1 Negative binomial (NB)	
NB2	
NB1	
NB-C (canonical)	
NB-H (Heterogeneous negative binomial)	
NB-P (variety of generalized NB)	
2 Zero-adjusting models	
Zero-truncated NB	
Zero-inflated NB	
NB with endogenous stratification (G)	
Hurdle NB models	
NB-logit hurdle	// geometric-logit hurdle
NB-probit hurdle	// geometric-probit hurdle
NB-cloglog hurdle	// geometric-cloglog hurdle
3 Censored and truncated NB	
Censored NB-E: econometric parameterization	
Censored NB-S: survival parameterization	
Truncated NB-E: econometric parameterization	
4 Sample selection NB models	
5 Models that handle endogeneity	
Latent class NB models	
Two-stage instrumental variables approach	
Generalized method of moments (GMM)	
NB2 with an endogenous multinomial treatment variable	
Endogeneity resulting from measurement error	
6 Panel NB models	
Unconditional fixed-effects NB	
Conditional fixed-effects NB	
Random-effects NB with beta effect	
Generalized estimating equations	
Linear mixed NB models	
Random-intercept NB	
Random-parameter NB	
7 Response adjustment	
Finite mixture NB models	
Quantile count models	
Bivariate count models	
8 Exact NB model	
9 Bayesian NB with (a) gamma prior, (b) beta prior	

on a simplification of the full maximum likelihood method. IRLS is intrinsic to the estimation of generalized linear models (GLMs), as well as to certain extensions of the generalized linear model algorithm. Poisson, standard NB2, and canonical negative binomial (NB-C) regressions are GLM models and can be estimated using either IRLS or FMLE. We examine the details of both methods of estimation in the following chapter. However, it should be mentioned that, for many software applications, Poisson, NB2, and NB-C regression can only be estimated from within a generalized linear models IRLS algorithm; no FMLE methods are provided.

When the negative binomial model, conceived as a Poisson–gamma mixture model, is estimated by maximum likelihood the software typically uses a modification of the traditional Newton–Raphson estimating algorithm. A commonly used variety is the Marquardt (1963) modification, which has itself been amended for use in leading commercial packages, e.g. Stata and SAS. Such a method allows the estimation of the negative binomial ancillary or overdispersion parameter, which we refer to as α.

The negative binomial was first supported in generalized linear models software by Hilbe (1994a), who incorporated it into the GLM procedures of both XploRe and Stata. So doing allowed use of the full range of GLM fit and residual capabilities. However, since the standard IRLS GLM algorithm allows estimation of only the mean parameter, μ, or $\exp(x\beta)$, the ancillary or heterogeneity parameter must be entered into the GLM algorithm as a known constant. It is not itself estimated. The scale parameter for all GLM count models is defined as 1, and does not enter into the estimation process.

There are methods to point estimate α (Breslow, 1984; Hilbe, 1993a) based on an iterative covering algorithm with an embedded IRLS that forces the Pearson dispersion statistic to a value of 1.0. But the resulting estimated value of α typically differs somewhat from the value estimated using maximum likelihood. Moreover, no standard error is obtained for the estimated α using the cited IRLS approaches. Typically though, the difference in estimated α between the two methods is not very much. Both will be examined in detail in Chapter 7, together with recommendations of how they can be used together for a given research project.

Recently SAS, R, and Stata have enhanced their GLM algorithm for the estimation of the negative binomial by allowing the heterogeneity parameter, α, to be estimated in a subroutine outside of GLM, with the resulting value entered back into the GLM algorithm as a constant. Users of these software applications may also choose to select their own preferred value for α rather than to have the software estimate it. Reasons for doing this will be discussed later in Section 10.1. Regardless, this mechanism now allows GLM-based negative binomial models to provide full maximum likelihood estimates of coefficients and α, without the limitations imposed on estimation when using older GLM facilities.

3.3 Fit considerations

Fit statistics usually take two forms: so-called goodness-of-fit statistics and residual statistics. With respect to the negative binomial, goodness-of-fit generally relates to the relationship of the negative binomial model to that of a Poisson model on the same data. Assuming that the purpose of negative binomial modeling is to model otherwise Poisson overdispersed data, it makes sense to derive fit statistics aimed at determining whether a negative binomial model is statistically successful in that regard. Two of the commonly used fit statistics include a Score test, or Lagrange multiplier test, and a Vuong test (Vuong, 1989). These, as well as others, will later be examined in more detail.

Residual analyses of negative binomial models generally follow the model varieties constructed for members of the family of generalized linear models. These include unstandardized and standardized Pearson and deviance residuals, as well as the studentized forms for both. Additionally, Anscombe residuals have been developed for the NB2 negative binomial (Hilbe, 1994a; Hardin and Hilbe, 2001). In application, however, values of the Anscombe are similar to those of the standardized deviance residuals, with the latter being substantially less complicated to calculate. The majority of commercial GLM software packages now provide both standardized deviance residuals and Anscombe residuals as standard options. A commonly used residual fit analysis for negative binomial models plots standardized deviance residuals against the fitted values, μ.

Summary

Negative binomial models have been derived from two different origins. First, and initially, the negative binomial can be thought of as a Poisson–gamma mixture designed to model overdispersed Poisson count data. Conceived of in this manner, estimation usually takes the form of a maximum likelihood Newton–Raphson type algorithm. This parameterization estimates both the mean parameter, μ, as well as the ancillary or heterogeneity parameter, α. Extensions to this approach allow α itself to be parameterized (NB-H), as well as the negative binomial exponent (NB-P). Violations of distributional assumptions are addressed by various adjustments to the base negative binomial (NB2) model. Examples include models such as ZINB, zero-truncated negative binomial, and censored negative binomial regression.

Secondly, the negative binomial can be derived as a full member of the single parameter exponential family of distributions, and hence be considered as one of the generalized linear models. The value of this approach rests on the ability to evaluate the model using well-tested GLM goodness-of-fit statistics as well as to employ the host of associated GLM-defined residuals. Estimation in this case takes the form of Fisher scoring, or iteratively re-weighted least squares. Since the traditional GLM algorithm only allows estimation of location parameter θ (equation 4.1), which gives us the value of μ, the ancillary parameter α must be specified directly into the estimating algorithm as a known constant – it is not itself estimated. Although this is a drawback for its usefulness in modeling, the ability to assess fit in part offsets this problem. The fact, however, that popular software such as Stata, R, and SAS employ negative binomial estimating algorithms that also incorporate a maximum likelihood subroutine to estimate α, a researcher no longer needs to engage a two-part process. Estimation of a negative binomial (NB2) model is as easy as estimation of a Poisson model.

It is also important to remember that alternative parameterizations of the negative binomial require either maximum likelihood or quadrature estimation methods. Stata has a maximum likelihood negative binomial command called **nbreg**, which can be used to estimate traditional NB2 models, as well as NB1 models.

We shall next examine the two foremost methods of estimating negative binomial models, regardless of parameterization: full maximum likelihood and IRLS.

4
Methods of estimation

Two general methods are used to estimate count response models: (1) an iteratively re-weighted least squares (IRLS) algorithm based on the method of Fisher scoring, and (2) a full maximum likelihood Newton–Raphson type algorithm. Although the maximum likelihood approach was first used with both the Poisson and negative binomial, we shall discuss it following our examination of IRLS. We do this for strictly pedagogical purposes, which will become evident as we progress.

It should be noted at the outset that IRLS is a type or subset of maximum likelihood which can be used for estimation of generalized linear models (GLM). Maximum likelihood methods in general estimate model parameters by solving the derivative of the model log-likelihood function, termed the *gradient*, when set to zero. The derivative of the gradient with respect to the parameters is called the *Hessian* matrix, upon which model standard errors are based. Owing to the unique distributional structure inherent to members of GLM, estimation of model parameters and standard errors can be achieved using IRLS, which in general is a computationally simplier method of maximum likelihood estimation. Both methods are derived, described, and related in this chapter.

4.1 Derivation of the IRLS algorithm

The traditional generalized linear models (GLM) algorithm, from the time it was implemented in GLIM (generalized linear interactive modeling) through its current implementations in Stata, R, SAS, SPSS, GenStat, and other statistical software, uses some version of an IRLS estimating algorithm. This method arises from Fisher scoring, which substitutes the expected Hessian matrix for the observed Hessian matrix in a Taylor series-defined updating step for a solution of the estimating equation. The resulting Newton–Raphson or updating equation for the regression coefficients may be written in terms of ordinary least squares (OLS) owing to the simplification afforded by Fisher scoring. The reason for the initial development of GLM had much to do with the difficulty of modeling individual GLM models using full maximum likelihood algorithms. In the late 1960s and early 1970s, statistical software was limited to mainframe batch runs. That is, one wrote an estimating algorithm in a higher programming language such as FORTRAN, tied it together with data stored on cards, and submitted it to a mainframe, which usually resided at a remote site. If one desired to make changes, then the entire batch file required re-writing. Each submission to the mainframe had a cost applied to it, normally charged to the department of the user. Problems with initial values, difficulties with convergence, and other such difficulties resulted in a modeling project taking substantial time and money.

When Nelder and Wedderburn (1972) proposed an IRLS algorithm for estimating regression models based on the exponential family of distributions, specific models still had to be submitted on cards and to a mainframe. The difference was, however, that there were substantially fewer difficulties with convergence, and one algorithm could be used for all members of the class. All that required alteration from one model type to another (e.g. a logit model compared with a probit model, or a gamma compared with a Poisson), was a change in the specification of the link, variance, and family functions. The algorithm took care of the rest. Time savings transferred to money savings.

As previously mentioned, GLIM was the first commercial software implementation of GLM. When desktop computing became available on PCs starting in August 1981, Numerical Algorithms Group in the UK, the manufacturer of GLIM, quickly implemented a desktop version of the GLIM software. For the first time models such as logit, probit, Poisson, and gamma could be modeled in an interactive and inexpensive manner. An international GLIM user group emerged whose members designed macros to extend the basic offerings. The negative binomial was one of these macros, but was not published to the general user base.

The important point in this discussion is that the IRLS algorithm was designed to be a single covering algorithm for a number of related models. Also important to understand is that the IRLS algorithm is a simplified maximum likelihood algorithm and that its derivation is similar to that of the derivation of general Newton–Raphson type models. We turn to this demonstration next.

IRLS methodology, like maximum likelihood methodology in general, is ultimately based on a probability distribution or probability mass function. Generalized linear models software typically offers easy specification of variance functions defined by eight families of distributions, each of which is a probability function. These include:

GLM DISTRIBUTIONAL FAMILIES

Gaussian	Bernoulli
Binomial	Gamma
Inverse Gaussian	Poisson
Geometric	Negative binomial

Each of these probability functions represents a model or family within the framework of generalized linear models. The key concept is that they all are separate parameterizations of an underlying single-parameter exponential probability distribution that is commonly expressed as:

EXPONENTIAL FAMILY PDF

$$f(y;\theta,\phi) = \exp\left\{\frac{y_i\theta_i - b(\theta_i)}{\alpha_i(\phi)} + c(y_i;\phi)\right\} \tag{4.1}$$

where

θ_i is the canonical parameter or link function
$b(\theta_i)$ is the cumulant
$\alpha(\phi)$ is the scale parameter, set to one in discrete and count models
$C(y_i;\phi)$ is the normalization term, guaranteeing that the probability function sums to unity.

The exponential family form is unique in that the first and second derivatives of the cumulant, with respect to θ, produce the mean and variance functions, respectively. The important point to remember is that if one can convert a probability function into exponential family form, its unique properties can be easily used to calculate the mean and variance, as well as facilitate estimation of parameter estimates based on the distribution. All members of the class of generalized linear models can be converted to exponential form, with

$$b'(\theta_i) = \text{mean}$$
$$b''(\theta_i) = \text{variance}$$

How the exponential family characterizes the above relationships with respect to Poisson and negative binomial (NB2) regression will be addressed in the following chapters. We now turn to the derivation of the estimating equations which are used to estimate the parameters and associated statistics for generalized linear models. Thereafter we narrow the discusson to a focus on the algorithms used to estimate Poisson and negative binomial models.

We may express the GLM probability function as

$$f(y;\theta,\phi) \tag{4.2}$$

where y is the response, θ is the location or mean parameter, and ϕ is the scale parameter. Count models, by definition, set the scale to a value of 1. The outcome y, of course, has distributional properties appropriate to the family used in estimation. Probability functions determine properties of the response, given values of the mean and scale parameters. Maximum likelihood, on the other hand, bases estimation on the likelihood. The likelihood function is the complement of the probability function. Rather than data being determined on the basis of mean and scale values, the mean, and possibly scale, parameters are estimated on the basis of the given data. The underlying goal of a likelihood is to determine which parameters make the given data most likely. This parameterization can be characterized as

$$L(\theta,\phi;y) \tag{4.3}$$

Statisticians normally employ the natural log of the likelihood function in order to facilitate estimation. The prime reason is that the observations and their respective parameter estimates enter the likelihood function in a multiplicative manner. However, it is much easier to estimate parameters if their relationship is instead additive. In fact, for many modeling situations, using a likelihood function rather than a log-likelihood function would prevent the estimation process from getting off the ground. An excellent discussion of this topic, together with numeric examples, can be found in Gould *et al.* (2006) and Edwards (1972).

The log-likelihood function can be written as

$$\mathcal{L}(\theta,\phi;y) \tag{4.4}$$

The first derivative of the log-likelihood function is called the *gradient*; the second derivative is the *Hessian*. These functions play an essential role in the estimation process for both IRLS and traditional Newton–Raphson type algorithms.

Gradient – first derivative of $\mathcal{L}$
Hessian – second derivative of $\mathcal{L}$

Derivation of the IRLS algorithm is based on a modification of a two-term Taylor expansion of the log-likelihood function. We can use the Taylor expansion to find a value X_1 for which $f(X_1) = 0$. In this form, the Taylor expansion appears as

$$0 = f(X_0) + (X_1 - X_0)\, f'(X_0) \quad \cdot \\ + \frac{(X_1 - X_0)2}{2!} f''(X_0) + \frac{(X_1 - X_0)^3}{3!} f'''(X_0) + \cdots \tag{4.5}$$

The first two terms reduce to

$$0 = f(X_0) + (X_1 - X_0) f'(X_0) \tag{4.6}$$

which can be recast to

$$X_1 = X_0 - \frac{f(X_0)}{f'(X_0)} \tag{4.7}$$

The Newton–Raphson method of estimation adopts the above by using the score or gradient of the log-likelihood function as the basis of parameter estimation. The form is

$$\beta_r = \beta_{r-1} - \frac{\partial \mathcal{L}\,(\beta_{r-1})}{\partial \mathcal{L}'\,(\beta_{r-1})} \tag{4.8}$$

where

$$\partial \mathcal{L} = \frac{\partial \mathcal{L}}{\partial \beta} \quad \text{and} \quad \partial^2 L = \frac{\partial^2 \mathcal{L}}{\partial \beta\, \partial \beta'} \tag{4.9}$$

The first derivative of the log-likelihood function is called the Fisher *score* function, or the gradient. If the log-likelihood function is concave, one may calculate the maximum likelihood parameter estimates by setting the score function to zero and solving with respect to β.

The second derivative of the log-likelihood function is called the *Hessian matrix*. It indicates the extent to which the log-likelihood function is peaked rather than flat. Minus the inverse of the Hessian gives us the *variance–covariance matrix*. Parameter standard errors are based on the diagonal elements of this matrix, which is also called the *information matrix*.

In the traditional nomenclature, we let

$$\mathbf{U} = \partial \mathcal{L} \quad \text{and} \quad \mathbf{H} = \partial^2 \mathcal{L} \tag{4.10}$$

Then we may estimate parameter estimates, β_r, by employing a Newton–Raphson type updating algorithm, displayed as:

$$\beta_r = \beta_{r-1} - \mathbf{H}^{-1}\mathbf{U} \tag{4.11}$$

where

$$\mathbf{H} = \mathbf{H}_{r-1} \quad \text{and} \quad \mathbf{U} = \mathbf{U}_{r-1}$$

The Newton–Raphson algorithm estimates β_r, the model parameter estimates, by iteratively finding solutions for **H** and **U**, which define β_r; β_r resets itself to β_{r-1} in each subsequent iteration until some predefined threshold is reached. We shall see, however, that the matrix **H** used by the Newton–Raphson is the observed information matrix (OIM). IRLS, on the other hand, defines **H** as the expected information matrix (EIM). We next turn to how both methods define **U**.

4.1.1 Solving for $\partial\mathcal{L}$ or U – the gradient

In exponential family form, the log-likelihood function is expressed as

$$\mathcal{L}(\theta; y, \phi) = \sum_{i=1}^{n} \frac{y_i\theta_i - b(\theta_i)}{\alpha_i(\phi)} + c(y_i; \phi) \tag{4.12}$$

Solving for $\mathcal{L}$, with respect to β, can be performed using the chain rule

$$\frac{\partial\mathcal{L}}{\partial\beta_j} = \sum_{i=1}^{n} \left(\frac{\partial\mathcal{L}_i}{\partial\theta_i}\right)\left(\frac{\partial\theta_i}{\partial\mu_i}\right)\left(\frac{\partial\mu_i}{\partial\eta_i}\right)\left(\frac{\partial\eta_i}{\partial\beta_j}\right) \tag{4.13}$$

Solving for each term yields

$$\frac{\partial\mathcal{L}}{\partial\beta_j} = \sum_{i=1}^{n} \frac{y_i\theta_i - b'(\theta_i)}{\alpha_i(\phi)} + \sum_{i=1}^{n} \frac{y_i - \mu_i}{\alpha_i(\phi)} \tag{4.14}$$

We obtain the above formula by solving each of the terms of the chain. We have $b'(\theta_i) = \mu_i$

$$\frac{\partial\mu_i}{\partial\theta_i} = \frac{\partial b'(\theta_i)}{\partial\theta_i} = b''(\theta_i) = V(\mu_i), \quad \frac{\partial\theta_i}{\partial\mu_i} = \frac{1}{V(\mu_i)} \tag{4.15}$$

and

$$\frac{\partial\eta_i}{\partial\beta_j} = \frac{\partial(x_i\beta_j)}{\partial\beta_j} = x_{ij}, \quad \text{since } \eta_i = x_i\beta_j \tag{4.16}$$

and

$$\frac{\partial\mu_i}{\partial\eta_i} = [g^{-1}(\eta_i)]' = \frac{1}{\partial\eta_i/\partial\mu_i} = \frac{1}{g'(\mu_i)} \tag{4.17}$$

which is the derivative of the link function with respect to μ. Recall that η is the inverse link function, and for canonical models is the linear predictor.

Substitutions of expressions specify that the maximum likelihood estimator of β is the solution of the vector-based estimating equation

$$\sum_{i=1}^{n} \frac{(y_i - \mu_i)x_i}{\alpha_i(\phi)V(\mu_i)g'(\mu_i)} = \sum_{i=1}^{n} \frac{(y_i - \mu_i)x_i}{\alpha_i(\phi)V(\mu_i)}\left(\frac{\partial\mu_i}{\partial\eta_i}\right) = 0 \tag{4.18}$$

where y and μ are the response and fitted variables respectively, x is a $1\times$j row vector, and the resulting sum is a j$\times 1$ column vector.

The next step in the derivation takes two turns, based on a decision to use the observed or expected information matrix. Again, Newton–Raphson type maximum likelihood estimation uses the observed matrix; IRLS, or Fisher scoring, uses the expected. We first address the latter.

4.1.2 Solving for $\partial^2\mathcal{L}$

The traditional GLM algorithm substitutes I for H, the Hessian matrix of observed second derivatives. I is the second of two equivalent forms of Fisher information given by

$$\mathbf{I} = -\mathbf{E}\left[\frac{\partial^2\mathcal{L}}{\partial\beta_j\partial\beta_k}\right] = \mathbf{E}\left[\frac{\partial\mathcal{L}}{\partial\beta_j}\frac{\partial\mathcal{L}}{\partial\beta_k}\right] \tag{4.19}$$

Solving the above yields

$$\mathbf{I} = \frac{\partial}{\partial\beta_j}\left[\frac{(y_i-\mu_i)\,x_j}{\alpha_i\,(\phi)\,V\,(\mu_i)}\left(\frac{\partial}{\partial\eta}\right)_i\right] \times \frac{\partial}{\partial\beta_k}\left[\frac{(y_i-\mu_i)\,x_k}{\alpha_i\,(\phi)\,V\,(\mu_i)}\left(\frac{\partial}{\partial\eta}\right)_i\right] \tag{4.20}$$

$$\mathbf{I} = \frac{(y_i-\mu_i)^2\,x_jx_k}{\{\alpha_i\,(\phi)\,V\,(\mu_i)\}^2}\left(\frac{\partial\mu}{\partial\eta}\right)_i^2 \tag{4.21}$$

Since

$$(y_i-\mu_i)^2 = \alpha_i(\phi)V\,(\mu_i) \tag{4.22}$$

and letting

$$V(y_i) = \alpha_i(\phi)V(\mu_i) = (y_i-\mu_i)^2 \tag{4.23}$$

I therefore becomes formulated as

$$\mathbf{I} = \frac{x_jx_k}{V(y_i)}\left(\frac{\partial\mu}{\partial\eta}\right)_i^2 = \frac{x_jx_k}{V(y_i)g'^2} \tag{4.24}$$

Putting the various equations together we have

$$\beta_r = \beta_{r-1} - \left[\frac{x_jx_k}{V(y_i)}\left(\frac{\partial\mu}{\partial\eta}\right)_i^2\right]^{-1}\left[\frac{(y_i-\mu_i)\,x_k}{V(y_i)}\left(\frac{\partial\mu}{\partial\eta}\right)_i\right] \tag{4.25}$$

Multiplying both sides by $\mathbf{I}$ yields

$$\left[\frac{x_jx_k}{V(y_i)}\left(\frac{\partial\mu}{\partial\eta}\right)_i^2\right]\beta_r = \left[\frac{x_jx_k}{V(y_i)}\left(\frac{\partial\mu}{\partial\eta}\right)_i^2\right]\beta_{r-1} + \left[\frac{(y_i-\mu_i)\,x_k}{V(y_i)}\left(\frac{\partial\mu}{\partial\eta}\right)_i\right] \tag{4.26}$$

We next let weight **W** equal

$$\mathbf{W} = \frac{1}{V(y_i)}\left(\frac{\partial\mu}{\partial\eta}\right)_i^2 \tag{4.27}$$

with the linear predictor, η, given as

$$\eta_i = x_{ik}\beta_{r-1} \tag{4.28}$$

where η_i is evaluated at the $(n-1)$th iteration.

We next convert the above algebraic representation (equation 4.26) to matrix form. This can be done in parts. First, given the definition of **W** above, the following substitution may be made

$$\left[\frac{x_j x_k}{V(y_i)}\left(\frac{\partial\mu}{\partial\eta}\right)_i^2\right]\beta_r = \left[\mathbf{X}'\mathbf{W}\mathbf{X}\right]\beta_r \tag{4.29}$$

Secondly, recalling the definition of $V(y)$ and **W**, we have

$$\frac{(y_i-\mu_i)\,x_k}{V(y_i)}\left(\frac{\partial\mu}{\partial\eta}\right)_i = \frac{(y_i-\mu_i)\,x_k}{\frac{1}{W}\left(\frac{\partial\mu}{\partial\eta}\right)_i^2}\left(\frac{\partial\mu}{\partial\eta}\right)_i \tag{4.30}$$

Thirdly, since $\eta_i = x_{ik}\,\beta_{r-1}$, we have, in matrix form

$$\left[\frac{x_j x_k}{V(y_i)}\left(\frac{\partial\mu}{\partial\eta}\right)_i^2\right]\beta_{r-1} = \mathbf{X}'\mathbf{W}\boldsymbol{\eta}_i \tag{4.31}$$

Combining the terms involved, we have

$$[\mathbf{X}'\mathbf{W}\mathbf{X}]\boldsymbol{\beta}_r = \mathbf{X}'\mathbf{W}\boldsymbol{\eta}_i + \left[\frac{(y_i-\mu_i)\,x_k}{\frac{1}{W}\left(\frac{\partial\mu}{\partial\eta}\right)_i^2}\left(\frac{\partial\mu}{\partial\eta}\right)_i\right] \tag{4.32}$$

$$[\mathbf{X}'\mathbf{W}\mathbf{X}]\boldsymbol{\beta}_r = \mathbf{X}'\mathbf{W}\boldsymbol{\eta}_i + \left[X_k W\,(y_i-\mu_i)\left(\frac{\partial\eta}{\partial\mu}\right)_i\right] \tag{4.33}$$

Finally, letting z, the model working response, be defined as

$$z_i = \eta_i + (y_i-\mu_i)\left(\frac{\partial\eta}{\partial\mu}\right)_i \tag{4.34}$$

we have

$$[\mathbf{X}'\mathbf{W}\mathbf{X}]\boldsymbol{\beta}_r = \mathbf{X}'\mathbf{W}\mathbf{z} \tag{4.35}$$

so that, by repositioning terms, $\boldsymbol{\beta}_r$ is equal to

$$\boldsymbol{\beta}_r = [\mathbf{X}'\mathbf{W}\mathbf{X}]^{-1}\mathbf{X}'\mathbf{W}\mathbf{z} \tag{4.36}$$

which is a weighted regression matrix used to iteratively update estimates of parameter vector β_r, as well as values for μ, η, and the deviance function. Iteration typically culminates when the difference in deviance values between two iterations is minimal, usually 10^{-6}. Some software uses the minimization of differences in the log-likelihood function as the basis of iteration. Others use differences in parameter estimates. In either case, the results are statistically identical. However, since the deviance is itself used to assess the fit of a GLM model, as well as being a term in a version of the goodness-of-fit BIC statistic, it has enjoyed more use in commercial software implementations of GLM. The log-likelihood function is also used to assess fit, and is a term in both the AIC goodness-of-fit statistic and the most popular parameterizations of the BIC statistic. We address these fit tests at a later point in our discussion.

The use of deviance over log-likelihood is a matter of preference and tradition. We generally calculate both statistics. Some software iterates based on the deviance statistic, then calculates the log-likelihood function at the end of the iteration process from the final values of μ and η, thereby providing a wider range of post-estimation fit statistics.

Recall that the matrix form of the estimation relationship between parameter and data for ordinary least squares regression (OLS) is

$$\beta_r = [\mathbf{X}'\mathbf{X}]^{-1}\mathrm{Xy} \tag{4.37}$$

The formula we derived is simply a weighted version of the OLS algorithm. Since the IRLS algorithm is iterative, and cannot be solved in one step for models other than the basic Gaussian model, the response is redefined as a function of the linear predictor – hence the value of z rather than y. A consideration of the iteration or updating process is our next concern.

4.1.3 The IRLS fitting algorithm

The IRLS algorithm, using an expected information matrix, may take one of several forms. Not using subscripts, a standard schema is:

1 Initialize the expected response, μ, and the linear predictor, η, or $g(\mu)$.
2 Compute the weights as

$$W^{-1} = Vg'(\mu)^2 \tag{4.38}$$

where $g'(\mu)$ is the derivative of the link function and V is the variance, defined as the second derivative of the cumulant, $b''(\theta)$.

3 Compute a working response, a one-term Taylor linearization of the log-likelihood function, with a standard form of

$$z = \eta + (y - \mu)\, g'(\mu) \tag{4.39}$$

4 Regress z on predictors $X_1, \ldots, X_n$ with weights, W, to obtain updates on the vector of parameter estimates, β.
5 Compute η, or $x\beta$, based on the regression estimates.
6 Compute μ, or $E(y)$, as $g^{-1}(\eta)$.
7 Compute the deviance or log-likelihood function.
8 Iterate until the change in deviance or log-likelihood between two iterations is below a specified level of tolerance, or threshold.

Again, there are many modifications to the above scheme. However, most traditional GLM software implementations use methods similar to the above.

The GLM IRLS algorithm for the general case is presented in Table 4.1. The algorithm can be used for any member of the GLM family. We later demonstrate how substitution of specific functions into the general form for link, $g(\mu)$,

Table 4.1 *Standard GLM estimating algorithm (expected information matrix)*

```
Dev = 0
μ = (y + 0.5)/(m + 1)          // binomial
μ = (y + mean(y))/2            // non-binomial
η = g(μ)                       // linear predictor
WHILE (abs(ΔDev) > tolerance){
w = 1 / (Vg'²)
z = η + (y - μ)g' - offset
β = (X'wX)⁻¹X'wz
η = Xβ + offset
μ = g⁻¹(η)
Dev0 = Dev
Dev = Deviance function
ΔDev = Dev - Dev0
}
Chi2 = Σ(y - μ)² / V(μ)
AIC = (-2LL + 2p) /n        // AIC at times defined w/o n
BIC = Dev - (dof)ln(n)      // alternative def. exist
Where p = number of model predictors + const
n = number of observations in model
dof = degrees of freedom (n - p)
```

inverse link, $g^{-1}(\eta)$, variance, V, and deviance or log-likelihood function create different GLM models. All other aspects of the algorithm remain the same, hence allowing the user to easily change models. Typically, with parameter estimates being of equal significance, the preferred model is the one with the lowest deviance, or highest log-likelihood, as well as the lowest AIC or BIC statistic. AIC is the acronym for *Akaike Information Criterion*, which is based on the log-likelihood function; BIC represents *Bayesian Information Criterion*, which was originally based on the deviance, but is now usually formulated on the log-likelihood.

These will be discussed at greater length later in the text. For count response models, statistics reflecting overdispersion need to be considered as well.

Table 4.1 provides a schematic view of the IRLS estimating algorithm, employing the traditional GLM expected information matrix for the calculation of standard errors. The algorithm relies on change in the deviance (*Dev*) value as the criterion of convergence. We have also added formulae for calculating the Pearson *chi*2, the AIC, and BIC statistics.

Other terms needing explanation are m, the binomial denominator, g', the first derivative of the link, and the two variables: *Dev*0, the value of the deviance in the previous iteration, and ΔDev, the difference in deviances between iterations. When the difference reaches a small value, or tolerance, somewhere in the range of 10^{-6}, iterations cease and the resultant parameter estimates, standard errors, and so forth are displayed on the screen.

There are of course variations of the algorithm found in the table. But the variations largely deal with whether iteration takes the form of a DO-WHILE loop, an IF-THEN loop, or some other looping technique. The form of Table 4.1 is that used for Stata's original GLM algorithm.

4.2 Newton–Raphson algorithms

In this section we discuss the derivation of the Newton–Raphson type algorithm. Until recently, the only method used to estimate the standard negative binomial model was by maximum likelihood estimation using a Newton–Raphson based algorithm. All other varieties of negative binomial are still estimated using a Newton–Raphson based routine. We shall observe in this section, though, that the iteratively re-weighted least squares method we discussed in the last section, known as Fisher scoring, is a subset of the Newton–Raphson method. We conclude by showing how the parameterization of the GLM mean, μ, can be converted to $x'\beta$.

4.2.1 Derivation of the Newton–Raphson

There are a variety of Newton–Raphson algorithms. Few software programs use the base version, which is a simple root-finding procedure. Rather, statistical software uses one of several types of modified Newton–Raphson algorithm to produce maximum likelihood estimates of model parameters. A type of modified Marquardt algorithm is perhaps the most popular commercial software implementation. Moreover, complex models require methods other than Newton–Raphson or Marquardt. Quadrature methods are commonly used with random-effects and mixed models; simulation-based methods are employed when no other estimating algorithm appears to work, or when estimation takes an inordinate amount of time.

In any event, a marked feature of the Newton–Raphson approach is that standard errors of the parameter estimates are based on the observed information matrix. However, the IRLS algorithm can be amended to allow calculation of observed information-based standard errors. The theoretical rationale and calculational changes required for being able to do this is detailed in Hardin and Hilbe (2001). GLM software such as SAS, Stata, XploRe, and LIMDEP allows the user to employ either the expected or observed information matrix. Note, though, that the more complex observed matrix reduces to the expected when the canonical link is used for the model. The log link is canonical for the Poisson, but not for the negative binomial. Therefore, which algorithm, and which modification, is used will have a bearing on negative binomial standard errors, and on the displayed significance of parameter estimates. Differences between the two matrices are particularly notable when modeling small numbers of observations.

Derivation of terms for the estimating algorithm begin as a Taylor linearization and continue through the calculation of the gradient, or first derivative of the likelihood function. Fisher scoring, used as the basis of the GLM estimating algorithm, calculates the matrix of second derivatives based on the expected information matrix. Equation 4.19 provides the usual manner of expressing this form of information and derivatives.

Newton–Raphson methodology, on the other hand, calculates the second derivatives of the likelihood on the basis of the observed information matrix, which allows estimation of likelihood-based algorithms other than the more limited exponential family form

$$H = \left[\frac{\partial^2 L}{\partial \beta_j \beta_k}\right] = \sum_{i=1}^{n} \frac{1}{\alpha_i(\phi)} \left[\frac{\partial}{\partial \beta_k}\right] \left\{ \frac{y_i - \mu_i}{V(\mu_i)} \left(\frac{\partial \mu}{\partial \eta}\right)_i x_j x_k \right\} \tag{4.40}$$

Solved, the above becomes

$$
\begin{aligned}
&\underbrace{-\sum_{i=1}^{n} \frac{1}{\alpha_i\,(\phi)} \left[\frac{1}{V(\mu_i)} \left(\frac{\partial \mu}{\partial \eta}\right)_i^2 - (\mu_i - y_i)\right.}_{\text{EIM}} \\
&\underbrace{\times \left.\left\{ \frac{1}{V(\mu_i)^2} \left(\frac{\partial \mu}{\partial \eta}\right)_i^2 \frac{\partial V(\mu_i)}{\partial \mu_i} - \frac{1}{V(\mu_i)} \left(\frac{\partial^2 \mu}{\partial \eta^2}\right)_i \right\}\right]}_{\text{OIM}} x_{ji} x_{ki}
\end{aligned}
\tag{4.41}
$$

In the case of the canonical link, terms including and following the term $-(\mu - y)$ in the above formula cancel and reduce to the value of the expected information. Compare (4.41) with (4.21). A single line is drawn under the formula for the expected information matrix. The double line rests under the added terms required for the observed information. Table 4.2 provides a schema for the modified Newton–Raphson algorithm used for SAS's GENMOD procedure.

We note here that this algorithm uses a convergence method based on the elementwise absolute differences between the vector of parameter estimates. βn represents the new β, βc represents the previously calculated, but current β. The intercept, sometimes represented as α_0, is included in comparison vectors. Elements of the entire parameter vector, $\alpha_0 + \beta_1 + \beta_2 + \cdots + \beta_j$, must not change (much) from one iteration to another in order for the algorithm to converge.

The Newton–Raphson type algorithm takes the general form of Table 4.3. Initial values must be provided to the algorithm at the outset. Some software set initial values to all zeros or all ones. Others calculate a simpler model, perhaps an OLS regression, to obtain initial parameter estimates. Negative binomial algorithms typically use the parameter estimates from a Poisson model on the same data when initializing parameters.

The algorithm employs a maximum difference in log-likelihood functions as well as a maximum difference in parameter estimates as the criterion of convergence. The first terms in the algorithm are calculated by the means shown in our previous discussion. The observed information matrix is used to calculate H. Maximum likelihood parameter estimates are calculated in line four of the loop. The remaining terms deal with the updating process. One can observe the similarity in the algorithms presented in Table 4.2 and Table 4.3.

Table 4.2 *A Newton–Raphson algorithm*

```
g = g(μ) = link
g_ = g'(μ) = 1st derivative of link, wrt μ
g__ = g"  (μ) = 2nd derivative of link, wrt μ
V = V(μ) = variance
V_ = V(μ) = derivative of variance
m = binomial denominator
y = response
p = prior weight
φ = phi, a constant scale parameter
off = offset
μ = (y + mean(y))/2          // binomial = (y+0.5)/(m+1)
η = g
βn = 0
while MAX(ABS(βn - βc)) > tolerance {
  βc = βn
  z = p(y - μ)/(Vg'φ)                // a column vector
  s = X'z                             // gradient
  We = p/(φVg'²) // weight : expected IM
  Wo = We + p(y - μ){(Vg" + V'g')/(V²g'³φ)} // obs. IM
  Wo = diag(Wo) // diagonal Wo
  H = -X'WoX // Hessian
  βn = βc - H⁻¹s :==: βc + (X'WoX)⁻¹ X' (p(y - μ))
                :==: (X'WoX)⁻¹ X'W[η + (y - μ)g']
                :==: (X'WoX)⁻¹ X'Wz // if z=η + (y-μ)g'
  η = X'β + off                        // linear predictor
  μ = g⁻¹(η) // inverse link
}
```

Table 4.3 *Maximum likelihood: Newton–Raphson*

```
Initialize β
WHILE (ABS(βn - βo) > tol & ABS(Ln - Lo) > tol) {
    g = ∂L/∂β
    H = ∂2L/∂β²
    βo = βn
    βn = βo - H⁻¹g
    Lo = Ln
    Ln
    }
V(μ)
```

4.2.2 GLM with OIM

The GLM algorithm can be modified to accommodate the calculation of standard errors based on the observed information matrix, which is the inverse of the negative Hessian. The main feature of the alteration is an extension of the w term in the standard GLM. w is defined as

$$w_i = 1/\{V(\mu_i)g'(\mu_i)^2\} \tag{4.42}$$

where $V(\mu)$ is the variance function and $g'(\mu)^2$ is the square of the derivative of the link function. GLM terminology calls this a model weight. Weighting in terms of frequency weights are termed prior weights, and are entered into the algorithm in a different manner.

Returning to the modification of w necessary to effect an observed information matrix, w is amended and entered into the GLM IRLS algorithm as

$$w_{io} = w_i + (y_i - \mu_i)\frac{[V(\mu_i)g''(\mu_i) + V'(\mu_i)g'(\mu_i)]}{[V(\mu_i)^2 g'(\mu_i)^3]} \tag{4.43}$$

so that it reflects (4.42). The full working algorithm is presented in Table 4.4. In terms of the Hessian, w_o is defined as

$$H = \frac{\partial^2 \mathcal{L}}{\partial\beta_j\beta_k'} \tag{4.44}$$

4.2.3 Parameterizing from μ to $x'\beta$

One finds parameterization of the GLM probability and log-likelihood functions, as well as other related formulae, in terms of both μ and $x'\beta$. μ is defined as the fitted value, or estimated mean, $E(y)$, whereas $x'\beta$ is the linear predictor. GLM terminology also defines the linear predictor as η. Hence $x'\beta = \eta$.

Transforming a log-likelihood function from a parameterization with respect to μ to that of $x'\beta$ is fairly simple. Making such a transformation is at times required when one needs to estimate more complex count models, such as zero-truncated or zero-inflated models.

The method involves substituting $x'\beta$ for η and substituting the inverse link function of μ at every instance of μ in the formula. For an example, we use the Poisson model. The probability distribution function may be expressed as equation 3.1, with the mean parameter expressed as μ in place of λ.

$$f(y;\mu) = \frac{e^{-\mu_i}(\mu_i)^{y_i}}{y_i!} \tag{4.45}$$

Table 4.4 *Standard GLM estimating algorithm (observed information matrix)*

```
Dev = 0
μ = (y + 0.5)/(m + 1)              // binomial
μ = (y + mean(y))/2                // non-binomial
η = g(μ)                           // g; linear predictor
WHILE (abs(Dev) > tolerance) {
V = V(μ)
V' = 1st derivative of V
g' = 1st derivative of g
g" = 2nd derivative of g
w = 1/(Vg'^2)
z = η + (y - μ)g' - offset
W_o = w + (y - μ)(Vg" + V'g')/(V^2g'^3)
β = (X'W_oX)^-1X'W_oz
η = X'β + offset
μ = g^-1(η)
Dev0 = Dev
Dev = Deviance function
Dev = Dev - Dev0
}
Chi2 = Σ(y - μ)^2/V(μ)
AIC = (-2LL + 2p)/n
BIC = -2LL + ln(n)*k    // original ver: Dev-(dof)ln(n)

Where p = number of model predictors + const
k = # predictors : dof = degrees of freedom (n - p)
n = number of observations in model
```

The Poisson log-likelihood function may then be derived as

$$\mathcal{L}(\mu; y) = \sum_{i=1}^{n} \{y_i \ln(\mu_i) - \mu_i - \ln(y_i!)\} \tag{4.46}$$

Since the Poisson has a link defined as $\ln(\mu)$, the inverse link is

$$\mu_i = \exp(x_i'\beta) \tag{4.47}$$

Substituting into equation 4.47 yields

$$\mathcal{L}(\beta; y) = \sum_{i=1}^{n} \left\{y_i \left(x_i'\beta\right) - \exp\left(x_i'\beta\right) - \ln(y_i!)\right\} \tag{4.48}$$

We also see the above expressed as

$$\mathcal{L}(\beta; y) = \sum_{i=1}^{n} \{y_i (x_i'\beta) - \exp(x_i'\beta) - \ln\Gamma(y_i + 1)\} \tag{4.49}$$

where $y!$ can be calculated in terms of the log-gamma function, $\ln\Gamma(y + 1)$.

The first derivative of the Poisson log-likelihood function, in terms of $x'\beta$, is

$$\frac{\partial \mathcal{L}}{\partial \beta} = \sum_{i=1}^{n} \{y_i x_i - x_i \exp(x_i'\beta)\} \tag{4.50}$$

or

$$\frac{\partial \mathcal{L}}{\partial \beta} = \sum_{i=1}^{n} \{(y_i - \exp(x_i'\beta))\, x_i\} \tag{4.51}$$

Solving for parameter estimates, β, entails setting the above to zero and solving

$$\sum_{i=1}^{n} \{(y_i - \exp(x_i'\beta))\, x_i\} = 0 \tag{4.52}$$

This is the Poisson estimating equation and is used to determine parameter estimates β. Refer to equation 6.28.

All Newton–Raphson maximum likelihood algorithms use the $x'\beta$ parameterization. μ is normally used with GLM-based estimation models. Both parameterizations produce identical parameter estimates and standard errors when the observed information matrix is used in the IRLS GLM algorithm. A similar transformation can be performed with negative binomial models. Except for the base NB2 model, all estimating algorithms use the $x'\beta$ parameterization.

4.2.4 Maximum likelihood estimators

The maximum likelihood estimator, $\hat{\theta}$, is a consistent, unbiased, and efficient estimate of the true parameter, θ. Likewise, $\hat{\beta}$ is a consistent, unbiased and efficient estimator of the true model parameters, β. These qualities of maximum likelihood can be understood as:

1 Consistent: An estimator $\hat{\theta}$ is consistent when it improves with sample size; i.e. as sample size gets larger, $\hat{\theta}$ approaches the true value of θ. As $n \to \infty$, $|\hat{\theta} - \theta| \to 0$.

2 Unbiased: An estimator $\hat{\theta}$ is unbiased when the mean of its sampling distribution approaches the true value of θ; $E(\hat{\theta}) = \theta$. Any bias that exists in an estimator is reduced as sample size increases.

3 Efficient: An estimator $\hat{\theta}$ is asymptotically efficient when it has the minimal asymptotic variance among all asymptotically unbiased estimators. That is, maximum likelihood estimators are more precise than other possible estimators.

Summary

We have discussed the two foremost methods used to estimate count response models – Newton–Raphson and Fisher scoring. Both are maximum likelihood methods, using the likelihood function, or its derived deviance function, as the basis for estimation. That is, both methods involve the maximization of the likelihood score function in order to estimate parameter values. Standard errors are obtained from the Hessian, or, more correctly, from the information matrix, which is calculated as the second derivative of the likelihood function. We also mentioned that most software applications use some variation of the traditional or basic Newton–Raphson algorithm. For our purposes, we refer to these methods collectively as full maximum likelihood methods of estimation. These methods produce standard errors based on the observed information matrix.

Fisher scoring is typically based on an iteratively re-weighted least squares (IRLS) algorithm. It is the algorithm traditionally used for estimation of generalized linear models (GLM). Fisher scoring is a simplification of the full maximum likelihood method that is allowed due to the unique properties of the exponential family of distributions, of which all GLM members are instances. Standard errors produced by this method are generally based on the expected information matrix. In the case of canonically linked GLMs, the observed information matrix reduces to the expected (see equation 4.42), resulting in standard errors of the same value. For example, a log-linked Poisson, which is the canonical link, can be estimated using a form of the Newton–Raphson algorithm employing the observed information matrix, or by considering it as a member of the family of generalized linear models, using the expected information matrix. In either case the calculated standard errors will be identical, except for perhaps very small rounding errors. Non-canonical linked GLMs will produce standard errors that are different from those produced using full maximum likelihood methods.

It is possible, though, to adjust the GLM IRLS algorithm so that the observed information matrix is used to calculate standard errors rather than the expected. We showed how this can be accomplished, and how it results in standard errors that are the same no matter which of the two major methods of estimation are used.

5

Assessment of count models

In this chapter we address the basic methods used to assess the fit of count models. In many respects they are the same as for other regression models, but there are also some methods that are unique to the modeling of counts.

The chapter is divided into three sections. First we discuss some of the primary residual analyses that have been employed to assess model fit. Second we look at what are termed goodness-of-fit (GOF) tests. Traditional GOF tests include R^2, deviance, and likelihood ratio tests. We also address the more recent information criterion statistics, AIC and BIC. Finally we look at validation sampling and how it can assist in assuring a well-fitted model.

5.1 Residuals for count response models

When modeling, using either full Newton–Raphson maximum likelihood or IRLS, it is simple to calculate the linear predictor as

$$x'_i\beta = \eta_i = \alpha_0 + \beta_1 x_{i1} + \beta_2 x_{i2} + \cdots + \beta_n x_{in} \tag{5.1}$$

Each observation in the model has a linear predictor value. For members of the exponential family, an easy relationship can be specified for the fitted value based on the linear predictor. The normal or Gaussian regression model has an identity canonical link, that is $\eta = \mu$. The canonical Poisson has a natural log link, hence $\eta = \ln(\mu) = \ln(\exp(x'\beta)) = x'\beta$. The traditional form of the negative binomial (NB2) also has a log link, but it is not the canonical form. The linear predictor and fit are essential components of all residuals.

The basic or raw residual is defined as the difference between the observed response and the predicted or fitted response. When y is used to identify the response, $\hat{y}$ or μ is commonly used to characterize the fit. Hence

$$\text{Raw residual} = (y_i - \hat{y}_i) \quad \text{or} \quad (y_i - \mu_i) \quad \text{or} \quad (y_i - E(y_i))$$

Other standard residuals used in the analysis of count response models include:

Pearson:	$R^{\mathrm{p}} = (y - \mu)/\mathrm{sqrt}(V(\mu))$
Deviance:	$R^{\mathrm{d}} = \mathrm{sgn}(y - \mu)^{*}\mathrm{sqrt}(\mathrm{deviance})$ Note: $\Sigma(R^{\mathrm{d}})^2$ = model deviance statistic
Standardized residuals:	Divide residual by sqrt(1 − h), which aims to make its variance constant. $h = \mathrm{hat} = (\mathrm{stdp})^{2*}V(\mu)$ where stdp = standard error of the prediction. A scale-free statistic.
Studentized residuals:	Divide residual by scale, ϕ. (See McCullagh and Nelder, 1989, p. 396.)
Standardized–studentized:	Divide by both standardized and studentized adjustments; e.g. $R^{\mathrm{d}} : (y - \mu)/\{\phi V(\mu)^{*}\mathrm{sqrt}(1 - h)\}$. Unit asymptotic variance; used to check model adequacy and assumptions regarding error distribution.

In the above formulae we indicated the model distribution variance function as $V(\mu)$, the hat matrix diagonal as *hat* or **h**, and the standard error of the prediction as *stdp*. A scale value, ϕ, is user defined, and employed based on the type of data being modeled. See McCullagh and Nelder (1989, p. 396), for a detailed account of these residuals.

The hat matrix diagonal, **h**, is defined in terms of matrix algebra as:

$$\mathbf{h} = \mathbf{W}^{1/2}\mathbf{X}\left(\mathbf{X}'\mathbf{W}\mathbf{X}\right)^{-1}\mathbf{X}'\mathbf{W}^{1/2}$$

where

$$\mathbf{W} = \mathrm{diag}\left\{\frac{1}{\{V(\mu)\}} * (\partial\mu/\partial\eta)^2\right\}$$

The *hat* statistic, or *leverage,* is a measure of the influence of a predictor on the model. When h-values are graphed against standardized Pearson residuals, values on the horizontal extremes are high residuals. Covariates with high leverage, h, and low residuals, are interesting because they are not easily detected by usual analysis – they do not fit the model, and do not appear to be outliers. Values of high leverage and outside the range of ± 2 standardized Pearson residuals indicate poor model fit.

We mentioned earlier that the Anscombe residual (Anscombe, 1972) has values close to those of the standardized deviance. There are times, however, when this is not the case, and the Anscombe residual performs better than R^{d}. Anscombe residuals attempt to normalize the residual so that heterogeneity in the data, as well as outliers, become easily identifiable.

Anscombe residuals use the model variance function. The variance functions for the three primary count models are

$$\text{Poisson:} \quad V = \mu_i \tag{5.2}$$

$$\text{Geometric:} \quad V = \mu_i(1 + \mu_i) \tag{5.3}$$

$$\text{NB2, NB1, NB-C:} \quad V = \mu_i + \alpha\mu_i^2 \quad \text{or} \quad \mu_i\,(1 + \alpha\mu_i) \tag{5.4}$$

Anscombe defined the residual, which later became known under his name as

$$R^{\mathrm{A}} = \frac{A\,(y_i) - A\,(\mu_i)}{A'\,(\mu_i)\,\sqrt{V(\mu_i)}} \tag{5.5}$$

where

$$A\,(.) = \int_{y_i}^{\mu_i} \frac{d\mu_i}{V(\mu_i)^{1/3}} \tag{5.6}$$

The calculated Anscombe residuals for the three basic count models are

$$\text{Poisson:} \quad 3\left(y_i^{2/3} - \mu_i^{2/3}\right)/\left(2\mu_i^{1/6}\right) \tag{5.7}$$

$$\text{Geometric:} \quad \frac{\left[3\{(1+y_i)^{2/3} - (1+\mu_i)^{2/3}\} + 3\left(y_i^{2/3} - \mu_i^{2/3}\right)\right]}{2\left(\mu_i^{2/3} + \mu_i\right)^{1/6}} \tag{5.8}$$

$$\text{Negative binomial:} \quad \frac{[3/\alpha\{(1+\alpha y_i)^{2/3} - (1+\alpha\mu_i)^{2/3}\} + 3\left(y_i^{2/3} - \mu_i^{2/3}\right)]}{2\left(\alpha\mu_i^2 + \mu_i\right)^{1/6}} \tag{5.9}$$

The negative binomial Anscombe residual has also been calculated in terms of

$$[A(r/y, 2/3, 2/3) - A(\mu, 2/3, 2/3)]\mu^{\frac{1}{6}}\,(1 - \mu)^{\frac{1}{6}}$$

and in terms of the hypergeometric2F1 function

$$y_i^{2/3} H(2/3, 1/3, 5/3, y_i/\alpha) - \mu^{2/3} H(2/3, 1/3, 5/3, \mu_i/\alpha) \tag{5.10}$$

$$= 2/3B(2/3, 2/3)\{y_i - B_1(2/3, 2/3, \mu_i/\alpha)\} \tag{5.11}$$

where H is the hypergeometric2F1 function, B is the beta function, and B_1 is the incomplete beta function. Hilbe (1994a) and Hardin and Hilbe (2001) show that the two-term beta function has the constant value of 2.05339. It is also noted that the value of α is the negative binomial ancillary parameter. See Hilbe (1993b) and Hardin and Hilbe (2007)

5.2 Model fit tests

A useful maxim to remember is that, "a model is only as good as the results of its fit statistics." Far too many times a researcher will apply a regression algorithm to data based on the distribution of the response (e.g. a Poisson regression for count response models), and then simply check the p-values of the respective explanatory predictors. If all p-values are under 0.05, they then declare that the model is well fitted. Software does not always help; some applications do not display appropriate fit statistics, or have no options to test for fit. The problem is that predictor p-values may all be under 0.05, or may even all be displayed as 0.000, and yet the model can nevertheless be inappropriate for the data. A model that has not undergone an analysis of fit is, statistically speaking, useless.

In this section we address the general type of fit tests that are employed to understand if the data are well modeled by the methods applied by the researcher. A fit test typically results in a single statistic, with an associated p-value. How the p-value is interpreted depends entirely on the type of test. We shall first provide an overview of the more traditional fit tests, i.e. those tests that have been associated with generalized linear models software since the early 1980s.

I should note that count models typically fail to fit the data being modeled because of a problem known as overdispersion. Overdispersion is another term for excessive correlation in the data, meaning more correlation than is allowed for on the basis of the distributional assumptions of the model being employed. Overdispersion is a problem for both count and binomial models; Chapter 7 is entirely devoted to this topic. Here we discuss general fit tests that provide information on overall fit without regard to the source of the bias.

5.2.1 Traditional fit tests

Three goodness-of-fit tests have been traditionally associated with members of the family of generalized linear models (GLM): R^2 and pseudo-R^2 statistics, deviance statistics, and the likelihood ratio test. The first is an overall test of the model, whereas the other two are comparative tests between models. Insofar as a count model is a GLM, these tests are appropriate – to a point. We shall discuss each in turn.

5.2.1.1 R^2 and pseudo-R^2 Goodness-of-fit tests

R^2 is a well-known test associated with ordinary least squares regression. Referred to as the *coefficient of determination*, R^2 indicates the percentage of variation in the data explained by the model. The statistic ranges in value from 0 to 1, with higher values representing a better fitted model. However, this statistic is not appropriate for non-linear models, for example Poisson, negative binomial, and logistic regression.

The R statistic that is usually displayed with count model output is called a pseudo-R^2 statistic, of which a variety have been formulated. The most common is the expression:

$$R_{\mathrm{p}}^2 = 1 - \mathcal{L}_{\mathrm{F}}/\mathcal{L}_{\mathrm{I}} \tag{5.12}$$

with

$\mathcal{L}_{\mathrm{F}}$ the log-likelihood of the full model
$\mathcal{L}_{\mathrm{I}}$ the log-likelihood of the intercept-only model.

For an example we shall use the **azpro** dataset, with length of (hospital) stay (LOS) as the response, and three binary explanatory predictors:

sex: 1=male; 0=female
admit: 1=emergency/urgent admission; 0=elective admission
procedure: 1=CABG; 0=PTCA

CABG is the standard acronym for coronary artery bypass graft, where the flow of blood in a diseased or blocked coronary artery or vein has been grafted to bypass the diseased sections. PTCA, or percutaneous transluminal coronary angioplasty, is a method of placing a balloon in a blocked coronary artery to open it to blood flow. It is a much less severe method of treatment for those having coronary blockage, with a corresponding reduction in risk.

Here we evaluate the predictors' bearing on the length of hospital stay for these two types of cardiovascular surgeries, controlled by gender and type of admission. In order to calculate the pseudo-R^2 statistic, we must model both an intercept-only model and a full model.

R

```
data(azpro)
p5_2a <- glm(los ~ 1, family=poisson, data=azpro)
summary(p5_2a)
11a <- p5_2a$rank - p5_2a$aic/2     # see page 74
p5_2b <- glm(los ~ sex + admit + procedure, family=poisson, data=azpro)
summary(p5_2b)
11b <- p5_2b$rank - p5_2b$aic/2     # page 74
1 - 11b/11a                          # 1-(-11237.256)/(-14885.314)
```

INTERCEPT-ONLY MODEL

```
. poisson los, nolog
Poisson regression                          Number of obs   =       3589
                                            LR chi2(0)      =       0.00
                                            Prob > chi2     =          .
Log likelihood = -14885.314                 Pseudo R2       =      0.000
--------------------------------------------------------------------------
   los |    Coef.    Std. Err.       z    P>|z|    [95% Conf.   Interval]
-------+------------------------------------------------------------------
 _cons |2.178254    .0056171    387.79   0.000    2.167244       2.189263
--------------------------------------------------------------------------
```

FULL MODEL

```
. poisson los sex admit procedure
Poisson regression                          Number of obs   =       3589
                                            LR chi2(3)      =    7296.12
                                            Prob > chi2     =     0.0000
Log likelihood = -11237.256                 Pseudo R2       =     0.2451
--------------------------------------------------------------------------
      los |    Coef.   Std. Err.   z     p>|z|   [95% Conf. Interval]
----------+---------------------------------------------------------------
      sex |-.1302223 .0117938 -11.04 0.000 -.1533378    -.1071068
    admit | .3330722 .0121044  27.52 0.000  .309348      .3567964
procedure | .9573838 .0121785  78.61 0.000  .9335144     .9812532
    _cons | 1.491405 .0153897  96.91 0.000  1.461242     1.521568
--------------------------------------------------------------------------
```

The pseudo-R^2 statistic may be calculated using the two log-likelihood functions as in equation 5.12.

```
. di 1 - (-11237.256/-14885.314)
.24507767
```

Note that the value calculated is identical to that given in the full model output. Again, we cannot interpret the pseudo-R^2 statistic as we would R^2. Statisticians are varied in how the statistic is actually to be interpreted. All we can confirm is that very low values may indicate a lack of fit, whereas high values do not. But there are no set criteria for being more clear.

5.2.1.2 Deviance goodness-of-fit test

The deviance statistic is used with generalized linear models software, where it is normally displayed in the model output. The deviance is expressed as

$$D = 2\sum_{i=1}^{n}\{\mathcal{L}(y_i; y_i) - \mathcal{L}(\mu_i; y_i)\} \tag{5.13}$$

with $\mathcal{L}(y_i; y_i)$ indicating a log-likelihood function with every value of μ given the value y in its place. $\mathcal{L}(\mu_i; y_i)$ is the log-likelihood.

The deviance is a comparative statistic. The test is performed by using a *chi*2 test with the value of the deviance and the degrees of freedom as the two *chi*2() parameters. The *chi*2 degrees of freedom is the number of predictors in the model, including interactions. If the resulting *chi*2 *p*-value is less than 0.05, the model is considered well fit.

The deviance test has been used to compare models from different GLM families and links based on an analogy to variance. The idea is that a model with smaller deviance is preferred over a model with greater deviance. However, this type of use of the statistic is now rare in actual practice. The problem with the deviance test is that too many in-fact ill-fitted models appear as well fitted. McCullagh and Nelder (1989) and Hardin and Hilbe (2012) discuss this test in more detail.

The deviance is also used as a goodness-of-fit test of the predicted to observed counts. A *Chi*2 test is used with observations minus number of predictors as the degrees of freedom. In Stata the test is *chiprob(dof, dev); in R: 1-pchisq(dev, dof)*. *p*-values less than 0.05 indicate that model predictions significantly differ from the observed counts. It likely is a poorly fitted model.

The deviance is sometimes used with nested models. If the difference in the deviance statistic is minimal when adding a predictor or predictors, we may conclude that they will add little to the model, and therefore are not incorporated into the final model. The logic of this type of reasoning is well founded, but there is no well-established *p*-value to quantify the significance of the differences in deviances.

5.2.1.4 Likelihood-ratio test

The likelihood-ratio (LR) test is a commonly used comparative fit test. It is generally used for nested models, but has also been used to test different models (e.g. whether data are better modeled using a negative binomial or a Poisson). These types of special case will be discussed when the models themselves are addressed. The formula for the likelihood-ratio test, which will be frequently employed in this text, is:

$$\mathrm{LR} = -2\{\mathcal{L}_{\text{reduced}} - \mathcal{L}_{\text{full}}\} \tag{5.14}$$

The likelihood-ratio test is also at times used as a global test of model fit. Using the same log-likelihood functions that we employed for the pseudo-R^2 test above (Section 5.2.1.1), we have

```
. di -2*(-14885.314 - (-11237.256))
7296.116
```

The likelihood-ratio statistic given in the full model output is the same. However, few statisticians now give this statistic much credence. The statistic simply tells us that the model with predictors is superior to one with only an intercept. Most models fit better than an intercept-only model – even if both fail to fit the data.

The LR test is useful, though, when deciding whether to add a predictor to a model, or to add a set of predictors such as a multilevel factor variable. The LR test is superior to a Wald test for evaluating the significance of individual predictors to a model. The Wald test consists of the *p*-values typically associated with a predictor in displayed model output. Statisticians tend to call this type of test a Wald test owing to history. Several of the major software applications based the associated *p*-value on a Wald distribution, which is the same as *t*-squared. The results correspond to a *z* or normal based distribution.

5.2.2 Information criteria fit tests

Information criterion tests are comparative in nature, with lower values indicating a better fitted model. Two primary groups of information criterion tests are current in research: the Akaike Information Criterion (AIC), and the Bayesian Information Criterion (BIC). Each of these criteria consists of a number of alternative parameterizations, each aimed at attempting to determine a method to best assess model fit.

5.2.2.1 Akaike Information Criterion

The Akaike Information Criterion (AIC) was developed by Hirotsugu Akaike in 1974. However, it did not begin to enjoy widespread use until the twenty-first century. It is now one of the most, if not the most, commonly used fit statistic displayed in statistical model output.

The AIC statistic is generally found in two forms,

$$\text{AIC} = \frac{-2\mathcal{L} + 2k}{n} = -\frac{2(\mathcal{L} - k)}{n} \tag{5.15}$$

and

$$\text{AIC} = -2\mathcal{L} + 2k = -2(\mathcal{L} - k) \tag{5.16}$$

where $\mathcal{L}$ is the model log-likelihood, k is the number of predictors including the intercept, and n represents the number of model observations. In both parameterizations $2k$ is referred to as a penalty term which adjusts for the size and complexity of the model. Since more parameters make what is observed more likely, $-2\mathcal{L}$ becomes smaller as k increases. This bias is adjusted by adding the penalty term to $-2\mathcal{L}$.

Larger n also affects the $-2\mathcal{L}$ statistic. Equation 5.15 divides the main terms by n, thereby obtaining a per-observation contribution to the adjusted $-2\mathcal{L}$. All other conditions being equal, a smaller AIC indicates a better fitted model.

Several of the major commercial statistical applications use equation 5.16 as the definition of AIC under the assumption that the two parameterizations produce the same information, i.e. both provide equal assessment regarding comparative model fit. However, this is not necessarily the case, especially when the statistic is improperly used, e.g. with clustered data.

Care must be taken to know which parameterization of AIC is being displayed in model ouput. Low values compared with the number of model observations indicate that equation 5.15 is being used; high values equation 5.16.

Other AIC statistics have been used in research. The most popular – other than the two above primary versions – is the finite sample AIC, which may be defined as

$$\mathrm{AIC_{FS}} = -2\{\mathcal{L} - k - k(k+1)/(n-k-1)\}/\mathrm{n} \tag{5.17}$$

or

$$\mathrm{AIC_{FS}} = \mathrm{AIC} + \frac{2\mathrm{k}(\mathrm{k}+1)}{\mathrm{n}-\mathrm{k}-1} \tag{5.18}$$

where k is the number of parameters in the model. Note that $\mathrm{AIC_{FS}}$ employs a greater penalty for added parameters compared to the standard AIC statistic. Note also that $\mathrm{AIC_{FS}} \approx \mathrm{AIC}$ for models with large numbers of observations. Hurvich and Tsai (1989) first developed the finite sample AIC for use with time series and auto-correlated data, however others have considered it as preferred to AIC for models with non-correlated data as well, particularly for models with many parameters, and/or for models with comparatively few observations. A variety of finite sample AIC statistics have been designed since 1989. We will say no more about it here.

How do we decide if one value of the AIC is significantly superior to another, i.e. whether there is a statistically significant difference between two values of the AIC? Hilbe (2009) devised a table based on simulation studies that can assist in deciding if the difference between two AIC statistic values is significant. Table 5.1 is taken from that source, and is based on equation 5.16 (i.e. AIC without division by n).

Table 5.1 *AIC significance levels*

Difference between models A and B	Result if A < B
<0.0 & <=2.5	No difference in models
<2.5 & <= 6.0	Prefer A if $n > 256$
<6.0 & <= 9.0	Prefer A if $n > 64$
10+	Prefer A

Table 5.1 is a general guide only, with intermediate cases generally being interpreted as being undetermined. Other tests, like BIC, should be used to complement what is determined using the AIC statistic.

If the models being compared employ equation 5.15, users need to multiply the AIC statistics by n and subtract. If equaton 5.15 is indicated as AIC_n, the value to use in Table 5.1 is

$$n[\text{AIC}_n(\text{ModelA}) - \text{AIC}_n(\text{ModelB})]$$

Users of R should note that the AIC is generally given using the form of equation 5.16, i.e. without dividing the adjusted log-likelihood by the number of model observations. The **glm** function displays this result. The form of AIC given in equation 5.15, i.e. dividing the adjusted log-likelihood function by the number of model observations, can be given as

```
aic_n <- aic/(model$df.null + 1)
```

where *model* is used to represent the name of the **glm** model, and df.null + 1 is the number of observations in the model.

The log-likelihood value is not displayed in **glm** results, nor is it saved for post-estimation use. The log-likelihood can be calculated using the value of AIC, together with other saved results from **glm**. Again supposing that the **glm** model is named *model*, the log-likelihood may be calculated as:

```
ll <- model$rank - model$aic/2
```

where *rank* provides the number of predictors, including the constant. This value can then be used for other formulations of the AIC statistic, as well as for versions of the BIC statistic, the subject of discussion in the following subsection.

Mention should be made of another fit test that is used by some statisticians in place of the AIC. Called the Consistent AIC, or CAIC (Bozdogan, 1987), the statistic includes an added penalty for models having a greater number of parameters, k, given sample size n.

$$\mathrm{CAIC} = -2\mathcal{L} + k(\ln(n) + 1) \tag{5.19}$$

Note its similarity to the $\mathrm{BIC_L}$ statistic below. Because it is at times found in research reports, it is desirable to be familiar with the CAIC statistic. There is evidence that the CAIC does a better job at selection over the standard AIC when there are many predictors in the model (Nylund, 2007), but it appears to give no selection advantage in other modeling situations. Additional testing is required.

5.2.2.2 Bayesian Information Criterion

The second foremost contemporary comparative fit statistic for likelihood-based statistical models is the Bayesian Information Criterion (BIC). Again, this statistic has undergone a variety of parameterizations. The original formulation is from Gideon Schwartz (1978), who based his reasoning on Bayesian grounds. The statistic is now commonly known as the Schwartz Criterion (SC) in SAS and as BIC in other software. It is formulated as equation 5.21.

The original formulation of BIC for GLMs was given by the University of Washington's Adrian Raftery in 1986. Based on the deviance statistic, it is given as:

$$\mathrm{BIC_R} = D - (df)\ln(n) \tag{5.20}$$

where D is the model deviance statistic and df is the model residual degrees of freedom. The residual degrees of freedom is defined as the number of observations minus the number of predictors in the model including the intercept. Stata's **glm** command provides this parameterization of BIC statistic in its output. Most other applications, including R and other BIC statistics in Stata other than in **glm**, employ the formula in equation 5.22, or Schwartz Criterion, as the meaning of BIC. The statistic is defined as

$$\mathrm{BIC_L} = -2\mathcal{L} + k\ln(n) \tag{5.21}$$

with k indicating the number of predictors, including intercept, in the model. It is clear that $\mathrm{BIC_R}$ and $\mathrm{BIC_L}$ have very different values. BIC statistics usually give a higher value to the second term of the function than do AIC statistics.

It should be noted that both the AIC and BIC statistics have been designed to be used with non-nested models of the same general distributional type having the same number of observations. However, they are both robust tests that some statisticians have used to compare non-nested models with slightly different numbers of observations. We may engage the statistics accordingly, but only as a rough guide to model preference.

Let us use the **azpro** data which was employed in Section 5.2.1.1 to display pseudo-R^2 results. The GLM formulation of the model appears as

R

```
summary(p5_2b)  # AIC in output; divide by n=3589
bic <- 8968.864992  (3585 * log(3589)) # or
bic <- p5_2b$deviance (p5_2b$df.residual * log(p5_2b$df.null +1)
```

```
. glm los sex admit procedure, fam(poi)

Generalized linear models                 No. of obs          =       3589
Optimization       : ML                   Residual df         =       3585
                                          Scale parameter     =          1
Deviance           = 8968.864992          (1/df) Deviance     =   2.501775
Pearson            =  11588.0839          (1/df) Pearson      =    3.23238
                                          .   .   .
                                          AIC                 =   6.264283
Log likelihood     = -11237.25647         BIC                 =  -20376.61

------------------------------------------------------------------------------
             |                 OIM
         los |     Coef.   Std. Err.     z    P>|z|  [95% Conf.  Interval]
-------------+----------------------------------------------------------------
         sex | -.1302223  .0117938  -11.04  0.000  -.1533378   -.1071068
       admit |  .3330722  .0121044   27.52  0.000    .309348    .3567964
   procedure |  .9573838  .0121785   78.61  0.000   .9335144    .9812533
       _cons |  1.491405  .0153897   96.91  0.000   1.461242    1.521568
------------------------------------------------------------------------------

. estat ic

------------------------------------------------------------------------------
Model |   Obs   ll(null)    ll(model)    df      AIC        BIC
------+-----------------------------------------------------------------------
    . | 3589    .          -11237.26      4    22482.51   22507.26
------------------------------------------------------------------------------
          Note: N-Obs used in calculating BIC; see [R] BIC note
```

Using equation 5.20, BIC_R is calculated as

```
. di 8968.864992 - 3585 * ln(3589)
-20376.615
```

which is the value displayed in the model output. Note that the parameterization of AIC given in the model output is equation 5.15, where the adjusted log-likelihood is divided by the number of observations in the model. Multiplying that value gives what is displayed in **estat ic** results.

```
. di 6.264283 * 3589
22482.512
```

Table 5.2 BIC_R *model preference*

|Difference|	Degree of preference
0–2	Weak
2–6	Positive
6–10	Strong
10+	Very strong

BIC_L is calculated as

```
. di -2* -11237.25647 + 4*ln(3589)
22507.255
```

which is the value we observe in the **estat ic** output. Neither BIC statistic is preferred to the other. The only requirement is that when AIC and BIC are being used to compare models as to best fit, it is requisite that the models being evaluated use the same statistic.

Raftery developed a table of model preference to be used with his statistic. Presented in Table 5.2, it is not to be used with BIC_L. Unfortunately some authors have used it for BIC_L, not realizing that it was designed for another parameterization.

Models A and B:

If $BIC_A - BIC_B < 0$, then A is preferred.
If $BIC_A - BIC_B > 0$, then B is preferred.

Simply put, the model having a lower value of BIC_R is the better fitted model. This relationship maintains for any of the parameterizations of the BIC.

Another parameterization of the BIC, which is used in LIMDEP software, was developed by Hannan and Quinn (1979). It is given as

$$QIC_H = -2(\mathcal{L} - k^*\ln(k))/n \tag{5.22}$$

```
. di -2*(-11237.25647 - 4*ln(4))/3589
6.2651444
```

which appears more like an AIC than a BIC statistic. We shall address the QIC, or Quasi-likelihood Information Criterion when we discuss panel models later in the text. In particular, QIC statistics are typically used for GEE and quasi-least squares models.

A Stata command called **abic** (Hilbe) displays the values of the most used parameterizations of the AIC and BIC statistics. The BIC statistic given in the

Table 5.3 R: *Function duplicates author's Stata abic command*

```
# modelfit function to calc AIC and BIC statistics post
  estimation
modelfit <- function(x) {
obs     <- x$df.null + 1
aic     <- x$aic
xvars   <- x$rank
rdof    <- x$df.residual
aic_n   <- aic/obs
ll      <- xvars - aic/2
bic_r   <- x$deviance - (rdof * log(obs))
bic_l   <- -2*ll + xvars * log(obs)
bic_qh <- -2*(ll - xvars * log(xvars))/obs
c(AICn=aic_n, AIC=aic, BICqh=bic_qh, BICl=bic_l)
}
modelfit(x) # substitute fitted model name for x
```

lower left corner of the output is QIC_H, whereas the lower right BIC statistic is BIC_L.

```
. abic
AIC Statistic = 6.264283    AIC*n        = 22482.514
BIC Statistic = 6.265144    BIC(Stata) = 22507.256
```

Users of R need to calculate the log-likelihood as discussed in the previous section when desiring to calculate log-likelihood–based BIC statistics following a **glm** model. Given the model saved as **p5_2b**, the BIC statistic formulated in equation 5.21 can be calculated as:

```
ll <- model$rank - model$aic/2
bic_l <- -2*ll + p5_2b$rank*log(p5_2b$df.null+1)
```

The form of BIC used in LIMDEP, and displayed on the lower left of the above **abic** command results, is also known as the Hannan and Quinn QIC_H statistic. It can be calculated after **glm** as:

```
xvars <- p5_2b$rank
bic_qh <- -2*(ll - xvars * log(xvars))/(p5_2b$df.null + 1)
```

In the above R **modelfit** script the following line can also be used to display the four AIC and BIC statistics that are given with the **abic** command, which was written using Stata's programming language. However, it does not display as well as the code used in the script. I also provide code for calculating Raftery's original version of the BIC statistic, which is normally displayed in the Stata **glm** command output. It is stored in bic_r. One may add it to the final line of **modelfit** within the parenthesis as "BICr=bic_r."

```
return(list("AIC" = aic, "AICn" = aic_n, "BIC" = bic_l,
  "BICqh" = bic_qh))
```

A caveat should be given when using either AIC or BIC statistics, regardless of the parameterization. The statistics are intended to be used with observations that are independent of one another. Violations of this requirement are frequently found in statistical literature. However, simulation tests have shown that even though the AIC statistic may not always identify the truly best fitted model, it does so better than competing tests found in commercial software (Bartlett *et al.*, 2010).

5.3 Validation models

Researchers commonly use a validation data set to help confirm the generalizability of the model to a greater population. This is particularly the case when the data come from a sample of observations within a population. This notion of extractability is inherent in the frequency interpretation of statistics, upon which the models we are discussing are based. We shall discover that many times the methods used in practice for count models fall back on frequentist principles.

Validation data generally come in two varieties: (1) as a sample of data from the population from which the model data derive, but which are not included in the model, and (2) a sample taken from the model itself. Percentages differ, but validation samples typically consist of about 20% of the observations from the model being examined – although there is no hard rule as to the actual percentage used. If taken from the estimated model, the procedure is to take a sample from the fitted model and remodel it. The validation model coefficients and standard errors should be similar to those produced by the main model. A Hausman test (1978) on the equality of coefficients can be used to access the statistical difference in the two groups of observations.

A given percentage of data may also be withheld from the estimation of the primary model. Once the selected data have been modeled, the withheld validation data are subsequently modeled for comparison purposes. Using a test like Hausman, the coefficients are tested for similarity. If it appears that the two sets of data come from the same population (i.e. the data are not statistically different), then the sets of data may be combined and remodeled for a final fitted model.

Both of these methods of constructing and implementing a validation sample help prevent a researcher from developing an overfitted model. Overfitted models are such that they cannot be used for classification, or for prediction of non-model data from a greater population. An overfitted model is so well fitted that it is applicable only to the data being modeled. Coefficient values and other summary statistics may not be extrapolated to other data. For most research situations, such a model is of little value.

It is easy to overfit a model. When a statistician incorporates a large number of predictors into the model and then attempts to fine-tune it with highly specific transformations and statistical manipulations, it is likely that the resultant model is overfitted. Testing a model against validation data assists in minimizing the this possibility and helps assure us of the stability of the parameter estimates.

Remember that even if we are confident that a model is well-fitted, it is nevertheless dangerous to make predictions outside the sample space of the model. If the modeled data is truly representative of the population on which the modeled data is a sample, predictions made outside the model, but within the scope of the population, may be constructed. Care must be taken, though, that the population parameters do not change. As example of this situation is examined in Hardin and Hilbe (2007) and Hilbe (2009). Validation samples are examined in more detail in Hilbe (2009).

Summary

In this chapter we addressed the type of residuals that are used to evaluate the worth of count response models. Residuals based on GLM methodology provide the statistician with a rather wide range of evaluative capability. It is recommended that standardized deviance and Anscombe residuals be used with the fitted values, μ, to determine if the data are appropriate for the model used in estimation.

The AIC and BIC statistics are the most commonly used statistics employed for the purpose of assessing comparative fit. Both have been formulated in a variety of ways, so care is needed to establish which has been calculated for model output. The parameterizations of AIC and BIC which appear to be most consistent with one another are:

$$\text{AIC} = \frac{-2(\mathcal{L} - k)}{n}$$

and

$$\text{BIC} = \frac{-2(\mathcal{L} - k^{*}\ln(k))}{n}$$

This formulation of BIC is found in LIMDEP software and is also referred to as the QIC_H statistic. Several of the major commercial statistical software packages prefer to use an AIC that does not adjust for model observations; we have argued here that the use of the above parameterization is probably superior for comparative purposes.

6

Poisson regression

Poisson regression is the standard or base count response regression model.We have seen in previous discussions that other count models deal with data that violate the assumptions carried by the Poisson model. Since the model does play such a pivotal role in count response modeling, we begin with an examination of its derivation and structure, as well as a discussion of how it can be parameterized to model counts per unit time or area. Also discussed are the different interpretations of Poisson coefficients, marginal effects and discrete change, and graphics such as effect plots. We also show how to construct synthetic Poisson models in both Stata and R, but which can be developed in other languages using the same logic as described here.

6.1 Derivation of the Poisson model

Poisson regression, based on the Poisson probability distribution, is the fundamental method used for modeling count response data. As previously mentioned, however, the Poisson model can also be construed as a subset of the NB2 model. Poisson is an NB2 model with a heterogeneity parameter value of zero. This relationship is discussed in detail in Chapter 8.

In this section we discuss the derivation of the Poisson regression model. First, in Section 6.1.1 we demonstrate how the Poisson probability distribution can be derived from the binomial distribution. In Section 6.1.2, given the Poisson distribution first displayed in equation 3.1, we derive the Poisson regression model. The model is an expansion of the discussion given in Section 4.2.3, where we derived the maximum likelihood estimating equation for the Poisson probability distribution in terms of $x'\beta$ from the Poisson PDF. In this section we derive the various statistical components of the Poisson algorithm.

6.1.1 Derivation of the Poisson from the binomial distribution

Later we shall discuss the relationship between Poisson and logistic regression, finding that their structure is closer than many realize. The logit link is the canonical or natural form of the binomial distribution. That is, given the exponential family form of the binomial probability function, the natural parameter, θ, is expressed as $\ln(p/(1-p))$ – the logit function. In fact, in Chapter 2 we earlier developed the relationship between the odds and risk ratio, which underlies both logistic and Poisson regression. We next address not simply the relationship, but how the Poisson can be derived from the binomial.

There are several ways mathematicians have derived the Poisson from the binomial distribution. The presentation here represents a rather standard method.

First, we begin with the binomial probability distribution, or more properly the binomial probability mass function. A standard expression for the binomial PDF is:

BINOMIAL PDF

$$f(y; p, n) = \binom{n_i}{y_i} p_i{}^{y_i} (1-p_i)^{n_i - y_i} \tag{6.1}$$

The choose function, '*n* choose *y*', may be expanded and incorporated into the equation to appear as

$$f(y; p, n) = \frac{n_i!}{y_i!(n_i - y_i)!} p_i^{y_i} (1-p_i)^{n_i - y_i} \tag{6.2}$$

with y indicating the number of Bernoulli successes, n the number of trials, and p the probability of success, i.e. the probability of $y = 1$.

Without employing subscripts for the derivation, the binomial mean is

$$\lambda = np \tag{6.3}$$

which is also represented as

$$\mu = np \tag{6.4}$$

From the binomial mean, we have

$$p = \lambda/n \tag{6.5}$$

so that

$$f(y; \lambda, n) = \frac{n!}{y!(n-y)!} \left(\frac{\lambda}{n}\right)^y \left(1 - \frac{\lambda}{n}\right)^{n-y} \tag{6.6}$$

Re-expressing the left term, and letting n become very large, and p small, we can reformulate the binomial distribution as:

$$\lim_{n\to\infty} f(y;\lambda,n) = \frac{n\,(n-1)\ldots(n-y+1)}{y!}\frac{\lambda^y}{n^y}\left(1-\frac{\lambda}{n}\right)^n\left(1-\frac{\lambda}{n}\right)^{-y} \tag{6.7}$$

Exchanging denominator terms gives

$$= \frac{n\,(n-1)\ldots(n-y+1)}{n^y}\frac{\lambda^y}{y!}\left(1-\frac{\lambda}{n}\right)^n\left(1-\frac{\lambda}{n}\right)^{-y} \tag{6.8}$$

Recall from basic mathematics, $e_{\lim x\to\infty} = (1+1/x)^x$, and $e^k_{\lim x\to\infty} = (1+k/x)^x$. With the left term cancelling to equal 1, $(1-\lambda/n)^n$ being the form for $e^{-\lambda}$, and with n being very large in the final term, resulting in $\lim_{n\to\infty}(1-(\lambda/n \to 0)) = 1$, equation 6.8 becomes

$$\lim_{n\to\infty} f(y;\lambda,n) = (1)\left(\frac{\lambda^y}{y!}\right)\left(e^{-\lambda}\right)(1) \tag{6.9}$$

Combining terms, and using subscripts to show the individual contribution of observations to the distribution, we have,

$$f(y;\lambda) = \frac{\lambda_i^{\ y_i} e^{-\lambda_i}}{y_i!} \quad \text{or} \quad \frac{\mu_i^{\ y_i} e^{-\mu_i}}{y_i!} \tag{6.10}$$

which is the standard form of the Poisson probability distribution.

The above line of reasoning is consistent with having a series of Bernoulli trials without knowledge of the probability of success on a particular trial p, or the number of trials, n, in the series. All that is known is the mean number of successes in the series, which is termed λ or μ. The Poisson distribution gives the probability of y successes in the series, given the series is large and p is small.

In Chapter 2 we discussed the formulae for both odds and risk ratios. Given 2×2 and 2×k tables, it was quite simple to demonstrate their relationship, and show what must occur for an odds ratio to be reduced in value such that it approximates a risk ratio. Essentially we have done the same above, but with distributions rather than the terms in a table.

6.1.2 Derivation of the Poisson model

Maximum likelihood models, as well as the canonical form members of generalized linear models, are ultimately based on an estimating equation derived from a probability distribution. This is done by parameterizing the association between the mean parameter, μ, and explanatory predictors, x. In the case of the Poisson, and count models in general, μ is the mean rate parameter. The probability distribution or mass function may be expressed as previously displayed in equation 4.45 as

$$f(y;\mu) = \frac{e^{-\mu_i}\mu_i^{\ y_i}}{y_i!}$$

for $y = \{\ 0,\ 1, \ldots\}$ and $\mu > 0$. Again, μ is the predicted mean of count response, y.

The conversion of the Poisson PDF to log-likelihood form is accomplished by casting it as a member of the exponential family of distributions, previously given in equation 4.53.

EXPONENTIAL FAMILY PDF

$$f(y;\theta,\phi) = \exp\left\{\frac{y_i\theta_i - b(\theta_i)}{\alpha_i(\phi)} + c(y_i;\phi)\right\}$$

The likelihood function is a transformation of the probability function for which the parameters are estimated to make the given data most likely. A probability function predicts unknown outcomes, y, on the basis of known parameters, θ and ϕ. A likelihood function estimates the unknown parameters on the basis of known outcomes. Statistical models are, of course, instances of the latter.

The form of the likelihood function appropriate for members of the exponential family of probability distributions, of which the Poisson and negative binomial distributions are members, can be expressed, as

$$L(\theta;y,\phi) = \prod_{i=1}^{n} \exp\left\{\frac{y_i\theta_i - b(\theta_i)}{\alpha_i(\phi)} + c(y_i;\phi_i)\right\} \tag{6.11}$$

Note the change in what is being estimated, from y to θ. However, for the models we are describing, the parameter of interest is μ. As we have previously observed, and will demonstrate in another manner, θ can be expressed in terms of η, the linear predictor, or μ, the fitted value – which for count models is an expected count. Recalling that the scale, ϕ, of Poisson and negative binomial models is 1, we may express the Poisson likelihood function in exponential family form as

$$L(\mu;y) = \prod_{i=1}^{n} \exp\{y_i \ln(\mu_i) - \mu_i - \ln(y_i!)\} \tag{6.12}$$

Logging both sides of the equation results in the Poisson *log-likelihood function*. Note that the product symbol has been changed to a summation. Summing the log-likelihood value over each observation in the model is computationally easier than multiplying the various likelihoods. Logging both sides of the equation also drops the exponentiation function displayed in the likelihood.

POISSON LOG-LIKELIHOOD FUNCTION

$$\mathcal{L}(\mu; y) = \sum_{i=1}^{n} \{y_i \ln(\mu_i) - \mu_i - \ln(y_i!)\} \tag{6.13}$$

The canonical link and cumulant terms can then be abstracted from the above, as

LINK: $$\theta_i = \ln(\mu_i) = \eta_i = x_i'\beta \tag{6.14}$$

CUMULANT: $$b(\theta_i) = \mu_i \tag{6.15}$$

The inverse link is a re-interpretation of μ with respect to η, the linear predictor. Transformation yields

INVERSE LINK: $$\mu_i = \exp(\eta_i) = \exp(x_i'\beta) \tag{6.16}$$

Recalling that the exponential family mean is defined as the first derivative of the cumulant with respect to θ, and the variance as the second derivative with respect to θ, we calculate the Poisson mean and variance as

MEAN

$$b'(\theta_i) = \frac{\partial b}{\partial \mu_i}\frac{\partial \mu_i}{\partial \theta_i} = (1)(\mu_i) = \mu_i \tag{6.17}$$

VARIANCE

$$b''(\theta_i) = \frac{\partial^2 b}{\partial \mu_i^2}\left(\frac{\partial \mu_i}{\partial \theta_i}\right)^2 + \frac{\partial b}{\partial \mu_i}\frac{\partial \mu_i^2}{\partial \theta_i^2} = (0)(1) + (\mu_i)(1) = \mu_i \tag{6.18}$$

Note the equality of the Poisson mean and variance functions. I should mention here the Poisson distribution coefficient of variation is $\mu^{-0.5}$, or $1/\sqrt{\mu}$.

Since the derivative of the link is important to the estimating algorithm, we have

DERIVATIVE OF LINK: $$\frac{\partial \theta_i}{\partial \mu_i}\frac{\partial\{\ln(\mu_i)\}}{\partial \mu_i} = \frac{1}{\mu_i} \tag{6.19}$$

Recall that equation 4.47 in the previous chapter specified the Poisson mean, μ, as equal to $\exp(x'\beta)$. We can therefore make the following translation such that μ takes the value $\exp(x'\beta)$: Note that the $x'\beta$ parameterization is used in all full maximum likelihood estimating algorithms.

$$\frac{1}{\mu_i} = \frac{1}{\exp(x_i'\beta)} \tag{6.20}$$

The Poisson log-likelihood, parameterized in terms of $x'\beta$, is therefore given as

POISSON LOG-LIKELIHOOD FUNCTION ($x\beta$)

$$\mathcal{L}(\beta; y) = \sum_{i=1}^{n} \left\{ y_i(x_i'\beta) - \exp(x_i'\beta) - \ln(y_i!) \right\} \tag{6.21}$$

or

$$\mathcal{L}(\beta; y) = \sum_{i=1}^{n} \left\{ y_i(x_i'\beta) - \exp(x_i'\beta) - \ln\Gamma(y_i + 1) \right\} \tag{6.22}$$

where $\ln\Gamma()$ is the log-gamma function. In programming, the log-gamma function may be calculated using the *lngamma()* function in Stata and *log(gamma())* nested functions in R. However, some applications use the straightforward factorial functions. Stata employs the *lnfactorial()* function, which is equivalent to R's *log(factorial())* functions. Note that *log(gamma(6)) = log(factorial(5))*. The Stata command for the factorial is *exp(lnfactorial(x))*, with x being the factorial number. A more exact term for factorial, recommended by Stata, is *round(exp(lnfactorial(x)),1)* It should also be noted that, given the first term of equations 6.21 and 6.22, the sufficient statistic of β is $\sum_{i=1}^{n} y_i x_i$.

When the response has a value of zero, the log-likelihood function reduces to

$$\mathcal{L}(\beta; y = 0) = \sum_{i=1}^{n} \left\{ -\exp(x_i'\beta) \right\} \tag{6.23}$$

Returning to the traditional GLM parameterization of the mean as μ, the GLM deviance function is defined as

$$D = 2 \sum_{i=1}^{n} \left\{ \mathcal{L}(y_i; y_i) - \mathcal{L}(\mu_i; y_i) \right\} \tag{6.24}$$

The deviance is a measure of the difference between the saturated and model likelihoods. It is in effect a likelihood ratio test of the full to the model

Table 6.1 *Poisson regression algorithm*

```
dev=0
μ = (y + mean(y))/2
η = ln(μ)
WHILE(abs(Δdev)>tolerance) {
u = (y - μ)/μ
w = μ                              /// weight
z = η + u - offset                 /// working response
β= (X'wX)-1X'wz                    /// weighted regression
η = X'β + offset                   ///linear predictor
μ = exp(η)                         /// inverse link -> fit
oldDev = dev
dev = 2Σ{yln(y/μ) - (y-μ)}
Δdev = dev - oldDev
}
Chi2 =Σ(y-μ)²/μ
AIC = (-2Σ(y*ln(μ)-μ-lngamma(y + 1)) + 2*p)/n
BIC = 2Σ(y*ln(y/μ) - (y-μ)) - df*ln(n)
*   n = # observations; p = # predictors
*  df = model degrees of freedom
```

likelihoods. Traditionally, the deviance statistic has been used as the basis of convergence for GLM algorithms (see Table 6.1). It has also been used as a goodness-of-fit statistic, with lower positive values representing a better fitted model (see Section 5.2.1.2).

Substituting the appropriate Poisson terms into the saturated likelihood parameterization entails substituting the value of y for each instance of μ. This gives us

$$D = 2\sum_{i=1}^{n}\{y_i \ln(y_i) - y_i - y_i \ln(\mu_i) + \mu_i\} \tag{6.25}$$

$$= 2\sum_{i=1}^{n}\left\{y_i \ln(\frac{y_i}{\mu_i}) - (y_i - \mu_i)\right\} \tag{6.26}$$

We may recall Table 4.1, which schematized the generic IRLS algorithm. We can now substitute the above Poisson statistics into Table 6.1 to develop a paradigm IRLS-type Poisson regression.

The Poisson regression model is also considered as a non-linear regression to be estimated using maximum likelihood methods. But, in order to do so, we must calculate the derivatives of the log-likelihood function, which define the gradient and observed Hessian matrix.

The gradient vector, score, or first derivative of the Poisson log-likelihood function with respect to parameters β, is calculated as

$$\frac{\partial(\mathcal{L}(\beta; y))}{\partial\beta} = \sum_{i=1}^{n} \left(y_i - \exp\left(x_i'\beta\right)\right) x_i' \tag{6.27}$$

For an intercept-only model, equation 6.27 reduces to

$$\sum_{i=1}^{n} \left\{ \left(\frac{y_i}{\exp(x_i'\beta)} \right) - 1 \right\}$$

Setting equation 6.27 to zero, the Poisson estimating equation

$$\sum_{i=1}^{n} \left(y_i - \exp(x_i'\beta)\right) x_i' = 0 \tag{6.28}$$

provides for the solution of the parameter estimates.

The Hessian matrix is calculated as the second derivative of the log-likelihood function and is negative definite for β. For the Poisson, it may be expressed as

$$\frac{\partial^2(\mathcal{L}(\beta; y))}{\partial\beta\partial\beta'} = -\sum_{i=1}^{n} \left(\exp(x_i'\beta)\right) x_i x_i' \tag{6.29}$$

For an intercept-only model, equation 6.29 reduces to

$$\sum_{i=1}^{n} \left(\frac{-y_i}{\exp(x_i'\beta)} \right)$$

Estimation of the maximum likelihood variance–covariance matrix is based on the negative inverse of the Hessian, many times represented as Σ, given as

$$\sum = -H^{-1} = \left[\sum_{i=1}^{n} \left(\exp(x_i'\beta)\right) x_i x_j' \right]^{-1} \tag{6.30}$$

The square roots of the respective terms on the diagonal of the negative inverse Hessian are the values of parameter standard errors. A Newton–Raphson algorithm can be used for the maximum likelihood estimation of parameters

$$\boldsymbol{\beta}_{r+1} = \boldsymbol{\beta}_r - \mathbf{H}^{-1}\mathbf{g} \tag{6.31}$$

which is the standard form of the maximum likelihood estimating equation.

The tools have now been developed to construct a full maximum likelihood Poisson algorithm. The algorithm typically updates estimates based on the value of the log-likelihood function. When the difference between old and updated values is less than a specified tolerance level – usually 10^{-6} – iteration discontinues and the values of the various statistics are at their maximum likelihood estimated values. Other iteration criteria have been employed, but they all generally result in the same estimates. Refer to Table 4.3 for the structure of the estimation algorithm.

We next turn to the construction of synthetic Poisson data. Simulated data have a variety of uses, as will be discussed in the following section.

6.2 Synthetic Poisson models

This section is concerned with the construction of synthetic data and models. We address the use of random number generators for creating synthetic data and show examples of models that closely reflect the parameters and conditions specified by the statistician when developing a synthetic model. Section 6.2.2 provides a brief discussion of how summary statistics change, if at all, when values of the response and predictors change.

6.2.1 Construction of synthetic models

Statisticians employ synthetic data sets to evaluate the appropriateness of fit statistics as well as to determine the effect of modeling the data after making specific alterations to the data. Models based on synthetically created data sets have proved to be extremely useful in this respect, and appear to be used with increasing frequency in texts on statistical modeling.

The key to constructing synthetic data is the use of pseudo-random number generators. The data are given a specified number of observations, a seed value may optionally be set, and explanatory predictors are defined. In the code below, values of the model predictors are specified on the line defining the linear predictor. The following line defines the fitted values by means of the inverse link function. The Poisson inverse link is the exponential function, i.e. the exponentiation of the linear predictor, $\exp(x'\beta)$. For Poisson models the fitted values are estimated counts.

Finally, the fitted value is employed as the parameter of a pseudo-random Poisson generator, in this case **rpoisson**. The resultant random Poisson variate is the response variable of a synthetic Poisson regression model.

Table 6.2 *Synthetic Poisson data*

[With predictors $x1$ and $x2$, having parameters of 0.75 and −1.25 and an intercept of 2]

```
* poi_rng.do
clear
set obs 50000
set seed 4744
gen x1 = rnormal() // normally distributed:
gen x2 = rnormal() // " -4.5 - +4.5
gen xb = 2 + 0.75*x1 - 1.25*x2 // define coefficients
gen exb = exp(xb) // inverse link; def mean
gen py = rpoisson(exb) // random Poisson variate
glm py x1 x2, fam(poi) // model resultant data
--------------------------------------------------------------
Note: lines generating xb, exb, and py may be shortened to
gen py = rpoisson(exp(2 + 0.75*x1 - 1.25*x2))
```

The model output is given as:

```
Generalized linear models                   No. of obs      =      50000
Optimization      : ML                      Residual df     =      49997
                                            Scale parameter =          1
Deviance          = 52295.46204             (1/df) Deviance =   1.045972
Pearson           = 50078.33993             (1/df) Pearson  =   1.001627

                                            AIC             =   4.783693
Log likelihood    = -119589.3262            BIC             =    -488661
-------------------------------------------------------------------------
      |                OIM
   py |     Coef.  Std. Err.       z    P>|z|     [95% Conf. Interval]
------+------------------------------------------------------------------
   x1 |   .7488765  .0009798   764.35   0.000    .7469562    .7507967
   x2 |  -1.246898  .0009878 -1262.27   0.000   -1.248834   -1.244962
_cons |   2.002672  .0017386  1151.91   0.000    1.999265     2.00608
-------------------------------------------------------------------------
```

The parameter estimates closely approximate the user defined values. If we delete the seed line, add code to store each parameter estimate, and convert the *do* file to an *rclass ado* program, it is possible to perform a Monte Carlo simulation of the synthetic model parameters. The above synthetic Poisson data and model code may be amended to execute a simple Monte Carlo simulation using Stata. The code is in Table 6.3, with values stored for the two parameter estimates and intercept together with the Pearson and deviance based dispersion statistics.

Table 6.3 *Monte Carlo simulation of synthetic Poisson data*

```
program define poi_sim, rclass   /* poi_sim.ado */
version 10
drop _all
set obs 50000
gen x1 = rnormal()
gen x2 = rnormal()
gen py = rpoisson(exp(2 + 0.75*x1 - 1.25*x2))
glm py x1 x2, nolog fam(poi)
return scalar sx1 = _b[x1]          /// x1
return scalar sx2 = _b[x2]          /// x2
return scalar sc = _b[_cons]        /// intercept
return scalar ddisp = e(dispers_s)  /// deviance dispersion
return scalar pdisp = e(dispers_p)  /// Pearson dispersion
end
```

The following basic **simulate** command can be used for a Monte Carlo simulation, repeating an execution of the code in Table 6.3 500 times. The command builds a 500-observation data set in memory consisting of the values saved in the created variables *sx*1, *sx*2, *sc, ddisp,* and *pdisp* respectively. Other numbers of repetitions can be requested. Generally it is a good idea to initially create a 5–10-observation data set to ensure the code is doing what is expected. There is typically little difference in the values of these statistics for repetitions greater than 30. But larger numbers of repetitions guarantee more accurate values, with the limit being the *true* value of the parameters given the data. This limit is displayed for each variable or statistic as the mean of the 500 values. A run of the Monte Carlo simulation of the synthetic data is given below.

```
. simulate mx1 = r(sx1) mx2 = r(sx2) mcon = r(sc)
  mdd = r(ddisp) mpd = r(pdisp), reps(500) : poi_sim
  . . .
. summ

variable |  Obs      Mean      Std. Dev.      Min         Max
---------+---------------------------------------------------------
     mx1 |  500    .7500655   .0010005    .7475729     .752877
     mx2 |  500   -1.250047   .0009626   -1.252795   -1.246607
    mcon |  500    1.999863   .0017003    1.993828    2.005672
     mdd |  500    1.041547   .0063002    1.021438     1.06144
     mpd |  500    .9997811   .0066034    .9791458    1.020914
```

The Monte Carlo estimates are very close to the values we specified for the synthetic Poisson model given in Table 6.2. Employing a greater number of repetitions will result in mean values even closer to the user specified values. Standard errors could also have been included in the above simulation, which is of considerable value when testing the variability of the data.

Of particular interest here is the value of the two dispersion statistics. Traditionally the dispersion and fit tests have been based on the deviance dispersion. However, it is clear from the Monte Carlo simulation that the appropriate test should instead be based on the Pearson dispersion. In short, and as will be discussed in greater length in the following chapter, a model having values of the Pearson dispersion greater than 1.0 is overdispersed. Values under 1.0 are underdispersed. It is common to find Poisson models with a dispersion statistic ranging from 0.2 to greater than 10. An exception to using the Pearson dispersion as an indicator of possible overdispersion exists when survey or *importance* weights are employed in the model. Importance weights magnify the Pearson dispersion as the product of the dispersion of the frequency-weighted model and the mean of the variable used as the weight. A survey-weighted model may appear overdispersed given the dispersion statistic, when it is not.

Extra dispersion in a Poisson model results from excessive correlation in the data. We will later discuss the many causes of such correlation, but be aware for now that, when it exists, it biases the standard errors of the model.

When a Poisson model is well specified and appropriately fits the data, the Pearson-based dispersion statistic has a value approximating 1.0. The deviance-based dispersion in the model above is some 4% greater (1.04) in value. Goodness-of-fit statistics based on the deviance-dispersion statistic will therefore be biased by this amount. Other simulation runs with differing parameter values and more predictors can produce deviance-dispersion values farther from unity than 4%. For example, if a 3-predictor Poisson model is simulated with values of $x1 = 1.75$, $x2 = -2.25$, $x3 = 2.0$, and _cons $= 1$ for 100 repetitions, the two dispersion statistics are:

```
Variable  |  Obs       Mean     Std. Dev.       Min         Max
----------+-----------------------------------------------------
      .   |   .          .          .             .           .
    mdd   |  100    .9096086    .0051867     .8913224    .9168785
    mpd   |  100    .9994591    .0176827     .9498385    1.035536
```

The parameter values are identical to the specified values, but are not shown. Note that the Pearson dispersion is near unity, but the deviance-dispersion is some 9% off 1.0. In the simulations I have conducted, the Pearson-dispersion is always fixed at 1.0 while the deviance-dispersion varies rather widely based on the number of coefficients and their values.

Care must be taken, therefore, to know which dispersion statistic is being used to evaluate a Poisson model when presented in study results. It is clear that only the Pearson dispersion statistic is valid for fit analysis. We shall simply refer to it as the dispersion statistic. For a comprehensive discussion of binomial overdispersion, see Hilbe (2009).

Looking ahead, the traditional parameterization of the negative binomial model (NB2) is an attempt to accommodate the overdispersion found in an otherwise Poisson model. The NB2 heterogeneity parameter, α, is a measure of Poisson overdispersion, or extra correlation, and corresponds in direction to the value of the Pearson dispersion given in model results. Higher values of the Poisson dispersion statistic are reflected in a greater value of α in NB2 model results. We discuss this relationship in detail in Chapter 8.

An example of using R to effect the same synthetic model as above is shown in Table 6.4. However, the function has been enhanced to generic form. With the **poisson_syn** function in memory (to the single line), create *sim.data* using the coefficients you desire. The first number is the intercept. The number of coefficients typed will tell the function how many predictors you wish. Then model the data as shown. Default values are given at the start of the function.

Table 6.4 R: *Synthetic Poisson model: generic function*

```
require(MASS)                        # posyn.r
poisson_syn  <- function(nobs = 50000, off = 0, xv = c(1, -.5,  1))  {
  p <- length(xv) - 1
  X <- cbind(1, matrix(rnorm(nobs * p), ncol = p))
  xb <- X %*% xv
  exb <- exp(xb + off)
  py <- rpois(nobs, exb)
  out <- data.frame(cbind(py, X[,-1]))
  names(out) <- c("py", paste("x", 1:p, sep=""))
  return(out)
}
-------------------------
sim.data <- poisson_syn(nobs = 50000, xv = c(2, .75,-1.25))
mypo <- glm(py ~ ., family=poisson, data = sim.data)
summary(mypo)
```

Relevant output of the R code in Table 6.4 is displayed as:

```
              Estimate  Std. Error z value Pr(>|z|)
(Intercept)  2.0003206  0.0017192  1163.5   <2e-16 ***
x1           0.7523057  0.0009594   784.1   <2e-16 ***
x2          -1.2492210  0.0009957 -1254.6   <2e-16 ***
---
```

Simulation of the Pearson dispersion can be obtained using R with the code given in Table 6.5. The algorithm requests 100 iterations, requiring a few minutes of processing. The Monte Carlo estimated coefficients, including the intercept, and Pearson dispersion are calculated and displayed. For a quicker output, change the value of 100 (iterations) to 10 in the final two lines of code.

Table 6.5 R: *Monte Carlo simulation of Poisson data*

```
mysim <- function()
{
 nobs <- 50000
 x1 <-runif(nobs)
 x2 <-runif(nobs)
 py <- rpois(nobs, exp(2 +.75*x1 -- 1.25*x2))
 poi <- glm(py ~ x1 + x2, family=poisson)
   pr <- sum(residuals(poi, type="pearson")^2)
   prdisp <- pr/poi$df.residual
   beta <- poi$coef
   list(beta,prdisp)
}
B <- replicate(100, mysim())
apply(matrix(unlist(B[1,]),3,100),1,mean)
mean(unlist(B[2,]))
```

```
> apply(matrix(unlist(B[1,]),3,100),1,mean)
[1] 1.9998356 0.7507025 -1.2502162
> mean(unlist(B[2,]))
[1] 1.000437
```

It may be of interest to construct a graph of the synthetic model mean. Following a run of the model displayed in Table 6.4, we may summarize the predicted mean of *py*, then use that value as μ in a Poisson probability distribution. A graph of the distribution *ypoi*, where *py* is considered to be the response term, is found in Figure 6.1.

R

```
mpy <- mean(py)
ypoi <- (exp(-mpy)*mpy^py)/gamma(py+1)
plot(ypoi~ py, xlim=c(0,50), main="Synthetic Poisson Model: Mean=21")
```

```
. sum py
. gen double ypoi = (exp(-r(mean))*r(mean)^py)/exp(lngamma(py+1))
. graph twoway scatter ypoi py if py<50, title(Synthetic
Model: Mean = 21.35) xline(21.35) c(0) sort
```

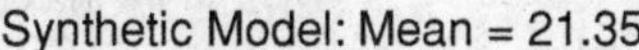

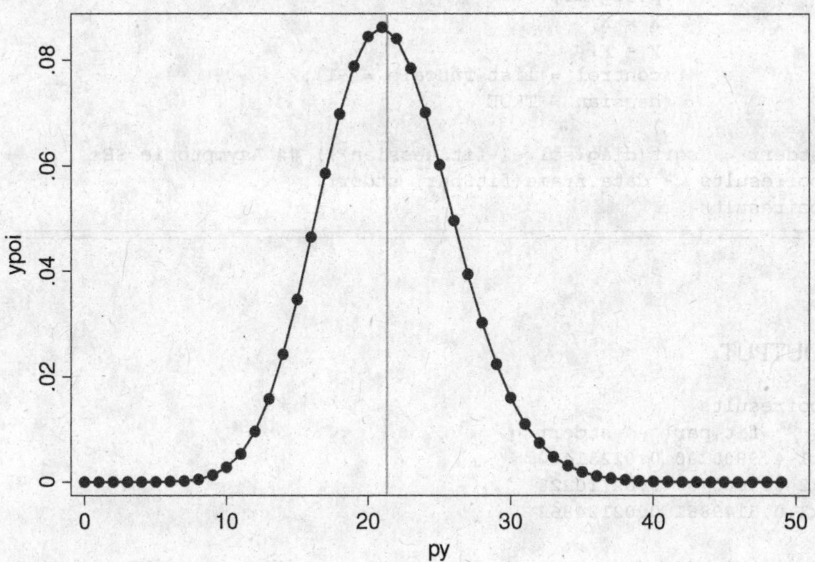

Figure 6.1 Graph of *py* from Table 6.2

As values of the mean of a Poisson distribution get larger, the distribution approaches normality. Values from 4 through 10 exhibit decreasing right skewness. Here the skew is 1.25 and kurtosis 3.04. Normality has values of 1.0 and 3.0 respectively. The Poisson distribution begins to appear normal when the mean is 4, with greater values more closely approximating normality.

Another more primitive method of estimating synthetic Poisson models using optimization can be given in the following table. Refer to Jones *et al.* (2009) for an exposition of optimization.

R: *Poisson optimization*

```
set.seed(3357)
b <-
c(5, 1, 0.5)                               ## Population parameters
n <- 10000
X <- cbind(1, rnorm(n), rnorm(n))          ## Design matrix
y <- rpois(n = n, lambda = X %*% b)
p.reg.ml <-
function(b.hat, X, y) {                    ## Joint Conditional LL
  sum(dpois(y, lambda = X %*% b.hat, log = TRUE))
}
p.0 <- lm.fit(X, y)$coef                   ## Obtain initial estimates
fit <- optim(p.0,                          ## Maximize JCLL
             p.reg.ml,
             X = X,
             y = y,
             control = list(fnscale = -1),
             hessian = TRUE
             )
stderr <- sqrt(diag(solve(-fit$hessian))) ## Asymptotic SEs
poiresults <- data.frame(fit$par, stderr)
poiresults
```

OUTPUT

```
poiresults
     fit.par     stderr
x1 4.9900700 0.02231480
x2 0.9600787 0.02110228
x3 0.5149861 0.02124862
```

A note should be given on what is meant by *pseudo-random* when referring to the generation of random numbers. The presumably random numbers that are generated by "random-number generators" in statistical software are not truly random. Computers are not able to generate truly random numbers. The pseudo-random numbers that are actually produced by computer algorithms are thoroughly deterministic, although they do have the appearance of being random. The key to determining that they are not in fact random is the use of a *seed*.

The uniform pseudo-random number generator underlies all other distribution generators. Symbolized as $U(0,1)$, the uniform generator produces a series of numbers between the values 0 and 1. The series of numbers is calculated by the use of an algorithm, with a starting value given by the computer clock.

However, the start can be specified by setting a seed, or initial value, which then generates the identical series of numbers with each repetition based on the algorithm used. That is, for a given seed, a deterministic series of numbers will be produced that is constrained to uniformly fit within the range 0 to1. Different seed numbers will produce different series of numbers based on the algorithm. Again, if a seed number is not specified, one is automatically produced based on the computer's clock.

An example will clearly demonstrate the deterministic nature of pseudo-random number generators. Reserve space for only four observations. Set a seed and generate two runs of the uniform function. Then clear memory and reserve space for eight observations. Set the same seed and execute the pseudo-random uniform number generator for one variate.

STATA R

```
. clear
. set obs 4                               > obs <- 1:4
. set seed 1357                           > set.seed(1357)
. gen a = runiform()                      > a <- runif(obs)
. gen b = runiform()                      > b <- runif(obs)
. clist                                   > list(a,b)
             a          b
  1.   .9032054   .0758667
  2.   .5338306   .7561663
  3.   .28684     .6649072
  4.   .577705    .1942068
. clear
. set obs 8                               > obs <- 1:8
. set seed 1357                           > set.seed(1357)
. gen a = runiform()                      > c <- runif(obs)
. clist                                   > list(c)
             a
  1.   .9032054
  2.   .5338306
  3.   .28684
  4.   .577705
  5.   .0758667
  6.   .7561663
  7.   .6649072
  8.   .1942068
```

Note that a in the eight-observation run is identical to the conjoined series of $a + b$ in the four-observation runs. It is easy to observe that, given a specified seed value, a deterministic series of numbers is produced. Other seed values produce other deterministic series, each mapped as a uniform distribution. The series of numbers produced is certainly not random but, with so many possible start values available, the appearance of randomness is evident. Refer to Jones *et al.* (2009) and Cameron and Trivedi (2010) for additional information.

6.2.2 Changing response and predictor values

In this section we look at how summary statistics and parameter values change when response and predictor values are varied by the application of a uniform mathematical operation. Doing so allows some insight into understanding how count models differ given changes in data.

Changes to the response

First create a synthetic Poisson data set with a single normal pseudo-random predictor. After modeling the random response, *y*, the response is multiplied by 10, creating a variable, *y*10. The new response is modeled on the same predictor. Observe the changes to the model statistics.

Construct a 50,000-observation synthetic Poisson model using the code below.

```
. clear
. set obs 50000
. gen x1= runiform()
. gen y = rpoisson(exp(1 + .5*x1))
. glm y x1, fam(poi)

Generalized linear models                   No. of obs       =      50000
Optimization        : ML                    Residual df      =      49998
                                            Scale parameter  =          1
Deviance            =  54127.23639          (1/df) Deviance  =   1.082588
Pearson             =  50111.16122          (1/df) Pearson   =   1.002263
. . .
                                            AIC              =   4.029168
Log likelihood      = -100727.1884          BIC              =    -486840
------------------------------------------------------------------------
     |                 OIM
   y |     Coef.   Std. Err.      z    P>|z|    [95% Conf. Interval]
-----+------------------------------------------------------------------
  x1 | .4966001   .0083048     59.80   0.000    .4803231    .5128771
_cons |  1.00199   .0050831    197.12   0.000    .9920273    1.011953
------------------------------------------------------------------------

. predict mu, mu
```

Next multiply the response, *y*, by 10, creating variable *y*10. Model with *y*10 as the response. Note that *y*10 is a constant that is applied to every value of *y*.

```
. gen y10 = y*10
. glm y10 x1, fam(poi)

Generalized linear models          No. of obs      =     50000
Optimization     : ML              Residual df     =     49998
                                   Scale parameter =         1
Deviance         =  541272.3639    (1/df) Deviance =  10.82588
Pearson          =  501111.6122    (1/df) Pearson  =  10.02263

                                   AIC             =  15.94285
Log likelihood   = -398569.2641    BIC             =  305.0892
------------------------------------------------------------------
      |             OIM
 px10 |     Coef.  Std. Err.     z     P>|z|  [95% Conf. Interval]
------+-----------------------------------------------------------
   x1 | .4966001  .0026262  189.09  0.000   .4914528    .5017473
_cons | 3.304575  .0016074  2055.83 0.000   3.301425    3.307726
------------------------------------------------------------------
. predict mu10, mu
```

Multiplying the Poisson response by 10 has the effect of multiplying the deviance and Pearson statistics, and their associated dispersion statistics, by 10. The coefficient, $x1$, remains unchanged. The square of the ratio of the two models is also 10.

$x1$

```
. di (.0083048 /.0026262)^2
10.000064
```

INTERCEPT

```
. di (.0050831 /.0016074)^2
10.000216
```

Finally, the predicted values also have a ten-fold difference in mean values.

```
. su mu*

Variable |   Obs      Mean    Std. Dev.      Min        Max
---------+--------------------------------------------------
      mu |  50000   3.52656    .5042567   2.723754   4.475327
    mu10 |  50000   35.2656    5.042567   27.23755   44.75327
```

A rule that can be abstracted from the above, and which is in fact applicable to real models, is that if the response is multiplied by x, then the Pearson and deviance statistics are also multiplied by x. Likewise, the deviance and Pearson dispersion statistics are multiplied by x. In the case of the standard errors, the ratio of the original and original*x standard errors is equal to the square root of x. Likewise, the mean predicted count of the original*x model is x times greater in value than the mean predicted count of the original model.

Changes to the predictor

What occurs when the value of a predictor changes instead of the response? This is certainly the more common situation. Using the same synthetic model for y as in the previous subsection, we create a model with $x1$ multiplied by 10.

```
. gen x10 = x1*10
. glm y x10, fam(poi)

Generalized linear models              No. of obs      =      50000
Optimization     : ML                  Residual df     =      49998
                                       Scale parameter =          1
Deviance         =  54127.23637        (1/df) Deviance =   1.082588
Pearson          =   50111.1612        (1/df) Pearson  =   1.002263
                                       . . .
                                       AIC             =   4.029168
Log likelihood   = -100727.1884        BIC             =    -486840
-------------------------------------------------------------------
     |                OIM
   y |   Coef.    Std. Err.      z     P>|z|   [95% Conf. Interval]
-----+-------------------------------------------------------------
 x10 |  .04966    .0008305    59.80    0.000    .0480323   .0512877
_cons | 1.00199   .0050831   197.12    0.000    .9920273   1.011953
-------------------------------------------------------------------

. predict mux10, mu
```

It should be noticed at the outset that the model summary statistics are identical to the original model. Moreover, the intercepts are the same, as are the predicted values.

```
. su mu mux10

Variable |    Obs        Mean   Std. Dev.        Min        Max
---------+----------------------------------------------------
      mu |  50000     3.52656    .5042567   2.723754   4.475327
   mux10 |  50000     3.52656    .5042567   2.723754   4.475327
```

The only difference in displayed output are the coefficient values. $x10$ is 10 times less than $x1$. That is, the coefficient of $x10$ times 10 equals the coefficient of $x1$.

This same relationship holds for any single predictor model. It does not matter if the data are synthetic or real – the same relationships hold.

This rule may be enhanced to the use of mathematical operations in general. Dividing by a constant value, or applying a square or square root, to a response or predictor result in the same logical relationships as we observed here for multiplication. R code for the above manipulations are given in the following inset.

Table 6.6 R: *Changes to univariable model*

```
# nbr2_6_2_2.r
nobs <- 50000
x1 <- runif(nobs)
y <-rpois(nobs, exp(1 + 0.5*x1))
poi <- glm(y ~ x1, family=poisson)
summary(poi)
mu <- predict(poi, type="response")
# change response y
y10 <- y*10
poi10 <- glm(y10 ~ x1, family=poisson)
summary(poi10)
mu10 <- predict(poi10, type="response")
# change predictor x1
x10 <- x1*10
poix <- glm(y ~ x10, family=poisson)
summary(poix)
mux10 <- predict(poix, type="response")
```

6.2.3 Changing multivariable predictor values

Changing response in a multivariable model

Here we briefly look at expanding the single predictor model to multivariable. We wish to determine if the same relationships are obtained as with the single predictor model. We do so by creating a new synthetic Poisson model with two predictors and an intercept.

```
. set obs 50000
. gen x1 = runiform()
. gen x2 = runiform()
. gen py = rpoisson(exp(1 + .5*x1 - .25*x2))
. glm py x1 x2, nolog fam(poi)
```

```
Generalized linear models                 No. of obs      =      50000
Optimization     : ML                     Residual df     =      49997
                                          Scale parameter =          1
Deviance         =  54537.48851           (1/df) Deviance =   1.090815
Pearson          =   49967.4882           (1/df) Pearson  =   .9994097
                                 . . .
                                          AIC             =    3.89607
Log likelihood   = -97398.74088           BIC             =    -486419
------------------------------------------------------------------------
       |                OIM
    py |      Coef.  Std. Err.      z  P>|z|    [95% Conf. Interval]
-------+----------------------------------------------------------------
    x1 |   .4894407   .0088071   55.57 0.000    .4721791     .5067023
    x2 | -.2586323   .0087254  -29.64 0.000  -.2757339    -.2415308
 _cons |   1.015795   .0068234  148.87 0.000    1.002421     1.029168
------------------------------------------------------------------------
```

Given that the model is based on pseudo-random variates, it is close to what we expect. We shall later discover using Monte Carlo techniques that this type of model generates without bias a synthetic model with "true" variates.

Multiply the response *py* by 10 (*py*10) and use it as the response in a new model.

```
. gen py10 = py*10
. glm py10 x1 x2, fam(poi)

Generalized linear models                 No. of obs      =      50000
Optimization     : ML                     Residual df     =      49997
                                          Scale parameter =          1
Deviance         =   545374.8851          (1/df) Deviance =   10.90815
Pearson          =    499674.882          (1/df) Pearson  =   9.994097
                                          . . .
                                          AIC             =   15.84373
Log likelihood   = -396090.1265           BIC             =    4418.43
------------------------------------------------------------------------
       |                OIM
  py10 |      Coef.  Std. Err.      z    P>|z|   [95% Conf. Interval]
-------+----------------------------------------------------------------
    x1 |   .4894407    .002785  175.74   0.000  .4839821      .4948993
    x2 | -.2586323   .0027592  -93.73   0.000 -.2640403     -.2532243
 _cons |    3.31838   .0021577 1537.90   0.000  3.314151      3.322609
------------------------------------------------------------------------
```

The relationship between the two multivariable models is identical to the relationship we observed with the single predictor model. Coefficients have the same values, and the deviance and Pearson statistics, associated dispersion statistics, and mean predicted counts are increased tenfold. The intercept, as before, differs, and the same relationship exist between standard errors.

Changing a predictor in a multivariate model

Next change the predictor $x1$, multiplying it by 10. The newly created variable is termed $x10$.

```
. gen x10 = x1*10
. glm py x10 x2, fam(poi)

Generalized linear models                   No. of obs      =      50000
Optimization     : ML                       Residual df     =      49997
                                            Scale parameter =          1
Deviance         =  54537.48851             (1/df) Deviance =   1.090815
Pearson          =   49967.4882             (1/df) Pearson  =   .9994097
                                            . . .
                                            AIC             =    3.89607
Log likelihood   = -97398.74088             BIC             =    -486419
------------------------------------------------------------------------
       |                 OIM
    py |     Coef.   Std. Err.      z    P>|z|    [95% Conf. Interval]
-------+----------------------------------------------------------------
   x10 |  .0489441   .0008807    55.57   0.000    .0472179    .0506702
    x2 | -.2586323   .0087254   -29.64   0.000   -.2757339   -.2415308
 _cons |  1.015795   .0068234   148.87   0.000    1.002421    1.029168
------------------------------------------------------------------------
```

As with the single predictor model, when a uniform change is made to a predictor, model summary statististics do not change. Only the coefficient of the amended predictor changes. Note that the original predictor, $x1$, is 10 times greater than the changed predictor, i.e. $x10$ is 10 times less $x1$.

A caveat needs to be made explicit at this point – these relationships hold only when a constant value is applied to a response or to a predictor(s). However, the exercise should demonstrate the relationship of the response and predictors to one another in the production of summary statistics.

We now discuss real data, and will detail the various ways in which parameter estimates may be constructed and interpreted. R code for this subsection is provided first.

Table 6.7 R: *Changes to multivariable model*

```
# nbr2_6_2_3.r
nobs <- 50000
x1m <- runif(nobs)
ym <-rpois(nobs, exp(1 + 0.5*x1 -.25*x2))
poim <- glm(ym ~ x1 + x2, family=poisson)
summary(poim)
```

```
mum <- predict(poim, type="response")
# change response ym
ym10 <- ym*10
poim10 <- glm(ym10 ~ x1 + x2, family=poisson)
summary(poim10)
mum10 <- predict(poim10, type="response")
# change predictor x1
xm10 <- x1*10
poix10 <- glm(ym ~ xm10 + x2, family=poisson)
summary(poix10)
mumx10 <- predict(poix10, type="response")
```

6.3 Example: Poisson model

Poisson models are typically used either to summarize predicted counts based on a set of explanatory predictors, or for the interpretation of exponentiated estimated slopes, indicating the expected change or difference in the incidence rate ratio of the outcome based on changes in one or more explanatory predictors. An example will demonstrate how each of these modeling concerns appear in fact. For pedagogical reasons both the coefficient and incident rate ratio parameterizations will be discussed separately.

6.3.1 Coefficient parameterization

This example comes from Medicare hospital length-of-stay data from the state of Arizona. The data are limited to only one diagnostic group (DRG 112). Patient data have been randomly selected from the original data to be part of the **medpar** data set. Relevant **medpar** variables for this example include:

RESPONSE

los: length of stay, a count of the days each patient spent in the hospital

PREDICTORS

hmo: Patient belongs to a Health Maintenance Organization (1), or private pay (0)

white: Patient identifies themselves as primarily Caucasian (1) in comparison to non-white (0)

type : A three-level factor predictor related to the type of admission. 1 = elective (referent), 2 = urgent, and 3 = emergency

Program code and results are displayed in both Stata and R.

```
. glm los hmo white type2 type3, fam(poi)

Generalized linear models                 No. of obs      =       1495
Optimization      : ML                    Residual df     =       1490
                                          Scale parameter =          1
Deviance          =  8142.666001          (1/df) Deviance =   5.464877
Pearson           =  9327.983215          (1/df) Pearson  =   6.260391
. . .
                                          AIC             =   9.276131
Log likelihood    = -6928.907786          BIC             =  -2749.057
------------------------------------------------------------------------
      |             OIM
  los |   Coef.   Std. Err.    z      P>|z|    [95% Conf. Interval]
------+-----------------------------------------------------------------
  hmo | -.0715493  .023944   -2.99    0.003    -.1184786   -.02462
white | -.153871   .0274128  -5.61    0.000    -.2075991   -.100143
type2 |  .2216518  .0210519  10.53    0.000     .1803908    .2629127
type3 |  .7094767  .026136   27.15    0.000     .6582512    .7607022
_cons | 2.332933   .0272082  85.74    0.000    2.279606    2.38626
------------------------------------------------------------------------
```

The variance covariance matrix can be displayed as

```
. matrix list e(V)

symmetric e(V)[5,5]
                 los:         los:         los:         los:         los:
                 hmo          white        type2        type3        _cons
  los:hmo     .00057331
los:white   -.00003093    .00075146
los:type2    .00003196    .00005636    .00044318
los:type3    .00007959    .00002412    .0001065     .00068309
los:_cons   -.00006879   -.00068476   -.00015703   -.0001355     .00074028
```

The standard errors are contained in the above matrix as the square root of the main diagonal elements. For example, the standard error of *hmo* is:

```
. di sqrt(.00057331)
.02394389
```

The coefficient and standard error of a predictor may also be abstracted from the table of parameter estimates. *hmo* statistics may be more nicely abstracted using the code:

```
. di "hmo: coefficient = " _b[hmo] " hmo: SE = " _se[hmo]
hmo: coefficient = -.07154931 hmo: SE =.02394396
```

As discussed in Chapter 2, the confidence intervals of the coefficient standard errors are determined by the formula given in equation 2.6. To repeat here for clarification, the confidence internals are

$$\beta_{\mathrm{k}} \pm z^{\alpha/2}\ \mathrm{SE}(\beta_{\mathrm{k}}) \tag{6.32}$$

where $z^{\alpha/2}$ is a quantile from the normal distribution. The $100(1-\alpha)\%$, or 95%, confidence interval has a value of 1.96, and is the same as $\alpha = 0.05$. β_k is the coefficient on the kth predictor, and $\mathrm{SE}(\beta_k)$ is its standard error.

Given the above formula, the 95% confidence intervals for *hmo* are

β_{hmo}	$\pm$	$z^{\alpha/2}$	*	$\mathrm{SE}(\beta_{hmo})$	=	Conf interval
−0.0715493	−	1.96	*	0.02394396	=	−0.11847946
−0.0715493	+	1.96	*	0.02394396	=	−0.02461914

which are the values given in the model output above. Other predictor standard errors are determined in the same manner. We shall find, however, that the methods discussed here cannot be used to determine the standard error for exponentiated coefficients, i.e. incidence rate ratios. IRR standard errors and confidence intervals will be addressed in the next section.

The R code for estimating the same Poisson model as above, together with relevant output, is given in Table 6.8.

The dependent or response variable of a Poisson model is a count. The natural log of the expected or fitted value, $\ln(\mu)$, is the canonical link function. Therefore, the Poisson model estimates the log of the expected count, given the values of the respective explanatory predictors. Coefficients represent the difference in the expected log-count of one level compared with another. For a binary predictor, we have $\ln(\mu_1) - \ln(\mu_0)$, which is the same as $\ln(\mu_1/\mu_0)$. Table 6.9 provides an interpretation of the Poisson binary and factor coefficients from the above displayed models.

Prior to viewing Table 6.9, it should be recalled that coefficients of any regression model are slopes, and are to be interpreted like the slopes of any linear equation. One of the basic concepts learned in beginning courses in algebra is the meaning of the slope of a line to describe the rate of change of a function. Given the simple linear equation,

$$y = \tfrac{3}{5}x + 5 \tag{6.33}$$

The slope is $\frac{3}{5}$ and the intercept is 5, giving the value of y when $x = 0$. The slope can be interpreted as: for each 5 units increase in x, y increases by 3. If

Table 6.8 R: *Poisson model on Medpar data*

```
data(medpar)
attach(medpar)
poi <- glm(los ~ hmo + white + factor(type), family=poisson,
  data=medpar)
summary(poi)
confint(poi)
vcov(poi)       # Variance-Covariance matrix; not displayed below
-----------------------------------------------------------------
 Coefficients:
             Estimate Std. Error z value Pr(>|z|)
(Intercept)  2.33293    0.02721  85.744  < 2e-16 ***
hmo         -0.07155    0.02394  -2.988  0.00281 **
white       -0.15387    0.02741  -5.613 1.99e-08 ***
type 2       0.22165    0.02105  10.529  < 2e-16 ***
type 3       0.70948    0.02614  27.146  < 2e-16 ***

                                     2.5 %      97.5 %
(Intercept)                      2.2792058  2.38586658
hmo                             -0.1187331 -0.02486953
white                           -0.2072290 -0.09976522
factor(type)Urgent Admit         0.1802390  0.26276400
factor(type)Emergency Admit      0.6579525  0.76041000
```

the slope is negative, then the slope would be interpreted as: for each 5 units increase in x, y decreases by 3. If the slope is 3 rather than $\frac{3}{5}$, a denominator of 1 is implied, so that the interpretation would be: for a unit increase in x, y increases by 3. In fact, it is this last interpretation that is the standard manner of interpreting regression coefficients. Converting the above slope, or coefficient, $\frac{3}{5}$ to 0.6, would then be interpreted as: for every one unit increase in x, y increases by 0.6. This interpretation is the standard when explaining regression coefficients.

The above linear model is that which is assumed for Gaussian or normal linear regression models. The linear predictor, $y|x$, or $x\beta$, is identical to the predicted fit, $E(y)$, $\hat{y}$, or μ. For models that require a link function to establish a linear relationship between the fit and linear predictor, such as the log link for Poisson models, the change in y given a change in x is based on the link. If equation 6.33 is a Poisson model, the slope is interpreted as: for a 1 unit change in x, y changes by $\log\left(\frac{3}{5}\right)$.

Recall that for a binary predictor, a coefficient reflects the change in value of y given $x = 1$ compared with $x = 0$. However, if we need to know the coefficient

Table 6.9 *Interpretation of Poisson coefficients*

hmo:	The difference in the log of the expected length of stay is expected to be 0.07 log-days lower for HMO than for private pay patients, with the remaining predictor values held constant.
white:	The difference in the log of the expected length of stay is expected to be 0.15 log-days lower for patients who identified themselves as white than for patients having another racial background, with the remaining predictor values held constant.
*type*2:	The difference in the log of the expected length of stay is expected to be 0.22 log-days higher for patients admitted as urgent compared with patients admitted as elective, with the remaining predictor values held constant.
*type*3:	The difference in the log of the expected length of stay is expected to be 0.71 log-days higher for patients admitted as emergency compared with patients admitted as elective, with the remaining predictor values held constant.
_cons:	The log of the expected length of stay is 2.33 log-days when a patient is private pay, non-white, and admitted to the hospital as an elective patient. That is, *hmo* and *white* are both 0 in value and *type*1 has a value of 1.

on $x = 0$ compared with $x = 1$, we simply reverse the signs. Therefore, for the model we have been using as an example, we can model *los* on non-white private pay patients, as well as on *type* of admission as before, by constructing the following modification.

Table 6.10 R: *Model with inversed binary predictors*

```
private <- medpar$hmo==0
nonwhite <- medpar$white==0
poi0 <- glm(los ~ private + nonwhite + factor(type),
   family=poisson, data=medpar)
summary(poi0)
confint(poi0)
```

```
. gen byte private = hmo==0
. gen byte nonwhite= white==0
. glm los hmo0 nonwhite type2 type3, fam(poi) nohead
```

```
------------------------------------------------------------------------------
             |                 OIM
         los |      Coef.   Std. Err.      z    P>|z|     [95% Conf. Interval]
-------------+----------------------------------------------------------------
     private |   .0715493    .023944     2.99   0.003       .02462    .1184786
    nonwhite |    .153871   .0274128     5.61   0.000      .100143    .2075991
       type2 |   .2216518   .0210519    10.53   0.000     .1803908    .2629127
       type3 |   .7094767    .026136    27.15   0.000     .6582512    .7607022
       _cons |   2.107513   .0222731    94.62   0.000     2.063858    2.151167
------------------------------------------------------------------------------
```

The coefficient and standard error values of *private* and *non-white* are identical to that of *hmo* and *white* in the earlier model – except that the signs of the coefficients are reversed. The value of the intercept has also changed, from 2.33 to 2.11. This fact allows for an easy interpretation of both levels of a binary predictor.

Continuous predictors have an interpretation similar to that of binary predictors, except that there is more than one level or units of the predictor to relate. The logic of the difference in log-count for continuous predictors is:

$$\beta_{\mathrm{K}} \;=\; \ln(\mu_{k+1}) - \ln(\mu_k) \tag{6.34}$$

where k is a given value of the continuous predictor and $k + 1$ is the next highest value.

Consider an example taken from the German health registry for the years 1984–8 (**rwm5yr**). A Poisson model can be constructed for the number of visits a patient makes to a doctor during the year (*docvis)* given their age (*age*), where *age* takes the values of 25 through 64. After a patient reaches 65 their visits are recorded in another registry. The logic of the coefficient on *age* is therefore:

$$\beta_{\mathrm{AGE}} = ln(\mu_{\mathrm{age}+1}) - \ln(\mu_{\mathrm{age}}) \tag{6.35}$$

for any two contiguous ages in the data. β_{AGE} is the coefficient on *age.*

The number of visits to the physician (*docvis*) is modeled on gender (*female*) and educational level. We wish to see if gender and education level have a bearing on the numbers of doctor visits during 1984–8.

```
Response   =  docvis    : count from 0-121
Predictor  =  female    :(1=female;0=male)
              edlevel1 : Not HS grad  edlevel2 : HS grad
              edlevel3 : Univ/Coll    edlevel4 : Grad school
panel id   =  id        : individual
time       =  year      : 1984-1988
```

```
DATA
     docvis |      Freq.     Percent        Cum.
------------+-----------------------------------------------------
          0 |      7,572       38.61       38.61
          1 |      2,582       13.17       51.78
          2 |      2,357       12.02       63.80
            |          .     .     .
         23 |         15        0.08       98.46
         24 |         37        0.19       98.65
         25 |         30        0.15       98.81
            |
     female |      Freq.     Percent        Cum.
------------+-----------------------------------------------------
          0 |     10,187       51.95       51.95
          1 |      9,422       48.05      100.00
------------+-----------------------------------------------------
      Total |     19,609      100.00
            |
       year |      Freq.     Percent        Cum.
------------+-----------------------------------------------------
       1984 |      3,874       19.76       19.76
       1985 |      3,794       19.35       39.10
       1986 |      3,792       19.34       58.44
       1987 |      3,666       18.70       77.14
       1988 |      4,483       22.86      100.00
------------+-----------------------------------------------------
      Total |     19,609      100.00
            |
    edlevel |      Freq.     Percent        Cum.
------------+-----------------------------------------------------
Not HS grad |     15,433       78.70       78.70      [Reference]
    HS grad |      1,153        5.88       84.58
  Coll/Univ |      1,733        8.84       93.42
Grad School |      1,290        6.58      100.00
------------+-----------------------------------------------------
      Total |     19,609      100.00
```

Ignoring the fact that there are multiple observations of each patient based on the year that visits were recorded, we have the following Poisson model.

Table 6.11 R: *German health data: 1984–8*

```
detach(medpar)
data(rwm5yr)
attach(rwm5yr)
rwmpoi <- glm(docvis ~ age, family=poisson, data=rwm5yr)
summary(rwmpoi)
confint(rwmpoi)
```

```
. glm docvis age, fam(poi) nohead

------------------------------------------------------------------------------
         |                 OIM
  docvis |      Coef.   Std. Err.      z    P>|z|     [95% Conf. Interval]
---------+--------------------------------------------------------------------
     age |    .023908   .0003616    66.12   0.000     .0231993    .0246168
   _cons |   .0728969   .0173841     4.19   0.000     .0388246    .1069691
------------------------------------------------------------------------------
```

The coefficient on *age* is 0.024, which can be interpreted as:

For a one-year increase in age, the difference in the log of the expected doctor visits increases by a factor of 0.024, or 2.4%. That is, for each one year of age, there is an increase in the expected log-count of doctor visits of 0.024.

The intercept for this model represents the value of the linear predictor, $x'\beta$, when $age = 0$, which is not in the data, and which makes little sense in the context of this study. Statisticians typically *center* a continuous predictor when a 0 value has little meaning. There are a number of ways that centering can be executed, but the most common is to use the mean value of the continuous predictor, subtracting it from the actual value of the predictor for a given observation. This is called *mean-centering*, or deviation scoring. The intercept of a mean-centered predictor is the value of the linear predictor at the mean value of the predictor. Other types of centering are occasionally used, like the median, but mean-centering is the most common. Centering is most commonly employed for interactions in which multicollinearity is a problem. However, not all statisticians agree on the value of centering for reducing multicollinearity.

By centering, the correlation between a multiplicative term (interaction or polynomial) and its constituent variables is lessened. However, it is simply wise to use centered continuous predictors when interpretation is made more meaningful as a result. A mean-centered *age* may be calculated as shown in Table 6.12.

Table 6.12 R: *Model with centered age*

```
mage <-mean(age)
cage <- age - mage
mage
mean(cage)
crwm <- glm(docvis ~ cage, family=poisson)
summary(crwm)
confint(crwm)
```

```
. qui sum age, meanonly     /// quietly summarize age
. gen mage = r(mean)        /// mage = mean of age, 43.78566
. gen cage = age  - mage    /// cage = centered value of age
. mean age cage

Mean estimation                      Number of obs   =   19609
--------------------------------------------------------------
     |        Mean    Std. Err.          [95% Conf. Interval]
-----+--------------------------------------------------------
 age |    43.78566     .0801855          43.62849     43.94283
cage |   -1.05e-07     .0801855         -.1571705     .1571703
--------------------------------------------------------------
```

Comparing the mean values of *age* and centered-*age* clearly indicates that the latter has a mean of zero, as we would expect. Modeling the mean-center *age* results in

```
. glm docvis cage, fam(poi) nohead

------------------------------------------------------------------
       |                 OIM
docvis |     Coef.   Std. Err.      z    P>|z|   [95% Conf. Interval]
-------+----------------------------------------------------------
  cage |   .023908   .0003616   66.12   0.000    .0231993   .0246168
 _cons |  1.119726   .0041508  269.76   0.000    1.111591   1.127861
------------------------------------------------------------------
```

Note the identical values of *age* and *cage*. Parameterizing *age* to its centered value has no bearing on the coefficient; however, the value of the intercept differs. Likewise, the coefficients of other non-centered predictors will stay the same as well. Only when there are interactions do the values of the main effects terms differ between centered and uncentered models. Noted also is the fact that the predicted counts for models with uncentered predictors compared with centered predictors are identical.

Interpretation of centered predictors relate to the deviation of the value from the mean. For the example given of *age* above, interpretation of the centered value of *age* can be given as:

> *For a one-year increase from the mean of age, the difference in the log of the expected doctor visits increases by a factor of 0.024, or 2.4%. That is, for each one-year increase from the mean of age, there is an increase in the expected log-count of doctor visits of 0.024.*

In data with a number of continuous predictors, all of which need to be centered, the following Stata program may prove useful. Execution of the code will create and label centered variables automatically. Centered variables will have a leading *c*, followed by the name of the original variable. In this instance, the three centered variables will have the names *cweight*, *cheight*, and *cage*. The code may easily be amended to name the variables in alternative ways.

```
. foreach var of varlist weight height age {
    summ `var', meanonly
    gen c`var' = `var' - r(mean)
    label var c`var' "centered `var'"
}
```

These three predictors may be more easily centered by using the user authored commands, **mcenter** (Simmons, 2004) and **center** (Jann, 2007) Type:

```
. ssc install mcenter
```

or

```
. scc install center
```

at the Stata command line to obtain the appropriate code.

To reiterate, recall that the value of the intercept is the value of the linear predictor when the values of all explanatory predictors are zero. When having a zero value for a continuous predictor makes little interpretative sense, then centering is the preferred method. The value of the mean-centered intercept is the value of the linear predictor at the mean value of the continuous predictor. For example, 1.12 is the linear predictor value of *age* at its mean, which is 43.79.

To see this, quietly run the non-centered model and calculate the linear predictor based on the mean value of age.

```
. qui glm docvis age, fam(poi) nohead /// quietly run
                                          noncentered model
. _b[_cons] + _b[age]*43.78566
```

or

```
. di.0728969 +.023908*43.78566
1.1197245
```

The value is the same as the intercept of the mean-centered model.

6.3.2 Incidence rate ratio parameterization

Log-counts are a bit difficult to handle for practical situations, therefore most statisticians prefer working with rate ratios, as described in Chapter 2. We shall use the **medpar** data set for examples in this section.

In the parameterization that follows, the coefficients are exponentiated to assess the relationship between the response and predictors as incidence rate ratios (IRR). For a model with a single predictor, the IRR is identical to the value we can calculate by hand using the methods described in Chapter 2. A factored predictor, like the levels of *type* below, would also have the same values that we can calculate using the methods of Chapter 2, given that it is the only predictor in the model. For models with two or more predictors, predictors are considered as adjusters for each other; For example, in the model above, *white* and the levels of *type* are kept at a constant value as adjusters for *hmo*; *hmo* and levels of *type* are adjusters for *white*, and so forth. As adjusters, their presence alters the value of the predictor of interest, particularly if the adjuster(s) are significant predictors. That is, statistically significant adjusters, or *confounders*, change the value of the predictor of interest from what it would be if no adjusters were present in the model. A statistical model is required to determine its value. Each significant predictor in the model may statistically be regarded as a predictor of interest. Correspondingly, dropping a highly non-significant predictor from a model only slightly changes the value of the coefficients of the remaining predictors.

The interpretation of a Poisson coefficient is different from when it is parameterized as a rate ratio. A table of incidence rate ratios and their corresponding standard errors and confidence intervals for our first model can be constructed using the following code:

```
rm(list=ls()) # Warning! Removes all objects in the current workspace.
data(medpar)
attach(medpar)
poi <- glm(los ~ hmo+white+type2+type3, family=poisson,
data=medpar)
```

```
. glm los hmo white type2 type3, fam(poi) nohead eform

------------------------------------------------------------------------------
             |                 OIM
         los |        IRR   Std. Err.      z    P>|z|     [95% Conf.  Interval]
-------------+----------------------------------------------------------------
         hmo |   .9309504   .0222906    -2.99   0.003     .8882708    .9756806
       white |   .8573826   .0235032    -5.61   0.000     .8125327     .904708
       type2 |   1.248137   .0262756    10.53   0.000     1.197685    1.300713
       type3 |   2.032927   .0531325    27.15   0.000     1.931412    2.139778
------------------------------------------------------------------------------
```

Using R, the rate ratios may be calculated from the previous R model fit by exponentiating the coefficients. Recall that *poi* was the name given the R model.

```
exp(coef(poi))
(Intercept)          hmo        white        type2        type3
 10.3081316    0.9309504    0.8573826    1.2481366    2.0329271
```

Note that the intercept is also exponentiated. Most software applications do not display this value since it is not a rate ratio. The intercept is not compared with any other level or value, and may therefore be ignored. The following command may be used to obtain the confidence intervals of the rate ratios,

```
exp(confint(poi))
                  2.5 %     97.5 %
(Intercept) 9.7689185 10.8684770
hmo         0.8880448  0.9754372
white       0.8128335  0.9050499
type2       1.1975035  1.3005198
type3       1.9308350  2.1391531
```

The similarity in values between Stata and R output, as would be expected. The interpretation of the rate ratios is much more intuitive than the interpretation of model coefficients; however, rate ratios are nevertheless based on the coefficient notion of the difference in the log of expected counts. Recalling simple algebra,

$$\beta_{\mathrm{K}} = \ln(\mu_1) - \ln(\mu_0) = \ln(\mu_1/\mu_0) \tag{6.36}$$

for binary predictors, and

$$\beta_{\mathrm{K}} = \ln(\mu_{k+1}) - \ln(\mu_k) = \ln(\mu_{k+1}/\mu_k) \tag{6.37}$$

for continuous predictors.

Note that the relationship of the two values of the predictor are in ratio form. Exponentiating the right most term in the two above equations gives us the rate ratio as understood in Chapter 2:

BINARY IRR

$$\exp(\beta_{\mathrm{K}}) = \mu_1/\mu_0 \tag{6.38}$$

CONTINUOUS IRR

$$\exp(\beta_{\mathrm{K}}) = \mu_{k+1}/\mu_k \tag{6.39}$$

Although we shall discuss this notion in more detail in Section 6.7, when counts are collected over a specified time period or area, we call the counts a *rate*. Hence the name rate ratio. The *incidence rate* is the rate at which counts enter a time period or area. Hence the term *incidence rate ratio*, which is also simply

Table 6.13 *Frequency interpretation of Poisson incidence rate ratios*

hmo:	HMO patients are expected to stay in the hospital 7% fewer days than are private pay patients, with the remaining predictor values held constant. (0.93–1)
white:	White patients are expected to stay in the hospital 14% fewer days than are non-white patients, with the remaining predictor values held constant. (0.857–1)
*type*2:	Urgent patients are expected to stay in the hospital 25% more days than are elective patients, with the remaining predictor values held constant. (1.248–1)
*type*3:	Emergency patients are expected to stay some 103% more days in the hospital than are elective patients, with the remaining predictor values held constant. (2.03–1)

a variety of risk ratio. Different names are given to the same process by various authors, but they mean the same. The area or time period may be specified as one, in which case the model is estimated without an offset. Again, we address this notion in more detail in Section 6.7.

In the case of our first example, counts are days stayed in the hospital; in the second, counts are numbers of visits to the doctor each year. The interpretation of the incidence rate ratios of the first model is given in Tables 6.13 and 6.14. Both frequency and expected count interpretations are given since both are commonly found in statistical literature. Which interpretation is used for a particular study depends on the nature of the study report, as well as on personal preference. Neither makes a statistical difference, although I tend to prefer the frequency interpretation.

Recall that we earlier discussed how the regression coefficient changes when the inverse of a binary predictor is estimated, compared with a model with $x = 1$. When, for example, we wish to estimate the coefficient of private pay (*hmo* $= 0$) rather than *hmo* (*hmo* $= 1$) using the present working example, we need only reverse the signs of the coefficient of *hmo*. However, since all exponentiated coefficients are positive, this same relationship does not obtain for values of IRR. Instead the inverse of binary predictors are estimated by inverting the value of IRR.

The IRR values of *hmo* and *white* in the example model are 0.9309504 and 0.8573826 respectively. Inverting these IRR values yields,

```
. di 1/exp(_b[hmo]) /// or 1/.9309504
1.0741711
and
. di 1/exp(_b[white]) /// or 1/.8573826
1.1663405
```

The model of *los* on *private*, *nonwhite*, and *type* of admission, with exponentiated coefficients, can be given as

R

```
private <- 1 - hmo
nonwhite <- 1 - white
poi0 <- glm(los ~ private+nonwhite+type2+type3, family=poisson,
  data=medpar)
exp(coef(poi0))
exp(confint(poi0))
```

```
. glm los private nonwhite type2 type3, fam(poi) nohead eform

----------------------------------------------------------------
             |                 OIM
         los |     IRR    Std. Err.     z    P>|z|  [95% Conf. Interval]
-------------+--------------------------------------------------
     private | 1.074171   .0257199    2.99   0.003   1.024926   1.125783
    nonwhite |  1.16634   .0319726    5.61   0.000   1.105329    1.23072
       type2 | 1.248137   .0262756   10.53   0.000   1.197685   1.300713
       type3 | 2.032927   .0531325   27.15   0.000   1.931412   2.139778
----------------------------------------------------------------
```

which verifies the inverse relationship of exponentiated coefficients of binary predictors. This relationship demonstrates that one can subtract, for example, the same IRR value that $x = 1$ is greater than 1.0 from 1.0 to obtain the IRR of $x = 0$.

For the present example, the IRR of *hmo* is 0.931, which is approximately 0.07 less than 1.0. Adding the same value of 0.07 to 1.0 produces 1.07, which is in fact the IRR value of *private*. Likewise for *white* − *non-white* (0.857–1.16), where the difference in IRR values is some $\pm$ 0.16. The IRR values of all binary predictors maintain the same relationship.

In order to examine the interpretation of a continuous predictor, we turn to the German health data example in **rwm5yr**, for which we earlier modeled the number of doctor visits on *age*, a continuous predictor. The incidence rate ratio of *age* may be interpreted by exponentiating its Poisson coefficient of 0.023908.

```
. di exp(.023908)
1.0241961
```

For a one-year increase in age, the rate of visits to the doctor increases by 2.4%, with the remaining predictor values held constant.

Table 6.14 *Expected count interpretation of Poisson incidence rate ratios*

hmo:	HMO patients have an expected decrease in the number of days in the hospital by a factor of 0.93 compared with private pay patients, with the remaining predictor values held constant.
white:	White patients have an expected decrease in the number of days in the hospital by a factor of 0.86 compared with non-white patients, with the remaining predictor values held constant.
*type*2:	Urgent patients have an expected increase in the number of days in the hospital by a factor of 0.25 compared with elective patients, with the remaining predictor values held constant.
*type*3:	Emergency patients have an expected increase in the number of days in the hospital by a factor of 2 compared with elective patients, with the remaining predictor values held constant. OR Emergency patients are expected to stay in the hospital twice as long as elective patients, other predictors being constant.

The rate ratio for a 10-year increase in age is obtained by exponentiating the incidence rate ratio 1.024 by 10.

```
. di 1.0241961^10
1.2700803
```

This same value may be directly calculated from the coefficient as:

```
. di exp(.023908*10)
1.2700801
```

Any non-incremental increase or decrease in the value of a continuous predictor uses the same logic for obtaining rate ratios.

The expected count interpretation of exponentiated Poisson coefficients, or incidence rate ratios, are given in Table 6.14.

The expected count interpretation of the incidence or risk ratio of *age* in predicting the number of doctor visits in the German health registry can be given as:

> *For a one-year increase in age, the expected number of visits to a doctor increases by a factor of 0.024 for a given year, with the other predictor values held constant. That is, for each additional year of age, patients are expected to have 0.024 more visits to a doctor, other predictor values being constant.*

On the surface, the Medicare model appears acceptable. However, the Pearson dispersion value, displayed above, is 6.26, far exceeding unity. Reiterating, the dispersion is defined as the ratio of the Pearson statistic to the degrees of freedom, or the number of observations less predictors. In this case we have

9327.983/1490 = 6.260. Such a value for a model consisting of some 1,500 observations is clearly excessive.

The dispersion statistic in the Medicare model output has been indicated with a "<" to the immediate right of the statistic. We will discover, however, that even if the dispersion statistic exceeds unity, the model may nevertheless be only apparently overdispersed, not overdispersed in fact. In this case, though, the model is indeed overdispersed. Lagrange multiplier and Z tests results indicate overdispersion (not shown). We discuss overdispersion, its varieties, and additional tests in the next chapter.

The standard errors of incidence rate ratios are not found in the variance-covariance matrix as they are for model coefficients. Rather, IRR standard errors are determined using the delta method. The method specifies that the standard error is calculated using the following formula:

$$\text{SE}_{\text{delta}} = \exp(\beta) * \text{SE}(\beta) \tag{6.40}$$

Employing the formula for the IRR of predictor *hmo*, we have

hmo IRR

```
. di exp(-.0715493)
.93095038
```

SE of *hmo* IRR

```
. di exp(-.0715493) *.02394396
.02229064
```

which are identical to the values displayed in the model output above.

What of the IRR confidence intervals? The values displayed in the above model output are 0.8882708 and 0.9756806. If we apply the same formula for determining the confidence intervals we used for model coefficients, we have:

```
. di.93095038 -- 1.96*.02229064
.88726073
```

and

```
. di.93095038 + 1.96*.02229064
.97464003
```

The values are close, and some software packages use this method. However, it is incorrect. It is generally considered preferable to simply exponentiate the coefficient confidence intervals. Doing so produces the following:

```
. di exp(-.1184786)
.88827082
. di exp(-.02462)
.9756806
```

Both values are identical to the model output. This same method is used for all exponentiated coefficient confidence intervals, including logistic regression odds ratios. The delta method for calculating standard errors is used in a variety of contexts – marginal effects standard errors being one.

6.4 Predicted counts

Expected or predicted counts (e.g. number of days of stay) can be calculated for a user defined set of predictor values. We may predict on the basis of the model that a non-white HMO patient entering the hospital as an urgent admission has an expected length of stay of 12 days.

This is particularly easy to calculate since all predictors are binary. Thus

intercept		hmo		urgent		linear predictor
β_0	+	$\beta_1 * 1$	−	$\beta_2 * 1$	=	xb
2.3329	−	0.07155	+	0.2217	=	2.48305

which is the value of the linear predictor ($x'\beta$). Next we apply the inverse link to determine the predicted count, μ; in this case exp(2.48305) = 11.977741 or 12 days.

Unfortunately the data are not well fitted, as we shall determine. This gives rise to the following observed values of *los*, given the above criteria:

R

```
head(medpar, n=10)
```

```
. l los hmo white type*  if hmo==1 & white==0 & type2==1, noob

------------------------------------------------------------------
     los        hmo        white          type1        type2        type3
------------------------------------------------------------------
       1          1            0              0            1            0
      14          1            0              0            1            0
       3          1            0              0            1            0
      19          1            0              0            1            0
------------------------------------------------------------------
```

The four observed values of *los* are 1, 3, 14, and 19. The mean of these counts is $(1 + 3 + 14 + 19)/4 = 9.25$, which is substantially lower than the predicted value of 11.98, or 12.

Of interest to many researchers is the relationship of observed and expected or predicted counts. Various programs exist to automate the calculation and graphing of the relationship. I show a method described in Long and Freese (2006). The authors use the **poisson** command, which saves specific model statistics that can be subsequently used to create specialized tables and graphs. The command **ereturn list** displays the various saved statistics following Stata estimations. The **glm** command produces saved statistics as well, but many of the same statistics are saved using different names. In any case, using the Medicare model as described above, a researcher first models the data using the **poisson** command. We may create the model without displaying it on screen by using the **quietly**, abbreviated **qui**, command prefix:

R

```
glm(los ~ hmo + white + type2 + type3, family=poisson, data=medpar)
```

```
. qui poisson los hmo white type2 type3
```

Next we may develop a table of mean predicted vs observed counts for each day of *los*. For the **medpar** data, the number of days after 24 drops off sharply, with each *los* having only a few instances. We shall therefore develop a table of observed versus predicted proportions for days 0 through 25, followed by a graph of the relationship. With a mean value of the count, *los*, at 9.85, we expect a fairly normal-appearing predicted distribution.

For the table of observed versus predicted days, we shall multiply the percentage values we obtain by 100, thereby rescaling the values to units. We also calculate the difference between the two values to more easily observe the discrepancy of observed to predicted days. The code for developing the table of observed versus predicted LOS, their difference and Figure 6.2 is contained in the Stata do-file, **medpar_obspred.do**. It can easily be amended to create the same type of table and graph for other data situations. Note that a standalone command named **prcounts**, authored by Long and Freese (2006), can be used to develop a similar table and graph. One may use the following command to effect similar results: `. prcounts psn, plot max(25)`

Readers may prefer using it to employing the code shown here for developing both the table and graph. However, I find the *do file* easier to use and amend for the examples in the text.

Table 6.15 R: *Observed vs predicted counts*

```
rm(list=ls())
data(medpar)
attach(medpar)
mdpar <- glm(los ~ hmo+white+type2+type3, family=poisson, data=medpar)
poi.obs.pred(len=25, model=mdpar)
```

```
* Code to create table Observed v Predicted LOS
* medpar_obspred.do   : also includes code for Fig 6.2
. predict mu
. local i 0
. local newvar "pr`i'"
*: Predicted probability at each los
. while `i' <=25 {
  2. local newvar "pr`i'"
  3. qui gen `newvar' =  exp(-mu)*(mu^`i')/exp(lnfactorial(`i'))
  4. local i = `i' + 1
  5. }
. quietly gen cnt = .
. quietly gen obpr = .
. quietly gen prpr = .
. local i 0
*: Observed and predicted los
. while `i' <=25 {
  2. local obs = `i' + 1
  3. replace cnt = `i' in `obs'
  4. tempvar obser
  5. gen `obser' = (los==`i')   /* generic = `e(depvar)' */
  6. sum `obser'
  7. replace obpr = r(mean) in `obs'
  8. sum pr`i'
  9. replace prpr = r(mean) in `obs'
 10. local i = `i' + 1
 11. }
*: Preparation for table
. gen obsprop = obpr*100      /*  outcomes equal to # */
. gen preprop = prpr*100      /* average predicted prob */
. gen byte count = cnt
. gen diffprop = obsprop - preprop
. format obsprop preprop diffprop %8.3f
. l count obsprop preprop diffprop in 1/21
```

	count	obsprop	preprop	diffprop
1.	0	0.000	0.012	-0.012
2.	1	8.428	0.108	8.320
3.	2	4.749	0.471	4.278
4.	3	5.017	1.381	3.636
5.	4	6.957	3.047	3.910
6.	5	8.227	5.402	2.825
7.	6	6.488	8.024	-1.536
8.	7	7.759	10.280	-2.521
9.	8	6.154	11.612	-5.458
10.	9	4.950	11.766	-6.816
11.	10	5.953	10.852	-4.899
12.	11	4.682	9.231	-4.549
13.	12	4.682	7.337	-2.654
14.	13	2.876	5.523	-2.647
15.	14	3.278	4.001	-0.723
16.	15	2.742	2.842	-0.099
17.	16	2.876	2.020	0.856
18.	17	1.940	1.465	0.475
19.	18	1.538	1.095	0.443
20.	19	1.605	0.845	0.761
21.	20	1.271	0.664	0.607

Rather severe underprediction occurs until day 6, at which time overprediction occurs until day 15. An interesting test may be given to determine if the difference between the observed and predicted probabilities by count, weighted by the number of observations per count, is statistically significant. A formatted graph of this same relationship can be given using the following code. Note that if the observed counts have extremely high values compared to the distributional mean, the fit at the extremes will be poor. Recall that for a given distributional mean, values far from it will have increasingly lower *p*-values.

R: *Graph of predicted versus observed LOS*

```
plot(0:25, avgp, type="b", xlim=c(0,25),
main = "Observed vs Predicted Days",
xlab = "Days in Hospital", ylab = "Probability of LOS")
lines(0:25, c(0,propObsv), type = "b", pch = 2)
legend("topright", legend = c("Predicted Days","Observed Days"),
       lty = c(1,1), pch = c(1,2))
```

```
. label var prpr "Predicted days"
. label var obpr "Observed days"
. label var count "Days in Hospital"
. twoway scatter prpr obpr count, c(l l) ms(T d) ///
    title(Observed vs Predicted days) ytitle(Probability of LOS)
```

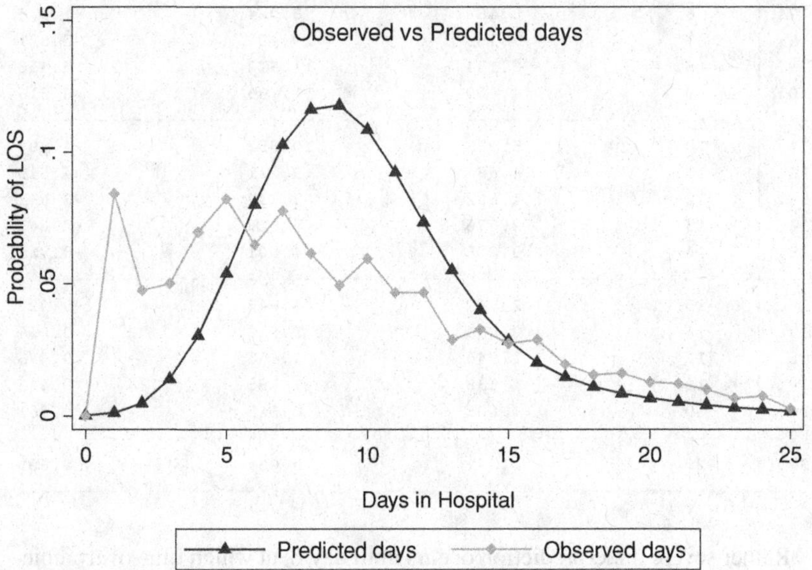

Figure 6.2 Observed versus predicted hospital days

There is a symbol at each day on the two lines in the graph. Again, note the substantial disparity of the days until day 15 is reached. The graph clearly displays the fact that the model underpredicts days until the sixth day, when it begins to overpredict for the next nine days. Predicted days greater than 15 fit well with the observed values from the data. This lack of fit can be expected when there is statistical evidence of overdispersion.

It is evident that the predicted Poisson days appear to approximate a normal or Gaussian distribution. We mentioned that this would be expected given the mean of *los* nearing 10. Poisson distributions with means from 4 to 10 and above appear Gaussian, with a right skew. A mean of 10 has only a slight skew, but has a lower kurtosis than a standard Gaussian distribution. A graphical representation should help visualize the various distributional shapes.

Table 6.16 R: *Poisson distributions*

```
m<- c(0.5,1,3,5,7,9) #Poisson means
y<- 0:19              #Observed counts
layout(1)
for (i in 1:length(m)) {
   p<- dpois(y, m[i]) #poisson pmf
   if (i==1) {
   plot(y, p, col=i, type='l', lty=i)
   } else {
   lines(y, p, col=i, lty=i)
   }
}
```

```
. clear
. set obs 20
. gen y = _n-1
. gen mu =.
. gen mu0_5 = (exp(-.5)*.5^y)/exp(lngamma(y+1))
. forvalues i = 1(2)8 {
    gen mu`i' = (exp(-`i')*`i'^y)/exp(lngamma(y+1))
    }
. graph twoway connected mu0_5 mu1 mu3 mu5 mu7 mu9 y, //
    title("Poisson Distributions")     /* Figure 6.3 */
```

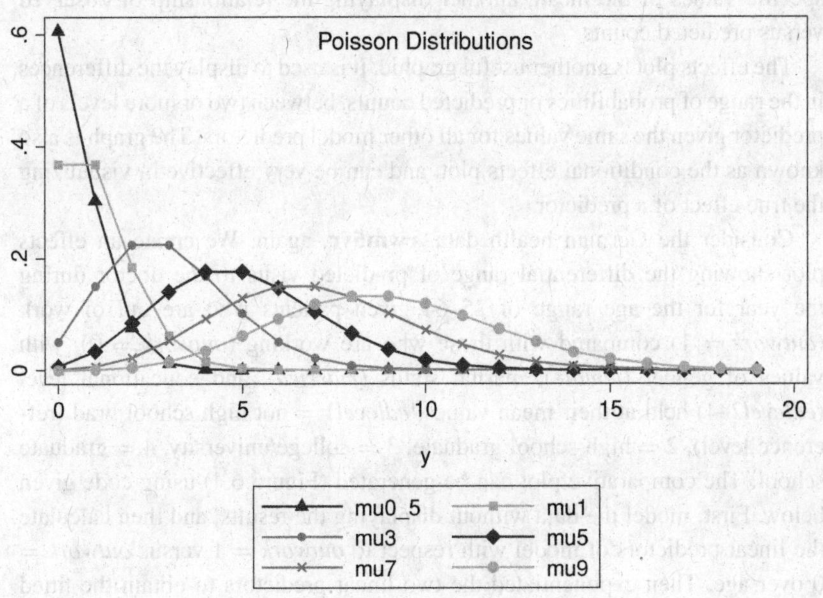

Figure 6.3 Poisson distributions with means at 0.5, 1, 3, 5, 7, 9

A Poisson mean of under 1 produces a negative exponential slope, with a high probability (~0.6) of a count being 0. Other means produce a distribution with its value at the point of maximum probability. The code used to determine the various probabilities entails the use of the Poisson PDF. The same relationship holds for the **medpar** study data. For a given observation:

Linear predictor

$$X'b = b_0 + b_1' X_1 + \cdots + b_n' X_n \tag{6.41}$$

Expected mean

$$\mu_i = \exp(x_i' \beta) \tag{6.42}$$

Probability

$$p_i = \frac{\exp(-\mu_i)\mu_i^{y_i}}{\exp(\ln\Gamma(y_i + 1))} \tag{6.43}$$

6.5 Effects plots

In the last section we used the relationships of expected mean and probability to construct two types of graphs, one displaying the Poisson distributions at specific values of the mean, another displaying the relationship of observed versus predicted counts.

The effects plot is another useful graphic. It is used to display the differences in the range of probabilities or predicted counts, between two or more levels of a predictor given the same values for all other model predictors. The graph is also known as the conditional effects plot, and can be very effective in visualizing the true effect of a predictor.

Consider the German health data, **rwm5yr**, again. We create an effects plot showing the differential range of predicted visits to the doctor during the year for the age range of 25–64 given patients who are out of work (*outwork* = 1) compared with those who are working (*outwork* = 0), with values of gender (*female*), marital status (*married*), and educational level (*edlevel*2–4) held at their mean values. *edlevel*1 = not high school grad (reference level), 2 = high school graduate, 3 = college/university, 4 = graduate school. The comparative plot can be generated (Figure 6.4) using code given below. First, model the data without displaying the results, and then calculate the linear predictors of model with respect to *outwork* = 1 versus *outwork* = 0 over age. Then exponentiated the two linear predictors to obtain the fitted counts.

Table 6.17 R: *Effect plot of doctor visits on working status*

```
rm(list=ls())
data(rwm5yr)
attach(rwm5yr)
eppoi <- glm(docvis ~ outwork+age+female+married+edlevel2+ edlevel3+
  edlevel4, family=poisson, data=rwm5yr)
rest <- eppoi$coef[4]*mean(rwm5yr$female) +
  eppoi$coef[5]*mean(rwm5yr$married) + eppoi$coef[6]*mean(rwm5yr$edlevel2) +
  eppoi$coef[7]*mean(rwm5yr$edlevel3) + eppoi$coef[8]*mean(rwm5yr$edlevel4)
out0 <- eppoi$coef[1] + eppoi$coef[3]*rwm5yr$age + rest
out1 <- eppoi$coef[1] + eppoi$coef[2]*1 + eppoi$coef[3]*rwm5yr$age + rest
eout1 <- exp(out1)
eout0 <- exp(out0)
matplot(cbind(rwm5yr$age, rwm5yr$age), cbind(eout0, eout1), pch=1:2,
  col=1:2, xlab='Count', ylab='Frequency')
matplot(cbind(rwm5yr$age, rwm5yr$age), cbind(eout0, eout1),
  type='l', lty=1:2, col=1:2, xlab='Doctor visits', ylab='Frequency')
```

```
. use rwm5yr, clear
. qui poisson docvis outwork age female married edlevel2-edlevel4
. gen rest = _b[_cons] + _b[female]*.4804937 + _b[married]*.7736244 +
  _b[edlevel2]*.0587995 + _b[edlevel3]*.0883778 + _b[edlevel4]*.0657861
. gen L1 = _b[outwork]*1 + _b[age]*age + rest
. gen L0 = _b[outwork]*0 + _b[age]*age + rest
. gen eout1 = exp(L1)
. gen eout0 = exp(L0)
. lab var eout1 "Not Working"
. lab var eout0 "Working"
. graph twoway connected eout1 eout0 age, c(l l) ms(T d)
    title("Predicted visits to doctor: age on outwork levels") sort
```

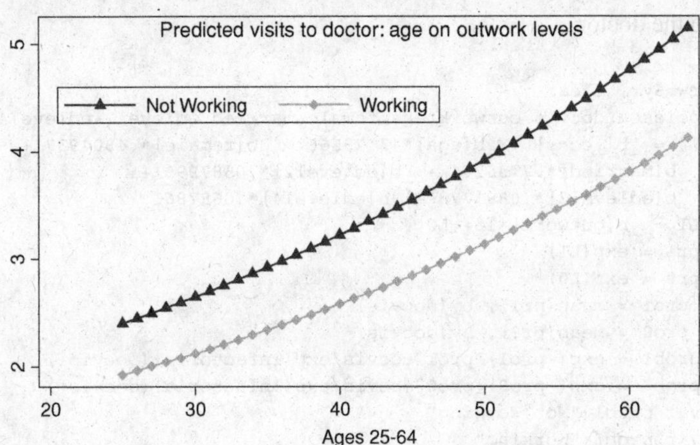

Figure 6.4a Predicted visits to doctor: age on outwork levels

The lower of the two ranges of counts is for working patients. Note that patients who are out of work, for each average age, visit the doctor more frequently. Moreover, older patients visit the doctor more frequently than younger patients. This is consistent with the model coefficients.

A program authored by Scott Long called **prgen** in Long and Freese (2006) can be used to automate the creation of conditional effects plots, but with the predicted probabilities being displayed rather than the predicted counts. After modeling the data as shown above, the following commands can be used to develop a graph of probabilities of counts by *outwork* over *age*.

```
. prgen age, x(outwork=0) rest(mean) from(25) to(64) gen(pr1) n(40)
. prgen age, x(outwork=1) rest(mean) from(25) to(64) gen(pr2) n(40)
```

The prefices *pr*1 and *pr*2 are user-assigned labels for the probabilities. Label the appropriate created variables and plot the difference in probabilities (not shown).

```
. lab var pr1p0 "Not Working"
. lab var pr2p0 "Working"
. lab var pr2x "Age"
. graph twoway connected pr1p0 pr2p0 pr2x, ///
title(Probability of doctor visits: work vs not working)
```

A straightforward differential probability plot can be constructed for the probability of visiting a doctor for those patients who are not working compared with those who are working using the following code. R code is listed in *Negative Binomial Regression Extensions*. The key is to convert the predicted counts to probabilities using the Poisson PDF. Values are at their mean per number of visits to the doctor.

```
. use rwm5yr, clear
. qui poisson docvis outwork age female married edlevel2-edlevel4
. gen L0 = _b[_cons] + _b[age]*43.78566 + _b[female]*.4804937 +
      _b[married]*.7736244 + _b[edlevel2]*.0587995 +
      _b[edlevel3]*.0883778 + _b[edlevel4]*.0657861
. gen L1 = _b[outwork]*1 + L0
. gen pr1 = exp(L1)
. gen pr0 = exp(L0)
. egen pro1 = mean(pr1), by(docvis)
. egen pro0 = mean(pr0), by(docvis)
. gen prob1 = exp(-pro1)*pro1^docvis/exp(lnfactorial(docvis))
. gen prob0 = exp(-pro0)*pro0^docvis/exp(lnfactorial(docvis))
. lab var prob1 "Not Working"
. lab var prob0 "Working"
```

```
. graph twoway connected prob1 prob0 docvis if docvis<16,
    sort ms(t 1) xlabel(1(2)16) title(Probability of doctor
    visits: work vs not working)
```

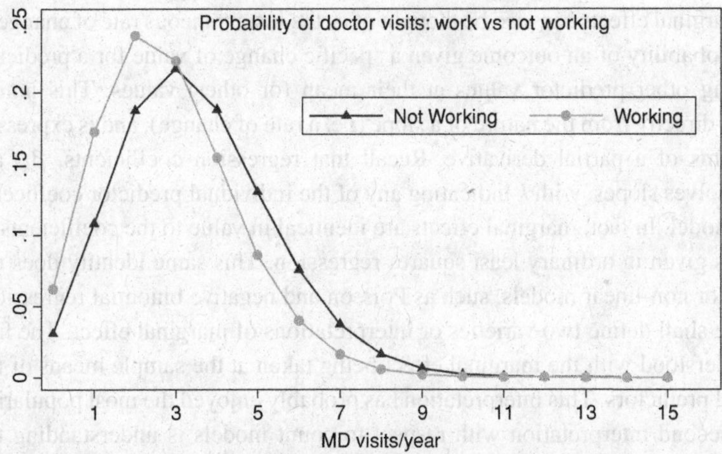

Figure 6.4b Probability of doctor visits: work versus not working

Each of the above described Figures can be calculated for both Poisson and negative binomial models.

6.6 Marginal effects, elasticities, and discrete change

Econometricians commonly find it helpful to interpret regression model coefficients in terms of either *marginal effects* or *discrete change*. Marginal effects are used when a predictor is continuous; "discrete change" is used for binary or factor variables. Discrete change is also referred to as *finite difference*. We shall use the former term in this text given that it involves a change of values between discrete units of a categorical variable.

Although it is rare to see these interpretations used in disciplines outside economics, they nevertheless can be used quite effectively to better understand relationships studied in health and medical analysis, transportation studies, fisheries and other environmental domains, as well as in most other areas of research.

6.6.1 Marginal effects for Poisson and negative binomial effects models

Essentially, a marginal effect is used to understand the relationship of an estimated predictor and its predicted probability, with other predictor values held at a specified value, usually at their means or medians. Specifically, marginal effects are used to quantify the relationship of how the probability of an outcome changes given a unit (or units) change in the value of the predictor.

Marginal effects can also be thought of as the instantaneous rate of change in the probability of an outcome given a specific change of value for a predictor, holding other predictor values at their mean (or other) values. This notion stems directly from the nature of a slope (i.e. a rate of change), and is expressed in terms of a partial derivative. Recall that regression coefficients, β_k, are themselves slopes, with k indicating any of the individual predictor coefficents in a model. In fact, marginal effects are identical in value to the coefficients or slopes given in ordinary least squares regression. This same identity does not hold for non-linear models, such as Poisson and negative binomial regression.

We shall define two varieties or interpretations of marginal effect. The first is understood with the marginal effect being taken at the sample means of the model predictors. This interpretation has probably enjoyed the most popularity. The second interpretation with respect to count models is understanding the marginal effect taken at the average of the predicted counts in the model. When related to a binary predictor, the average marginal effect is also termed a *partial effect*, although some statisticians allow the term to be used for any form of predictor. We shall examine each of these two interpretations, beginning with the marginal effect at mean approach.

Marginal effects at the mean

Being essentially non-linear models, the marginal effects of a continuous predictor in a Poisson or negative binomial model vary across observations. The logic of the effect for a specific continuous predictor can be given as:

$$\frac{\partial E\left(y_i|x_i\right)}{\partial x_i} = \frac{\partial \mu_i}{\partial x_i} \tag{6.44}$$

$$= \frac{\partial e^{x_i'\beta}}{\partial x_i} \tag{6.45}$$

$$= \frac{\partial e^{x_i'\beta}}{\partial x_i'\beta}\frac{\partial x_i'\beta}{\partial x_i} \tag{6.46}$$

$$= e^{x_i'\beta}\beta_k = \exp(x_i'\beta)\beta_k \tag{6.47}$$

For any continuous predictor in a Poisson or negative binomial (NB2) model, the marginal effect may be expressed as

$$\frac{\partial E(y_i|x_i)}{\partial x_{ik}} = E\left(y_i|x_i\right)\beta_k = \exp(x_i'\beta_k)\beta_k \tag{6.48}$$

Using the derivative nature of the marginal effect, I shall demonstrate how to calculate it employing a simple Poisson example. Using the German health data, **rwm5yr**, with the number of visits a patient makes to a doctor during a given year as the response, and work status and *age* as predictors, the marginal effect of *age* may be calculated using the following logic. First we model the data using a Poisson regression, but not displaying header information.

Table 6.18 R: *Marginal effects*

```
rwm <- rwm5yr
rwm1 <- glm(docvis ~ outwork + age, family=poisson, data=rwm)
meanout <- mean(rwm$outwork)
meanage <- mean(rwm$age)
xb <- coef(rwm1)[1] + coef(rwm1)[2]*meanout + coef(rwm1)[3]
  *meanage
dfdxb <- exp(xb) * coef(rwm1)[3]
mean(dfdxb)    # answer =  0.06043214
```

```
. use rwm5yr
. keep docvis outwork age
. glm docvis outwork age, nolog fam(poi) nohead

------------------------------------------------------------------
 docvis |  Coef.    Std. Err.   z      P>|z|  [95% Conf. Interval]
--------+---------------------------------------------------------
outwork | .3256733 .0083798  38.86   0.000  .3092493    .3420973
    age |  .019931  .0003697  53.91   0.000  .0192064    .0206555
  _cons |  .124546  .0171297   7.27   0.000  .0909724    .1581197
------------------------------------------------------------------

. qui sum outwork               /// summarize outwork
. qui replace outwork=r(mean)   /// change outwork to its mean
. qui sum age                   /// summarize age
. qui replace age=r(mean)       /// change age to its mean
. predict xb, xb                /// calculate linear predictor
                                    on basis of mean values
. gen dfdxb = exp(xb)*_b[age]   /// exp(xb)*b
. l dfdxb in 1                  /// display the ME

     +----------+
     |   dfdxb  |
     |----------|
  1. | .0604319 |
     +----------+
```

Stata's **margins** command calculates the marginal effect at the mean of *age* by using the following command. Observe that the displayed value for *age* is identical to what we calculated by hand.

The marginal effect of *age* may be interpteted as: at the mean values of the predictors in the model (sample mean), the number of visits to the doctor increases by 0.06, or 6%, for every year increase in *age*.

```
. margins, dydx(age) atmeans

Conditional marginal effects              Number of obs    = 19609
Model VCE      : OIM
Expression     : Predicted number of events, predict()
dy/dx w.r.t. : age
at             : outwork          =    .3439237 (mean)
                 age              =    43.78566 (mean)
-------------------------------------------------------------------
    |         Delta-method
    |      dy/dx   Std. Err.      z    P>|z|    [95% Conf. Interval]
----+--------------------------------------------------------------
age | .0604319    .001097    55.09   0.000    .0582818      .062582
-------------------------------------------------------------------
```

Average marginal effects

Marginal effects can also be taken as *average marginal effects*, calculated as

$$\hat{\beta}_k \bar{y}$$

The command to calculate the average marginal effect is:

R

```
mean(rwm$docvis) * coef(rwml)[3]
   age
0.0633046
```

```
. margins, dydx(age)

Average marginal effects                     Number of obs   =    19609
Model VCE      : OIM

Expression     : Predicted mean docvis, predict()
dy/dx w.r.t. : age
--------------------------------------------------------------------
    |        Delta-method
    |      dy/dx   Std. Err.      z       P>|z|   [95% Conf. Interval]
----+---------------------------------------------------------------
age | .0633046    .0012013     52.70    0.000    .0609502      .065659
--------------------------------------------------------------------
```

The average marginal effect for *age* specifies that, for each additional year of *age*, there are an average of 0.0633 additional visits to the doctor. With the coefficient of *age* given in the coefficient table above as 0.019931, and the mean of *docvis* as 3.176195, we may apply the formula to calculate the average marginal effect by hand.

```
. di 3.176195*.019931
.06330474
```

The calculated value is identical to the value displayed in the results. Also note that the standard error is calculated using the delta method.

Confidence intervals are generated as:

```
. di.0633046 - 1.96*.0012013
.06095005

. di.0633046 + 1.96 *.0012013
.06565915
```

Elasticities

Elasticities are commonly used in econometrics and relate to percentages of change. Elasticities are defined as ME * x/y, where ME is the interpretation of marginal effects defined earlier, x is the predictor and y the response. Continuing the example we have been using, the elasticity may be interpreted as: a 1% increase in *age* is associated with a 0.873% increase in visits to the doctor.

R: *Elasticity*

```
sel <- mean(dfdxb)
mage <- mean(rwm5yr$age)
sel * mage/3.032
[1] 0.8727114
```

```
. margins, eyex(age) atmeans noatlegend

Conditional marginal effects                   Number of obs   =   19609
Model VCE     : OIM
Expression    : Predicted number of events, predict()
ey/ex w.r.t. : age
at            : outwork        =    .3439237 (mean)
                age            =    43.78566 (mean)
------------------------------------------------------------------------
    |          Delta-method
    |    ey/ex   Std. Err.      z    P>|z|     [95% Conf. Interval]
----+-------------------------------------------------------------------
age | .8726901   .0161868    53.91   0.000    .8409646     .9044155
------------------------------------------------------------------------
```

Here x is 43.786 and y is the value of the fit at *age* 43.786, or 3.032. Given the marginal effect at the mean as calculated earlier, 0.604319, we have (with rounding error)

```
. di.0604319 * 43.78566 /3.032
.87270799
```

which is the same value as given in the output.

A **semi-elasticity** may be calculated as well. It may be interpreted as: for every one year increase in *age*, there is a 2% increase in visits to the doctor. If this interpretation appears familiar, it should. Owing to the exponential nature of the Poisson and negative binomial mean, the Poisson/negative binomial coefficient is a semi-elasticity. This identity does not occur with NB-C, or with models in which the mean value is not exponential, e.g. $\mu = \exp(x'\beta)$.

R: *Semi-elasticity*

```
se1/3.032
[1] 0.01993144
```

```
. margins, eydx(age) atmeans noatlegend

Conditional marginal effects                   Number of obs   =   19609
Model VCE     : OIM
Expression    : Predicted mean docvis, predict()
ey/dx w.r.t. : age

------------------------------------------------------------------------------
    |          Delta-method
    |    ey/dx   Std. Err.      z      P>|z|     [95% Conf. Interval]
----+-------------------------------------------------------------------------
age | .019931    .0003697   53.91    0.000     .0192064     .0206555
------------------------------------------------------------------------------
```

This is the same value that we calculated for *age* in the initial Poisson model. Exponentiating produces: exp(0.019931) = 1.0201309.

Elasticities may also be calculated for average marginal effects. The interpretations are analogous to the relationship of the marginal effects at mean and elasticities we defined above. Note that the Poisson/NB-semi-elasticity is only interpreted "at the mean."

6.6.2 Discrete change for Poisson and negative binomial models

Discrete change is used to calculate change in the predicted probability of a binary predictor, x, as $x = 0$ changes to $x = 1$. The *at means* and *average* interpretations obtain with discrete change as well as for marginal effects. We begin by examining the 'discrete change at means' interpretation. The relationship may be formalized as:

$$\frac{\Delta \Pr(y_i|(x_i = 1|x_i = 0))}{\Delta x_k} \tag{6.49}$$

where Δ symbolizes change. Discrete change is also referred to as *finite differences* (Cameron and Trivedi, 2010). We use the former terminology, but keep in mind that either is used in the literature discussing it.

The predicted values of both levels of x are calculated at the mean values of the other predictors in the model. In the case of the example being used, only the mean of *age* need be calculated.

R

```
mu0 <- exp(.124546 + .019931*meanage)              # outwork=0
mu1 <- exp(.124546 + .3256733 + .019931*meanage)   # outwork=1
dc <- mu1 - mu0
mean(dc)   # answer = 1.04355  Discrete Change
```

```
. sum age

Variable |     Obs        Mean    Std. Dev.      Min        Max
---------+--------------------------------------------------------
     age |   19609    43.78566    11.22855        25         64

. replace age=r(mean)     /// change age to its mean
. replace outwork=0       /// change outwork to 0
. predict p0              /// exp(xb) when outwork==0
. replace outwork=1       /// change outwork to 1
. predict p1              /// exp(xb) when outwork==1
. di p1-p0                /// subtract each predicted prob
1.0435493
```

The value calculated, 1.0435, is identical to that displayed in the marginal effects table.

We first construct a Poisson model with factor variables specifically indicated as such. The **margins** command then recognizes that a discrete change calculation is required.

```
. glm docvis i.outwork age, fam(poi) nohead

------------------------------------------------------------------------------
             |                 OIM
      docvis |      Coef.   Std. Err.      z    P>|z|     [95% Conf. Interval]
-------------+----------------------------------------------------------------
   1.outwork |   .3256733   .0083798    38.86   0.000     .3092493    .3420973
         age |    .019931   .0003697    53.91   0.000     .0192064    .0206555
       _cons |    .124546   .0171297     7.27   0.000     .0909724    .1581197
------------------------------------------------------------------------------

. margins, dydx(outwork) atmean

Conditional marginal effects                      Number of obs   =      19609
Model VCE    : OIM
Expression   : Predicted mean docvis, predict()
dy/dx w.r.t. : 1.outwork
at           : 0.outwork       =    .6560763 (mean)
               1.outwork       =    .3439237 (mean)
               age             =    43.78566 (mean)
------------------------------------------------------------------------------
             |            Delta-method
             |      dy/dx   Std. Err.      z    P>|z|     [95% Conf. Interval]
-------------+----------------------------------------------------------------
   1.outwork |   1.043549   .0281831    37.03   0.000     .9883113    1.098787
------------------------------------------------------------------------------
Note: dy/dx for factor levels is the discrete change from the
      base level
```

The key to calculating a marginal effect for members of the generalized linear models family is the inverse link function (i.e. the function which converts the linear predictor to a fitted value). Both Poisson and the traditional parameterization of the negative binomial (NB2) have a log-link, which entails that their inverse link is $\exp(\eta)$. In the case of Poisson, $\exp(\eta) = \exp(x'\beta)$.

Average discrete change: partial effects

Average discrete change, which is referred to by many statisticians as partial effects, measures the average effect of a change in the predictor on the probability of the response. Using the same data as in the previous subsections, the code for calculating the average discrete change of *outwork* as it changes from a value of 0 to 1 is given as:

R

```
bout <- coef(rwm1$outwork)
mu <- fitted.values(rwm1)
xb <- linear.predictors(rwm1)
```

```
pe_out <- 0
pe_out <- ifelse(rwm1$outwork==0,exp(xb+bout) - exp(xb),.)
pe_out <- ifelse(rwm1$outwork==1, exp(xb)- exp(xb-bout),.)
mean(pe_out)
```

```
. qui poisson docvis outwork age
. predict mu, n
. predict xb, xb
. scalar bout= _b[outwork]
.gen pe_out=0
. replace pe_out=exp(xb+bout) - exp(xb) if outwork==0
. replace pe_out=exp(xb)- exp(xb-bout) if outwork==1
. sum pe_out

Variable |    Obs         Mean     Std. Dev.        Min        Max
---------+--------------------------------------------------------
  pe_out |  19609      1.06999     .2399167    .7176391     1.5613

. margins, dydx(outwork)

Average marginal effects                  Number of obs    =   19609
Model VCE    : OIM
Expression   : Predicted number of events, predict()
dy/dx w.r.t. : 1.outwork
---------------------------------------------------------------------
         |       Delta-method
         |    dy/dx  Std. Err.      z    P>|z|   [95% Conf. Interval]
---------+-----------------------------------------------------------
1.outwork | 1.06999  .0286055   37.41   0.000  1.013924     1.126056
---------------------------------------------------------------------
Note: dy/dx for factor levels is the discrete change from the
     base level.
```

Note the identity of the mean value of *pe_out* calculated by hand and the results of **margins**. A more general application of partial effects relates to any one-unit change in a predictor. An extensive discussion of marginal effects and discrete change, emphasizing logistic models, is given in Hilbe (2009). Cameron and Trivedi (2010) present a view similar to that expressed here for models in general.

It should be understood that there are a number of alternative statistics that econometricians, social statisticians, and others have derived in order to better understand the relationship of a predictor to the expected count, as well as to the expected probability of observations. For example, a type of elasticity has been constructed that calculates the percentage change in the expected count for a standard deviation increase, or decrease, in the value of a predictor. Texts such as Winkelmann (2008), Long and Freese (2006), Cameron and Trivedi (1998, 2010), and Greene (2008) provide additional information.

6.7 Parameterization as a rate model

6.7.1 Exposure in time and area

We briefly addressed the rate parameterization of the Poisson model in Chapter 1. Although μ is an intensity or rate parameter, it is usually conceived as such when it is in conjunction with the constant coefficient, *t*. The rate parameterization of the Poisson PDF may be expressed as

$$f(y;\mu) = \frac{e^{-t_i\mu_i}(t_i\mu_i)^{y_i}}{y_i!} \tag{6.50}$$

where *t* represents the length of time, or exposure, during which events or counts occur; *t* can also be thought of as an area in which events occur, each associated with a specific count. For instance, when using a Poisson model with disease data, $t_i\mu_i$ can be considered as the rate of disease incidence in specified geographic areas, each of which may differ from other areas in the population. Again, the incidence rate of hospitalized bacterial pneumonia patients may be compared across counties within the state. A count of such hospitalizations divided by the population size of the county, or by the number of total hospitalizations for all diseases, results in the incidence rate ratio (IRR) for that county. *Person-time* is also a common application for *t*. When $t = 1$, the model is understood as applying to individual counts without a consideration of size or period. This is the parameterization that has been considered in the previous sections of this chapter.

Many commercial software applications indicate exponentiated Poisson coefficients as incidence rate ratios. IRR is also used with exponentiated negative binomial coefficients. Other software refers to the same ratio as simply a rate ratio, or as relative risk. Specifically, rate ratio refers to counts of events entering a period of time or area, even if the latter is reduced to 1. Relative risk is a more general term, that includes rate ratios as well as binary response relationships, as discussed in Chapter 2. All refer to the same mathematical relationship.

The Poisson mean or rate parameter, μ, adjusted by the time or area, *t*, in which counts, *y*, occur, may be expressed as

$$E(y_i/t_i) = \mu_i/t_i \tag{6.51}$$

The Poisson regression model for the expected count rate per unit of t may be expressed as

$$\log(\mu_i/t_i) = \alpha + x_i'\beta \tag{6.52}$$

or

$$\log(\mu_i) - \log(t_i) = \alpha + x_i'\beta \tag{6.53}$$

or

$$\log(\mu_i) = \alpha + x_i'\beta + \log(t_i) \tag{6.54}$$

which is the standard manner for which the rate parameterized Poisson is symbolized.

With intercept α incorporated within $x_i'\beta$,

$$\log(\mu_i/t_i) = x_i'\beta \tag{6.55}$$

The term $\log(t)$ is an adjustment term, or *offset*, to the linear predictor whereby observations of counts may have a different value of t, with t representing various time periods or areas. Again, when $t = 1$, we have the standard Poisson model, where the time periods, e.g. person-time, or areas are set at 1 for all counts.

From equation 6.52 it follows that μ has the values given in equations 6.56 and 6.57, which clearly show the multiplicative effect of the terms. Exponentiating both sides of equation 6.52 gives us, for a single observation:

$$\mu/t = e^{\alpha}e^{x'\beta} \tag{6.56}$$

Multiplying both sides by t yields

$$\mu = te^{\alpha}e^{x'\beta} \tag{6.57}$$

or

$$\mu = te^{\alpha+x'\beta} = te^{x'\beta} \tag{6.58}$$

where the intercept, α, is incorporated into $x'\beta$ in the rightmost term. With subscripts representing multiple observations, we have

$$\mu_i = \exp(x_i'\beta + \ln(t_i)) \tag{6.59}$$

which is also observed to be the exponentiation of both sides of 6.54. Equations 6.57 through 6.59 are identical, and are found in the literature of Poisson regression as the definition of μ.

To summarize, Poisson models are rate models, estimating the number of counts in, or entering, given time periods or areas. When these periods or areas have a value of 1, we have the traditional Poisson model. When they have positive values greater than 1, they are employed in the model as an offset, which reflects the area or time over which the count response is generated. Since the natural log is the canonical link of the Poisson model, the offset must be logged prior to entry into the estimating algorithm, and entered into the model as a constant. In fact, the offset is an exposure variable with a coefficient constrained to a value of 1.0.

6.7.2 Synthetic Poisson with offset

It may be helpful to construct a synthetic Poisson regression model with an offset. Recall from Table 6.1 that the offset is added to the calculation of the linear predictor, but is subtracted from the working response just prior to the iterative estimation of parameters. In constructing a synthetic rate-parameterized Poisson model, however, the offset is only added to the calculation of the linear predictor.

A synthetic offset may be randomly generated, or may be specified by the user. For this example I will create an area offset having increasing values of 100 for each 10,000 observations in the 50,000-observation data set. The shortcut code used to create this variable is given in the first line below. I have commented code that can be used to generate the same offset as in the single line command that is used in the algorithm. The extended code better demonstrates what is being done.

We expect that the resultant model will have approximately the same parameter values as the earlier model, but with different standard errors. Modeling the data without using the offset option results in similar parameter estimates to those produced when an offset is employed, but with an inflated value of the intercept. The COUNT **poisson_syn** function can also be used.

Table 6.19 *Synthetic rate Poisson data*

```
* poio_rng.do
clear
set obs 50000
set seed 4744
gen off = 100+100*int((_n-1)/10000) // creation of offset
* gen off=100 in 1/10000 // lines duplicate single line above
* replace off=200 in 10001/20000
* replace off=300 in 20001/30000
* replace off=400 in 30001/40000
* replace off=500 in 40001/50000
```

```
gen loff = ln(off) // log offset prior to model entry
gen x1 = rnormal()
gen x2 = rnormal()
gen py = rpoisson(exp(2 + 0.75*x1 - 1.25*x2 + loff))
glm py x1 x2, fam(poi) off(loff) // added offset option
```

The results of the rate parameterized Poisson algorithm given in Table 6.19 (R-Table 6.20) above is displayed below:

```
Generalized linear models                 No. of obs        =      50000
Optimization      : ML                    Residual df       =      49997
                                          Scale parameter   =          1
Deviance          =  49847.73593          (1/df) Deviance   =   .9970145
Pearson           =  49835.24046          (1/df) Pearson    =   .9967646

                                          AIC               =   10.39765
Log likelihood    = -259938.1809          BIC               =  -491108.7
-------------------------------------------------------------------------
       |              OIM
    py |     Coef.  Std. Err.      z     P>|z|   [95% Conf. Interval]
-------+-----------------------------------------------------------------
    x1 |  .7500656  .0000562 13346.71   0.000   .7499555    .7501758
    x2 | -1.250067  .0000576 -2.2e+04   0.000   -1.25018   -1.249954
 _cons |  1.999832  .0001009 19827.16   0.000   1.999635     2.00003
  loff |  (offset)
-------------------------------------------------------------------------
```

The same code logic is employed when using R to effect the same model result.

Table 6.20 R: *Poisson with rate parameterization using R*

```
oset <- rep(1:5, each=10000, times=1)*100
loff <- log(oset)
sim.data <- poisson_syn(nobs = 50000, off = loff, xv = c(2, .75, -1.25))
poir <- glm(py ~ . + loff, family=poisson, data = sim.data)
summary(poir)
confint(poir)
```

Relevant model R output is displayed as:

```
Coefficients:
             Estimate Std. Error z value Pr(>|z|)
(Intercept)  2.0002055  0.0002715    7367   <2e-16 ***
x1           0.7501224  0.0003612    2077   <2e-16 ***
x2          -1.2507453  0.0003705   -3375   <2e-16 ***
---
```

A display of the offset table is shown as:

```
. table(off)

off
  100   200   300   400   500
10000 10000 10000 10000 10000
```

The resultant parameter estimates are nearly identical to the specified values. We expect that the parameter estimates of the model with an offset or rate parameterized model will closely approximate those of the standard model. However, we also expect that the standard errors will differ. Moreover, if the rate model were modeled without declaring an offset, we would notice a greatly inflated intercept value.

6.7.3 Example

An example of a Poisson model parameterized with an offset is provided below. The data are from the Canadian National Cardiovascular Disease registry called, FASTRAK. They have been grouped by covariate patterns from individual observations and stored in a file called **fasttrakg**. The response is *die*, which is a count of the number of deaths of patients having a specific pattern of predictors. Predictors are *anterior*, which indicates if the patient has had a previous anterior myocardial infarction in difference to having had an inferior site infarct; *hcabg*, if the patient has a history of having had a CABG (coronary artery bypass graft) procedure compared with having had a PTCA (percutaneous transluminal coronary angioplasty); and *killip* class, a summary indicator of the cardiovascular health of the patient, with increasing values indicating increased disability. The number of observations sharing the same pattern of covariates is recorded in the variable *case*. This value is log-transformed and entered into the model as an offset.

Table 6.21 R: *Poisson model of Fasttrak data*

```
rm(list=ls())
data(fasttrakg)
attach(fasttrakg)
lncases <- log(cases)
poioff <- glm(die ~ anterior + hcabg + kk2 + kk3 + kk4 +
   offset(lncases), family=poisson, data=fasttrakg)
exp(coef(poioff))
exp(confint(poioff))
```

```
. glm die anterior hcabg kk2-kk4,fam(poi) eform lnoffset(cases)

Generalized linear models                    No. of obs      =        15
Optimization     : ML                        Residual df     =         9
                                             Scale parameter =         1
Deviance         =  10.93195914              (1/df) Deviance =  1.214662
Pearson          =  12.60791065              (1/df) Pearson  =  1.400879
                         . . .
                                             AIC             =   4.93278
Log likelihood   = -30.99584752              BIC             = -13.44049
----------------------------------------------------------------------
             |                 OIM
         die |        IRR   Std. Err.      z    P>|z|  [95% Conf. Interval]
-------------+--------------------------------------------------------
    anterior |   1.963766   .3133595    4.23   0.000    1.436359     2.6848
       hcabg |   1.937465   .6329708    2.02   0.043    1.021282   3.675546
         kk2 |   2.464633   .4247842    5.23   0.000     1.75811   3.455083
         kk3 |   3.044349   .7651196    4.43   0.000     1.86023   4.982213
         kk4 |   12.33746   3.384215    9.16   0.000    7.206717   21.12096
       cases | (exposure)
----------------------------------------------------------------------
```

The Pearson dispersion has a relatively low value of 1.40. Given a total observation base of 5,388, the added 40% overdispersion may represent a lack of model fit. We shall delay this discussion until the next chapter where we deal specifically with models for overdispersed data.

Table 6.22 *Interpretation of rate parameterized Poisson*

anterior:	A patient is twice as likely to die within 48 hours of admission if they have sustained an anterior rather than in inferior infarct, holding the other predictor values constant.
hcabg:	A patient is twice as likely to die within 48 hours of admission if they have a previous history of having had a CABG rather than a PTCA, holding the other predictor values constant.
*kk*2:	A patient rated as Killip level 2 is some two and a half times more likely to die within 48 hours of admission than is a patient rated at Killip level 1, holding the other predictor values constant.
*kk*3:	A patient rated as Killip level 3 is some three times more likely to die within 48 hours of admission than is a patient rated at Killip level 1, holding the other predictor values constant.
*kk*4:	A patient rated as Killip level 4 is over 12 times more likely to die within 48 hours of admission than is a patient rated at Killip level 1, holding the other predictor values constant.

Summary

The Poisson model is the traditional basic count response model. We discussed the derivation of the model and how the basic Poisson algorithm can be amended to allow estimation of rate models, i.e. estimation of counts per specified defined areas or over various time periods. The rate parameterization of the Poisson model is also appropriate for modeling counts that are weighted by person years.

A central distributional assumption of the Poisson model is the equivalence of the Poisson mean and variance. This assumption is rarely met with real data. Usually the variance exceeds the mean, resulting in what is termed as *overdispersion*. Underdispersion occurs when the variance is less than the nominal mean, but this also rarely occurs in practice. Overdispersion is, in fact, the norm, and gives rise to a variety of other models that are extensions of the basic Poisson model.

Negative binomial regression is nearly always thought of as the model that is to be used instead of Poisson when overdispersion is present in the data. Because overdispersion is so central to the modeling of counts, we next address it, and investigate how we determine if it is real or only apparent.

7
Overdispersion

This chapter can be considered as a continuation of the previous one. Few real-life Poisson data sets are truly equidispersed. Overdispersion to some degree is inherent to the vast majority of Poisson data. Thus, the real question deals with the amount of overdispersion in a particular model – is it statistically sufficient to require a model other than Poisson? This is a foremost question we address in this chapter, together with how we differentiate between real and apparent overdispersion.

7.1 What is overdispersion?

Not all overdispersion is real; apparent overdispersion may sometimes be identified and the model amended to eliminate it. We first address the difference between real and apparent overdispersion, and what can be done about the latter.

1 What is overdispersion?
  Overdispersion in Poisson models occurs when the response variance is greater than the mean.
2 What causes overdispersion?
  Overdispersion is caused by positive correlation between responses or by an excess variation between response probabilities or counts. Overdispersion also arises when there are violations in the distributional assumptions of the data, such as when the data are clustered and thereby violate the likelihood independence of observations assumption.
3 Why is overdispersion a problem?
  Overdispersion may cause standard errors of the estimates to be deflated or underestimated, i.e. a variable may appear to be a significant predictor when it is in fact not significant.

4 How is overdispersion recognized?
A model may be overdispersed if the value of the Pearson (or χ^2) statistic divided by the degrees of freedom (dof) is greater than 1.0. The quotient of either is called the dispersion.
Small amounts of overdispersion are of little concern; however, if the dispersion statistic is greater than 1.25 for moderate sized models, then a correction may be warranted. Models with large numbers of observations may be overdispersed with a dispersion statistic of 1.05.
5 What is apparent overdispersion; how may it be corrected?
Apparent overdispersion occurs when:
(a) the model omits important explanatory predictors;
(b) the data include outliers;
(c) the model fails to include a sufficient number of interaction terms;
(d) a predictor needs to be transformed to another scale;
(e) the assumed linear relationship between the response and the link function and predictors is mistaken, i.e. the link is misspecified.

7.2 Handling apparent overdispersion

We can show the impact of the various causes of apparent overdispersion, delineated in 5(a)–(e) above, by creating simulated data sets. Each constructed data set will entail a specific cause for the overdispersion observed in the display of model output.

We shall first create a 50,000-observation Poisson data set consisting of three uniform distributed predictors.

7.2.1 Creation of a simulated base Poisson model

Create a data set with the following constructed predictors, and associated parameters:

```
intercept == 1.00 x1 == 0.50
x2 == -0.75       x3 == 0.25
```

Stata code to create the simulated data consists of the following:

```
. clear
. set obs 50000
. set seed 4590
. gen x1  = runiform()
. gen x2  = runiform()
. gen x3  = runiform()
. gen py  = rpoisson(exp(1 + 0.5*x1 - 0.75*x2  + 0.25*x3))
. tab py                            // tabulate values of py
```

Table 7.1 R: *Poisson synthetic data set*

```
# nbr2_7_1.r
library(MASS)           # use for all R sessions
nobs <- 50000
x1 <- runif(nobs)
x2 <- runif(nobs)
x3 <- runif(nobs)
py <-rpois(nobs, exp(1 + 0.5*x1 - 0.75*x2 + 0.25*x3))
cnt <- table(py)
df <- data.frame( prop.table( table(py) ) )
df$cumulative <- cumsum( df$Freq )
dfall <- data.frame(cnt, df$Freq *100, df$cumulative *100)
dfall
```

Table 7.2 *Frequency distribution of synthetic Poisson model*

py	Freq.	Percent	Cumulative
0	3,855	7.71	7.71
1	9,133	18.27	25.98
2	11,370	22.74	48.72
3	9,974	19.95	68.66
4	7,178	14.36	83.02
5	4,266	8.53	91.55
6	2,328	4.66	96.21
7	1,115	2.23	98.44
8	478	0.96	99.39
9	195	0.39	99.78
10	71	0.14	99.93
11	22	0.04	99.97
12	11	0.02	99.99
13	4	0.01	100.00

The created response variable, *py*, has an observed count distribution displayed in Table 7.2.

We could have created random normal numbers for each of the three predictors rather than random uniform numbers. The range of values would extend from 0 to 74 rather from 0 to 13 (there is variability if no seed number is specified), but the Poisson model results would be identical.

py is next modeled on the three pseudo-randomly generated predictors, given as:

R

```
poy1 <- glm(py ~ x1 + x2 + x3, family=poisson)
summary(poy1)
confint(poy1)
poy1$aic/(poy1$df.null+1)
```

```
. glm py x1 x2 x3, fam(poi)

Generalized linear models                 No. of obs      =      50000
Optimization     : ML                     Residual df     =      49996
                                          Scale parameter =          1
Deviance         = 53978.55675            (1/df) Deviance =   1.079658
Pearson          = 50016.11952          =>(1/df) Pearson  =   1.000402
                                 .   .   .
                                          AIC             =   3.732233
Log likelihood   = -93301.81687           BIC             = -486967.1
----------------------------------------------------------------------
     |      OIM
  py |     Coef.  Std. Err.      z     P>|z|    [95% Conf. Interval]
-----+----------------------------------------------------------------
  x1 |  .5005752 .0021681    230.88   0.000     .4963258    .5048247
  x2 | -.7494748 .0021955   -341.37   0.000    -.7537779   -.7451717
  x3 |   .249406 .0021796    114.43   0.000     .2451341    .2536778
_cons|  .9998925  .002987    334.74   0.000      .994038    1.005747
----------------------------------------------------------------------
```

Although the Stata data is randomly generated, having a set seed value guarantees that the same random data will be generated any time the program code is run using the same seed.

No seed value was used with the R script, therefore each time the script is run, slightly different simulated data sets will be created. Resultant model statistics will therefore have slightly different values. If we run several hundred simulated models, taking the mean value of parameters and other statistics of interest, we would find that the mean estimates would equal the values we assigned them, and that the Pearson dispersion statistic, defined as the Pearson statistic divided by the model degrees of freedom, would equal 1.0. This procedure is called Monte Carlo simulation, and was discussed in the last chapter (Table 6.3). Note the Pearson dispersion statistic in the above model is 1.000402, with the parameter estimates approximating the values we specified.

Regardless, we now have a synthetic data set with "true" parameter values. These values can be used to compare against the values that result when we make changes to the data in certain ways.

7.2.2 Delete a predictor

Next model the above data as before, but exclude predictor $x1$.

R

```
poy2 <- glm(py ~ x2 + x3, family=poisson)
summary(poy2)
confint(poy2)
```

```
. glm py x2 x3, fam(poi)

Generalized linear models                   No. of obs      =      50000
Optimization     : ML                       Residual df     =      49997
                                            Scale parameter =          1
Deviance         =  107161.7793             (1/df) Deviance =   2.143364
Pearson          =  110609.0344             (1/df) Pearson  =   2.212313

                                            AIC             =   4.795857
Log likelihood   = -119893.4282             BIC             =  -433794.7
--------------------------------------------------------------------------
        |               OIM
     py |     Coef.  Std. Err.      z    P>|z|  [95% Conf. Interval]
--------+-----------------------------------------------------------------
     x2 | -.753503   .0022008  -342.37   0.000  -.7578165   -.7491895
     x3 | .2467799   .0021803   113.19   0.000   .2425066    .2510532
  _cons | 1.122657   .0027869   402.83   0.000   1.117194    1.128119
--------------------------------------------------------------------------
```

Parameter estimates deviate from those defined in the base data set – but not substantially. What has notably changed is the dispersion statistic. It has nearly doubled to a value of 2.1. Given a data set of 50,000 observations, the dispersion statistic correctly indicates that the data are overdispersed. The AIC and BIC statistics are also inflated. These fit statistics are commonly used when comparing models; those with lower AIC and BIC statistics are better-fitted.

7.2.3 Outliers in data

We create 50 outliers out of the 50,000 values of the response, *py*. This represents one-tenth of 1 percent of the observations. The synthetic values of *y* that we created range from 0 to 110, although the 49,997th largest number is 99, and 49,950th largest is 52. That is, 99/1,000 of the values are less than or equal to 52. The mean is 4.2 and median 3.0. Two sets of outliers will be generated. One set will add 10 to the first 50 values of *py* in the data, which have been randomized. The second will add 100 to *y* of the first observation. Code showing the creation of the outliers, together with a listing of the first 15 values of *py* and outlier values is given below:

R

```
y10_50 <- py
y10_50[1:50] <- y10_50[1:50] + 10
poy3 <- glm(y10_50 ~ x1 + x2 + x3, family=poisson)
summary(poy3)
confint(poy3)
```

```
. gen y10_50 = py
. replace y10_50 = y10_50 + 10 in 1/50          //+10 to 1st 50 obs
. glm y10_50 x1 x2 x3, fam(poi)

Generalized linear models                 No. of obs       =      50000
Optimization      : ML                    Residual df      =      49996
                                          Scale parameter  =          1
Deviance          =  55544.10589          (1/df) Deviance  =   1.110971
Pearson           =  57208.26572          (1/df) Pearson   =   1.144257
                                          .
                                          AIC              =   3.765705
Log likelihood    = -94138.63697          BIC              =  -485401.5
------------------------------------------------------------------------
        |              OIM
 y10_50 |    Coef.  Std. Err.    z     P>|z|   [95% Conf. Interval]
--------+---------------------------------------------------------------
     x1 |  .4988582  .002165  230.41  0.000   .4946148    .5031016
     x2 | -.7465734 .0021923 -340.54  0.000  -.7508702   -.7422765
     x3 |  .2486545 .0021765  114.25  0.000   .2443887    .2529203
  _cons |  1.005949 .0029776  337.83  0.000   1.000113    1.011785
------------------------------------------------------------------------
```

The parameter estimates are nearly identical to the synthetic model having a response of *py*, i.e. *py* with the first 50 responses having 10 added to the value *py*. The Pearson dispersion statistic, however, has increased from 1.000 to 1.144. Given the large number of observations, a value of 1.14 indicates overdispersion – 14% extra correlation. Of course, we know that the source of the overdispersion results from the 50 outliers.

Suppose that we add 100 to a single observation out of the 50,000.

R

```
y100_1 <- py
y100_1[1] <- y100_1[1] + 100
print py[1] y100_1[1]
```

```
. gen y100_1 = py
. replace y100_1 = y100_1 + 100 in 1
```

We chose to add 100 to the first value of *py*, which is 0. The new value is therefore 100. All other values remain the same.

```
. l y y100_1 in 1
     +-------------+
     | py  y100_1 |
     +-------------+
  1. | 0      100 |
     +-------------+
```

Modeling the amended data results in

R

```
poy4 <- glm(y100_1 ~ x1 + x2 + x3, family=poisson)
summary(poy4)
confint(poy4)
```

```
. glm y100_1 x1 x2 x3, fam(poi) nolog

Generalized linear models                    No. of obs      =      50000
Optimization      : ML                       Residual df     =      49996
                                             Scale parameter =          1
Deviance          =  54968.60634             (1/df) Deviance =    1.09946
Pearson           =   87936.2096             (1/df) Pearson  =   1.758865
. . .
                                             AIC             =   3.752163
Log likelihood    = -93800.06403             BIC             =    -485977
---------------------------------------------------------------------------
         |                OIM
  y100_1 |    Coef.   Std. Err.    z     P>|z|     [95% Conf. Interval]
---------+-----------------------------------------------------------------
      x1 |  .5001942 .0021676  230.76   0.000     .4959458    .5044426
      x2 | -.7475691 .0021949 -340.59   0.000    -.7518711   -.7432671
      x3 |  .2497153  .002179  114.60   0.000     .2454444    .2539862
   _cons |  1.001911 .0029838  335.78   0.000     .9960625    1.007759
---------------------------------------------------------------------------
```

The dispersion statistic is 1.76, representing some 75% extra dispersion in the data. Changing one value by a rather substantial amount substantially alters the dispersion statistic. The values of the parameter estimates are closely the same as in the original model. The log-likelihood and therefore the AIC and BIC statistics have changed only a little. Only the dispersion gives evidence that something is wrong. Typos frequently result in inflated dispersion statistics.

Dropping the first observation from the model by changing the value of *y*100_1 to missing results in the extra dispersion being eliminated.

R

```
y100_1 <- py
y100_1[1] <- NA
poy5 <- glm(y100_1 ~ x1 + x2 + x3, family=poisson)
summary(poy5)
confint(poy5)
```

```
. replace y100_1=. in 1
. glm y100_1 x1 x2 x3, fam(poi)

Generalized linear models                No. of obs      =      49999
Optimization     : ML                    Residual df     =      49995
                                         Scale parameter =          1
Deviance         =  53978.03793          (1/df) Deviance =   1.079669
Pearson          =  50015.59649          (1/df) Pearson  =   1.000412
                         .     .     .
                                         AIC             =   3.732297
Log likelihood   = -93301.55747          BIC             =  -486955.8
----------------------------------------------------------------------
        |                 OIM
 y100_1 |     Coef.   Std. Err.     z     P>|z|   [95% Conf. Interval]
--------+-------------------------------------------------------------
     x1 |   .5005742  .0021681   230.88   0.000    .4963248   .5048237
     x2 |  -.7494699  .0021955  -341.36   0.000    -.753773  -.7451667
     x3 |   .2494068  .0021796   114.43   0.000    .2451349   .2536786
  _cons |   .9998977   .002987   334.75   0.000    .9940432   1.005752
----------------------------------------------------------------------
```

This example provides good evidence of the importance of checking for model outliers in the presence of apparent overdispersion. Once the outliers are corrected the Pearson dispersion reduces to near 1.0. In fact, in some cases where outliers have been identified, but we do not have information as to how they are to be amended, it may be preferable to simply drop them from the model. Outliers can be typos, which may be corrected, or they may be accurate data. How we deal with them will depend on the nature of the data, and what we are attempting to determine. Of course, with real data, there is likely to be at least some residual overdispersion which cannot be accommodated by identifying outliers, adding a predictor, and so forth. We shall address this subject in Section 7.3.

An interesting approach might be to jackknife the model and then show the change in the Pearson dispersion for each observation as a means of identifying outliers.

7.2.4 Creation of interaction

We next consider a Poisson model having an interaction term. We create it in the same manner as we did the base model. The interaction term, created by *x*2 and *x*3, is represented by the predictor *x*23. We furthermore specify a parameter value of 0.2 for the interaction term. The data set is created with main effects predictors and the interaction:

R

```
x23 <- x2*x3
yi <-rpois(nobs, exp(1 + 0.5*x1 - 0.75*x2 + 0.25*x3 + 0.2*x23))
poy6 <-glm(yi ~ x1 + x2 + x3 + x23, family=poisson)
summary(poy6)
confint(poy6)
```

```
. gen x23 = x2*x3
. gen yi = rpoisson(exp(1 + 0.5*x1 - 0.75*x2 + 0.25*x3 + 0.2*x23))
```

Model the synthetic data including the main effects and interaction. The values should appear nearly the same in the model results, as specified in the above code.

```
. glm yi x1 x2 x3 x23, fam(poi)

Generalized linear models                 No. of obs      =     50000
Optimization     : ML                     Residual df     =     49995
                                          Scale parameter =         1
Deviance         =  53851.76409           (1/df) Deviance =  1.077143
Pearson          =  50017.58421           (1/df) Pearson  =  1.000452
                          . . .
                                          AIC             =   3.72901
Log likelihood   = -93220.24239           BIC             = -487083.1
-----------------------------------------------------------------------
       |                OIM
    yi |   Coef.    Std. Err.     z     P>|z|  [95% Conf. Interval]
-------+---------------------------------------------------------------
    x1 |  .5015653  .0021757   230.53   0.000   .4973009    .5058296
    x2 | -.7501484  .0021679  -346.03   0.000 -.7543974   -.7458994
    x3 |  .2499188  .0026366   100.89   0.000   .1955398    .2032875
   x23 |  .1994136  .0019765   100.89   0.000   .1955398    .2032875
 _cons |  .9987994  .0029977   333.19   0.000    .992924    1.004675
-----------------------------------------------------------------------
```

All parameter estimates appear as expected, including the interaction. Again, the dispersion statistic approximates 1.0. The AIC and BIC statistics are very close to those of the base model.

Now, model the data without the interaction:

R

```
poy7 <-glm(yi ~ x1 + x2 + x3, family=poisson)
summary(poy7)
confint(poy7)
```

```
. glm yi x1 x2 x3, fam(poi)

Generalized linear models                    No. of obs      =      50000
Optimization     : ML                        Residual df     =      49996
                                             Scale parameter =          1
Deviance         =  63322.68972              (1/df) Deviance =   1.266555
Pearson          =   60648.7788              (1/df) Pearson  =   1.213073
                                             . . .
                                             AIC             =   3.918388
Log likelihood   = -97955.70521              BIC             = -477622.9
------------------------------------------------------------------------
     |              OIM
  yi |     Coef.   Std. Err.      z    P>|z|    [95% Conf. Interval]
-----+------------------------------------------------------------------
  x1 |   .5048273   .0021767   231.93   0.000    .5005612    .5090935
  x2 | -.7322403   .0021595 -339.08   0.000   -.7364728   -.7280077
  x3 |   .0997768   .0021682    46.02   0.000    .0955272    .1040263
_cons |  1.030957   .0029284   352.06   0.000    1.025218    1.036697
------------------------------------------------------------------------
```

The dispersion rises 21%, AIC is higher by 0.19, and the BIC statistic increases by 9,344; $x3$ is also quite different from the true model having the interaction term – changing from 0.25 to 0.10. $x3$ is one of the two terms from which the interaction has been created. This model is apparently overdispersed. Of course, the overdispersion can be accommodated by creating the proper interaction. On the other hand, if we are modeling data, not knowing beforehand the parameter values of the model, the only indication that an interaction may be required is the errant dispersion statistic.

7.2.5 Testing the predictor scale

We next construct a Poisson data set where $x1$ has been transformed to $x1$-squared with a parameter value of 0.50. Other predictors are given the same parameter values as before.

R

```
x1sq <- x1*x1
pysq  <- rpois(nobs, exp(1 + .5*x1sq -.75*x2 +.25*x3))
poy8 <- glm(pysq ~ x1sq + x2 + x3, family=poisson)
summary(poy8)
confint(poy8)
```

```
. gen x1sq = x1*x1                          // square x1
. gen xbsq = 1 +.5*x1sq -.75*x2 +.25*x3     // exchange x1sq for x1
. gen exbsq = exp(xbsq)
. gen pysq  = rpoisson(exbsq)

. glm pysq x1sq x2 x3, fam(poi)

Generalized linear models             No. of obs      =      50000
Optimization     : ML                 Residual df     =      49996
                                      Scale parameter =          1
Deviance         =  52753.58396       (1/df) Deviance =   1.055156
Pearson          =  49146.16442       (1/df) Pearson  =   .9830019
                                      AIC             =   4.252571
Log likelihood   = -106310.2636       BIC             =  -488192.1
------------------------------------------------------------------
       |              OIM
  pysq |    Coef.  Std. Err.     z    P>|z|  [95% Conf. Interval]
-------+----------------------------------------------------------
  x1sq | .4998989   .000215  2325.38  0.000   .4994776   .5003203
    x2 | -.7496343 .0013599  -551.24  0.000  -.7522997  -.7469689
    x3 | .2500206  .0011969   208.89  0.000   .2476746   .2523665
 _cons | 1.000495  .0021954   455.72  0.000   .9961922   1.004798
------------------------------------------------------------------
```

Parameter estimates all approximate the synthetically assigned values, and the dispersion statistic is close to 1. We model the data as with the base model, except for the new *py*, which we call *ysq*.

R

```
poy9 <- glm(pysq ~ x1 + x2 + x3, family=poisson)
summary(poy9)
confint(poy9)
```

```
. glm pysq x1 x2 x3, fam(poi)

Generalized linear models             No. of obs      =      50000
Optimization     : ML                 Residual df     =      49996
                                      Scale parameter =          1
```

```
Deviance          =  2657739.466   .   (1/df) Deviance =  53.15904
Pearson           =  92511911.96       (1/df) Pearson  =  1850.386
                                       AIC             =  56.35229
Log likelihood    = -1408803.205       BIC             =   2116794
------------------------------------------------------------------
       |                 OIM
  pysq |      Coef.   Std. Err.      z    P>|z|   [95% Conf. Interval]
-------+----------------------------------------------------------
    x1 |   .366116   .0012164    300.98  0.000    .3637319    .3685001
    x2 |  -.660156   .0012297   -536.84  0.000   -.6625662   -.6577458
    x3 |  .2255974   .0012219    184.63  0.000    .2232026    .2279922
 _cons |  2.284213   .0015569   1467.20  0.000    2.281161    2.287264
------------------------------------------------------------------
. save odtest  ///  save simulated datasets in one file
```

Parameter estimates now differ greatly from the true values. The dispersion statistics are both extremely high (Pearson dispersion = 1,850), as are the AIC and BIC statistics. The model is highly overdispersed. Note the difference created by not taking into account the quadratic nature of $x1$. Squaring $x1$, of course, results in the correct model.

7.2.6 Testing the link

The final correction we address in this chapter relates to a misspecification of link. Link specification tests are typically applied to members of the binomial family, e.g. evaluating if the logit, probit, complementary loglog, or loglog is the most appropriate link to use with a Bernoulli or binomial family GLM. Misspecification for these sets of models is examined in length in Hilbe (2009). The test is valid, however, for any single equation model. Termed a Tukey–Pregibon test in Hilbe (2009), the test is constructed by fitting the model, calculating the hat diagonal statistic, and running the model again with *hat* and *hat-squared* as the sole predictors. If *hat-squared* is significant, at $p < 0.05$, the model is misspecified. Another link may be more appropriate.

Three links have been used with Poisson and negative binomial count models. These include those listed in Table 7.3.

We shall delay specification tests for negative binomial models until later; for now we shall observe how data that are appropriate for an identity-linked Poisson are affected when estimated using a standard log-linked Poisson. Recall that the natural log-link is the Poisson canonical link, deriving directly from the Poisson PDF.

When the coefficients of an identity-linked Poisson are exponentiated, they are estimates of an *incidence rate difference* in two contiguous levels of a

Table 7.3 *GLM count model links*

Link	Form	Models
log	$\ln(\mu)$	Poisson; NB2
canonical NB	$-\ln(1/(\alpha\mu)+1)$	NB-C
identity	μ	Poisson; NB-I

Table 7.4 *Identity Poisson GLM algorithm*

```
dev=0
μ = (y + mean(y))/2
η = μ
WHILE (abs(Δdev) > tolerance) {
    w = μ
    z = (y-μ)/μ
    β = (X'wX)^-1 X'wz
    μ = X'β
    oldDev = dev
    dev = 2S{yln(y/μ) - (y-μ)}
    Δdev = dev - oldDev
}
```

predictor with respect to the count response. The notion is based on the relative risk difference that we discussed with respect to a binary response in Section 2.4. The logic is the same, except that it is now applied to counts. Rate difference models are rarely discussed in statistical literature, but they are nevertheless viable models. When estimated using GLM software, the model setup is identical to that of a standard Poisson regression, with the exception that the link is $\eta = xb = \mu$, and the inverse link is simply the linear predictor. Generally the algorithm may be simplified owing to the identity of the linear predictor and fitted value. An abbreviated estimating algorithm for the identity Poisson model can be given as displayed in Table 7.4.

Using the same data as in the previous analyses, including the same value of *xb*, the linear predictor, an identity Poisson model may be constructed by using xb as the parameter to the *rpoisson()* function. The randomly generated counts, *piy*, will reflect the identity Poisson structure of the data. It is then used as the response term with the *x*1, *x*2, and *x*3 predictors as before.

R

```
xb <- 1 + 0.5*x1 - 0.75*x2 + 0.25*x3
piy  <- rpois(nobs, xb)
poy10 <- glm(piy ~ x1+x2+x3, family=poisson(link=identity))
summary(poy10)
confint(poy10)
```

```
. gen piy  = rpoisson(xb)
. glm piy x1 x2 x3, fam(poi) link(iden)

Generalized linear models                  No. of obs       =      50000
Optimization       : ML                    Residual df      =      49996
                                           Scale parameter =          1
Deviance           =  56186.38884           (1/df) Deviance =   1.123818
Pearson            =  49682.03785          ( 1/df) Pearson  =   .9937203
                                             .   .   .
                                           AIC              =   2.567087
Log likelihood     = -64173.16593          BIC              = -484759.2
---------------------------------------------------------------------
           |                 OIM
       piy |     Coef.   Std. Err.     z     P>|z|   [95% Conf. Interval]
-----------+---------------------------------------------------------
        x1 |   .5123143  .0148118    34.59   0.000    .4832837   .5413449
        x2 | -.7595172   .0148764   -51.06   0.000   -.7886744    -.73036
        x3 |   .2483579  .0148463    16.73   0.000    .2192597   .2774561
     _cons |  1.000588   .0137555    72.74   0.000    .9736275   1.027548
---------------------------------------------------------------------
```

The parameter estimates are close to those we specified. The distribution of counts for the identity Poisson model are displayed in Table 7.5.

This sharp decrease in the frequency of counts is characteristic of the identity linked Poisson model. Note that the dispersion statistic approximates 1.0, as we would expect.

Table 7.5 R: *Table with Count, Freq, Prop, cumProp*

```
piy <- poy10$piy
myTable <- function(x) {
myDF <- data.frame(table(x))
myDF$Prop <- prop.table(myDF$Freq)
myDF$CumProp <- cumsum(myDF$Prop)
myDF
}
myTable(piy)
```

```
. tab piy
        piy |      Freq.     Percent        Cum.
------------+-----------------------------------
          0 |     18,997       37.99       37.99
          1 |     17,784       35.57       73.56
          2 |      8,848       17.70       91.26
          3 |      3,197        6.39       97.65
          4 |        917        1.83       99.49
          5 |        199        0.40       99.88
          6 |         49        0.10       99.98
          7 |          8        0.02      100.00
          8 |          1        0.00      100.00
------------+-----------------------------------
      Total |     50,000 100.00
```

Now, suppose that we were given a model with the above distribution of counts, together with the adjusters $x1$, $x2$, and $x3$. It is likely that we would first attempt a standard Poisson model. Modeling it thus appears as:

R

```
poy11 <- glm(piy ~ x1 + x2 + x3, family=poisson))
summary(poy11)
confint(poy11)
```

```
. glm piy x1 x2 x3, fam(poi)

Generalized linear models                  No. of obs      =      50000
Optimization     : ML                      Residual df     =      49996
                                           Scale parameter =          1
Deviance         =  56261.22688            (1/df) Deviance =   1.125315
Pearson          =   49706.1183            (1/df) Pearson  =   .9942019

                                           AIC             =   2.568583
Log likelihood   = -64210.58495            BIC             =  -484684.4
------------------------------------------------------------------------
             |                 OIM
         piy |      Coef.   Std. Err.      z    P>|z|   [95% Conf. Interval]
-------------+----------------------------------------------------------
          x1 |   .5214119   .0155631   33.50   0.000    .4909088    .551915
          x2 |  -.7706432   .0156961  -49.10   0.000   -.8014069   -.7398794
          x3 |   .2563454   .0155219   16.52   0.000    .2259231    .2867677
       _cons |  -.0410198   .0142529   -2.88   0.004   -.0689549   -.0130847
------------------------------------------------------------------------
```

The dispersion statistic is still close to 1.0, and the parameter estimates are somewhat close to those produced given the correct model. The difference in intercept values is, however, significant.

We may test the link using what is termed the Tukey–Pregibon link test (Hilbe, 2009). The test is performed by first modeling the data (e.g. as a Poisson model), then calculating the *hat* matrix diagonal statistic. The *hat* statistic is squared and the response term is modeled on *hat* and *hat-squared*. If the value of *hat-squared* is statistically significant, the link of the original model is not well specified. That is, the model is well specified if the *hat-squared* is not significant.

The test is automated in Stata as the **linktest** command. Employed on the log-linked Poisson with identity-linked data, we clearly observe a misspecified link.

Table 7.6 R: *Tukey-Pregibon link test*

```
poitp <- glm(piy ~ x1+x2+x3, family=poisson))
hat <-hatvalues(poitp)
hat2 <- hat*hat
poy12 <- glm(piy ~ hat + hat2, family=poisson)
summary(poy12)
confint(poy12)
```

```
. qui poisson piy x1 x2 x3
. linktest

Poisson regression                          Number of obs   =      50000
                                            LR chi2(2)      =    3911.35
                                            Prob > chi2     =     0.0000
Log likelihood = -64173.639                 Pseudo R2       =     0.0296
------------------------------------------------------------------------
     piy |     Coef. Std. Err.      z  P>|z|   [95% Conf.  Interval]
---------+--------------------------------------------------------------
    _hat |  1.015117 .0167228  60.70  0.000    .9823409    1.047893
  _hatsq | -.4355458 .0512416  -8.50  0.000   -.5359776   -.3351141
   _cons |  .0323105 .0058467   5.53  0.000    .0208511    .0437699
------------------------------------------------------------------------
```

With the dispersion statistic approximating 1.0, and significant predictors, we would normally believe that the model would appropriately fit the data. It does not. But we only know this by using a link specification test. Here the remedy is to employ an alternative link.

These examples show how apparent overdispersion may be corrected. The caveat here is that one should never employ another model designed for overdispersed count data until the model is evaluated for apparent overdispersion. A model may in fact be a well-fitted Poisson or negative binomial model once appropriate transformations have taken place. This is not always an easy task, but necessary when faced with indicators of overdispersion. Moreover, until overdispersion has been accommodated either by dealing with the model as above, or by applying alternative models, one may not simply accept seemingly significant *p*-values.

Although it has not been apparent from the examples we have used, overdispersion does often change the significance with which predictors are thought to contribute to the model. Standard errors may be biased either upwards or downwards. A model may appear well fitted yet be incorrectly specified. All of these checks for apparent overdispersion should be traversed prior to declaring a model well fitted, or needing adjustment because of real overdispersion.

7.3 Methods of handling real overdispersion

The possible remedies that can be made to a model when faced with apparent overdispersion may be summarized by the following:

OVERDISPERSION ONLY APPARENT

1 Add appropriate predictor
2 Construct required interactions
3 Transform predictor(s)
4 Transform response
5 Adjust for outliers
6 Use correct link function

When faced with indicators of overdispersion, we first check for the possibility of apparent overdispersion. If overdispersion persists, there are a variety of methods that statisticians have used to deal with it – each based on addressing a reason giving rise to overdispersion.

MODELS DEALING WITH POISSON OVERDISPERSION

1 Scale SEs *post hoc*; deviance, *chi*2 dispersion
2 Scale SEs iteratively; scale term
3 Use robust variance estimators
4 Use bootstrap or jackknife SE
5 Multiply the variance by a specified constant
6 Multiply the variance by a parameter to be estimated : quasi-Poisson
7 Use generalized Poisson
8 Use negative binomial: NB2
9 Use heterogeneous negative binomial
10 Use NB-P; generalized negative binomial; Waring; Faddy; COM

OVERDISPERSION DUE TO ZERO COUNTS

1 Zero-inflated model
2 Zero-truncated model
3 NB with endogenous stratification
4 Heterogeneous NB with endogenous stratification
5 Hurdle models

OVERDISPERSION RESULTING FROM CENSORING/ TRUNCATION

1 Censored model: econometric
2 Censored model: survival
3 Truncated models

OVERDISPERSION RESULTING FROM PANEL STRUCTURE OF DATA

1 Generalized estimating equations (GEE) Poisson and NB
2 Unconditional fixed-effects Poisson and NB
3 Conditional fixed-effects Poisson and NB
4 Random effects
5 Random intercept
6 Mixed effects

OVERDISPERSION DUE TO ENDOGENEITY

1 Finite mixture model: multiple distributions inherent in response
2 Endogenous predictors
3 Selection models
4 Latent class models
5 Quantile count models

OTHER SOURCES OF OVERDISPERSION

Bivariate responses: Poisson and negative binomial

7.3.1 Scaling of standard errors / quasi-Poisson

Scaling of standard errors was the first method used within the GLM framework to deal with overdispersion in binomial and count response models. The method replaces the **W**, or model weight, in

$$\beta = (\mathbf{X}'\mathbf{W}\mathbf{X})^{-1}\ \mathbf{X}'\mathbf{W}_{\mathrm{Z}}$$

with the product of the model standard error and the square root of the dispersion. Scaling by the Pearson-based dispersion, applying the transformation, then running one additional iteration of the algorithm, but as

$$\beta = (\mathbf{X}'\mathbf{W}_d\mathbf{X})^{-1}\mathbf{X}'\mathbf{W}_{dZ} \tag{7.1}$$

Scaling in effect adjusts the model standard errors to the value that would have been calculated if the dispersion statistic had originally been 1.0. McCullagh and Nelder (1989) recommend that deviance-based scaling be used with discrete response models, while Pearson-based scaling be used with continuous response models. They assert that both deviance and Pearson scaling should produce similar standard errors if the model is well fitted. However, simulation studies have demonstrated that Pearson *chi*2-based scaling of count models is preferred over deviance-based scaling. Refer to Chapter 6 and Hilbe (2010c).

In Table 7.7 we see an IRLS Poisson algorithm showing both offsets and how scaling is calculated. An example will demonstrate how an overdispersed model can have the standard errors adjusted, providing the user with a more accurate indication of the true standard errors.

Table 7.7 *Poisson algorithm: scaled by chi2 dispersion*

```
μ = (y + mean(y))/2
η = ln(μ)
WHILE(abs(Δdev) > tolerance) {
u = (y - μ)/μ
w = μ
z = η + u - offset
β = (X'wX)^-1 X'wz
η = X'β + offset
μ = exp(η)
oldDev = dev
dev = 2Σ{yln(y/μ) - (y - μ)}
Δdev = dev - oldDev
}
/* After convergence, calculate */
dof = n - pred - 1
sc = chi2/dof
w = 1/sqrt(sc)              /* se(β)*sqrt(sc) */
w = se(β)*sqrt(sc)
```

A non-scaled model using the **medpar** data is given in Table 7.8.

Table 7.8 R: *Medpar data – Poisson estimates*

```
data (medpar)
attach (medpar)
poi <- glm(los ~ hmo + white + factor(type), family=poisson,
   data=medpar)
summary(poi)
confint(poi)
```

```
. glm los hmo white type2 type3, fam(poi)

Generalized linear models              No. of obs      =       1495
Optimization     : ML                  Residual df     =       1490
                                       Scale parameter =          1
Deviance         = 8142.666001         (1/df) Deviance =   5.464877
Pearson          = 9327.983215         (1/df) Pearson  =   6.260391
                                       . . .
                                       AIC             =   9.276131
Log likelihood   = -6928.907786        BIC             =  -2749.057

-----------------------------------------------------------------------
      |                OIM
  los |      Coef.  Std. Err.      z    P>|z|    [95% Conf. Interval]
------+----------------------------------------------------------------
  hmo | -.0715493    .023944    -2.99   0.003   -.1184786      -.02462
white |  -.153871   .0274128    -5.61   0.000   -.2075991     -.100143
type2 |  .2216518   .0210519    10.53   0.000    .1803908     .2629127
type3 |  .7094767    .026136    27.15   0.000    .6582512     .7607022
_cons |  2.332933   .0272082    85.74   0.000    2.279606      2.38626
-----------------------------------------------------------------------
```

The Pearson *chi*2, or χ^2, dispersion is extremely high at 6.26, especially considering the relatively large number of observations. For example, based on the original standard error for *hmo* (0.023944), we may calculate a scaled standard error as sqrt(6.260391) $*0.023944 = 0.0599097$. Note the calculated value in the model output below. For later discussion I should mention that the standard errors of IRR estimates are calculated using the delta method. IRR confidence intervals are calculated as the exponentiation of the model confidence intervals; they are not based on the IRR standard error.

The R code below first gives code to calculate the Pearson statistic as well as the Pearson dispersion. Following this is code to calculate the scaled SEs of the model coefficients using the quasi-Poisson family.

R

```
pr <- sum(residuals(poi,type="pearson")^2) # Pearson statistic
dispersion <- pr/poi$df.residual          # dispersion
dispersion
sse <- sqrt(diag(vcov(poi))) * sqrt(dispersion) # model SE
sse
poiQL <- glm(los ~ hmo + white + factor(type), family=quasipoisson,
  data=medpar)
coef(poiQL)
confint(poiQL)                             # scaled SEs
```

```
. glm los hmo white type2 type3, fam(poi) scale(x2)

Generalized linear models                 No. of obs      =       1495
Optimization     : ML                     Residual df     =       1490
                                          Scale parameter =          1
Deviance         =  8142.666001           (1/df) Deviance =   5.464877
Pearson          =  9327.983215           (1/df) Pearson  =   6.260391

                                          AIC             =   9.276131
Log likelihood   = -6928.907786           BIC             =  -2749.057
-----------------------------------------------------------------------
       |                OIM
   los |     Coef.   Std. Err.      z    P>|z|     [95% Conf. Interval]
-------+---------------------------------------------------------------
   hmo |  -.0715493  .0599097   -1.19   0.232   -.1889701    .0458715
 white |  -.153871   .0685889   -2.24   0.025   -.2883028   -.0194393
 type2 |   .2216518  .0526735    4.21   0.000    .1184137    .3248899
 type3 |   .7094767  .0653942   10.85   0.000    .5813064    .837647
 _cons |  2.332933   .0680769   34.27   0.000    2.199505    2.466361
-----------------------------------------------------------------------
(Standard errors scaled using square root of Pearson X2-based
dispersion.)
```

R output, from the above inset code is displayed as below. Note that the *summary* function following the use of the *quasipoisson* family displays the (Pearson) dispersion statistic.

PEARSON DISPERSION

```
dispersion
[1] 6.260391 Quasipoisson
```

SCALED SE

```
sse
  (Intercept)          hmo       white factor(type)2 factor(type)3
   0.06807690   0.05990964   0.06858885   0.05267347    0.06539395
```

QUASI-POISSON ESTIMATES AND SE

```
Coefficients:
                              Estimate Std. Error t value Pr(>|t|)
(Intercept)                    2.33293    0.06808  34.269  < 2e-16 ***
hmo                           -0.07155    0.05991  -1.194    0.233
white                         -0.15387    0.06859  -2.243    0.025 *
factor(type)Urgent Admit       0.22165    0.05267   4.208 2.73e-05 ***
factor(type)Emergency Admit    0.70948    0.06539  10.849  < 2e-16 ***
---
Signif. codes:  0 '***' 0.001 '**' 0.01 '*' 0.05 '.' 0.1 ' ' 1

(Dispersion parameter for quasipoisson family taken to be 6.260405)
```

QUASI-POISSON CONFIDENCE INTERVALS

```
> confint(poiQL)
Waiting for profiling to be done...
                                 2.5 %      97.5 %
(Intercept)                  2.1969663  2.46392933
hmo                         -0.1905868  0.04433228
white                       -0.2860215 -0.01703740
factor(type)Urgent Admit     0.1174482  0.32397344
factor(type)Emergency Admit  0.5794062  0.83584722
```

It needs to be emphasized that the parameter estimates remain unaffected, and only the standard errors are scaled. Overdispersion biases the standard errors such that *hmo* appears to contribute significantly to the model, and our consequent understanding of *los*, when in fact it does not. Scaling adjusts the model to what the standard errors would be if there were no overdispersion in the data.

Although the coefficients and standard errors are the same between Stata and R output, the values of the confidence intervals differ slightly. The difference rests with the probability functions upon which the *p*-values are based. GLM *p*-values in Stata are based on the *z*-statistic, or normal distribution. The R **glm** function employs a *t*-statistic instead. Although *t*-statistics are appropriate for Gaussian models, most statisticians prefer the normal distribution, *z*, or the Wald statistic, for non-normal GLM models. The Wald statistic, t^2, is used in SAS for non-Gaussian GLM models, producing confidence intervals similar to Stata output. The *t*-distribution approximates the normal when there are more than 30 observations. However, confidence intervals based on *t* will differ slightly from those based on *z* or Wald.

When standard errors have been scaled, the model is no longer a likelihood model since standard errors are not based on either the expected or observed information matrix. It is classified as a variety of quasi-likelihood model. We discuss the quasi-likelihood modeling of Poisson data further in the next section.

7.3.2 Quasi-likelihood variance multipliers

Quasi-likelihood (QL) methods were first developed by Wedderburn (1974). The method is based on GLM principles, but allows parameter estimates to be calculated based only on a specification of the mean and variance of the model observations without regard to those specifications originating from a member of the single-parameter exponential family of distributions. Further generalizations to the quasi-likelihood methodology were advanced by Nelder and Pregibon (1987). Called extended quasi-likelihood (EQL), these methods were designed to evaluate the appropriateness of the QL variance in a model. EQL models take us beyond the scope of our discussion, but quasi-likelihood models are important to understanding extensions to the Poisson and negative binomial models we consider in this text.

Quasi-likelihood methods allow one to model data without explicit specification of an underlying log-likelihood function. Rather, we begin with a mean and variance function, which are not restricted to the collection of functions defined by single-parameter exponential family members, and abstract backward to an implied log-likelihood function. Since this implied log-likelihood is not derived from a probability function, we call it quasi-likelihood or quasi-log-likelihood instead. The quasi-likelihood, or the derived quasi-deviance function, is then used in an IRLS algorithm to estimate parameters just as for GLMs when the mean and variance function are those from a specific member of the single-parameter exponential family.

Derived from equation 4.18, quasi-likelihood is defined as

$$Q(\mu; y) = \int_{y_i}^{\mu_i} \frac{y_i - \mu_i}{\phi V(\mu_i)} d\mu_i \tag{7.2}$$

and the quasi-deviance as

$$QD(\mu; y) = 2\int_{\mu_i}^{y_i} \frac{y_i - \mu_i}{V(\mu_i)} d\mu_i \tag{7.3}$$

In an enlightening analysis of leaf-blotch data, the quasi-likelihood was applied by Wedderburn (1974) using the logit link and a "squared binomial" variance function $\mu^2(1-\mu)^2$, with a corresponding quasi-likelihood given as:

$$Q\left[\mu_i^2(1-\mu_i)^2\right] = \sum_{i=1}^{n}(2y_i - 1)\ln\left(\frac{\mu_i}{1-\mu_i}\right) - \frac{y_i}{\mu_i} - \left(\frac{1-y_i}{1-\mu_i}\right) \quad (7.4)$$

The above quasi-likelihood function can be converted to a quasi-deviance using equation 6.24. Both functions behave like ordinary log-likelihood and deviance functions for discrete response models, regardless of not being directly derived from an underlying probability distribution.

In the case of the Poisson, we see that the solution of equation 7.2 with a variance of μ_i is identical to the Poisson log-likelihood, but without the final $\ln(y!)$ normalizing term, i.e. $\Sigma\{y\ln(\mu) - \mu\}$. The normalizing term is what ensures that the sum of the probabilities over the probability space adds to unity. The negative binomial (NB2) log-likelihood function can be similarly abstracted using the variance function $\mu + \alpha\mu^2$, but without its three characteristic log-gamma terms. It can be expressed as:

QUASI-LIKELIHOOD NEGATIVE BINOMIAL

$$Q_{\text{NB2}} = \sum_{i=1}^{n}\left\{y_i\ln(\mu_i) - \frac{1+\alpha y_i}{\alpha}\ln(1+\alpha\mu_i)\right\} \quad (7.5)$$

The manner in which quasi-likelihood methodology is typically brought to bear on overdispersed Poisson data is to multiply the variance μ by some constant scale value, indicated as ψ. A quasi-deviance Poisson regression algorithm is shown in Table 7.9. The quasi-likelihood can replace the quasi-deviance function without affecting the estimated model statistics.

The fact that the variance function is multiplied by a constant changes the likelihood, or, in this case, the deviance statistic, by dividing it by the scale. It is the next stage in which we discuss amending the Poisson variance function, and log-likelihood/deviance, to accommodate or adjust for overdispersion. We present an example using the same **medpar** data as used in the previous section. In this case we enter the Pearson dispersion statistic from the base model as the variance multiplier.

We may obtain the Pearson dispersion statistic from the standard Poisson model of the **medpar** example data as given in Section 7.2.1. The value 6.260391 is then inserted in the appropriate dispersion option of GLM software. The *dispersion*() option is used with Stata, but because it is a quasi-likelihood model, the *irls* option must be specified. The quasi-likelihood Poisson model is displayed below in Table 7.10.

Table 7.9 *Quasi-deviance Poisson regression algorithm variance multiplier*

```
μ = (y + mean(y))/2
η = ln(μ)
WHILE(abs(Δdev) >tolerance) {
u = (y - μ)/μ
w = μ*ψ
z = η + u - offset
β = (X'wX)^-1 X'wz
η = X'β + offset
μ = exp(η)
oldDev = dev
dev = [2Σ{yln(y/μ) - (y - μ)}]/ψ
Δdev = dev - oldDev
}
```

Table 7.10 R: *Quasi-likelihood Poisson standard errors*

```
poiQL <- glm(los ~ hmo + white + type 2 + type 3,
  family=poisson, data=medpar)
summary(poiQL)
pr <-sum(residuals(poiQL, type="pearson")^2)
disp <- pr/poiQL$df.residual          # Pearson dispersion
se <-sqrt(diag(vcov(poiQL)))
QLse <- se/sqrt(disp)
QLse
```

```
. glm los hmo white type2 type3, fam(poi) irls disp(6.260391)

Generalized linear models                No. of obs      =      1495
Optimization      : MQL Fisher scoring   Residual df     =      1490
                    (IRLS EIM)           Scale parameter = 6.260391
Deviance          =  1300.664128         (1/df) Deviance = .8729289
Pearson           =  1490.00008          (1/df) Pearson  =         1
                          .      .      .
Quasi-likelihood model with dispersion: 6.260391 BIC     = -9591.059
--------------------------------------------------------------------
      |                 EIM
  los |     Coef.   Std. Err.      z     P>|z|    [95% Conf. Interval]
------+-------------------------------------------------------------
  hmo | -.0715493   .0095696    -7.48   0.000   -.0903054   -.0527932
white |  -.153871   .010956    -14.04   0.000   -.1753444   -.1323977
type2 | .2216518   .0084138    26.34   0.000    .2051611    .2381424
type3 | .7094767   .0104457    67.92   0.000    .6890035    .7299499
_cons | 2.332933   .0108742   214.54   0.000     2.31162    2.354246
--------------------------------------------------------------------
```

Extra variation is dampened from the variance by multiplying it by the value of the Pearson dispersion. Note that the Pearson-dispersion value of this quasi-likelihood model is now 1.0. Note also the tight values of the standard errors.

One may calculate the quasi-likelihood standard errors, which also results in a type of quasi-likelihood model, by hand using the following formula

modelSE/sqrt(dispersion)

For example, the coefficient standard error of *hmo* in the standard Poisson model is 0.023944. If we divide this value by the square root of the dispersion, the result is the same as the value of adjusted standard error given in the model output above.

```
. di 0.023944/sqrt(6.260391)
.00956965
```

Each standard error may be calculated in the same manner. The same relationship obtains, of course, for exponentiated coefficients. For some purposes it may be simpler to put the vector of model predictor standard errors in a loop and divide each by the square root of the dispersion.

The same quasi-likelihood model may be obtained by employing the Pearson dispersion statistic as an importance weight. The importance weight, as defined in Stata, operates on the standard errors in the same manner as does the dispersion option. We have, therefore,

```
. gen pearsond = 6.260391
. glm los hmo white type2 type3 [iw=pearsond], fam(poi)

Generalized linear models                  No. of obs      =       1495
Optimization      : ML                     Residual df     =       1490
                                           Scale parameter =          1
Deviance          =  50976.27487           (1/df) Deviance =   34.21227
Pearson           =  58396.82436           (1/df) Pearson  =    39.1925

                                            AIC            =   58.03702
Log likelihood    = -43377.67358            BIC            =   40084.55
---------------------------------------------------------------------------
       |              OIM
   los |     Coef.  Std. Err.     z     P>|z|     [95% Conf. Interval]
-------+-------------------------------------------------------------------
   hmo | -.0715493 .0095696    -7.48   0.000    -.0903054    -.0527932
 white |   -.153871  .010956   -14.04   0.000    -.1753444    -.1323977
 type2 |  .2216518 .0084138    26.34   0.000     .2051611     .2381424
 type3 |  .7094767 .0104457    67.92   0.000     .6890035     .7299499
 _cons |  2.332933 .0108742   214.54   0.000      2.31162     2.354246
---------------------------------------------------------------------------
```

The standard errors as well as confidence intervals are identical in both manners of adjusting the standard errors. However, the summary model statistics of the weighted model differ considerably. Most notably the deviance and Pearson statistics differ between the standard, multiplier, and weighted models. The value of the Pearson dispersion in the above model is 39.1925, the square of the original dispersion statistic. This reflects the underlying difference in how each algorithm arrived at the same values for the adjusted standard errors.

It is important to keep in mind that, when using importance weights with Stata, how they are to be interpreted is based on the type of model being estimated. The manner in which they are used here is standard across GLM based models.

Compare the summary statistics of this model with the standard Poisson model applied to the same data. The quasi-likelihood model employing the dispersion option does not display an AIC statistic since the AIC is defined by a likelihood value. A likelihood function is not calculated or used in the model. The BIC parameterization that is used is that of the original GLM version developed by Raftery. Defined in equation 5.21, we refer to it as BIC_R. Stata's **glm** command uses BIC_R rather than other likelihood-based parameterizations so that a more appropriate fit statistic is displayed with quasi- and pseudo-likelihood GLM models.

The BIC statistic is substantially less than that of the standard (and scaled) model, indicating a better fit. The deviance statistic is also substantially less than that of the standard models.

	Standard	QL model
Deviance	8142.67	1300.66
BIC	−2749.10	−9591.06

Ordinarily one would employ a robust variance estimator with quasi- and pseudo-likelihood models, or any model in which the distributional assumptions of the model have been violated. We shall discuss that type of variance adjustment in the next subsection. However, since a scale was directly applied to the standard errors for the purpose of adjusting them to ameliorate overdispersion, applying yet another adjustment to the quasi-likelihood model, as understood here, in addition serves no statistically useful purpose. In fact, applying robust adjustments simply negates the effect of scaling. That is, multiplying the variance by a scale, and applying a robust estimator simultaneously produces the same standard errors as a model with no scaling, but only robust standard errors. We next turn to this discussion.

7.3.3 Robust variance estimators

Unlike the standard variance estimator, $-H(\beta)^{-1}$, the robust estimator does not need $L(\beta; x)$ to be based on the correct distribution function for x. Robust variance estimators have also been referred to as *sandwich* variance estimators. Associated standard errors are sometimes called *Huber standard errors* or *White standard errors*. Huber (1967) was the first to discuss this method, which was later independently discussed by White (1980) in the field of econometrics. Robust estimators are implemented in a post-estimation procedure according to the schema outlined in Table 7.11. Mention should be given as to why the modification of model-based standard errors are termed *sandwich* standard errors. The robust or sandwich estimator is derived from both the first and second derivatives of the model likelihood. Recall that these derivatives are known as the gradient or score and Hessian respectively. The outer product of the gradient is enclosed between two instances of the Hessian, **H***g***H**, analogous to a piece of meat between slices of bread. An abbreviated estimator termed OPG for *outer product gradients* has also been employed as the basis of standard errors. Readers interested in a more complete exposition can refer to Hardin and Hilbe (2003) or in the forthcoming Hilbe and Robinson (2011).

We shall find that robust variance estimators are quite robust, hence the name, when modeling overdispersion in count response models. They also play an important role when interpreting the Poisson or negative binomial parameter estimates as rate ratios. The robust score equations for three primary count response models are listed in Table 7.12.

Table 7.11 *Implementation of robust variance estimators*

1 Estimate the model
2 Calculate the linear predictor, $x\beta$
3 Calculate score vector: $g' = g(\beta;x) = x\partial L(x\beta)/\partial x\beta) = ux$
4 Calculate dof adjustment: $n(n - 1)$
5 Combine terms: $V(\beta) = V\{n/(n - 1)\ \Sigma u^2 x'x\}V$
6 Replace model Variance-Covariance matrix with robust estimator: an additional iteration with new matrix.

Table 7.12 *Robust score equations*

Poisson	:	$y - \exp(x\beta)$
Geometric (log)	:	$(y - \exp(x\beta))/(1 + \exp(x\beta))$
Negative binomial (log)	:	$(y - \exp(x\beta))/(1 + \alpha\exp(x\beta))$

An example using the same **medpar** data is displayed in Table 7.13.

Table 7.13 R: *Robust SEs of Medpar model*

```
library(sandwich)
poi <- glm(los ~ hmo + white + factor(type), family=poisson,
  data=medpar)
vcovHC(poi)
sqrt(diag(vcovHC(poi, type="HC0"))) # final HC0 = H-C-zero
```

```
. glm los hmo white type2 type3, fam(poi) robust

Generalized linear models                    No. of obs      =      1495
Optimization     : ML                        Residual df     =      1490
                                             Scale parameter =         1
Deviance         =  8142.666001              (1/df) Deviance =  5.464877
Pearson          =  9327.983215              (1/df) Pearson  =  6.260391
                                             . . .
                                             AIC             =  9.276131
Log pseudolikelihood = -6928.907786          BIC             = -2749.057
------------------------------------------------------------------------
      |             Robust
  los |      Coef.  Std. Err.      z    P>|z|    [95% Conf. Interval]
------+-----------------------------------------------------------------
  hmo | -.0715493   .0517323   -1.38   0.167   -.1729427    .0298441
white |  -.153871   .0833013   -1.85   0.065   -.3171386    .0093965
type2 |  .2216518   .0528824    4.19   0.000    .1180042    .3252993
type3 |  .7094767   .1158289    6.13   0.000    .4824562    .9364972
_cons |  2.332933   .0787856   29.61   0.000    2.178516     2.48735
------------------------------------------------------------------------
```

When robust variance estimators are applied to this type of quasi-likelihood model, we find that the effect of the robust variance overrides the adjustment made to the standard errors by the multiplier. It is as if the initial quasi-likelihood model were not estimated in the first place.

Robust variance estimators can also be applied to models consisting of clustered or longitudinal data. Many data situations take this form. For instance, when gathering treatment data on patients throughout a county, it must be assumed that treatments given by individual providers are more highly correlated within each provider than between providers. Likewise, in longitudinal data, treatment results may be recorded for each patient over a period of time. Again it must be assumed that results are more highly correlated within each patient record than between patients. Data such as these are usually referred to as *panel* data. Robust variance adjustments of some variety must be applied to the data because observations are not independent.

Modified sandwich variance estimators or robust-cluster variance estimators provide standard errors that allow inference that is robust to within group correlation, but assume that clusters of groups are independent. The procedure to calculate this type of robust estimate begins by summing the scores within each respective cluster. The data set is thereupon collapsed so that there is only one observation per cluster or panel. A robust variance estimator is then determined in the same manner as in the non-cluster case, except n is now the number of clusters and u consists of cluster sums (see Table 7.11). A thorough discussion of robust panel estimators is found in Hardin and Hilbe (2003).

The **medpar** data provide the hospital provider code with each observation. Called *provnum*, it is entered as an option to obtain the modified sandwich variance estimator. Unlike scaling and variance multipliers, robust estimators may be used with any maximum likelihood algorithm, not only GLM-based algorithms.

POISSON: CLUSTERING BY PROVIDER

R

```
poi <- glm(los ~ hmo+white+factor(type),family=poisson,
  data=medpar)
library(haplo.ccs)
sandcov(poi, medpar$provnum)
sqrt(diag(sandcov(poi, medpar$provnum)))
```

```
. glm los hmo white type2 type3, fam(poi) cluster(provnum)

Generalized linear models                   No. of obs       =      1495
Optimization     : ML                       Residual df      =      1490
                                            Scale parameter  =         1
Deviance         =  8142.666001             (1/df) Deviance  =  5.464877
Pearson          =  9327.983215             (1/df) Pearson   =  6.260391
                                   . . .
                                            AIC              =  9.276131
Log pseudolikelihood = -6928.907786         BIC              = -2749.057
                          (Std. Err. adjusted for 54 clusters in provnum)
---------------------------------------------------------------------------
       |              Robust
   los |     Coef.   Std. Err.      z    P>|z|    [95% Conf. Interval]
-------+-------------------------------------------------------------------
   hmo | -.0715493   .0527299   -1.36   0.175   -.1748979    .0317993
 white |  -.153871   .0729999   -2.11   0.035   -.2969482   -.0107939
 type2 |  .2216518   .0609139    3.64   0.000    .1022626    .3410409
 type3 |  .7094767    .202999    3.49   0.000     .311606    1.107347
 _cons |  2.332933   .0669193   34.86   0.000    2.201774    2.464093
---------------------------------------------------------------------------
```

Standard errors are produced by adjusting for the clustering effect on providers – that is, we suppose that the relationship between length of stay (*los*) and predictors is more highly correlated within a provider than between providers. This is a reasonable supposition. Only *hmo* fails to be contributory. Note again that all summary statistics are the same as in the unadjusted model. Also note that the model parameters are not adjusted for clustering. It is only the standard errors that are adjusted.

7.3.4 Bootstrapped and jackknifed standard errors

Bootstrap and jackknife are two additional methods that are used to adjust standard errors when they are perceived to be overdispersed. Non-parametric bootstrapping makes no assumptions about the underlying distribution of the model. Standard errors are calculated based on the data at hand. Samples are repeatedly taken from the data (with replacement), with each sample providing model estimates. The collection of vector estimates for all samples is used to calculate a variance matrix from which reported standard errors are calculated and used as the basis for calculating confidence intervals. Such confidence intervals can be constructed from percentiles in the collection of point estimates or from large sample theory arguments. The example below uses 1,000 samples of 1,495 observations; each sample provides an estimated coefficient vector from which standard errors are calculated. The number of samples may be changed. This method is primarily used with count data when the data are not Poisson or negative binomial, and the model is overdispersed.

Table 7.14 R: *Bootstrap standard errors*

```
poi <- glm(los ~ hmo + white + factor(type), family=poisson,
  data=medpar)
summary(poi)
t <- function (x, i) {
xx <- x[i,]
 bsglm <-
glm(los ~ hmo + white + factor(type), family=poisson,
  data=medpar) return(sqrt(diag(vcov(bsglm))))
 }
bse <- boot(medpar, ffit, R=1000)
sqrt(diag(vcov(poi)))
apply(bse$t,2, mean)
```

```
. glm los hmo white type2 type3, fam(poi) vce(bootstrap)
    reps(1000) saving(bsmedpar))
(running glm on estimation sample)

Bootstrap replications (1000)
----+--- 1 ---+--- 2 ---+--- 3 ---+--- 4 ---+--- 5
..................................................    50
..................................................  1000

Generalized linear models             No. of obs      =       1495
Optimization     : ML                 Residual df     =       1490
                                      Scale parameter =          1
Deviance         =  8142.666001       (1/df) Deviance =   5.464877
Pearson          =  9327.983215       (1/df) Pearson  =   6.260391
                              . . .
                                      AIC             =   9.276131
Log likelihood   = -6928.907786       BIC             =  -2749.057
------------------------------------------------------------------------
     |  Observed  Bootstrap                        Normal-based
 los |     Coef.  Std. Err.      z    P>|z|  [95% Conf. Interval]
-----+------------------------------------------------------------------
  hmo | -.0715493  .0527865  -1.36   0.175  -.1750089    .0319103
white |  -.153871  .0806069  -1.91   0.056  -.3118576    .0041155
type2 |  .2216518  .0537124   4.13   0.000   .1163775    .3269261
type3 |  .7094767   .117573   6.03   0.000   .4790378    .9399156
_cons |  2.332933   .076994  30.30   0.000   2.182028    2.483838
------------------------------------------------------------------------
```

NOTE

The same results as displayed above may be obtained using the command:

```
. bootstrap, reps(1000): glm los hmo white type2 type3, fam(poi)
```

Standard errors again indicate that *hmo* and *white* are problematic.

Jackknifed standard errors are used for the same purpose as standard errors calculated from bootstrapping. The model is estimated as many times as there are observations in the data – in this case 1,495. Each iteration excludes one observation. The collection of estimated coefficient vectors is used to calculate a variance matrix from which the standard errors are reported in the output, together with (large-sample based) confidence intervals.

Table 7.15 R: *Jackknifed standard errors*

```
poi <- glm(los ~ hmo + white + factor(type), family=poisson,
       data=medpar)
summary(poi)
library(bootstrap)
```

```
theta <- function (i) {
  xx <- medpar[i,]
  jkglm <- glm(los ~ hmo + white + factor(type), family=poisson,
       data=medpar)
  return(sqrt(diag(vcov(jkglm))[iCoef]))
}
sqrt(diag(vcov(poi)))
for (iCoef in 1:5) {
  #repeat these time-consuming calculations for each coef
  jJE <- jackknife(1:nrow(medpar), theta)
  #perform jackknife for
  coef[iCoef] cat(iCoef, jSE$jack.se, "\n")
}
```

```
. glm los hmo white type2 type3, fam(poi) vce(jack)
(running glm on estimation sample)
Jackknife replications (1495)
----+--- 1 ---+--- 2 ---+--- 3 ---+--- 4 ---+--- 5
..................................................    50
..................................................  1450
.............................................

Generalized linear models                  No. of obs      =      1495
Optimization     : ML                      Residual df     =      1490
                                           Scale parameter =         1
Deviance         =  8142.666001            (1/df) Deviance =  5.464877
Pearson          =  9327.983215            (1/df) Pearson  =  6.260391

                                           AIC             =  9.276131
Log likelihood   = -6928.907786            BIC             = -2749.057
------------------------------------------------------------------------
     |              Jackknife
 los |      Coef.   Std. Err.      t    P>|t|     [95% Conf. Interval]
-----+------------------------------------------------------------------
  hmo| -.0715493   .0520596    -1.37   0.170    -.173667     .0305684
white| -.153871    .0854792    -1.80   0.072    -.321543     .0138009
type2|  .2216518   .0532418     4.16   0.000     .1172152    .3260883
type3|  .7094767   .1188121     5.97   0.000     .4764205    .9425329
_cons|  2.332933   .0808175    28.87   0.000     2.174405    2.491461
------------------------------------------------------------------------
```

NOTE

The same results as displayed above may be obtained using the command:

```
. jackknife: glm los hmo white type2 type3, fam(poi)
```

Jackknifing yields standard errors similar to those obtained from bootstrapping. Again, *hmo* and *white* are questionable contributors to the Poisson model. To compare standard errors between models on the same data, see Table 7.15.

It is clear that the **medpar** data we modeled as Poisson are overdispersed. The standard model suggests that all predictors are statistically significant. However, when we employ adjustments to the variance in order to accommodate any overdispersion in our inference, we find that *hmo* and *white* are now questionable contributors to the model, and that the adjusted standard errors are much the same for all but *type*3 (emergency admissions). Actually, interpretation of coefficients using robust, bootstrapped, or jackknifed standard errors are similar for *type*3 – only Pearson-scaling and clustering on *provider* differ.

Table 7.16 *Comparison of standard errors: Medpar Poisson model*

	EIM/OIM	P-SCALE	QL	ROBUST	CLUSTER	BOOT	JACK
hmo	.0239	.0599	.0096	.0517	.0527	.0528	.0521
white	.0274	.0686	.0110	.0822	.0730	.0806	.0855
type2	.0211	.0527	.0084	.0529	.0609	.0537	.0532
type3	.0261	.0654	.0104	.1158	.2030	.1176	.1188

7.4 Tests of overdispersion

The concept of overdispersion is central to the understanding of both Poisson and negative binomial models. Nearly every application of the negative binomial is in response to perceived overdispersion in a Poisson model. We have previously addressed the problem of ascertaining whether indicators of overdispersion represent real overdispersion in the data, or only apparent. Apparent overdispersion can usually be accommodated by various means in order to eradicate it from the model. However, real overdispersion is a problem that affects the reliability of both the model parameter estimates and fit in general.

We showed one manner in which overdispersion could be detected in a Poisson model – observing the value of the Pearson dispersion statistic. Values greater than 1 indicate Poisson overdispersion. The problem with this method is in having no specific criteria for evaluating how far over 1 the statistic must be to qualify it as significantly overdispersed over random variation within the Poisson model. For example, a 50-observation Poisson model with a dispersion statistic of 1.25 does not provide support that significant overdispersion exists in the data. Everything else being equal, such a model appears well modeled using Poisson regression. Likewise, a 100-observation Poisson model with a dispersion statistic of 1.1 is likely equidispersed, except for random variation in the data upon which the model is based. On the other hand, such a dispersion statistic accompanying a 100,000-observation Poisson model may well indicate real overdispersion.

There are three commonly used statistics to quantify the amount of overdispersion in the data. The tests provide a *p*-value upon which the researcher can more clearly base a decision on whether a model is Poisson or requires modeling by a method such as negative binomial regression. We first discuss the *Z*-score and Lagrange multiplier tests, which are related. We then follow with an examination of a boundary likelihood ratio test that has been designed for assessing Poisson overdispersion.

7.4.1 Score and Lagrange multiplier tests

A score test first developed by Dean and Lawless (1989) to evaluate whether the amount of overdispersion in a Poisson model is sufficient to violate the basic assumptions of the model may be defined as:

Z-TESTS

$$Z_i = \frac{(y_i - \mu_i)^2 - y_i}{\mu_i \sqrt{2}} \tag{7.6}$$

The test is *post-hoc*, i.e. it is performed subsequent to modeling the data. Using the **medpar** data set as earlier delineated, we first model the data using a Poisson regression. Since there is no need to display the model results, the data are modeled but not shown. The interest here relates to the testing of the model for Poisson overdispersion. The *Z*-statistic may then be calculated using the following code (Table 7.17). The **predict** code tells the software to calculate the expected counts.

Table 7.17 R: *Z-score test*

```
poi <- glm(los ~ hmo + white + factor(type), family=poisson,
    data=medpar)
mu <-predict(poi, type="response")
z <- ((los - mu)^2 - los)/ (mu * sqrt(2))
zscore <- lm(z ~ 1)
summary(zscore)
```

```
. qui glm los hmo white type2 type3, fam(poi)
. predict mu
. gen double z=((los-mu)^2-los)/ (mu*sqrt(2))
. regress z, nohead
------------------------------------------------------------------
    z |     Coef.   Std. Err.     t     P>|t|   [95% Conf. Interval]
------+-----------------------------------------------------------
_cons | 3.704561   .3947321    9.39    0.000    2.930273    4.478849
------------------------------------------------------------------
```

The Z-score test is 3.7, with a t-probability of less than 0.0005. Z-tests the hypothesis that the Poisson model is overdispersed. In practice, it tests whether the data should be modeled as Poisson or negative binomial. This example indicates that the hypothesis of no overdispersion is rejected, i.e. it is likely that real overdispersion does exist in the data. A negative binomial model is preferred to a Poisson model. Of course, this information corresponds to the large Pearson dispersion statistic we know accompanies the model.

The score test given in equation 7.6 assumes a large model sample size, and a value of Z that is normally distributed. Winkelmann (2008) derives a similar Score test, with the numerator being the same as in Z, but with 2 in the denominator. This form of the test originates from Dean and Lawless (1989), and is motivated by assuming a large value for μ. Another variant of the Z-test is one that appears the same as Z, but without the corrective $\sqrt{2}$ in the denominator. Only μ is given as a correction. This version is presented in Cameron and Trivedi (1998), and is proportional in value to the Z-test as we have characterized it.

Other forms of the score test for determining if the model data are Poisson or negative binomial have appeared in the literature as well. I see little reason to use them over the test described here. Be certain to know which formulation of score test is used when evaluating the research results of others. The Lagrange multiplier test is evaluated using a *chi*2 test rather than on the t-test probability required for the score test.

LAGRANGE MULTIPLIER TEST

$$\mathrm{LM} = \frac{\left(\sum_{i=1}^{n} \mu_i^2 - n\bar{y}_i\right)^2}{2\sum_{i=1}^{n} \mu_i^2} \tag{7.7}$$

Again, using Stata commands to calculate the statistic, we have:

Table 7.18 R: *Lagrange multiplier test*

```
# continue from above R code
obs <- nrow(medpar)
mmu <- mean(mu)
nybar <- obs*mmu
musq <- mu*mu
mu2 <- mean(musq)*obs
chival <- (mu2 - nybar)^2/(2*mu2)
chival
pchisq(chival,1,lower.tail = FALSE)
```

```
. summ los, meanonly /* solving for Lagrange Multiplier */
. scalar nybar = r(sum)
. gen double musq = mu*mu
. summ musq, meanonly
. scalar mu2 = r(sum)
. scalar chival = (mu2-nybar)^2/(2*mu2)
. display "LM value =" chival _n "P-value ="
     chiprob(1,chival)
LM value = 62987.861
P-value = 0
```

With one degree of freedom, the test appears to be significant – the hypothesis of no overdispersion is again rejected. It should be noted that for a *chi*2 test with one degree of freedom, a *chi*2 value of 3.84 has a *p*-value of 0.05. Higher values of *chi*2 result in lower *p*-values. A *chi*2 value of 5.024 is 0.025.

7.4.2 Boundary likelihood ratio test

The Poisson model is identical to an NB2 negative binomial with an ancillary or heterogeneity parameter of 0. The Poisson is the lower limiting case of the NB2 model, which we describe in detail in the next chapter.

We earlier mentioned that, for a Poisson model, values of the Pearson dispersion in excess of 1.0 indicate possible overdispersion. Overdispersion represents more correlation in the data than is allowable on the basis of the distributional assumptions of the model. In the case of Poisson, with its assumption of equidispersion, the mean and variance functions should have equal values. In terms of the NB2 model, this distributional situation is represented by having a heterogeneity parameter value of 0. That is, when a Poisson model of given data has a Pearson dispersion of 1, the corresponding heterogeneity parameter of an NB2 model of the same data is 0. Poisson dispersion statistics greater than 1 correspond to NB2 heterogeneity parameter values of greater than 0. The relationship between the two is not linear, but there is a positive correlation in the two values. We shall refer to the negative binomial heterogeneity parameter as α.

A modeling problem exists when the Poisson dispersion is less than 1, indicating possible underdispersion. We shall later discuss this subject. For the present, it is simply necessary to understand that NB2 values of α cannot be lower than 0. As a result, NB2 models cannot properly estimate underdispersed Poisson models. Models such as generalized Poisson regression can be used to deal with underdispersed data. Binomial models, and positive-only continuous response models such as 2-parameter log-gamma and log-inverse Gaussian regression can also be used for this purpose. The subject of underdispersion, therefore, is not addressed by negative binomial models, and as such is not elaborated upon in this text.

A Poisson model can be underdispersed, equidispersed, or overdispersed. An NB2 model can be used to accommodate Poisson dispersion to a lower limit of the Pearson dispersion statistic equal to 1.0. Therefore this is the boundary of the interface between Poisson and NB2 regression, i.e. the test is on the boundary of the parameter space of the negative binomial. To decide if a value of α is small enough to warrant the estimation of data using a Poisson model rather than NB2, we may use a likelihood ratio test with a boundary at $\alpha = 0$, which corresponds to a Poisson dispersion equal to 1.0.

It may be argued that a researcher should simply model count data using an NB2 model instead of Poisson. The reasoning might be that if the model is truly Poisson, it can be estimated equally well using NB2 regression. The problem with this line of thinking is that the data may in fact be underdispersed, in which case the NB2 model will not converge. Secondly, if the true value of α is 0, current NB2 software will fail to converge. The reasons are numeric and entail division by 0. Poisson may be the limit of an NB2 model, but NB2 can only estimate data that are Poisson overdispersed, no matter how slightly.

The recommendation is to first model count data using Poisson regression. If the model is checked for apparent overdispersion and is found to truly be overdispersed, then employ an alternative. We shall discover that standard negative binomial regression (NB2) is only one of many ways to deal with overdispersed data. However, it is the most popular method, and for most purposes is satisfactory.

The boundary likelihood ratio test is used to evaluate if the value of α is significantly greater than 0, or negative binomial. The difference in this test and the standard likelihood ratio test rests in how the p-value is determined. The p-value is one-half the probability that a *chi*2 value, with one degree of freedom, is greater than the obtained likelihood ratio statistic. The likelihood ratio is the same as described in Section 5.2.1.4, with the full model being the same as the negative binomial. We have, therefore

$$\mathrm{LR} = -2(\mathcal{L}_{\mathrm{P}} - \mathcal{L}_{\mathrm{NB}}) \tag{7.8}$$

Supposing a Poisson log-likelihood of −4849.0395 and an NB2 model log-likelihood of −7264.0456, the likelihood ratio value is 4830. The test probability is determined by

R

```
pchisq(4830,1,lower.tail = FALSE)/2
```

```
. di chiprob(1, 4830)/2
0
```

This method of assessing Poisson overdispersion is perhaps the most popular, and simplest, of alternatives. However, it is applicable only to the relationship of Poisson and NB2 models. Fortunately it may also be used when comparing other pairs of count models, e.g. zero-truncated Poisson and zero-truncated negative binomial, censored Poisson and censored negative binomial.

7.4.3 R_{p}^2 and R_{pd}^2 tests for Poisson and negative binomial models

Fridstrøm *et al.* (1995) proposed a method to assess Poisson goodness-of-fit. Based on the logic of the linear model's R^2 coefficient of determination, the equation is partitioned into two components. The numerator consists of the traditional R^2 statistic,

$$R^2 = 1 - \frac{\sum_i (y_i - \mu_i)^2}{\sum_i (y_i - \bar{y})^2} \tag{7.9}$$

The denominator is given as

$$P^2 = 1 - \frac{\sum_i \mu_i}{\sum_i (y_i - \bar{y})^2} \tag{7.10}$$

which is the most amount of variation that can be explained in a equidispersed Poisson model. The ratio of the two statistics provides the proportion of potentially explainable systematic variation in the data: $R_{\mathrm{p}}^2 = R^2 P^2$ (see Vogt and Bared, 1998).

Fridstrøm *et al.* (1995) also proposed a log-likelihood R^2 for the negative binomial, where the heterogeneity or dispersion parameter is considered. Based on the negative binomial model deviance statistic, the equation is similar to the Poisson test. It is composed of two components, R_{p}^2 and P_{p}^2. The test statistic, R_{pd}^2, is the ratio of the two, which are defined respectively as,

$$R_{\mathrm{d}}^2 = 1 - \frac{D^{\mathrm{m}} / (n - k - 1)}{D^{\mathrm{o}} / (n - 2)} \tag{7.11}$$

and

$$P_{\mathrm{d}}^2 = 1 - \frac{D_{\mathrm{E}}^{\mathrm{m}} / (n - k)}{D^{\mathrm{o}} / (n - 2)} \tag{7.12}$$

where D^{o} is the model, deviance for an intercept-only negative binomial (NB2) model, n is the number of observations and k is the number of predictors. D^{m} is the deviance of the full NB2 model, and $D_{\mathrm{E}}^{\mathrm{m}}$ is the deviance of a Poisson model on the same data (i.e. $\alpha = 0$). The R^2_{pd} statistic is a measure of the total variation explained by the negative binomial model over the Poisson.

Another R^2 test for negative binomial models was developed by Miaou (1996). It may be expressed as

$$R^2_{\text{M}} = 1 - \frac{\alpha}{\alpha_0} \tag{7.13}$$

where α is the value of the model heterogeneity parameter for the full model, and α_0 is the value of α for an intercept-only version of the model. Miaou states that an interesting feature of the statistic rests in the fact that when one adds predictors of equal importance to the model, the R^2_{M} statistic increases in value by an absolute amount, regardless of the order of entry into the model. Moreover, he claims that the value of the intercept has no bearing on the statistic.

These tests have not found support in the literature, but they are interesting attempts to determine statistics that evaluate the fit of a count model independent of a comparative model.

7.5 Negative binomial overdispersion

Although the subject is rarely discussed, it is implicit in what we have been discussing that negative binomial (NB2) models may also be overdispersed. Poisson overdispersion occurs when its observed distributional variance exceeds the mean. That is, the Poisson distribution assumes equidispersion. We have mentioned several reasons that give rise to Poisson overdispersion in count data.

Given a specified, as well as calculated, value of the mean, μ, we may define negative binomial overdispersion as occurring when the calculated model variance exceeds $\mu + \alpha\mu^2$. That is, a count model may be both Poisson and negative binomial overdispersed if the variance produced by the estimated model exceeds the nominal negative binomial variance. Likewise, a model may be Poisson overdispersed yet negative binomial underdispersed, or even equidispersed. How can we know? By observing the negative binomial (NB2) Pearson dispersion.

We may simulate Poisson and negative binomial responses by generating random variates for both distributions. We begin by generating an intercept or constant value of 3, and using it to calculate a Poisson and a negative binomial random variate with a mean approximating 20, i.e. exp(3).

POISSON

R

```
library(MASS)
xb  <- 3
exb <- exp(xb)
yp <-rpois(50000, exp(xb))
p3 <-glm(yp ~ 1, family=poisson)
summary(p3)
exb
```

```
. set obs 50000
. gen xb = 3
. gen exb = exp(xb)
. gen yp= rpoisson(exp(3))
. di exb
20.085537

. su yp, detail              /// partial output
     Mean      =  20.08708 <=
     Variance =  20.12094 <=
```

Taking into account the random nature of the data, we may take the mean and variance as equal. The data are next modeled using a Poisson regression. Given the specified value of 3 that we assign for the value of the intercept ($xb = 3$), we expect a value of 3 for the Poisson parameter estimate. Exponentiating 3 produces 20.08.

```
. glm yp, fam(poi)

Generalized linear models                 No. of obs      =      50000
Optimization     : ML                     Residual df     =      49999
                                          Scale parameter =          1
Deviance         =  50591.43033           (1/df) Deviance =   1.011849
Pearson          =  50083.28007           (1/df) Pearson  =   1.001686
                                          .
                                          AIC             =   5.832425
Log likelihood   = -145809.6298           BIC             =  -490386.7
----------------------------------------------------------------------
     |                 OIM
  yp |     Coef.   Std. Err.      z    P>|z|     [95% Conf. Interval]
-----+----------------------------------------------------------------
_cons | 3.000077   .0009978  3006.60   0.000    2.998121    3.002033
----------------------------------------------------------------------
```

The same random variates are now used to construct negative binomial data with α assigned a value of 0.333.

NEGATIVE BINOMIAL (NB2)

R

```
a <- .33333333
ia <- 1/.33333333
xg <- rgamma(50000, a, a, ia)
xbg <-exb*xg
ynb <- rpois(50000, xbg)
nb3 <-glm.nb(ynb ~ 1)
summary(nb3)
-------------------------------------
# results obtained from a run
exp(3.00240) + .3351*exp(3.00240)^2
[1] 155.9733
```

```
. gen xg = rgamma(3, .3333333)
. gen xbg = exb*xg
. gen ynb = rpoisson(xbg)
. sum ynb, detail
      Mean       = 20.01352
      Variance = 155.5534

. di 20.01352 + .3351*20.01352^2        /// NB2 variance
154.2348                                /// obs var ~ pre var

. glm ynb, fam(nb ml)

Generalized linear models                   No. of obs      =      50000
Optimization      : ML                      Residual df     =      49999
                                            Scale parameter =          1
Deviance         =  52956.44357             (1/df) Deviance =    1.05915
Pearson          =  50424.30764             (1/df) Pearson  =   1.008506
                                             .  .  .
Variance function: V(u) = u+(.3351)u^2

                                            AIC             =   7.629868
Log likelihood    = -190745.6955            BIC             =  -488021.7
---------------------------------------------------------------------------
       |              OIM
   ynb |     Coef.  Std. Err.      z     P>|z|    [95% Conf. Interval]
------+--------------------------------------------------------------------
_cons |  2.996408  .0027752  1079.71  0.000    2.990969    3.001847
---------------------------------------------------------------------------
Note: Negative binomial parameter estimated via ML and treated
      as fixed once estimated.
```

Repeating the simulation will produce slightly different results, but the values for the respective variances will cluster around the appropriate values. The values for the regression statistics will vary as well.

Overdispersion for both Poisson and negative binomial models is at least in part indicated by the value of the Pearson *chi*2 dispersion statistic, which reflects the underlying variability in the model data, i.e. the variance. As previously discussed (Section 6.2), the equation for the dispersion is $chi2 = \Sigma_i\,(y_i - \mu_i)^2/V(\mu_i)$.

Note that for both the synthetic Poisson and negative binomial (NB2) models displayed above, the dispersion approximates 1.0. The observed and predicted values of the mean and variance are also closely the same. Like Poisson overdispersion, negative binomial overdispersion is indicated if the model dispersion is in excess of 1.0. Also like the Poisson, the NB2 model can also be thought of as a member of a covering model of greater scope – a generalized negative binomial model. We shall discuss this type of model later in the text, but be aware that there are several parameterizations of a generalized NB2 model. A problem with most of these algorithms, however, is that of bias. Save for one that we will examine in some detail, the NB-P model, most versions have been shown to lead to biased statistics.

Summary

Whether count model overdispersion is significant depends on: (1) the value of the Pearson dispersion statistic, (2) the number of observations in the model, and (3) whether apparent overdispersion can be remedied by performing various operations on the data, e.g. adding appropriate predictor(s), constructing required interactions, transforming predictor(s) or the response, adjusting for outliers, and using the correct link function for the model. Data that are highly unbalanced, as well as data in clustered or longitudinal (panel) format also give rise to overdispersion.

After examining the theory, nature, and applications of the standard negative binomial model in the following two chapters, the remainder of the text is devoted to discussing extensions to the Poisson and negative binomial which are aimed at ameliorating real overdispersion. Enhanced negative binomial models (e.g. ZINB, zero-truncated negative binomial) all attempt to accommodate negative binomial overdispersion just as enhanced Poisson models attempt to accommodate Poisson overdispersion. The difference is that negative binomial regression is itself one of the models used for overdispersed Poisson data. It is

Table 7.19 *Methods to directly adjust the variance*

1 Scaling SE's by Pearson-dispersion; quasipoisson
2 Adjusting SE's by a robust or empirical estimator
3 Multiplying V by a constant
  QL = $V\phi$ (with ϕ a constant)
4 Bootstrapping or jackknifing SE's
5 Variance multiplied by ancillary parameter
  NB-1 = $V\phi = \mu(1 + \alpha) = \mu + \alpha\mu$
4 Geometric
  $V\phi = \mu(1 + \mu) = \mu + \mu^2$
7 Negative binomial
  NB-2 = $V\phi = \mu(1 + \alpha\mu) = \mu + \alpha\mu^2$
8 Heterogeneous negative binomial
  NB-H = $V(1 + (\alpha\gamma)\mu)$ (with α parameterized by γ)
  $\mu + (\alpha\gamma)\mu^2$
9 Negative binomial P
  NB-P = $\mu + \alpha\mu^{\rho}$
10 Generalized Estimating Equations (GEE)
  GEE = V[correlation matrix]V'

important to keep in mind the relationships between the various models, and what it is that each is constructed to do.

In this chapter we have outlined various methods employed to determine whether a Poisson model is subject to real or to only apparent overdispersion. Remedies for apparent overdispersion were detailed, and methods for adjusting Poisson model standard errors were specified. We now turn to negative binomial regression, viewing it both as a means to ameliorate Poisson overdispersion, and as a model unto itself, independent of a relationship to the Poisson.

Table 7.19 provides a listing of direct adjustments that are commonly applied to the Poisson variance. Most are formatted as products to the basic Poisson mean, μ. The first four are discussed in this chapter.

8

Negative binomial regression

In this and subsequent chapters we shall discuss the nature and utility of some 25 varieties of negative binomial regression that are useful for modeling count response data. In addition, we examine certain models that are related to the negative binomial family of models. This chapter will primarily be devoted to an examination of the derivation of the negative binomial model and to the two foremost methods of its estimation. We also consider how the probabilities generated from a negative binomial model differ from the Poisson, as well as how they differ among various negative binomial models based on both mean and ancillary parameters.

8.1 Varieties of negative binomial

I mentioned that the basic negative binomial model can be enhanced to allow for the modeling of a wide range of count response situations. The Poisson can likewise be enhanced to adjust for data that violate its essential distributional assumptions. In fact, many of the same distributional problems face both Poisson and negative binomial models. We therefore find similar approaches to the handling of such data for both the Poisson and negative binomial. These include models such as zero-inflated Poisson (ZIP), which is directly related to the zero-inflated negative binomial (ZINB). All of the models allow for an estimated variance that exceeds the mean – the principal assumption of the Poisson regression model being that of equidispersion. With respect to the negative binomial, allowance of extra variation involves: (1) an adjustment to the NB2 variance function, or (2) a modification to the NB2 probability distribution, resulting in an amended log-likelihood function. The NB2 variance function, it may be recalled, is expressed as $\mu + \alpha\mu^2$.

Canonical linked models maintain both the μ-parameterized likelihood and variance functions of the NB2 model, but modify the link function. It can be argued, however, that it is the NB2 model that is an alteration of the basic canonical form, which derives directly from the negative binomial probability function. This is in fact the case when NB2 is modeled within the GLM framework. On the other hand, we have indicated that the negative binomial distribution can be derived as a Poisson–gamma mixture. However, as we discovered earlier, when the negative binomial PDF is understood within the context of GLM, it should be thought of as being characterized as the probability of successes and failures in a series of *iid* Bernouilli trials. Specifically, the probability distribution underlying the GLM-based negative binomial model is based on the distribution of y failures before the rth success in a series of $1/0$ Bernoulli trials. It can also be considered as the distribution of y successes before the rth failure. Parameterized in either of these ways, both y and r are integers. When r is limited to integer-only values, the distribution is termed a Pascal distribution. When r is allowed to take any positive real value, the distribution is known as a Polya distribution. Both are forms of the more general negative binomial distribution. We employ the first of the above two parameterizations of success–failure relationships in this monograph.

It is clear that allowing r to be positive real gives more pliability to the model. As actually employed in a statistical model, r indicates the amount of excess correlation in the data. Limiting it to integer values would dismiss the continuous nature of the range of correlation in data.

The point here is that the negative binomial statistical model is not based on one derivation. It can be derived as a Poisson–gamma mixture, as a series of Bernoulli trials, as a Polya–Eggenberger urn model, or as a type of inverse binomial distribution. Other derivations have been devised as well, each leading to the statistical model we call the negative binomial. In fact, Boswell and Patil (1970) identified twelve specific ways in which to derive the negative binomial distribution. The relationship between the negative binomial and binomial will be briefly examined in Section 8.2.2. This will help provide a meaning to the *negative* in negative binomial.

Table 8.1 lists 22 varieties of negative binomial regression for modeling count response data. Keep in mind that there are variations of these models, and that I have not identified models that have no commercial or well-used freeware support, e.g. R.

Note that the geometric model is a variety of negative binomial, i.e. a negative binomial with $\alpha = 1$. The Poisson may also be included as a negative binomial, i.e. $\alpha = 0$. We exclude them, however, since each belongs to the parent negative binomial of which it is entailed. Note that, as a negative binomial distribution

Table 8.1 *22 varieties of negative binomial model*

0	NB2 $V = \mu + a\mu^2$
1	NB1 $V = \mu + a\mu$
2	NB-C <Canonical>
3	Truncated NB; NB1; NB-C
5	Zero-inflated NB (ZINB)
8	Censored NB
9	NB-hurdle
10	NB-P
11	NB-H <Heterogeneous>/ w ES
12	Finite mixture models
13	Conditional fixed effects
14	Random effects
15	Mixed and multilevel effects
16	NB w endogenous stratification
17	Sample selection NB
18	Latent class NB
19	Quantile count
20	Generalized method of moments
21	Instrumental variables
22	Bivariate NB

or model, α can only approach zero, never reach it. On the other hand, as we have observed in practice, a negative binomial model with a value of α close to zero is statistically indistinguishable from a Poisson model. In this sense the Poisson is a variety of negative binomial. Table 8.1 simply details the specific varieties of negative binomial that will be discussed in this and coming chapters.

8.2 Derivation of the negative binomial

We have previously mentioned that the standard negative binomial regression model is also termed the NB2 model owing to the quadratic nature of its variance function, $\mu + \alpha\mu^2$. NB2 is to be distinguished from the NB1 model, which has a variance appearing as $\mu + \alpha\mu$. We shall consider the NB1 model in a later chapter.

In this section we describe the derivation of the NB2 model as either a Poisson–gamma mixture model, or as a member of the exponential family of distributions which serve as the basis of generalized linear models. In deriving the NB2 as a GLM, we must first derive what can be called the canonical negative binomial, or NB-C. An NB2 model is created by converting the canonical link and inverse link in the GLM derived NB-C to log form. There are some complications that must be addressed, but we shall take those up when appropriate.

We shall first address the derivation of the Poisson–gamma mixture model negative binomial.

8.2.1 Poisson–gamma mixture model

The negative binomial PDF can be derived from the specification of a count outcome characterized by

$$f(y;\lambda,u) = \frac{e^{-\lambda_i u_i}(\lambda_i u_i)^{y_i}}{y_i!} \tag{8.1}$$

which can be thought of as a Poisson model with gamma heterogeneity where the gamma noise has a mean of 1. The gamma mixture accommodates overdispersed or correlated Poisson counts.

The distribution of y, conditioned on x and u, is Poisson with the conditioned mean and variance given by μ (8.1). The conditional mean of y under gamma heterogeneity is thereby expressed as λu rather than as only λ. As a result, the unconditional distribution of y is derived from the following expression:

$$f(y;x,u) = \int_0^\infty \frac{e^{-(\lambda_i u_i)}(\lambda_i u_i)^{y_i}}{y_i!} g(u_i)\partial u_i \tag{8.2}$$

The unconditional distribution of y is specified by how we define $g(u)$. For this model a gamma distribution is given $u = \exp(\varepsilon)$ where $\ln(\mu) = x\beta + \varepsilon$. Assigning a mean of 1 to the gamma distribution, we have

$$f(y;x,u) = \int_0^\infty \frac{e^{-(\lambda_i u_i)}(\lambda_i u_i)^{y_i}}{y_i!} \frac{\nu^\nu}{\Gamma(\nu)} u_i^{\nu-1} e^{-\nu u_i} du_i \tag{8.3}$$

The gamma nature of u is evident in the derivation from equations 8.3 to 8.4.

$$= \frac{\lambda_i^{y_i}}{\Gamma(y_i+1)} \frac{\nu^\nu}{\Gamma(\nu)} \int_0^\infty e^{-(\lambda_i+\nu)u_i} u_i^{(y_i+\nu)-1} du_i \tag{8.4}$$

We carry the derivation further by moving

$$\frac{\lambda_i^{y_i}}{\Gamma(y_i+1)}\frac{\nu^\nu}{\Gamma(\nu)}\frac{\Gamma(y_i+\nu)}{(\lambda_i+\nu)^{y_i+\nu}}$$

to the left of the integral, with the remaining terms under the integral equaling 1.

We continue as

$$= \frac{\lambda_i^{y_i}}{\Gamma(y_i+1)}\frac{\nu^\nu}{\Gamma(\nu)}\Gamma(y_i+\nu)\left(\frac{\nu}{\lambda_i+\nu}\right)^\nu \frac{1}{\nu^\nu}\left(\frac{\lambda_i}{\lambda_i+\nu}\right)^{y_i}\frac{1}{\lambda_i^{y_i}}$$

$$= \frac{\Gamma(y_i+\nu)}{\Gamma(y_i+1)\,\Gamma(\nu)}\left(\frac{\nu}{\lambda_i+\nu}\right)^\nu\left(\frac{\lambda_i}{\lambda_i+\nu}\right)^{y_i}$$

$$= \frac{\Gamma(y_i+\nu)}{\Gamma(y_i+1)\,\Gamma(\nu)}\left(\frac{1}{1+\lambda_i/\nu}\right)^\nu\left(1-\frac{1}{1+\lambda_i/\nu}\right)^{y_i} \tag{8.5}$$

Inverting ν, the gamma scale parameter, yields α, the negative binomial heterogeneity or overdispersion parameter. We also equate λ and μ. Doing so, we then recognize the resulting negative binomial probability mass function

$$f(y;\mu,\alpha) = \frac{\Gamma(y_i+1/\alpha)}{\Gamma(y_i+1)\,\Gamma(1/\alpha)}\left(\frac{1}{1+\alpha\mu_i}\right)^{\frac{1}{\alpha}}\left(1-\frac{1}{1+\alpha\mu_i}\right)^{y_i} \tag{8.6}$$

Equation 8.7 is a commonly observed form of the negative binomial PMF. Given y, it follows that $\Gamma(y+1) = y!$, $\Gamma(y+1/\alpha-1) = (y+1/\alpha)!$, and $\Gamma(1/\alpha) = (1/\alpha-1)!$

$$\frac{\Gamma(y_i+1/\alpha)}{\Gamma(y_i+1)\,\Gamma(1/\alpha)} = \frac{(y_i+\frac{1}{a})!}{y_i!\,(1/\alpha-1)!} = \binom{y_i+\frac{1}{\alpha}-1}{\frac{1}{\alpha}-1} \tag{8.7}$$

The left term with gamma functions may be therefore re-structured to the form of a combination, and the rightmost term may be converted to a single fraction, resulting in another popular expression of the negative binomial probability distribution.

$$f(y;\mu,\alpha) = \binom{y_i+\frac{1}{\alpha}-1}{\frac{1}{\alpha}-1}\left(\frac{1}{1+\alpha\mu_i}\right)^{\frac{1}{\alpha}}\left(\frac{\alpha\mu_i}{1+\alpha\mu_i}\right)^{y_i} \tag{8.8}$$

I should reiterate that the above expression of the negative binomial PMF is found in different forms. Some authors prefer to leave the gamma scale parameter, ν, as it is, perhaps using the symbol r or k to represent the negative binomial ancillary parameter, which we term the negative binomial heterogeneity parameter. In this form the heterogeneity parameter is inversely related to the amount of Poisson overdispersion in the data. Most contemporary statisticians, however, prefer a direct relationship. Hence the form of $1/\alpha$ that is displayed in the above equations.

It should also be noted that when dealing with combinations, as we have in deriving this parameterization of the negative binomial, y and α are assumed to be integers. But this assumption does not have to obtain when it is used as the distributional basis of a regression model. As a count model, the negative binomial response, y, should consist of non-negative integer values, but α need not. The only restriction placed on α is that it take positive rational values, rarely above 4.

Moreover, do not confuse the negative binomial heterogeneity (ancillary) parameter with the dispersion statistic of the model. At times some statisticians refer to the heterogeneity parameter as the dispersion parameter since it relates to the amount of overdispersion in the data. However, the dispersion statistic that we discussed in previous chapters is not a parameter, but rather a statistic generated by dividing the Pearson *chi*2 statistic by the model residual degrees of freedom. This is to be distinguished from the parameter which is an integral term of the negative binomial probability distribution.

The likelihood function is a re-expression of the negative binomial probability function displayed in equation 8.8, given as

$$L(\mu; y, \alpha) = \prod_{i=1}^{n} \exp\left\{ y_i \ln\left(\frac{\alpha\mu_i}{1+\alpha\mu_i}\right) - \frac{1}{\alpha}\ln(1+\alpha\mu_i) + \ln\Gamma\left(y_i + \frac{1}{\alpha}\right) - \ln\Gamma(y_i+1) - \ln\Gamma\left(\frac{1}{\alpha}\right)\right\} \tag{8.9}$$

The log-likelihood is obtained by taking the natural log of both sides of the equation. As with the Poisson models, the function becomes additive rather than multiplicative.

$$\mathcal{L}(\mu; y, \alpha) = \sum_{i=1}^{n} y_i \ln\left(\frac{\alpha\mu_i}{1+\alpha\mu_i}\right) - \frac{1}{\alpha}\ln(1+\alpha\mu_i) + \ln\Gamma\left(y_i + \frac{1}{\alpha}\right) - \ln\Gamma(y_i+1) - \ln\Gamma\left(\frac{1}{\alpha}\right) \tag{8.10}$$

The negative binomial log-likelihood, parameterized in terms of β, the model coefficients, can be expressed as:

$$\mathcal{L}(\beta_j; y, \alpha) = \sum_{i=1}^{n} y_i \ln\left(\frac{\alpha \exp(x_i'\beta)}{1+\alpha \exp(x_i'\beta)}\right) - \frac{1}{\alpha}\ln\left(1+\alpha\exp(x_i'\beta)\right)$$
$$+ \ln\Gamma\left(y_i + \frac{1}{\alpha}\right) - \ln\Gamma(y_i+1) - \ln\Gamma\left(\frac{1}{\alpha}\right) \qquad (8.11)$$

Maximum likelihood principles define estimating equations as the derivatives of the log-likelihood function. We discovered in Chapter 4 that maximum likelihood estimates of the model parameters are determined by setting the first derivative of the log-likelihood with respect to β, *the gradient,* to zero, and solving. A more accurate representation is that if there are more than one parameter to be estimated, the gradient, or score, is the vector of first partial derivatives of the log-likelihood. If the negative binomial heterogeneity parameter is also estimated, we take the derivative of the log-likelihood with respect to α, set to zero, and solve. If α is entered into the GLM algorithm as a constant, α is not estimated, unless it is estimated outside GLM and subsequently entered into the GLM algorithm.

The matrix of second partial derivatives of the log-likelihood is commonly known as the Hessian. The observed information matrix is defined as the negative of the Hessian, evaluated at the maximum likelihood estimate. The expected information is the expected value of the negative Hessian. The derivation of these two types of information is given in Chapter 4. Recall, however, for canonical GLM models, e.g. Poisson and NB-C, the observed and expected information matrices are identical. For non-canonical models like NB2, the observed and expected information matrices differ. Model standard errors are obtained as the square root of the diagonal terms of the variance–covariance matrix, which is the inverse of the information. Therefore, model standard errors are the square root of the diagonal terms of the negative inverse Hessian matrix. Generally, except for models with few observations, standard errors from the expected information matrix are close to those produced from the observed. Recent studies have shown that for non-canonical models observed standard errors are asymptotically less biased than expected standard errors.

Newton–Raphson based maximum likelihood estimation entails an observed information matrix, whereas the traditional GLM IRLS algorithm uses the expected. As we shall observe in this chapter, it is fairly simple to amend GLM IRLS estimation so that it incorporates the observed information matrix as the basis of model standard errors.

With $\mu = \exp(x'\beta)$, the first and second partials of the log-likelihood function with respect to β and α are given in equations 8.12 through 8.16.

NB GRADIENT – β

$$\frac{\partial \mathcal{L}}{\partial \beta} = \sum_{i=1}^{n} \frac{x_i(y_i - \mu_i)}{1+\alpha\mu_i} \tag{8.12}$$

Note, owing to the importance of the parameterization, the gradient in terms of μ is

$$\frac{\partial \mathcal{L}}{\partial \mu} = \sum_{i=1}^{n} \frac{y_i - \mu_i}{\mu_i(1+\alpha\mu_i)} \tag{8.13}$$

NB GRADIENT – α

$$\frac{\partial \mathcal{L}}{\partial \alpha} = \sum_{i=1}^{n}\left[\frac{1}{\alpha^2}\left(\ln(1+\alpha\mu_i) + \frac{\alpha(y_i-\mu_i)}{1+\alpha\mu_i}\right) + \psi\left(y_i + \frac{1}{\alpha}\right) - \psi\left(\frac{1}{\alpha}\right)\right] \tag{8.14}$$

NB – HESSIAN – β

$$\frac{-\partial^2 \mathcal{L}}{\partial \beta \partial \beta'} = \sum_{i=1}^{n} \frac{\mu_i(1+\alpha y_i)}{(1+\alpha\mu_i)^2} x_i x_i' \tag{8.15}$$

or

$$\frac{\partial^2 \mathcal{L}}{\partial \beta_j' \partial \beta_j} = \sum_{i=1}^{n}\left[-x_{ij}x_{ij}' \frac{\mu_i(1+\alpha y_i)}{(1+\alpha\mu_i)^2}\right] \tag{8.16}$$

NB – HESSIAN – β; α

$$\frac{\partial^2 \mathcal{L}}{\partial \beta \partial \alpha} = E\left[-\sum_{i=1}^{n} \frac{\mu_i(y_i-\mu_i)x_{ij}}{(1+\alpha\mu_i)^2}\right] \tag{8.17}$$

NB – HESSIAN – α

$$\frac{\partial^2 \mathcal{L}}{\partial \alpha^2} = \sum_{i=1}^{n}\left[-\frac{1}{\alpha^3}\left(\frac{\alpha(1+2\alpha\mu_i)(y_i-\mu_i) - \alpha\mu_i(1+\alpha\mu_i)}{(1+\alpha\mu_i)^2} + 2\ln(1+\alpha\mu_i)\right) + \psi'\left(y_i+\frac{1}{\alpha}\right) - \psi'\left(\frac{1}{\alpha}\right)\right] \tag{8.18}$$

The digamma function is the derivative of the log-gamma function, lnΓ(), and the trigamma is the second derivative of the lnΓ(). Fairly good approximations can be made to them as:

$$\text{DIGAMMA} = \Psi_1 = \Psi(x) = \frac{\ln\Gamma(x+0.0001) - \ln\Gamma(x-0.0001)}{0.0002}$$

or
$$\frac{\{\ln\Gamma((y_i+1/\alpha)+0.0001) - \ln\Gamma((y_i+1/\alpha)-0.0001)\}}{0.0002} \tag{8.19}$$

and

$$\text{TRIGAMMA} = \Psi_2 = \Psi'(\text{x}) = \frac{-\ln\Gamma(x+0.002) + 16^* \ln\Gamma(x+0.001) - 30\ln\Gamma(x) + 16\ln\Gamma(x-0.001) - \ln\Gamma(x-0.002)}{0.000012}$$

or
$$\frac{\{\ln\Gamma((1/\alpha)+0.0001) - \ln\Gamma((1/\alpha)-0.0001)\}}{0.0002} \tag{8.20}$$

8.2.2 Derivation of the GLM negative binomial

Two major forms of the negative binomial may be derived from the negative binomial probability function. Both forms may be considered as members of the exponential family of distributions, and modeled under the framework of generalized linear models. One is the canonical, NB-C, being derived directly from the negative binomial PDF. The other is a conversion from the canonical form to the log-linked form. The latter is the traditional negative binomial regression (NB2) model we have discussed in the previous pages. Utilizing the log-link allows a direct comparison with the Poisson model, which is NB2 with $\alpha = 0$.

We can describe the negative binomial PDF as the probability of observing y failures before the rth success in a series of Bernoulli trials. Under such a description r is a positive integer. However, as discussed earlier, there is no compelling mathematical reason to limit this parameter to integers.

Although the negative binomial may be parameterized differently, it is always possible to convert terms to produce the final form derived here. Nevertheless, the form with which we begin is expressed in equation 8.21, which is the same as equation 8.8 with $\alpha = 1/r$.

Notice that this form is roughly similar to the binomial PDF, appearing as $\binom{n}{y} p^y (1-p)^{n-y}$, where p is the probability of success. For negative binomial models, p is rather thought of as the probability of r successes and y the number of failures before the rth success. It is the difference in how successes and failures are distributed that differentiates between binomial, geometric, and negative binomial models. The binomial distribution describes the number of successes in n trials, the geometric distribution describes the number of failures before the first success, r, and the negative binomial distribution describes the number of failures before the rth success. Given that the binomial describes successes, and the negative binomial describes failures before a specified number of successes, it is clear why the latter was given the name *negative* binomial.

Another view, based on the binomial distribution arising from the "binomial" expansion of $(q + p)^n$ where $0 < p < 1$ and $q = 1 - p$, is given in Johnson *et al.* (2005, Chapter 5). They show that the negative binomial arises from the same expansion, but with the signs reversed. That is, $(q - p)^{-n}$, where $p > 0$ and $q = 1 + p$. Again, the *negative* in negative binomial is clear.

As we discussed in Section 6.1, the Poisson may also be considered in relationship to a series of Bernoulli trials. However, it is generated by knowing the mean number of successes in a series of Bernoulli trials, λ, but without knowledge of the number of trials in the series, n, or the probability of success on any single trial, p. The Poisson distribution, given this understanding, is the probability of y number of successes in the series provided that n is large and p small. Bernoulli trials are therefore fundamental to all four distributions: Poisson, binomial, geometric, and negative binomial.

The formulation of the negative binomial PDF below is identical to equation 8.8, with $r = 1/\alpha$, and $p = 1/(1 + \alpha\mu)$ and $(1 - p) = \alpha\mu/(1 + \alpha\mu)$.

The usual manner of deriving the probability distribution of equation 8.21 from the context of a sequence of independent Bernoulli trials is expressed as follows. If we consider y as the number of failures before the rth success, consider the situation when the rth success is on the xth trial. It follows then that the previous $r - 1$ successes can occur any time in the previous $x - 1$ trials. Expressed as a combination, we have

$$\binom{x-1}{r-1}$$

ways that we can obtain the rth success on trial x. Each of the ways given in the combination has a probability specified as $p^r(1 - p)^{x-r}$ – giving us the probability of $x - r$ failures before the rth success. Note that $x - r$ is identical to how we first defined y. That being the case, we have $x = y + r$, producing equation 8.21.

It should be emphasized that a negative binomial regression model has a negligible tie to how the underlying PDF is derived. When such a model is being used to accommodate Poisson overdispersion, or to estimate predicted counts, it matters little how many failures have occurred before a specific number of successes. On the other hand, because the negative binomial is based on a sequence of Bernoulli trials, as are the binomial and geometric distributions, we shall discover some interesting relationships between the models. For example, the geometric model is a negative binomial with $\alpha = 1$. The relationship between the binomial models – with the logistic model being canonical – and the negative binomial model is more subtle, and will be discussed in Section 9.3.

NEGATIVE BINOMIAL PDF

$$f(y; p, r) = \binom{y_i + r - 1}{r - 1} p_i^r (1 - p_i)^{y_i} \tag{8.21}$$

or

$$f(y; p, r) = \frac{(y_i + r - 1)!}{y_i!(r - 1)!} p_i^r (1 - p_i)^{y_i} \tag{8.21a}$$

Converting the negative binomial PDF into exponential family form results in:

EXPONENTIAL FAMILY FORM

$$f(y; p, r) = \exp\left\{ y_i \underbrace{\ln(1 - p_i)}_{\text{link}} + \underbrace{r\ln(p_i)}_{\text{cumulant}} + \ln\binom{y_i + r - 1}{r - 1} \right\} \tag{8.22}$$

From our earlier discussion we found that the canonical link and cumulant can easily be abstracted from a PDF when it is expressed in exponential family form.

LINK, CUMULANT, SCALE

$$\theta_i = \ln(1 - p_i) \Rightarrow p_i = 1 - \exp(\theta_i) \tag{8.23}$$

$$b(\theta_\mathrm{i}) = -r\ln(p_i) \Rightarrow -r(1 - \exp(\theta_i)) \tag{8.24}$$

$$\alpha_i(\emptyset) \text{ (scale)} = 1$$

The first and second derivatives, with respect to θ, respectively yield the mean and variance functions.

NEGATIVE BINOMIAL MEAN

$$b'(\theta_i) = \frac{\partial b}{\partial p_i}\frac{\partial p_i}{\partial \theta_i} = -\frac{r}{p_i}(-(1-p_i)) = \frac{r(1-p_i)}{p_i} = \mu_i \tag{8.25}$$

NEGATIVE BINOMIAL VARIANCE

$$\begin{aligned} b''(\theta_i) &= \frac{\partial^2 b}{\partial p_i^2}\left(\frac{\partial p_i}{\partial \theta_i}\right)^2 + \frac{\partial b}{\partial p_i}\frac{\partial^2 p_i}{\partial \theta_i^2} = \frac{r}{p_i^2}(1-p_i)^2 + \frac{-r}{p_i}(-(1+p_i)) \\ &= \frac{r(1-p_i)}{p_i^2} \end{aligned} \tag{8.26}$$

$V(\mu)$ therefore equals $r(1-p)/p^2$. We now parameterize p and r in terms of μ and α.

$$\begin{aligned} (1-p_i)/(\alpha p_i) &= \mu_i \\ (1-p_i)/p_i &= \alpha\mu_i \\ p_i &= 1/(1+\alpha\mu_i) \end{aligned} \tag{8.27}$$

where $\alpha = 1/r$.

Given the defined values of μ and α, we may re-parameterize the negative binomial PDF such that

$$f(y;\mu,\alpha) = \binom{y_i + \frac{1}{\alpha} - 1}{\frac{1}{\alpha} - 1}\left(\frac{1}{1+\alpha\mu_i}\right)^{\frac{1}{\alpha}}\left(\frac{\alpha\mu_i}{1+\alpha\mu_i}\right)^{y_i} \tag{8.28}$$

This is identical to equation 8.8, which we derived via a Poisson–gamma mixture.

Re-expressed in terms of the log-likelihood, equation 8.28 yields

$$\begin{aligned} \mathcal{L}(\mu; y, \alpha) = \sum_{i=1}^{n}\Bigg\{ & y_i \ln\left(\frac{\alpha\mu_i}{1+\alpha\mu_i}\right) - \left(\frac{1}{\alpha}\right)\ln(1+\alpha\mu_i) + \ln\Gamma\left(y_i + \frac{1}{\alpha}\right) \\ & - \ln\Gamma(y_i + 1) - \ln\Gamma\left(\frac{1}{\alpha}\right)\Bigg\} \end{aligned} \tag{8.29}$$

or

$$\begin{aligned} \mathcal{L}(\mu; y, \alpha) = \sum_{i=1}^{n}\Bigg\{ & y_i \ln(\alpha\mu_i) - \left(y_i + \frac{1}{\alpha}\right)\ln(1+\alpha\mu_i) \\ & + \ln\Gamma\left(y_i + \frac{1}{\alpha}\right) - \ln\Gamma(y_i + 1) - \ln\Gamma\left(\frac{1}{\alpha}\right)\Bigg\} \end{aligned} \tag{8.30}$$

which we derived using the Poisson–gamma mixture approach.

The GLM deviance function is derived from both the saturated and fitted log-likelihood functions. The saturated function consists of replacing the value of y for each value of μ.

DEVIANCE

$$D = 2\sum_{i=1}^{n}\{\mathcal{L}(y_i; y_i) - \mathcal{L}(\mu_i; y_i)\}$$

Substituting the log-likelihood function as specified in either equation 8.29 or 8.30, we have

$$D_{\mathrm{NB}} = 2\sum_{i=1}^{n}\left\{y_i \ln\left(\frac{y_i}{\mu_i}\right) - \left(\frac{1}{\alpha} + y_i\right)\ln\left(\frac{1+\alpha y_i}{1+\alpha\mu_i}\right)\right\} \tag{8.31}$$

Calculating the terms required for the IRLS algorithm, we have

LINK (NB-C)

$$g(\mu_i) = \theta_i = \ln\left(\frac{\alpha\mu_i}{1+\alpha\mu_i}\right) = -\ln\left(\frac{1}{\alpha\mu_i} + 1\right) \tag{8.32}$$

INVERSE LINK (NB-C)

$$g^{-1}(\theta_i) = \mu_i = \frac{1}{\alpha(e^{-\theta_i} - 1)} = \frac{1}{\alpha(\exp(-\theta_i) - 1)} \tag{8.33}$$

CUMULANT

$$\begin{aligned} b(\theta) &= 1/\alpha \ln(1/(1+\alpha\mu_i)) \\ &= -1/\alpha \ln(1+\alpha\mu_i) \end{aligned} \tag{8.34}$$

MEAN, VARIANCE, AND DERIVATIVE

$$b'(\theta_i) = \mu_i = \frac{\partial b}{\partial\mu_i}\frac{\partial\mu_i}{\partial\theta_i} = \frac{1}{1+\alpha\mu_i}\mu_i(1+\alpha\mu_i) = \mu_i \tag{8.35}$$

$$b''(\theta_i) = V(\mu_i) = \frac{\partial^2 b}{\partial\mu_i^2}\left(\frac{\partial\mu_i}{\partial\theta_i}\right)^2 + \frac{\partial b}{\partial\mu_i}\frac{\partial^2\mu_i}{\partial\theta_i^2} = \mu_i + \alpha\mu_i^2 \tag{8.36}$$

$$g'(\theta_i) = \frac{\partial\theta_i}{\partial\mu_i} = \frac{\partial\{\ln(\alpha\mu_i/(1+\alpha\mu_i))\}}{\partial\mu_i} = \frac{1}{\mu_i + \alpha\mu_i^2} \tag{8.37}$$

IRLS algorithms normally are parameterized in terms of the fit statistic μ rather than xb, the linear predictor. Maximum likelihood algorithms such as Newton–Raphson or Marquardt are parameterized as $x\beta$. In Section 4.2.3 we showed how to convert between the two parameterizations, which is to substitute the inverse link function for μ. We do that for the negative binomial log-likelihood function.

Given the inverse link function

$$\frac{1}{\alpha(e^{-\theta_i}-1)} = \frac{1}{\alpha\left(\exp\left(-x_i'\beta\right)-1\right)}$$

the right term is substituted for every instance of μ in the negative binomial log-likelihood function. Although equation 8.29 is in exponential family form, we use a restructured version, equation 8.30, to do the substitution. Since the three log-gamma function terms do not include μ, I shall collectively refer to them as C. They are the normalization terms of the re-parameterized negative binomial PDF.

$$\mathcal{L}(\beta; y, \alpha) = \sum_{i=1}^{n}\left\{y_i \ln(1/(\exp\left(-x_i'\beta\right)-1)) - \left(y_i + \frac{1}{\alpha}\right)\right. \\ \left. \times \ln(1 + 1/(\exp\left(-x_i'\beta\right)-1)) + C\right\}$$

Collecting terms and incorporating the log-gamma terms, we have

NB-C LOG-LIKELIHOOD – $x\beta$

$$\mathcal{L}\left(\beta; y, \alpha\right) = \sum_{i=1}^{n}\left\{y_i\left(x_i'\beta\right) + \left(\frac{1}{\alpha}\right)\ln\left(1-\exp(x_i'\beta)\right) + \ln\Gamma\left(y_i + \frac{1}{\alpha}\right)\right. \\ \left. - \ln\Gamma\left(y_i+1\right) - \ln\Gamma\left(\frac{1}{\alpha}\right)\right\} \tag{8.38}$$

The NB-C model will be discussed in a later chapter, where we discuss its modeling capability, particularly with reference to other negative binomial models.

The NB2 log-likelihood substitutes the inverse log-link, $\exp(x)$, in place of μ in equation 8.29. I use equation 8.29 since it is the exponential family form and can also be more easily substituted than the canonical inverse link. The following form, or restructured version, is found in maximum likelihood parameterizations of the NB2 model.

NEGATIVE BINOMIAL (NB2) LOG-LIKELIHOOD – $x\beta$

$$\mathcal{L}(\beta; y, \alpha) = \sum_{i=1}^{n} \left\{ y_i \ln\left(\frac{\alpha \exp(x_i'\beta)}{1+\alpha\exp(x_i'\beta)}\right) - \left(\frac{1}{\alpha}\right)\ln\bigl(1+\alpha\exp(x_i'\beta)\bigr) + \ln\Gamma\left(y_i + \frac{1}{\alpha}\right) - \ln\Gamma(y_i+1) - \ln\Gamma\left(\frac{1}{\alpha}\right)\right\} \quad (8.39)$$

Given this parameterization, the NB2 link is $\ln(\alpha\exp(x'\beta)/(1+\alpha\exp(x'\beta)))$ and the cumulant is $\ln(1+\alpha\exp(x'\beta))/\alpha$. Maximum likelihood estimation of α can be expressed as (Piegorsch, 1990)

$$\mathcal{L}(\alpha; \mu) = \frac{1}{n}\sum_{i=1}^{n}\left\{\frac{\Gamma(y_i+1/\alpha)}{\Gamma(1/\alpha)} + \bar{y}\ln(\mu_i) - \left(\bar{y}+\frac{1}{\alpha}\right)\ln(1-\alpha\mu_i)\right\} \quad (8.40)$$

8.3 Negative binomial distributions

Figures 8.1–8.11 illustrate negative binomial distributions for various values of both the mean and α. Note that, when $\alpha = 1$, all distributions take the form of a geometric distribution, which is the discrete correlate of the continuous negative exponential distribution. Note also that as the mean increases, the probability of a zero decreases, and the more the shape approximates a Gaussian distribution. Again, the negative binomial distribution with $\alpha = 0$ is Poisson.

Range of mean (0.5, 1, 2, 5, 10) for each alpha(0, 0.33, 0.67, 1.0, 1.5, 3.0)
Range of alpha (0, 0.33, 0.67, 1.0, 1.5, 3.0) for each mean (0.5, 1, 2, 5, 10)

The code for Figures 8.1 through 8.13 can be found in *Negative Binomial Regression Extensions*, which can be downloaded from the book's website. Both Stata and R code are provided. R scripts for figures in the book can be found in the COUNT package.

Two sets of distributions that may be of particular interest involve graphs of NB2 distributions with a mean of 10 and α-values of 0, 0.1, 0.3, 0.5 and of 0.6, 0.8, 1, 1.2. An NB2 model with $\alpha = 0$, as you recall, is identical to a Poisson, without considering those cases where α is greater than, but not significantly different from, zero. Remember also that as the Poisson mean gets larger it approaches normality. The negative binomial distribution does not respond to larger mean values in that manner. Note in Figure 8.12 that

TOP TO BOTTOM ON VERTICAL AXIS; MEAN = 0.5, 1, 2, 5, 10

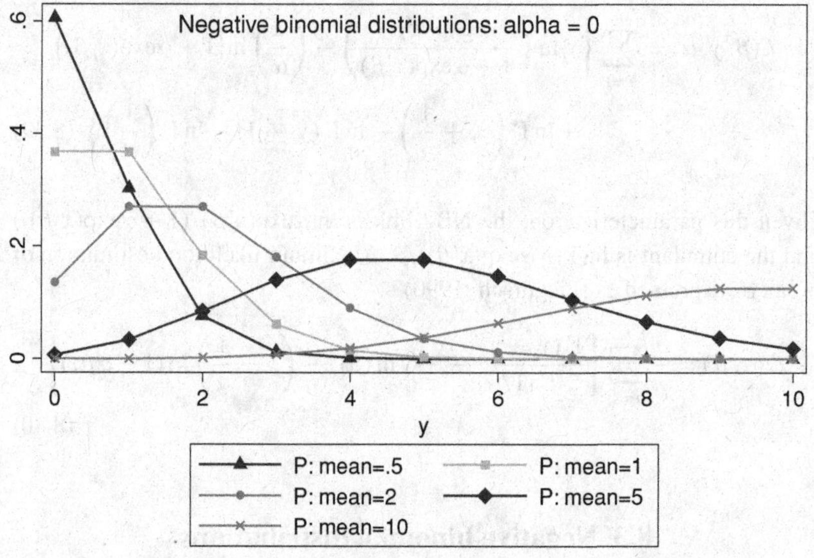

Figure 8.1 Negative binomial distributions: $\alpha = 0$

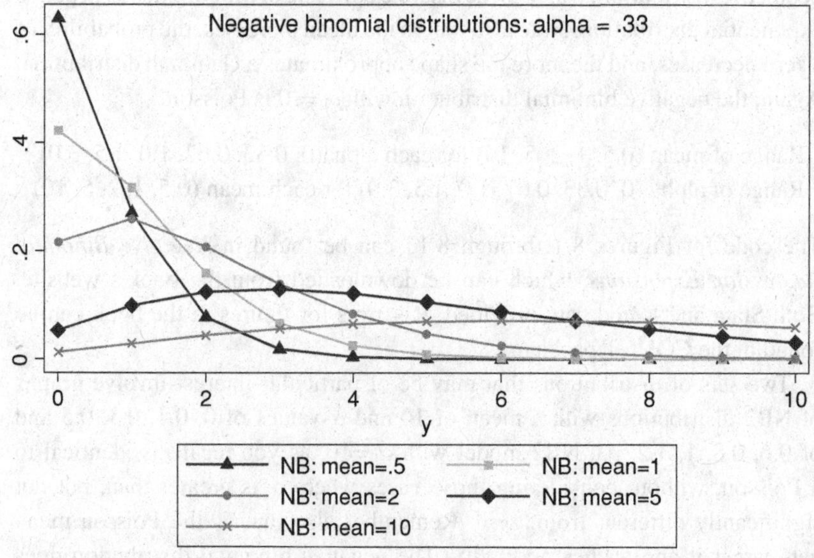

Figure 8.2 Negative binomial distributions: $\alpha = 0.33$

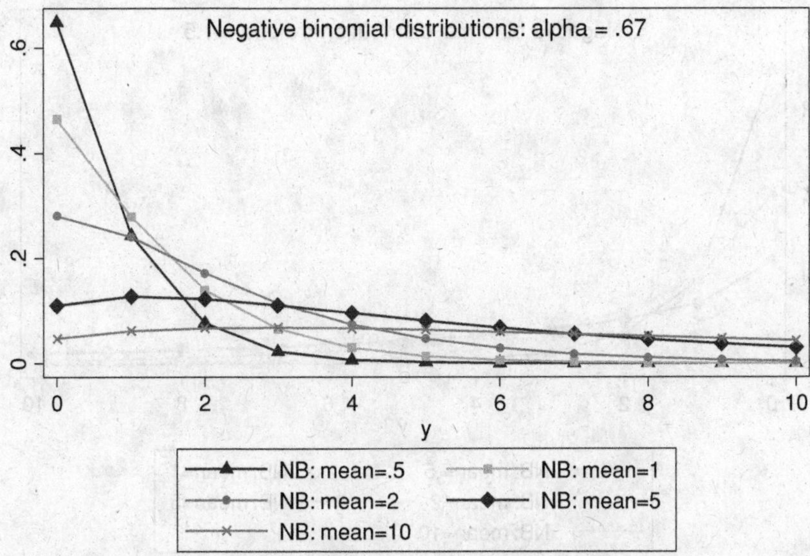

Figure 8.3 Negative binomial distributions: $\alpha = 0.67$

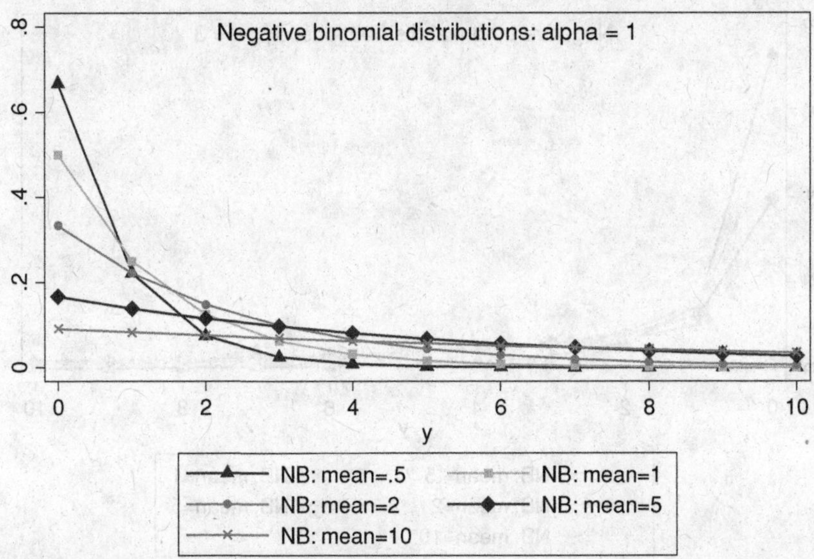

Figure 8.4 Negative binomial distributions: $\alpha = 1$

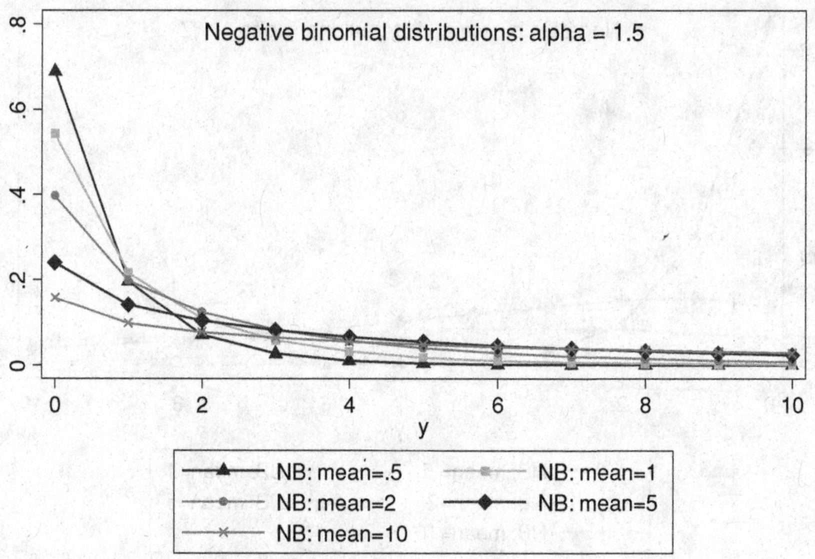

Figure 8.5 Negative binomial distributions: $\alpha = 1.5$

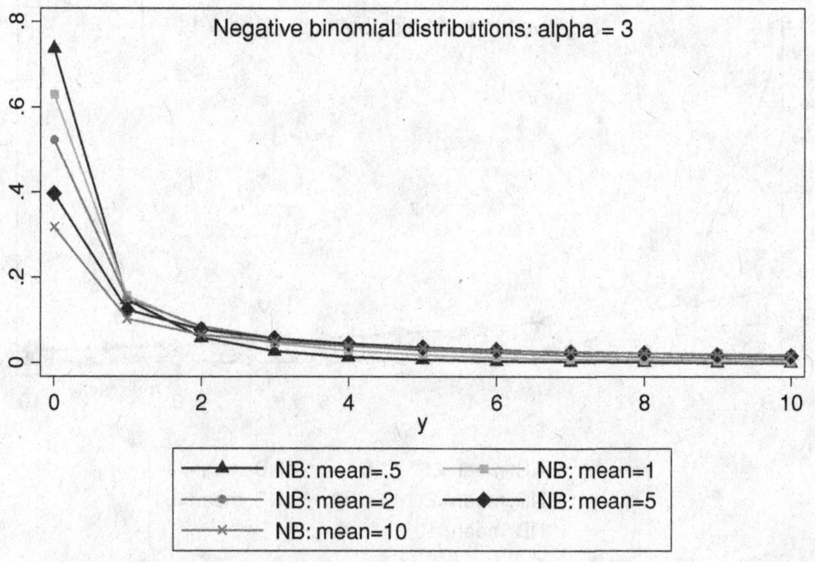

Figure 8.6 Negative binomial distributions: $\alpha = 3.0$

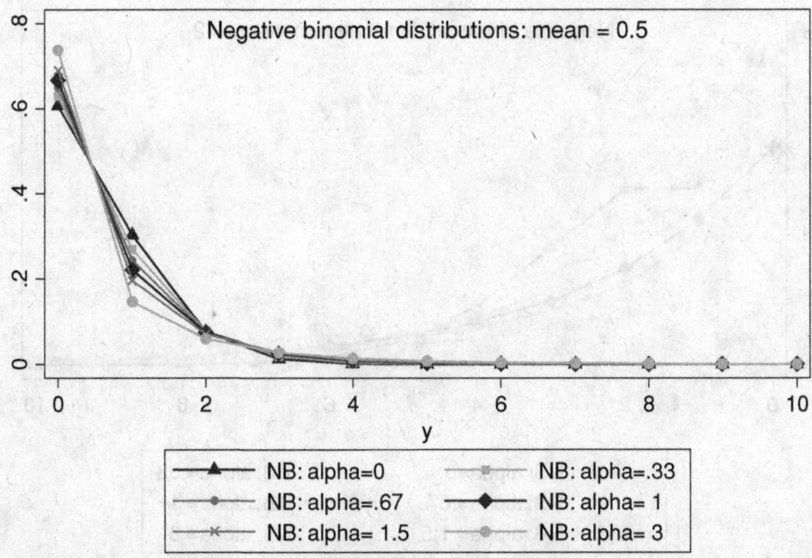

Figure 8.7 Negative binomial distributions: mean = 0.5

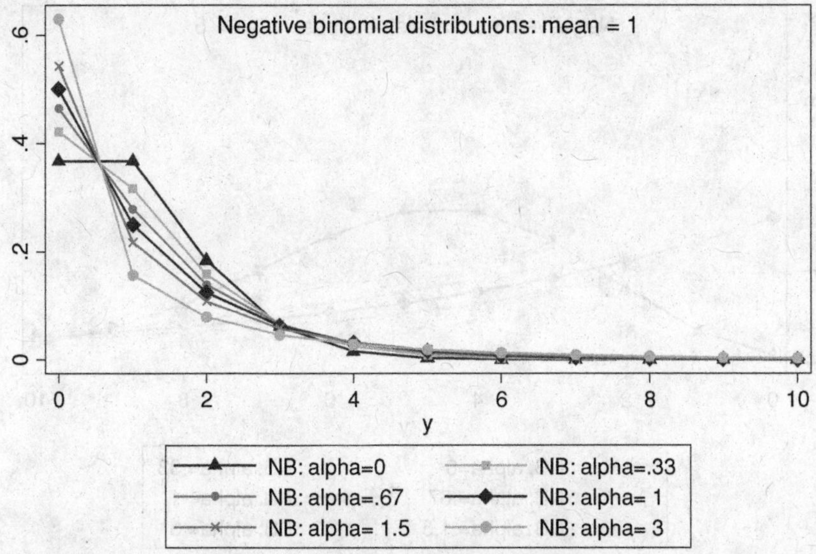

Figure 8.8 Negative binomial distributions: mean = 1

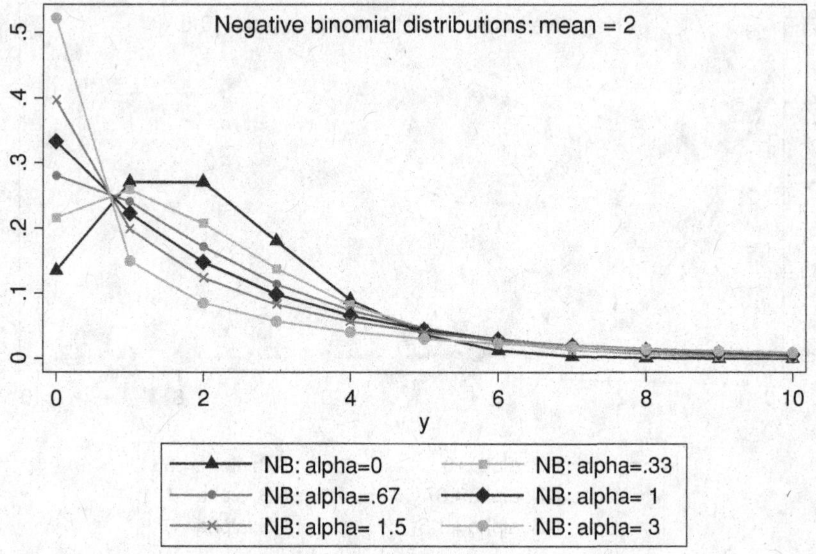

Figure 8.9 Negative binomial distributions: mean = 2

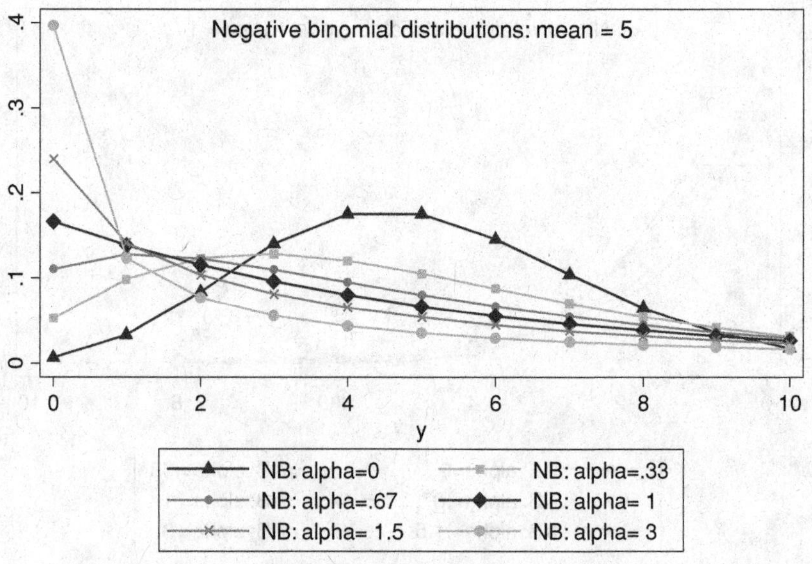

Figure 8.10 Negative binomial distributions: mean = 5

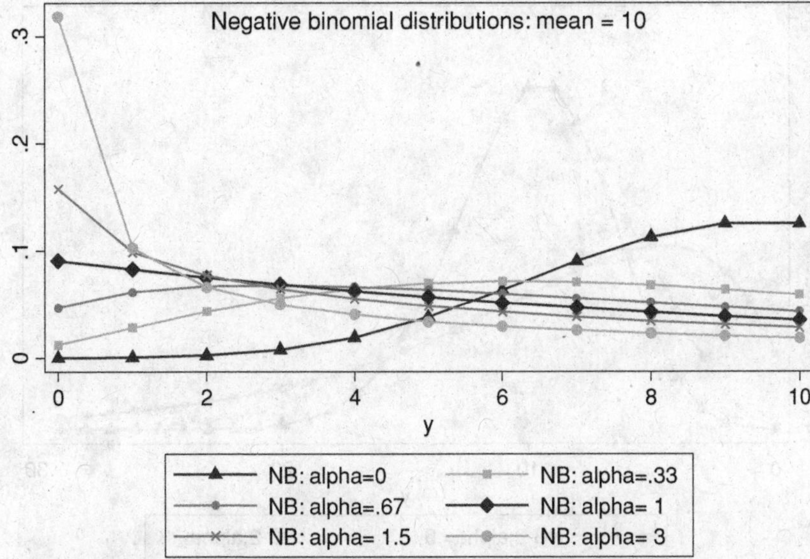

Figure 8.11 Negative binomial distributions: mean = 10

the Poisson distribution is the tallest Gaussian-approaching curve. The negative binomial distributions get lower and shift to the left as α moves from 0.1 to 0.5. What is also the case is that, given a constant mean, as α gets larger, the probability of a zero count gets larger. The mass of the distribution shifts left, raising the probability of obtaining smaller counts. This fact is clear from all of the values of the mean given in the above sequences of distributions.

We also observe from the above sequence of distributions that, for a constant value of α, increasing values of the mean reflect a decrease in the percentage of zero counts.

Figure 8.13 shows the NB2 distributions with a mean of 10 and α-values increasing by 0.2 starting with 0.6. The graph should be thought of as a continuation of Figure 8.12. With higher values of α come higher probabilities of zero. When $\alpha = 1$, we have a smooth negative exponential-like distribution curve. With higher values of α, the curve persists, but with higher probabilities for zero. Values of α greater than 1.2 appear quite similar, with gradually increasing initial heights.

The following inset summarizes the relationship of the negative binomial (NB2) mean and value of α:

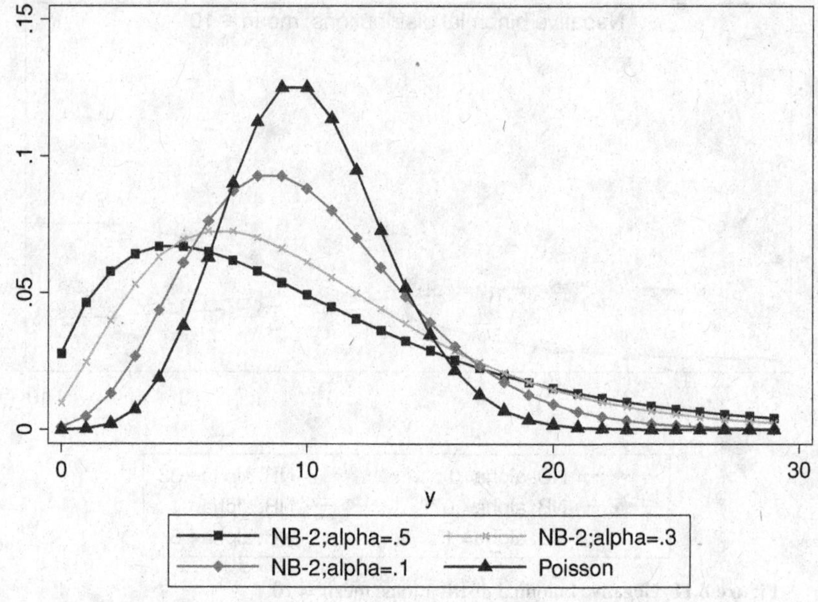

Figure 8.12 Mean = 10; $\alpha = \{0, 0.1, 0.3, 0.5\}$

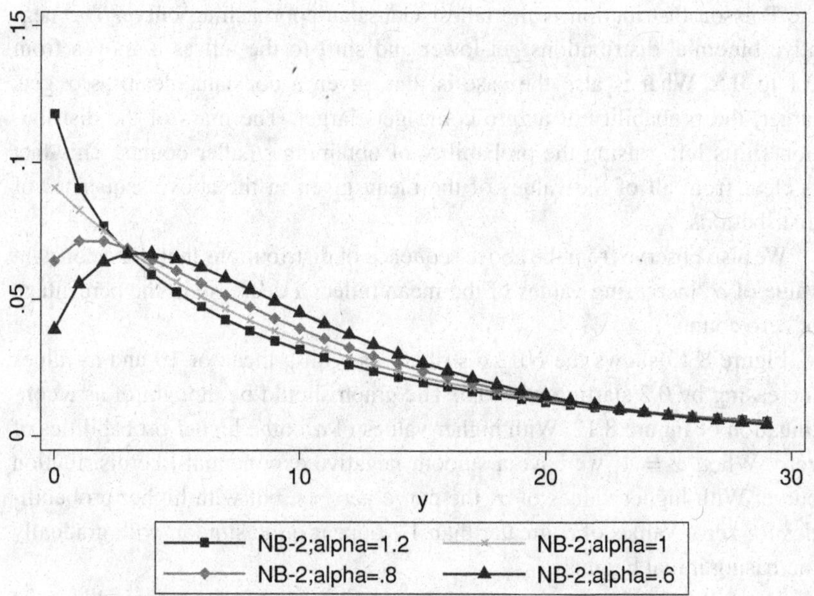

Figure 8.13 Mean = 10; $\alpha = \{0.6, 0.8, 1, 1.2\}$

Constant mean:	$> \alpha$	=	> percentage of zero counts
Constant α:	$> \mu$	=	< percentage of zero counts

In order to demonstrate how the number of zero counts grows with an increasing value of α, I created a series of negative binomial data sets all with 50,000 observations, a seed of 4,321, and specified parameters of $x1 = 0.75$, $x2 = -1.25$, and an intercept of 2.

Synthetic NB2: *Change in percentage of zero counts with differing* α

Alpha	# 0 counts	percent
0	4,156	8.31
0.05	4,324	8.65
0.10	4,472	8.94
0.25	5,187	10.37
0.50	6,408	12.82
0.75	7,904	15.81
1.00	9,331	18.66
1.50	12,131	24.26
2.00	14,496	29.00
4.00	21,547	43.78
6.00	23,981	50.98
8.00	24,517	55.40

The percentage of zero counts range from 8.31 for a Poisson model with $\alpha = 0$ to 55.4% for an NB2 model with $\alpha = 8$.

8.4 Negative binomial algorithms

In Chapter 4 we developed the logic and structure of both the Newton–Raphson type maximum likelihood algorithm as well as the traditional IRLS algorithm that is used in generalized linear models (GLM). The IRLS algorithm may be used for both the NB-C and NB2 parameterizations of the negative binomial model, but not for other parameterizations, e.g. NB1, censored, truncated, zero inflated, hurdle, and finite mixture negative binomial models. NB-C is the parameterization of the negative binomial that is directly derived from the negative binomial PDF. It is not appropriate, however, as a model to accommodate overdispersed Poisson data. It is a count model in its own right, independent of the Poisson model. We discuss the algorithm for this model in the next subsection.

NB2 is the standard form of negative binomial used to estimate data that are Poisson-overdispersed, and is the form of the model which most statisticians understand by negative binomial. NB2 is typically the first model we turn to when we discover that a Poisson model is overdispersed.

As a traditional GLM model, the negative binomial is estimated using the expected information matrix, which is used to calculate model standard errors. The NB-C model also generally employs the expected information matrix but, as a canonical model, the expected and observed information are the same. For the NB2 model, however, they are not the same. An adjustment must be made to the IRLS estimating algorithm so that standard errors are based on observed information. Doing so allows the IRLS estimation of the NB2 model to have identical standard errors to a full maximum likelihood estimation of the data.

In this section we shall discuss the NB-C algorithm, followed by an overview of the IRLS estimating algorithms for NB2 models incorporating both the expected information and the observed information. In the following chapter we put these algorithms to use by estimating both simulated and real data situations.

Table 4.1 provides a schematic for the GLM IRLS algorithm when the standard errors are calculated based on the expected information matrix. Recall the earlier discussion of information matrices in Chapter 4 and Section 8.2.1. Both forms will be examined in terms of coding for a schematic estimation algorithm. Also recall that the NB-C and NB2 models may be estimated using a full maximum likelihood (ML) procedure. Other methods of estimation exist as well. Regardless, we discussed ML estimation in Chapter 4 and earlier in this chapter, and have no reason to elaborate on it further. As earlier mentioned, GLM algorithms exist that call a ML negative binomial to calculate the heterogeneity parameter, α, and then used it as a constant in the GLM algorithm. SAS, Stata, and R provide this capability. Non-GLM full ML commands to estimate the NB2 model currently exist only in Stata, LIMDEP, and MATLAB software.

8.4.1 NB-C: canonical negative binomial

We have derived the IRLS functions required to construct the canonical form of the negative binomial, which we have referred to using the acronym NB-C. When we derived the Poisson–gamma mixture model, the resultant PDF, log-likelihood, cumulant, and so forth, were all appropriate for the traditional negative binomial or NB2 model. It is for this reason that statisticians have

Table 8.2 *Negative binomial: canonical NB-C parameterization*

```
μ = (y - mean(y))/2                   /// initialization
η = -ln(1/αμ + 1)                     /// canonical link
WHILE(ABS(ΔDev) > tolerance {
w = μ + αμ^2                          /// variance
z = η + (y - μ)/w - offset
β = (X'wX)^-1 X'wz
η = X'α + offset                      /// linear predictor
μ = 1/(α(exp(-η) - 1))                /// inverse link
oldDev = Dev
Dev = 2Σ{yln(y/μ) - (y + 1/α)ln((1 + αy)/(1 + αμ))}
ΔDev = Dev - oldDev
}
```

tended to think that this form of the negative binomial is basic. As a mixture model, it is; as a model directly derived from the negative binomial PDF, it is not. This has caused some confusion among those using negative binomial models.

Using the IRLS algorithm displayed in Table 4.1, we can substitute the NB-C equations derived in Section 8.2.2 for the creation of a GLM NB-C algorithm. Table 8.2 presents this algorithm, using the negative binomial deviance function. A likelihood function appropriate for calculating a post-estimation AIC or BIC fit statistic may be obtained using the estimated model values of μ and α.

Comparing Table 8.2 with Table 8.3, we find that the only difference in the traditional GLM NB2 model (8.3) and NB-C (8.2) canonical paramerization are the lines for the link and inverse link functions, i.e. lines 2 and 8 respectively. Code for using GLM to estimate an NB2 model incorporating an observed information matrix requires an adjustment of the weighting, as given in Table 8.6. Regardless, the algorithmic differences between the NB2 and NB-C are minimal, but the interpretation of the estimates differ considerably. Here we emphasize the algorithms, leaving model construction and interpretation until the next chapter.

I have used the deviance as the basis for the convergence criterion. The log-likelihood function could have been used as well. The deviance statistic has been the commonly used form since the initial release of the GLIM software package in the early 1980s. A trend is developing, however, to use the log-likelihood instead (when appropriate), calculating the deviance function after the parameters have been estimated. The deviance statistic can be derived directly from the final values of μ and α, and used as a term in the BIC goodness-of-fit statistic. The deviance is also used as a goodness-of-fit statistic in its own right, with lower values indicating a comparatively preferable model. However, most statisticians now prefer the use of the AIC, BIC, and other model-specific fit statistics to the deviance.

Note that, when modeled as a GLM, the negative binomial heterogeneity parameter α enters the algorithm as a constant. Unlike the traditional NB2 model, the canonical-linked algorithm incorporates α into the link, and as a consequence the inverse link. Having α as a term in the link and inverse link resulted in convergence difficulties with older estimating algorithms. Convergence seemed to be particularly tedious when estimated via Newton–Raphson type maximum likelihood. However, most of the current optimization code is sufficiently sophisticated to handle canonically linked models without difficulty. Maximum likelihood NB-C software accompanies this book, and can be downloaded from the book's website. We use it in the next chapter for comparison with the GLM-based NB-C model.

The canonical model has not been used for any research project of which I am aware. However, this need not be the case. It is a viable parameterization, and is directly derived from the negative binomial probability and likelihood functions. Using it with various example data has at times resulted in a better fit than when modeled with NB2 or NB1. In addition, exact statistics algorithms can be developed for canonically linked GLM-type models. However, exact negative binomial models have not yet been developed.

Cytel's LogXact has the capability of calculating exact *p*-values and confidence intervals for logit and Poisson regression models. The logit link is the canonical form derived from the binomial and Bernoulli distributions. The natural log link is canonical for Poisson. Exact statistics developed for negative binomial models can utilize the canonical form, but not NB2 or other parameterizations.

Examples of NB-C modeling will be given in Section 10.3.

8.4.2 NB2: expected information matrix

To convert the canonically linked GLM model to a non-canonical natural log link, change the initial values of the link and inverse link.

Link: $\eta = \ln(\mu)$
Inverse link: $\mu = \exp(\eta)$

Table 8.3 *Fisher scoring: NB2 expected information matrix*

```
μ = (y - mean(y))/2
η = ln(μ)
WHILE(ABS(ΔDev) > toleration {
w = μ + αμ^2
z = η + (y - μ)/w - offset
β = (X'wX)^-1 X'wz
η = X'β + offset
μ = exp(η)
oldDev = Dev
Dev = 2Σ{yln(y/μ) - (y + 1/α)ln((1 + αy)/(1 + αμ))}
ΔDev = Dev - oldDev
}
```

When we substitute the non-canonical log link into the GLM algorithm, the standard errors change from being generated on the basis of the observed information matrix to the expected information matrix, unless the appropriate adjustment has been made to the software being used. If no adjustment is made, standard errors will differ, particularly in smaller and unbalanced data sets. However, the differences are not usually enough to change the apparent significance level of the model predictors. On the other hand, when those values are near the edge of a pre-assigned level of significance, like 0.05, apparent significance may change. In medium and large data sets this situation is not usually of concern.

The value of using a Fisher scoring method of estimation, which uses the expected information matrix for production of standard errors, is the simplification of the Newton–Raphson steps to a sequence of weighted ordinary least squares model fits. Furthermore, models can be easily changed from one to another within a distributional family by simply changing the link and inverse link functions. GLMs can be interchanged between families by changing the variance and deviance functions, as well as the link and inverse link functions. Thus, all GLMs are specified through four functions. Creating an overall IRLS algorithm for the estimation of GLMs is thus a relatively simple matter, and it affords a great deal of modeling flexibility. The IRLS algorithm can be schematized as shown in Table 8.3.

The problem of using Fisher scoring for modeling the log-negative binomial is the necessity of entering α into the algorithm as a constant. Alternative values of α result in different parameter estimates and standard errors.

Table 8.4 *Negative binomial regression (log-linked) with iterative estimation of α via chi2 dampening*

```
Poisson y < predictors >
Chi2 = Σ(y - μ)^2/μ
Disp = Chi2/df
φ = 1/disp
j = 1
WHILE(ABS(ΔDisp) > tolerance {
oldDisp = Disp
NB y < predictors >, α = φ
Chi2 = Σ{(y - μ)^2/(μ + αμ^2)}
Disp = Chi2/df
φ = Disp*φ
ΔDisp = Disp - oldDisp
j = j + 1
}
```

Aside from employing a method of calling a maximum likelihood outside the GLM algorithm to estimate α, a fairly accurate point estimate of α can be obtained by searching for the value of α which results in the Pearson *chi*2 (χ^2) dispersion statistic approximating 1.0. So doing indicates that Poisson overdispersion has been dampened from the model. This value is also close to that produced using maximum likelihood methods, which directly estimate α as a parameter. Breslow (1984) was the first to develop this idea, but used the deviance as the basis of dampening. Hilbe (1993a) developed an algorithm to iteratively search for the optimal value of α based on the Pearson dispersion. In fact the algorithm allowed the user to select which mechanism they preferred for dampening, but the default was a Pearson–dispersion. The algorithm was made into a SAS macro based on the SAS/STAT GENMOD procedure. It was also implemented into Stata (Hilbe, 1993c) and Xplore software (Hilbe and Turlach, 1995).

The logic of the updating algorithm is simple. The algorithm begins by estimating a Poisson model. The inverse of the Pearson *chi*2 dispersion statistic is calculated, and is given the value of a constant, ϕ; ϕ is equated with α, and entered into the GLM negative binomial model. After estimation of the negative binomial, another *chi*2 dispersion statistic is calculated. This time, however, the value of ϕ is multiplied by the dispersion, resulting in an updated value of ϕ. Convergence is based on minimizing the difference between old and new dispersion statistics. Once convergence is achieved, the value of α is recorded. It is the optimal value of α produced by this dampening algorithm. Table 8.4 schematizes the algorithm.

Again, the algorithm, called **negbin**, estimates the negative binomial (NB2) model, and heterogeneity parameter, α, by iteratively forcing the Pearson *chi*2 dispersion to 1. We use the **medpar** data set to demonstrate how the algorithm works, and to check its accuracy compared with a maximum likelihood model. Note that as the dispersion reduces to 1.0, the model deviance and α statistics converge to approximate the maximum likelihood parameter estimates.

```
. negbin los hmo white type2 type3
1:     Deviance:  3110  Alpha (k):     .1597  Disp:  2.298
2:     Deviance:  1812  Alpha (k):      .367  Disp:  1.274
3:     Deviance:  1512  Alpha (k):     .4677  Disp:  1.048
4:     Deviance:  1458  Alpha (k):     .4902  Disp:  1.008
5:     Deviance:  1449  Alpha (k):     .4942  Disp:  1.001
6:     Deviance:  1447  Alpha (k):     .4949  Disp:      1
7:     Deviance:  1447  Alpha (k):      .495  Disp:      1
8:     Deviance:  1447  Alpha (k):      .495  Disp:      1

                                          No obs.        =     1495
Poisson Dp =       6.26                   Poisson Dev =        8143
Alpha (k)  =       .495                   Neg Bin Dev =        1447
                                          Prob > Chi2 =           0
Chi2          =    1490.01                Deviance       =  1447.139
Prob>chi2     =   .4950578                Prob>chi2      =  .7824694
Dispersion    =   1.000006                Dispersion     =  .9712345

Log negative binomial regression
--------------------------------------------------------------------
   los |    Coef.  Std. Err.     z     P>|z|   [95% Conf. Interval]
-------+------------------------------------------------------------
   hmo | -.0679255  .0555534  -1.22   0.221   -.1768082   .0409573
 white | -.1287343  .0716316  -1.80   0.072   -.2691297   .0116611
 type2 |  .2212414  .0528523   4.19   0.000    .1176529     .32483
 type3 |  .7061285  .0797535   8.85   0.000    .5498145   .8624426
 _cons |  2.309978    .07101  32.53   0.000    2.170801   2.449155
--------------------------------------------------------------------
Loglikelihood = -4800.2524
```

A comparison to a full maximum likelihood model via the GLM algorithm is given in Table 8.5.

It appears that the model parameter estimates and standard errors are close. The deviance, Pearson, and log-likelihood values are close, but they do differ. However, they differ in proportion to the amount that the dispersion statistic

Table 8.5 R: *Medpar NB2 model*

```
rm(list=ls())
# alternate: library(gamlss)
data (medpar)
attach (medpar)
mednb2 <- glm.nb(los ~ hmo+ white + factor(type), data=medpar)
# alternate: mednb2 <- gamlss(los ~white+factor(type),
    data=medpar, family=NBI)
summary(mednb2)
```

```
. glm los hmo white type2 type3, fam(nb ml)

Generalized linear models                  No. of obs      =      1495
Optimization     : ML                      Residual df     =      1490
                                           Scale parameter =         1
Deviance         =  1568.14286              (1/df) Deviance = 1.052445
Pearson          = 1624.538251              (1/df) Pearson  = 1.090294
Variance function: V(u) = u+(.4458)u^2 [Neg. Binomial]
Link function    : g(u) = ln(u)             [Log];
                                           AIC             =  6.424718
Log likelihood   = -4797.476603            BIC             = -9323.581
----------------------------------------------------------------------
     |                OIM
 los |     Coef.   Std. Err.     z    P>|z|   [95% Conf. Interval]
-----+----------------------------------------------------------------
 hmo |  -.0679552  .0532613   -1.28   0.202   -.1723455    .0364351
white | -.1290654  .0685416   -1.88   0.060   -.2634046    .0052737
type2 |   .221249  .0505925    4.37   0.000    .1220894    .3204085
type3 |  .7061588  .0761311    9.28   0.000    .5569446    .8553731
_cons |  2.310279  .0679472   34.00   0.000    2.177105    2.443453
----------------------------------------------------------------------
Note: Negative binomial parameter estimated via ML and treated
      as fixed once estimated.
```

exceeds 1. Recall that the point-estimate algorithm (PE) has the dispersion iteratively converge to one. But the true negative binomial is NB-overdispersed, given the dispersion of 1.09. That is, the negative binomial variance exceeds nominal variance.

Scaling the standard errors of the GLM model results in standard errors that closely approximate the values of the PE model. This demonstrates that the dampening algorithm provides the value of α given a negative binomial with a dispersion of 1.0.

R: *Negative binomial model scaled standard errors*

```
se <- sqrt(diag(vcov(mednb2)))
pr <- sum(residuals(mednb2, type="pearson")^2)
dispersion <- pr/mednb2$df.residual
sse <- sqrt(dispersion)*se
sse
```

```
. glm los hmo white type2 type3, fam(nb ml) scale(x2)

------------------------------------------------------------------------
       |                OIM
   los |     Coef.   Std. Err.      z    P>|z|   [95% Conf. Interval]
-------+----------------------------------------------------------------
   hmo | -.0679552    .055614    -1.22   0.222   -.1769566    .0410461
 white | -.1290654   .0715692    -1.80   0.071   -.2693385    .0112076
 type2 |   .221249   .0528273     4.19   0.000    .1177094    .3247886
 type3 |  .7061588   .0794939     8.88   0.000    .5503536    .8619641
 _cons |  2.310279   .0709486    32.56   0.000    2.171222    2.449336
------------------------------------------------------------------------
```

The Poisson dispersion statistic is shown in PE as 6.26. The GLM negative binomial model appropriately accommodates the Poisson overdispersion, but the data have an extra 9 percent correlation over that provided by a perfectly fitted negative binomial (NB2) model. Note that multiplication of model standard errors by the square root of the Pearson statistic (1624.538) yields scaled standard errors. Moreover, dividing the Pearson statistic of 1624.538 by the Pearson-dispersion of 1.09 yields

```
. di 1624.528065 /1.090287
1490.0004
```

which is identical in value to the Pearson *chi*2 statistic of the PE model. The PE technique is therefore as accurate a point estimation of α as is the fit of the underlying GLM negative binomial model, based on the Pearson dispersion. Using the algorithm on synthetically created NB2 data yields near identical results to a full maximum likelihood or GLM negative binomial model.

8.4.3 NB2: observed information matrix

Finally we parameterize the IRLS log-negative binomial (NB2) by transforming the weight function so that standard errors are based on an observed rather than an expected information matrix. This subject was initially addressed in Chapters 2 and 4.

To effect this conversion, various terms need to be calculated and introduced into the estimating algorithm. Common to both Newton–Raphson and Fisher scoring are:

LINK	$\eta = g(\mu) = \ln(\mu)$
hence	$g'(\mu) = 1/\mu$
and	$g''(\mu) = -1/\mu^2$
VARIANCE	$V(\mu) = \mu + \alpha\mu^2$
hence	$V'(\mu) = 1 + 2\alpha\mu$
and	$V^2 = (\mu + \alpha\mu^2)^2$

DEFINING w

$$\begin{aligned} u &= (y - \mu)g\prime(\mu) = (y - \mu)/\mu \\ w^{-1} &= V\{g'(\mu)\}^2 = (\mu) + \alpha\mu^2)/\mu^2 = (1 + \alpha\mu)/\mu \\ w &= \mu/(\alpha\mu) \end{aligned}$$

DEFINING w_o

The observed information matrix adjusts the weights, w, such that

$$\begin{aligned} w_o &= w + \frac{(y - \mu)\{V(\mu)g''(\mu) + V'(\mu)g'(\mu)\}}{\{V^2 g'(\mu)^3\}} \\ &= \frac{\mu}{(1 + \alpha\mu)} + \frac{(y - \mu)\{[-(\mu + \alpha\mu^2)/\mu^2] + [(1 + 2\alpha\mu)/(\mu]\}}{[(\mu + \alpha\mu^2)/\mu^3]} \\ &= \frac{\mu}{(1 + \alpha\mu)} + (y - \mu)\left\{\frac{\alpha\mu}{(1 + 2\alpha\mu + \alpha^2\mu^2)}\right\} \end{aligned}$$

DEFINING z_o

A revised working variate, z_o, is defined as

$$\begin{aligned} z_o &= \eta + w_o^{-1} wu \\ &= \eta + \frac{(y - \mu)}{\{w_o(1 + \alpha\mu)\}} \end{aligned}$$

with w and w_o representing diagonal weight matrices. Substituting w_o and z_o into the log-negative binomial algorithm provides the proper adjustment.

Table 8.6 *NB2 negative binomial with observed information matrix*

```
μ = (y - mean(y))/2
η = ln(μ)
WHILE (ABS(ΔDev) > toleration {
w = (μ/αμ) + (y - μ){αμ/(1 + 2αμ + α^2 + μ^2)}
z = {η + (y - μ)/(w(1 + αμ))} - offset
β = (X'wX)^-1 X'wz
η = X'β + offset
μ = exp(η)
oldDev = Dev
Dev = 2Σ{yln(y/μ) - (y + 1/α)ln((1 + αy)/(1 + αμ))}
ΔDev = Dev - oldDev
}
```

Table 8.6 schematizes the IRLS estimating algorithm, which is adjusted such that standard errors are produced from the observed information matrix. This IRLS algorithm will produce the same estimates and standard errors as that of a full maximum likelihood algorithm, which also estimates α. However, the GLM IRLS algorithm only allows α to be entered as a constant. This is a rather severe limitation of the GLM approach, although at times it is a valuable capability. As we have observed, however, an updating algorithm synthesized into the IRLS algorithm can approximate the maximum likelihood value of α. Secondly, an algorithm can estimate α outside GLM and insert it into the algorithm as a constant. Third, one may not have GLM software that calls an outside maximum likelihood routine, but may have a traditional GLM program and a separate full maximum likelihood procedure (e.g. Stata's **nbreg** command). A tactic that can be taken is to first estimate α using a full maximum likelihood command or function, then substitute the value of α obtained in estimation into a GLM-based algorithm that is adjusted to calculate an observed information matrix as in Table 8.6. Typically the algorithm schematized in Table 8.6 uses a log-likelihood function as the basis of convergence rather than the deviance. Either method produces the same result.

The values required for creating an observed information matrix based negative binomial were derived earlier in this subsection. A schematic GLM with an observed information matrix was presented in Table 4.4. How it appears for the NB2 model is displayed in Table 8.6.

We next address issues related to the development of well-fitted negative binomial models and how they relate to Poisson and binomial models. We shall limit our discussion to the NB2, or log-negative binomial. Subsequent chapters will deal with alternative parameterizations and extensions to the base model.

8.4.4 NB2: R maximum likelihood function

It is perhaps instructive to display an R NB2 function, which is general in that it will estimate almost any overdispersed Poisson count data submitted to it. As an example, paste the **ml.nb2** function into R's script editor and run. Then load the **medpar** data into memory and declare *type* as a factor variable. Execute the function. Estimates and standard errors are displayed below Table 8.7.

Table 8.7 R: *NB2 maximum likelihood function* ml.nb2

```
ml.nb2 <- function(formula, data, start = NULL, verbose = FALSE) {
 mf <- model.frame(formula, data)
 mt <- attr(mf, "terms")
 y <- model.response(mf, "numeric")
 nb2X <- model.matrix(formula, data = data)
 nb2.reg.ml <- function(b.hat, X, y) {
   a.hat <- b.hat[1]
   xb.hat <- X %*% b.hat[-1]
   mu.hat <- exp(xb.hat)
   r.hat <- 1 / a.hat
   sum(dnbinom(y,
            size = r.hat,
            mu = mu.hat,
            log = TRUE))
   }
if (is.null(start))
  start <- c(0.5, -1, rep(0, ncol(nb2X) - 1))
fit <- optim(start,
          nb2.reg.ml,
          X = nb2X,
          y = y,
          control = list(
            fnscale = -1,
            maxit = 10000),
          hessian = TRUE
          )
if (verbose | fit$convergence > 0) print(fit)
beta.hat <- fit$par
se.beta.hat <- sqrt(diag(solve(-fit$hessian)))
results <- data.frame(Estimate = beta.hat,
                      SE = se.beta.hat,
                      Z = beta.hat / se.beta.hat,
                      LCL = beta.hat - 1.96 * se.beta.hat,
                      UCL = beta.hat + 1.96 * se.beta.hat)
rownames(results) <- c("alpha", colnames(nb2X))
results <- results[c(2:nrow(results), 1),]
return(results)
}
```

```
data(medpar)
medpar$type <- factor(medpar$type)
med.nb2 <- ml.nb2(los ~ hmo + white + type, data = medpar)
med.nb2
               Estimate         SE         Z        LCL         UCL
(Intercept)  2.31214519 0.06794358 34.030372  2.1789758 2.445314604
hmo         -0.06809686 0.05323976 -1.279060 -0.1724468 0.036253069
white       -0.13052184 0.06853619 -1.904422 -0.2648528 0.003809103
type2        0.22049993 0.05056730  4.360524  0.1213880 0.319611832
type3        0.70437929 0.07606068  9.260754  0.5553003 0.853458231
alpha        0.44522693 0.01978011 22.508817  0.4064579 0.483995950
```

Note the close similarity of values to Stata and **glm.nb** output.

Summary

In this chapter we introduced the both the canonical and traditional forms of the negative binomial model. The traditional form of the model is a Poisson–gamma mixture model in which the gamma distribution is used to adjust the Poisson in the presence of overdispersion. The original manner of expressing the negative binomial variance clearly shows this mixture relationship: $\mu + \mu^2/\nu$, where μ is the Poisson variance and μ^2/ν the two-parameter gamma distribution variance. We inverted the gamma scale parameter, ν, to α, producing the negative binomial variance, $\mu + \alpha\mu^2$. This form of the variance provides a direct relationship of α to the amount of overdispersion in the otherwise Poisson model; α is sometimes referred to as simply the overdispersion parameter.

The traditional negative binomial model, as defined above, is many times referred to as the NB2 model, with 2 indicating the degree of the exponential term. We also derived the negative binomial model directly from the negative binomial probability function. This is the normal method used to derive any of the members from the family of generalized linear models.

When the negative binomial is derived from its probability function, the canonical form differs from the Poisson–gamma mixture model version of NB2. We showed the canonical linked algorithm, defining its primary components. We also showed how the canonical form of the model can be amended to the traditional of NB2 form. We do this by converting the canonical form to a log-linked form, which is the same link as the canonical Poisson. We see then that the GLM log-negative binomial is the same as the NB2 model that is traditionally estimated using full maximum likelihood methods. As a non-canonical linked model, however, it must be further amended to allow estimation of standard errors based on the observed rather than expected information matrix. All of the relevant derivations are shown in this chapter.

By equating the traditional Poisson–gamma mixture model parameterization with the amended GLM log-negative binomial model, we find that either version can be used to estimate parameters and standard errors. The only drawback with the GLM version is that the heterogeneity parameter, α, is not estimated, but rather has to be entered into the GLM model as a constant. Algorithms have been developed using point-estimate methods that provide a close approximation of the value of α compared with the value estimated using full maximum likelihood methods. GLM software also exists that calls an outside routine that estimates α and subsequently inserts it into the standard GLM algorithm.

I believe that GLM, with a call to a subroutine estimating α via ML, is the optimal approach to the estimation of NB2. The value of doing this rests with the associated suite of GLM residuals and fit statistics that accompany GLM models. Full maximum likelihood methods typically provide minimal associated fit statistics, and residuals must be created specifically for the model. Using the GLM approach for estimating NB2, as well as NB-C, allows the user to more easily assess model fit. It is also easier to employ an alternative modeling strategy with the data – comparing NB2 with Poisson and varieties of binomial models. We next turn to the task of describing these differences.

9

Negative binomial regression: modeling

In this chapter we describe how count response data can be modeled using the NB2 negative binomial regression. NB2 is the traditional parameterization of the negative binomial model, and is the one with which most statisticians are familiar. For this chapter, then, any reference to negative binomial regression will be to the NB2 model unless otherwise indicated.

9.1 Poisson versus negative binomial

We have earlier stated that, given the direct relationship in the negative binomial variance between α and the fitted value, μ, the model becomes Poisson as the value of α approaches zero. A negative binomial with $\alpha = 0$ will not converge because of division by zero, but values close to zero allow convergence. Where α is close to zero, the model statistics displayed in Poisson output are nearly the same as those of a negative binomial.

The relationship can be observed using simulated data. The code below constructs a 50,000 observation synthetic Poisson model with an intercept value of 2 and parameter values of $x1 = 0.75$ and $x2 = 1.25$. Each predictor is generated as a synthetic random normal variate. The Poisson data are then modeled using a negative binomial where the value of α is estimated prior to the calculation of parameter estimates and associated statistics.

Recall that the simulated data, and model statistics, will change slightly with each run. If α, the negative binomial heterogeneity or dispersion parameter, were exactly 0, the model would not converge. As it is, the 0.16% difference displayed in the simulated Poisson model between the dispersion parameter and the true Poisson dispersion value of 1.0 allows convergence as a negative binomial. The values of the various statistics between the Poisson and negative binomial will vary due to random error only.

Table 9.1 R: *Synthetic Poisson modeled as Poisson and NB2*

```
library(MASS)
nobs <- 50000
x1 <- runif(nobs)
x2 <- runif(nobs)
py <-rpois(nobs, exp(2 + 0.75*x1 - 1.25*x2))
poi9_1 <- glm(py ~ x1 + x2, family=poisson)
summary(poi9_1)
nb9_1 <- glm.nb(py ~ x1 + x2)
summary(nb9_1)
```

```
. set obs 50000
. gen x1 = rnormal()
. gen x2 = rnormal()
. gen py = rpoisson(exp(2 + 0.75*x1 - 1.25*x2))
```

POISSON

```
. glm py x1 x2, fam(poi)

Generalized linear models                   No. of obs        =     50000
Optimization        : ML                    Residual df       =     49997
                                            Scale parameter   =         1
Deviance            =  52084.19216          (1/df) Deviance   =  1.041746
Pearson             =  50074.77135          (1/df) Pearson    =  1.001556
                                   . . .
                                            AIC               =  4.770061
Log likelihood      = -119248.5131          BIC               = -488872.3
-------------------------------------------------------------------------
      |                OIM
   py |    Coef.   Std. Err.      z    P>|z|   [95% Conf. Interval]
------+------------------------------------------------------------------
   x1 |  .7499734  .0009792   765.89   0.000    .7480542   .7518926
   x2 | -1.251626  .0009883 -1266.45   0.000   -1.253563  -1.249689
_cons |  1.998462  .0017292  1155.73   0.000    1.995073   2.001851
-------------------------------------------------------------------------
```

NEGATIVE BINOMIAL (NB2)

```
. glm py x1 x2, fam(nb ml)

Generalized linear models                   No. of obs        =     50000
Optimization        : ML                    Residual df       =     49997
                                            Scale parameter   =         1
```

```
Deviance         =  52052.46267       (1/df) Deviance =  1.041112
Pearson          =  50043.36076       (1/df) Pearson  =  1.000927
Variance function: V(u) = u+(0)u^2  [Neg. Binomial]
Link function    : g(u) = ln(u)     [Log]
                                     AIC              =  4.770059
Log likelihood   = -119248.4687      BIC              =   -488904

------------------------------------------------------------------
      |               OIM
   py |      Coef.   Std. Err.      z    P>|z| [95% Conf. Interval]
------+-----------------------------------------------------------
   x1 |   .7499666  .0009824   763.42  0.000   .7480412   .751892
   x2 |  -1.251636  .0009925 -1261.09  0.000  -1.253582 -1.249691
_cons |   1.998456  .0017318  1153.95  0.000   1.995061   2.00185
------------------------------------------------------------------
Note: Negative binomial parameter estimated via ML and treated
      as fixed once estimated.
```

The negative binomial model displays an estimated value of $\alpha = 0$ `(u + (0)u^2)`. This is due to rounding variation. Values of α closer to a true 0 will not converge. Nevertheless, it is clear that these are near identical models.

The negative binomial model will always have a variance larger than that of the Poisson. With the Poisson mean and variance being equal, μ, and the negative binomial mean and variance as μ and $\mu + \alpha\mu^2$ respectively, the difference in values between the negative binomial mean and variance varies based on the specific relationship of α and μ. The larger the combined values of α and μ, the greater is the percentage difference between μ and $\mu + \alpha\mu^2$. Table 9.2 displays the variance given various values of both α and μ. These theoretical values may be compared with the variance produced by the model which is used to access overdispersion.

Table 9.2 *Negative binomial mean–variance relationship*

	MEAN						
ALPHA	.1	.5	1	2	4	10	100
.001	.10001	.50025	1.001	2.004	4.016	10.1	110
.01	.1001	.5025	1.01	2.04	4.16	11	200
.1	.101	.525	1.1	2.4	5.6	20	1100
.5	.105	.625	1.5	4	12	60	5100
1	.11	.75	2	6	20	100	10100
2	.12	1	3	10	36	210	20100
4	.14	1.5	5	18	68	410	40100

All values of the variance are greater than the mean, with the exception of when both the mean and α are very low. With the mean and α having a value of 0.001, the mean and variance are nearly identical. Nevertheless, the variance is still greater than the mean, although rounding may not make this clear.

```
. di .001+.001*.001*.001
.001
```

A boundary likelihood ratio test is a preferred manner of testing whether the data being modeled by a negative binomial are in fact Poisson. See Section 7.4.2 for details.

Run the code used to create a synthetic Poisson model that we employed in the previous example. Being pseudo-random, the parameter estimates will differ slightly. We shall then use Stata's **nbreg** command, a full maximum likelihood procedure, to estimate results. A likelihood ratio test is displayed under the table of estimates, with the null hypothesis being $\alpha = 0$.

```
. nbreg py x1 x2

Negative binomial regression                Number of obs   =       50000
                                            LR chi2(2)      =   155697.16
Dispersion      = mean                      Prob > chi2     =      0.0000
Log likelihood = -119532.37                 Pseudo R2       =      0.3944
------------------------------------------------------------------------------
      py |      Coef.   Std. Err.      z    P>|z|     [95% Conf. Interval]
---------+--------------------------------------------------------------------
      x1 |   .7492835   .0009454   792.55   0.000     .7474305    .7511365
      x2 |  -1.250556   .0009921 -1260.52   0.000    -1.252501   -1.248612
   _cons |   1.999388   .0017058  1172.14   0.000     1.996045    2.002732
---------+--------------------------------------------------------------------
/lnalpha |  -9.211381   .9551375                     -11.08342   -7.339346
---------+--------------------------------------------------------------------
   alpha |   .0000999   .0000954                      .0000154    .0006495
------------------------------------------------------------------------------
Likelihood-ratio test of alpha=0:   chibar2(01) =     1.44
Prob>=chibar2 = 0.115
```

The likelihood ratio test *chi*2 value of 1.44 is not significant at the 0.05 level, indicating that α is not significantly different from 0. Normally, a *chi*2 value of 3.84 with 1 degree of freedom is the criteria for a *p*-value of 0.05. However, as discussed in Section 7.4.2, the statistic is halved. This results in a criterion value of 2.705 being the cut-point determining significance at the 0.05 level. The *p*-value based on the model is calculated as:

```
. di chi2tail(1,1.44)/2
.11506967
```

We now turn to the construction of synthetic negative binomial models. Their generation is not as straightforward as with Poisson data.

9.2 Synthetic negative binomial

In Section 6.2 we detailed how to construct and manipulate synthetic Poisson models, including those with an offset. In this section we do the same for synthetic negative binomial (NB2) models. Synthetic NB-C and NB1 models will be developed in sections specifically devoted to them in Chapter 10.

Recall that the NB2 model is typically derived as a Poisson–gamma mixture. We exploit this derivation in the construction of synthetic NB2 models. Another term that must be introduced which was not included in the Poisson model is the heterogeneity parameter, α. The construction of synthetic NB2 data is therefor rather complicated in comparison with Poisson models.

R and Stata code to construct synthetic NB2 data are provided in Tables 9.3 and 9.4 respectively. The model has an intercept of 2, and two explanatory predictors with coefficients of 0.75 and -1.25; α is specified with a value of 0.5. The R **nb2_syn** function is generic, as was the synthetic Poisson model of Table 6.4. With the function in memory, create *sim.data* (below single line) using the coefficients desired, in this case an intercept of 2 and $x1$, $x2$ values of 0.75 and -1.25 respectively. We set *alpha* at 0.5. Default values are assigned at the top of the function.

Table 9.3 R: *Synthetic negative binomial (NB2): generic function*

```
require(MASS)
nb2_syn  <- function(nobs = 50000, off = 0,
                     alpha = 1,
                     xv = c(1, 0.75, -1.5))  {
  p <- length(xv) - 1
  X <- cbind(1, matrix(rnorm(nobs * p), ncol = p))
  xb <- X %*% xv
  a <- alpha
  ia <- 1/a
  exb <- exp(xb + off)
  xg <- rgamma(nobs, a, a, ia)
  xbg <-exb*xg
  nby <- rpois(nobs, xbg)
  out <- data.frame(cbind(nby, X[,-1]))
  names(out) <- c("nby", paste("x", 1:p, sep=""))
  return(out)
```

```
}
# -------------------------------------------------
sim.data <- nb2_syn(nobs = 50000, alpha=.5, xv = c(2, .75,
-1.25))
mynb <- glm.nb(nby ~ ., data = sim.data)
summary(mynb)
```

Table 9.4 *Synthetic negative binomial (NB2)*

```
* nb2_rng.do
* x1=.75, x2=-1.25, _cons=2, alpha=0.5
clear
set obs 50000
set seed 4321
gen x1 = rnormal()
gen x2 = rnormal()
gen xb = 2 + 0.75*x1 - 1.25*x2
gen a = .5
gen ia = 1/a
gen exb = exp(xb)
gen xg = rgamma(ia, a)
gen xbg = exb * xg
gen nby = rpoisson(xbg)
glm nby x1 x2, nolog fam(nb ml)
```

Note the creation of a random gamma variate, *xg*, which is multiplied by the exponentiated Poisson linear predictor, *exb*, to produce an exponentiated negative binomial linear predictor, *xbg*. This value is given as the mixture mean to the Poisson random number generator which we used in Chapters 6 and 7. A model of the synthetic negative binomial data is given below.

```
. glm nby x1 x2, fam(nb ml)

Generalized linear models              No. of obs      =     50000
Optimization     : ML                  Residual df     =     49997
                                       Scale parameter =         1
Deviance         =  54197.68583        (1/df) Deviance =  1.084019
Pearson          =  50137.35328        (1/df) Pearson  =  1.002807
Variance function: V(u) = u+(.4962)u^2  [Neg. Binomial]
Link function    : g(u) = ln(u)         [Log]
                                       AIC             =  6.153603
Log likelihood   = -153837.0815        BIC             = -486758.8
```

```
------------------------------------------------------------------------------
             |                 OIM
         nby |      Coef.   Std. Err.      z    P>|z|     [95% Conf. Interval]
-------------+----------------------------------------------------------------
          x1 |   .7508486   .0038458   195.24   0.000      .743311    .7583862
          x2 |  -1.246021   .0040671  -306.37   0.000    -1.253992   -1.238049
       _cons |   2.004094   .0039238   510.76   0.000     1.996404    2.011784
------------------------------------------------------------------------------
Note: Negative binomial parameter estimated via ML and treated
      as fixed once estimated.
```

The values of the coefficients and of α approximate the values specified in the algorithm. These values may of course be altered by the user. If these same data were modeled using a Poisson regression, the resultant dispersion statistic would be 11.5. The data are therefore Poisson overdispersed, but NB2 equidispersed, as we expect.

A Monte Carlo simulation may be performed to verify that the model is not biased and that the true model parameters inherent in the data are as we specified. The method runs a given number of randomly generated models, saving important model statistics in a separate file. At the conclusion we check the mean value of each statistic of interest, comparing them with the values that were specified at the outset.

Performing a Monte Carlo simulation of the NB2 model is similar to the Monte Carlo simulation we performed for the Poisson model in Section 6.2. However, we add α to the statistics that are saved and summarized at the conclusion of the procedure. If a researcher does not have GLM software that estimates α using a maximum likelihood subroutine, but does have access to a standalone maximum likelihood procedure or function, Monte Carlo can proceed using the following method. During each iteration of the Monte Carlo algorithm, the value of α is estimated using a separate procedure. This value is then passed as a constant to the GLM command or function that provides estimation of the dispersion statistics as well as parameter estimates of the NB2 model. The value of α must be stored in a manner suitable for the software being used, and is entered into the GLM algorithm using its stored designation. For example, in Stata, if the **nbreg** command is used to calculate α, it can be entered into the subsequent **glm** command by use of the option, `fam(nb `= e(alpha)')`. If it is saved using **glm**, then the Monte Carlo algorithm uses the **glm** storage name for α, '*e(a)*'. It is easier, however, to use the Stata **glm** command with the *fam(nb ml)* option. The estimated value of α is given as *e(a)*.

In the Monte Carlo algorithm given in Table 9.6, the statistics for each parameter estimate and intercept is stored, together with both dispersion statistics and α. Note how the statistics we wish to use in the Monte Carlo simulation program are saved.The code in Table 9.6 is called from Stata's **simulate** command, which traverses code listed in the program, saving what has been designated at the bottom in a separate file.

The R script in Table 9.5 is a 100-iteration Monte Carlo simulation of the NB2 coefficients, Pearson-dispersion statistic, and α. It takes a few minutes to run. Output for each of these statistics, and the coefficient vector, is displayed directly underneath the three lines at the bottom of the code. The order corresponds to the list created on the final line of the simulation function. In order to speed calculation, one may reduce the number of iterations, e.g. to 10. To do so, replace 100 with 10 in the **replicate** function and in the last line of code.

Table 9.5 R: *Synthetic Monte Carlo negative binomial*

```
library(MASS)
mysim <- function()
{
 nobs <- 50000
 x1 <-runif(nobs)
 x2 <-runif(nobs)
 xb <- 2 + .75*x1 - 1.25*x2
  a <- .5
ia <- 1/.5
exb <- exp(xb)
xg <- rgamma(nobs, a, a, ia)
xbg <-exb*xg
nby <- rpois(nobs, xbg)
nbsim <-glm.nb(nby ~ x1 + x2)
   alpha <- nbsim$theta
   pr <- sum(residuals(nbsim, type="pearson")^2)
   prdisp <- pr/nbsim$df.residual
   beta <- nbsim$coef
   list(alpha,prdisp,beta)
}
B <- replicate(100, mysim())
mean(unlist(B[1,]))
mean(unlist(B[2,]))
apply(matrix(unlist(B[3,]),3,100),1,mean)
```

Table 9.6 Stata: *Synthetic Monte Carlo negative binomial (NB2)*

```
* nb2ml_sim.ado
* x1=.75, x2=-1.25, _cons=2, alpha=0.5
program define nb2ml_sim, rclass
version 10
clear
set obs 50000
gen x1 = rnormal()
gen x2 = rnormal()
gen xb = 2 + 0.75*x1 - 1.25*x2
gen a = .5
gen ia = 1/a
gen exb = exp(xb)
gen xg = rgamma(ia, a)
gen xbg = exb * xg
gen nby = rpoisson(xbg)
glm nby x1 x2, nolog fam(nb ml)
return scalar sx1 = _b[x1]
return scalar sx2 = _b[x2]
return scalar sxc = _b[_cons]
return scalar pd  = e(dispers_p)
return scalar dd  = e(dispers_s)
return scalar _a = `e(a)'
end
```

The command below iterates 100 times, calling the code stored in the program named **nb2ml_sim** at each iteration. The results are summarized afterwards.

```
* simulate mx1= r(sx1) mx2= r(sx2) mxc= r(sxc) pdis= r(pd)
  ddis = r(dd) alpha= r(_a), reps(100): nb2ml_sim

. su

    Variable |       Obs        Mean    Std. Dev.       Min        Max
-------------+--------------------------------------------------------
         mx1 |       100    .7493749    .0036255   .7400422   .7596162
         mx2 |       100   -1.249798    .0036624  -1.257861  -1.242815
         mxc |       100    2.000098    .0038215   1.989675   2.009035
        pdis |       100    1.000225    .0052843    .980517   1.011739
        ddis |       100    1.083838     .001822   1.079629   1.089549
-------------+--------------------------------------------------------
       alpha |       100    .4991269    .0042018   .4895445    .509799
```

The values we specified in the algorithm are very closely matched in the simulation. If we use 500 iterations the values would be even closer to the true parameter value specified in the algorithm. We also reinforce the fact that extra-dispersion is identified using the Pearson dispersion statistic, *pdis*.

It should be emphasized that synthetic models do not require that coefficients for predictors have random uniform or normal values. We could have constructed a synthetic model with binary or categorical predictors just as well. An example will clarify.

We specify two random categorical variables based on the approximate percentage of observations in each level. Given 50,000 observations, $x1$ will be defined to have three levels, the first level having 50% of the observations, the second level the next 30%, and the third the final 20%. A second binary predictor is defined with 60% of the observations in level 0 and 40% in level 1. $x1$ is further factored into three indicator levels, with $x1\_1$ as the reference.

The parameter values are assigned as follows: intercept = 1; $x1\_2 = 2$; $x1\_3 = 3$; $x2 = -2.5$. The code provides synthetic data for both Poisson and negative binomial (NB2) data. Both model results are displayed; header statistics are not provided for the Poisson model.

Table 9.7 R: *Synthetic Poisson and NB2 with categorical predictors*

```
library(MASS)
library(stats)
nobs <- 1:50000
x1 <- ifelse(nobs<25000   , 1, NA)
x1 <- ifelse( nobs>= 25000 & nobs<40000 , 2, x1)
x1 <- ifelse( nobs>=40000 , 3, x1)
x2 <- ifelse( nobs<=30000, 0, NA)
x2 <- ifelse( nobs> 30000, 1, x2)
cntDta <-data.frame(nobs,x1,x2)
x1_2 <- ifelse(x1==2, 1, 0)
x1_3 <- ifelse(x1==3, 1, 0)
x2_2 <- ifelse(x2==1, 1, 0)
xb <- 1 + 2*x1_2 + 3*x1_3 - 2.5*x2_2
a  <- .5
ia <- 1/.5
exb <- exp(xb)
py <- rpois(nobs, exb)
xg <-  rgamma(nobs, a, a, ia)
xbg <- exb*xg
nby <- rpois(nobs, xbg)
poi <-glm(py ~ x1_2 + x1_3 + x2_2, family=poisson)
summary(poi)
nb2 <- glm.nb(nby ~ x1_2 + x1_3 + x2_2)
summary(nb2)
```

Table 9.8 *Synthetic Poisson and NB2 with categorical predictors*

```
. clear
. set obs 50000
. set seed 12345
. gen byte x1 = irecode(runiform(), 0, 0.5, 0.8, 1)
. gen byte x2 = irecode(runiform(),  0.6, 1)
. tab x1, gen(x1)
. gen py = rpoisson(exp(1 + 2*x1_2 + 3*x1_3 - 2.5*x2))
. gen a = .5
. gen ia = 1/.5
. gen xg = rgamma(ia, a)
. gen xbg = exb * xg
. gen nby = rpoisson(xbg)        /// NB2 variates
```

Tabulations of the two predictors are given below. The percentages very closely approximate what was specified.

R

```
t1 <- table(x1)
t1
cat(format(t1/sum(t1), digits=3), '\n')
t2 <- table(x2)
t2
cat(format(t2/sum(t2), digits=3), '\n')
```

```
. tab x1

         x1 |      Freq.     Percent        Cum.
------------+-----------------------------------
          1 |     25,011       50.02       50.02
          2 |     14,895       29.79       79.81
          3 |     10,094       20.19      100.00
------------+-----------------------------------
      Total |     50,000      100.00

. tab x2

         x2 |      Freq.     Percent        Cum.
------------+-----------------------------------
          0 |     29,869       59.74       59.74
          1 |     20,131       40.26      100.00
------------+-----------------------------------
      Total |     50,000      100.00
```

Next, factor the categorical predictor, $x1$, into three indicator variables.

POISSON

```
. tab x1, gen(x1_)
. glm py x1_2 x1_3 x2, fam(poi) nohead

------------------------------------------------------------------------
       |                OIM
    py |     Coef.   Std. Err.      z     P>|z|    [95% Conf. Interval]
-------+----------------------------------------------------------------
  x1_2 |  2.000533   .0053642    372.94   0.000     1.99002    2.011047
  x1_3 |  3.002113   .0051354    584.59   0.000    2.992048    3.012178
    x2 | -2.500408   .0059163   -422.63   0.000   -2.512003   -2.488812
 _cons |  .9981614   .0048582    205.46   0.000    .9886394    1.007683
------------------------------------------------------------------------
```

NEGATIVE BINOMIAL – NB2

```
. glm nby x1_2 x1_3 x2, fam(nb ml)

Generalized linear models                  No. of obs      =      50000
Optimization     : ML                      Residual df     =      49996
                                           Scale parameter =          1
Deviance         =  50978.58711            (1/df) Deviance =   1.019653
Pearson          =  49752.14355            (1/df) Pearson  =   .9951225
Variance function: V(u) = u+(.4996)u^2 [Neg. Binomial]
Link function    : g(u) = ln(u)            [Log]
                                           AIC             =   4.918204
Log likelihood   = -122951.0991            BIC             =    -489967
------------------------------------------------------------------------
       |                 OIM
   nby |     Coef.    Std. Err.     z     P>|z|    [95% Conf. Interval]
-------+----------------------------------------------------------------
  x1_2 |  2.002228   .0100081    200.06   0.000    1.982613    2.021844
  x1_3 |  3.001993   .0107295    279.79   0.000    2.980963    3.023022
    x2 | -2.482167   .0101532   -244.47   0.000   -2.502067   -2.462267
 _cons |  .9954278   .0073059    136.25   0.000    .9811085    1.009747
------------------------------------------------------------------------
Note: Negative binomial parameter estimated via ML and treated as
      fixed once estimated.
```

We may construct a categorical predictor where the levels are defined by counts, i.e. each level has a specified number of observations. Suppose we wish to create a categorical variable with 25,000 observations in the first level, 15,000 observations in the second level, and 10,000 in the third. The code is

```
. set obs 50000   /* if not already declared */
. gen x1=_n
. replace x1 = inrange(_n, 1, 25000)*1 + ///
               inrange(_n, 25001,40000)*2 + ///
               inrange(_n,40001,50000)*3
```

```
. tab x1, gen(x1_)
         x1 |      Freq.     Percent        Cum.
------------+-----------------------------------
          1 |     25,000       50.00       50.00
          2 |     15,000       30.00       80.00
          3 |     10,000       20.00      100.00
------------+-----------------------------------
      Total |     50,000      100.00
```

Randomly generated continuous, binary, and categorical predictors may be used when creating synthetic data for modeling. Most statisticians use continuous uniform or normal variates since it is easier.

It is important to mention that rate parameterizations of synthetic negative binomial models are constructed as were synthetic rate Poisson models. The key is to provide for a defined offset. Code for constructing the synthetic rate Poisson data and model is given in Table 6.19. We amend that code to produce a rate negative binomial model with an intercept specified at 0.7 and three predictors with coefficients specified as 1.2, −0.8, and 0.2 respectively. An offset is provided that generates values of 100, 200, . . . , 500 for each 10,000 observations. We can interpret the offset as an area or by a time period.

Table 9.9 R: *Synthetic NB2 with offset*

```
require(MASS)
nb2_syn  <- function(nobs = 50000, off = 0,
                      alpha = 1,
                      xv = c(1, 0.75, -1.5))  {
  p <- length(xv) - 1
  X <- cbind(1, matrix(rnorm(nobs * p), ncol = p))
  xb <- X %*% xv
  a <- alpha
  ia <- 1/a
  exb <- exp(xb + off)
  xg <- rgamma(nobs, a, a, ia)
  xbg <-exb*xg
  nby <- rpois(nobs, xbg)
  out <- data.frame(cbind(nby, X[,-1]))
  names(out) <- c("nby", paste("x", 1:p, sep=""))
  return(out)
}
# ---------------------------------------------------
```

```
off <- rep(1:5, each=10000, times=1)*100  # offset
loff <- log(off)                          # log of offset
sim.data <- nb2_syn(nobs = 50000, off = loff, alpha = .5,
      xv = c( 0.7, 1.2, -0.8, 0.2 ))
summary(glm.nb(nby ~ . + loff, data = sim.data))
```

Table 9.10 *Synthetic negative binomial with offset*

```
* nb2o_rng.do
clear
set obs 50000
set seed 7785
gen off = 100+100*int((_n-1)/10000)
gen loff = ln(off)
gen x1 = rnormal()
gen x2 = rnormal()
gen x3 = rnormal()
gen xb = .7 + 1.2*x1 -0.8*x2 + 0.2*x3 + loff
gen a = .3
gen ia= 1/a
gen exb = exp(xb)
gen xg = rgamma(ia, a)
gen xbg = exb * xg
gen nby = rpoisson(xbg)
glm nby x1 x2 x3, off(loff) fam(nb ml)
```

The results of the model are given as

```
Generalized linear models                No. of obs      =     50000
Optimization     : ML                    Residual df     =     49996
                                         Scale parameter =         1
Deviance         =  52478.29297          (1/df) Deviance =   1.04965
Pearson          =  49683.33408          (1/df) Pearson  =  .9937462

Variance function: V(u) = u+(.2984)u^2            [Neg. Binomial]
Link function    : g(u) = ln(u)                   [Log]
                                         AIC             =  13.95852
Log likelihood   = -348958.9907          BIC             = -488467.3
```

```
------------------------------------------------------------------------
             |                 OIM
         nby |      Coef.   Std. Err.      z    P>|z|   [95% Conf. Interval]
-------------+----------------------------------------------------------
          x1 |   1.198298    .002492   480.87   0.000    1.193414   1.203182
          x2 |  -.7939464   .0024874  -319.18   0.000   -.7988217  -.7890712
          x3 |   .2000177   .0024602    81.30   0.000    .1951959   .2048396
       _cons |   .7041177   .0024666   285.47   0.000    .6992834   .7089521
        loff |   (offset)
------------------------------------------------------------------------
Note: Negative binomial parameter estimated via ML and treated
      as fixed once
```

Another method of generating synthetic negative binomial data, which is then submitted to maximum likelihood estimation, with asymptotic standard errors being abstracted from the inverse Hessian, is given in Table 9.11. The algorithm is assigned parameter values of 5 (intercept), 2, 3, and 0.5. The maximum likelihood results follow the table.

Table 9.11 R: *Synthetic NB2 data and model*

```
set.seed(85132)
b <- c(5, 2, 3, 0.5)                  ## Population parameters
n <- 10000
X <- cbind(rlnorm(n), rlnorm(n)) ## Design matrix
y <- rnbinom(n = n,                   ## Choice of parameterization
           mu = b[1] + b[2] * X[,1],
           size = b[3] + b[4] * X[,2])
nb.reg.ml <- function(b.hat, X, y) {  ## JCLL
  sum(dnbinom(y,
            mu = b.hat[1] + b.hat[2] * X[,1],
            size = b.hat[3] + b.hat[4] * X[,2],
            log = TRUE))
}
p.0 <- c(1,1,1,1)                     ## initial estimates
fit <- optim(p.0,                     ## Maximize the JCLL
           nb.reg.ml,
           X = X,
           y = y,
           control = list(fnscale = -1),
           hessian = TRUE
           )
stderr <- sqrt(diag(solve(-fit$hessian))) ## Asymptotic SEs
nbresults <- data.frame(fit$par, stderr)
nbresults
```

```
nbresults
     fit.par      stderr
1 4.9705908 0.06270555
2 1.9829566 0.04401998
3 3.0486320 0.11324624
4 0.5405919 0.08001268
```

9.3 Marginal effects and discrete change

Marginal effects for negative binomial models NB2 and NB1 are calculated and interpreted in the same was as shown in Section 6.6. For continuous variables, marginal effects at the mean are calculated as

MARGINAL EFFECT AT MEAN

$$\exp\left(x_i'\beta_k\right)\beta_k$$

while average marginal effects are based on

AVERAGE MARGINAL EFFECT

$$\hat{\beta}_k \bar{y}$$

For binary and indicator levels of categorical predictors discrete change is determined by

$$\frac{\Delta \Pr(y_i|(x_i = 1|x_i = 0))}{\Delta x_k}$$

As described earlier for Poisson models, marginal effects depend on the value of x, as well as the values of the other model predictors. Formally, the marginal effect of a predictor is the expected instantaneous rate of change in the response as a function of the change in the specified predictor, maintaining the values of the other predictors at some constant value. That is, a marginal effect is calculated on a single predictor, with other predictors typically held to a specified constant value. The effect provides the change in the probability of y that will be produced given a 1-unit change in x_k, based either on the marginal effects being taken at the sample means of the model predictors (ME at means), or on the marginal effects taken at the average of the predicted counts in the model (Average ME).

Examples were given in Section 6.2 that are fully applicable here. Given the mean value of the Poisson, NB1 and NB2 distributions as $\exp(x'\beta)$, the definition and interpretations employ identical logic. I will give a brief demonstration of the similarity of Poisson and negative binomial marginal effects.

We shall use the same German data set as before, **rwm5yr**, but keeping only data for the year 1984, and modeling the number of visits a patient makes to a doctor during the year on their working status (*outwork* = 1 represents patients who are out of work) and *age* (ages range almost continuously from 25 to 64). To save space, regression results will not be displayed, only the marginal effect of *age* and discrete change on *outwork*. It is expected that their values will be somewhat different from when all fives years were included in the model data.

MARGINAL EFFECT AT MEAN: AGE

R: *Marginal effects*

```
data(rwm5yr)
attach(rwm5yr)
rwm <- rwm5yr
rwm1 <- glm.nb(docvis ~ outwork + age, data=rwm)
meanout <- mean(rwm$outwork)
meanage <- mean(rwm$age)
xb <- coef(rwm1)[1] + coef(rwm1)[2]*meanout + coef(rwm1)[3]*meanage
dfdxb <- exp(xb) * coef(rwm1)[3]
mean(dfdxb)    # answer =  0.0656549
```

```
. use rwm5yr
. keep if year==1984
. keep docvis age outwork
. glm docvis age outwork,fam(nb ml)
. qui sum outwork
. qui replace outwork = r(mean)
. qui sum age
. qui replace age = r(mean)
. predict xb, xb
. gen me = exp(xb) * _b[age]
. su me

Variable |       Obs        Mean    Std. Dev.       Min        Max
---------+--------------------------------------------------------
      me |      3874    .0656549           0   .0656549   .0656549
                      =======
```

The **margins** command will be used to calculate the marginal effect at the mean of *age,* and discrete change of *outwork*. Note that the same values are calculated.

```
. margins, dydx(age outwork) atmeans

Conditional marginal effects                  Number of obs    =   3874
Model VCE    : OIM
```

```
Expression    : Predicted mean docvis, predict()
dy/dx w.r.t. : age 1.outwork
at            : age              =    43.99587 (mean)
                0.outwork        =    .6334538 (mean)
                1.outwork        =    .3665462 (mean)
------------------------------------------------------------------
           |           Delta-method
           |     dy/dx  Std. Err.      z   P>|z|  [95% Conf. Interval]
-----------+------------------------------------------------------
       age |  .0656549  .0071768    9.15  0.000  .0515886     .0797211
 1.outwork |  1.308611  .1890565    6.92  0.000  .9380674     1.679155
------------------------------------------------------------------
Note: dy/dx for factor levels is the discrete change from the base
      level.
```

We use the same calculations to determine Poisson and NB2 marginal effects. The marginal effect of *age* at the mean of the predictors in the model reflects the increase in doctor visits by 6.5% for every year increase in age.

The discrete change calculations are given as

R

```
mu0 <- exp(.124546 + .019931*meanage)              # outwork=0
mu1 <- exp(.124546 + .3256733 + .019931*meanage)   # outwork=1
dc <- mu1 - mu0
mean(dc)  # answer =  1.3086 Discrete Change
```

```
. use rwm5yr
. keep if year==1984
. keep docvis outwork age
. qui sum age
. replace age=r(mean)
. replace outwork=0
. predict p0
. replace outwork=1
. predict p1
. di p1 - p0
1.3086114
```

Another way of calculating the marginal effects and discrete change in a single command is by the use of the **prchance** command (Long and Freese, 2006)

```
. prchange

nbreg: Changes in Rate for docvis

           min->max       0->1      -+1/2     -+sd/2  MargEfct
     age     2.6705     0.0251     0.0657     0.7399    0.0657
                                                        ======
outwork      1.3086     1.3086     1.2382     0.5934    1.2293
                        ======
```

The underlined values are the same as the marginal effect for *age* and discrete change from 0 to 1 for *outwork*.

Average marginal effects follow the same line of reasoning as demonstrated in Section 6.2. You are referred to that site for their construction and interpretation.

9.4 Binomial versus count models

The binomial mean-variance relationship is such that the variance is always less than the mean, and the relationship is symmetric about the value 0.5. This differs from the Poisson, where the mean and variance are assumed equal, or the negative binomial, where the variance is greater than the mean.

Table 9.12 *Bernoulli/binomial mean–variance values*

MEAN		VARIANCE	MEAN		VARIANCE
.025	.975	= .024375	.3	.7	= .21
.05	.95	= .0475	.35	.65	= .2275
.1	.9	= .09	.4	.6	= .24
.15	.85	= .1275	.45	.55	= .2475
.2	.8	= .16	.475	.525	= .249375
.25	.75	= .1875	.5	.5	= .25

There are times, however, when the count data are such that they are better modeled using a binomial model, e.g. a logistic or probit model. Recall that the binomial model response is a count of the number of successes over the number of observations with a given pattern of covariates. Recall also that we earlier derived the Poisson distribution from the binomial by letting the binomial denominator become very large as the probability of an event, p, becomes small. This relationship was also demonstrated in Chapter 2 when discussing the difference between odds and risk ratios in the context of $2 \times k$ tables.

If we do find that a binomial logit model fits the data better than a Poisson or negative binomial model, we must be certain that the question which generated the model can be appropriately answerered using a binomial rather than the original count model. Because the model may fit the data more tightly than another model does not guarantee that the better-fitted model answers the question which is the aim of the study. Binomial models estimate the probability of success for each observation in the model, whereas count models emphasize estimation of the counts in a specified period of time or area. Count models can also estimate the probability of counts, but that is not usually of primary interest. On the other hand, logistic models estimate the odds of one level of the predictor compared with another with respect to the response, whereas a count model is generally used to estimate the ratio of the number of times the response occurs, for levels of a predictor.

At times data come to us in the form of individual data that we wish to model as exposures. It is also the case that we may need to decide if data should be modeled as a logistic regression model in place of a count model, or the reverse. To draw out these relationships I shall begin with a noted data set that has traditionally been analyzed using a logistic model.

The data come to us from Hosmer and Lemeshow (2000). Called the *low birth weight* (**lbw**) data, the response is a binary variable, *low*, indicating whether the birth weight of a baby is under 2,500 g ($low = 1$), or over ($low = 0$). To simplify the example, only two predictors will be included in the model: *smoke* (1/0), and *race*, which enters the model as a three-level factor variable, with level 1 as the referent. I have also expanded the data threefold, i.e. I used three copies of each covariate pattern. Doing so allows the example to more clearly illustrate the relevant relationships. The initial data set consists of 189 observations; we now have 567.

The data are first modeled as a logistic regression. Parameter estimates have been exponentiated to odds ratios. Note also the low value of the AIC and BIC goodness-of-fit statistics.

Table 9.13 R: *Bernoulli logistic model – lbw data*

```
rm (list=ls())
data(lbw)
attach(lbw)
lr9_3 <- glm(low ~ smoke + factor(race), family=binomial,
  data=lbw)
summary(lr9_3)
exp(coef(lr9_3))
modelfit(lr9_3)
```

```
. glm low smoke race2 race3, fam(bin) eform

Generalized linear models                 No. of obs      =        567
Optimization     : ML                     Residual df     =        563
                                          Scale parameter =          1
Deviance         =  659.9241316           (1/df) Deviance =   1.172157
Pearson          =  553.7570661           (1/df) Pearson  =   .9835827
```

```
                                           AIC             = 1.177997
Log likelihood    = -329.9620658           BIC             = -2909.698
------------------------------------------------------------------------
       |              OIM
   low | Odds Ratio Std. Err.    z    P>|z|   [95% Conf.  Interval]
-------+----------------------------------------------------------------
 smoke | 3.052631  .6507382    5.24  0.000    2.010116    4.635829
 race2 | 2.956742  .8364414    3.83  0.000    1.698301    5.147689
 race3 | 3.030001  .7002839    4.80  0.000    1.926264    4.766171
------------------------------------------------------------------------

. abic
AIC Statistic  =    1.177997                AIC*n      = 667.92413
BIC Statistic  =    1.183447                BIC(Stata) = 685.28558
```

Observation level data can be converted to grouped format, i.e. one observation per covariate pattern, using the following Stata code. Users of other packages can easily apply the same logic to produce the same result.

Table 9.14 *Low birth weight data*

```
n = 567 data = lbwch6
-----------------------------------------------------------
variable name variable label
-----------------------------------------------------------
low              birthweight<2500g
smoke            smoked during pregnancy
race1            race = = white
race2            race = = black
race3            race = = other
```

```
/* Be certain that data consist of low, smoke, race1
race2 race3 only */
. egen grp = group(smoke-race3)
. egen cases = count(grp), by(grp)
. egen lowbw = sum(low), by(grp)
. sort grp
. by grp: keep if _n == 1 /// discard all but 1st of like CP's
.
```

The above code can be compressed into a single line using the **collapse** command in Stata (not available in R). However, the command changes the response value *low*. The previous code group employs a response with a different name.

We use the name *lowbw*.

```
. gen byte cases = 1
. collapse (sum) cases (sum) low, by(smoke race1 race2 race3)
. glm low smoke race2 race3, fam(bin cases) eform
```

As an aside, the **collapse** command can obtain the same results below by converting the data set to one that is frequency weighted. Twelve observations rather than the six result from using the previous commands.

```
. collapse (sum) cases, by(low smoke race2 race3)
. glm low smoke race3 race3 [fw=cases], fam(bin)
```

In grouped format, using the data created by the first method, the binomial logit model appears as shown in Table 9.15.

Table 9.15 R: *Grouped binomial logistic regression – **lbwgrp** data*

```
rm(list=ls())
data(lbwgrp)
attach(lbwgrp)
notlow <- (cases - lowbw)
lr9_3g <- glm(cbind(lowbw, notlow) ~ smoke + race2 + race3,
       family=binomial, data=lbwgrp)
summary(lr9_3g)
exp(coef(lr9_3g))
modelfit(lr9_3g)
```

```
. glm lowbw smoke race2 race3, fam(bin cases) eform

Generalized linear models                 No. of obs       =          6
Optimization     : ML                     Residual df      =          2
                                          Scale parameter =          1
Deviance         =  9.470810009           (1/df) Deviance =   4.735405
Pearson          =  9.354399596           (1/df) Pearson  =     4.6772
                                      .   .   .
                                          AIC              =   7.443406
Log likelihood   = -18.33021651           BIC              =   5.887291
--------------------------------------------------------------------
      |                OIM
lowbw | Odds Ratio Std. Err.    z    P>|z|   [95% Conf. Interval]
------+-------------------------------------------------------------
smoke | 3.052631  .6507383    5.24  0.000    2.010117    4.635829
race2 | 2.956742  .8364415    3.83  0.000    1.698301     5.14769
race3 | 3.030001   .700284    4.80  0.000    1.926264    4.766172

. abic
AIC Statistic    =    7.443406           AIC*n       =  44.660435
BIC Statistic    =    7.958465           BIC(Stata) =  43.827473
```

Run the R script, **modelfit.r,** to obtain the same values as displayed in the above **abic** command output.

Parameter estimates and associated standard errors and confidence intervals are identical, except for small differences resulting from estimation rounding errors. On the other hand, both AIC and BIC statistics have markedly risen.

Overdispersion is not applicable for binary response models in the sense that we have defined overdispersion. However, see Hilbe (2009) for an alternative meaning of binary response overdispersion. Regardless, the dispersion statistic displayed in the output of a binary logistic model does not indicate overdispersion, or, in this instance, equidispersion. For binomial models, overdispersion can only be assessed when the data are formatted as grouped. Here the binomial model has significant overdispersion: 4.677.

We could scale the model, estimate robust variance estimators, use bootstrapping and and jackknife techniques, or engage in specialized procedures such as the Williams Procedure (Collett, 1989) to handle the overdispersion. We should also look to see if the overdispersion is real or only apparent. In addition, we can model the data using an entirely different GLM family. Once the binary logistic data have been converted to grouped format, the binomial numerator can be modeled as Poisson. The binomial numerator is considered as a count (rather than a success) and the denominator is considered an exposure. Exposures enter the model as a log-transformed offset.

Table 9.16 R: *Poisson model – lbwgrp*

```
lncases <- log(cases)
p9_3 <- glm(lowbw ~ smoke + race2 + race3 + offset(lncases),
       family=poisson, data=lbwgrp)
summary(p9_3)
exp(coef(p9_3))
```

```
. glm lowbw smoke race2 race3, fam(poi) eform lnoffset(cases)

Generalized linear models                    No. of obs      =          6
Optimization     : ML                        Residual df     =          2
                                             Scale parameter =          1
Deviance         =  9.717852215              (1/df) Deviance =   4.858926
Pearson          =  8.863298559              (1/df) Pearson  =   4.431649
. . .
                                             AIC             =   7.954159
Log likelihood   = -19.86247694              BIC             =   6.134333
```

```
------------------------------------------------------------------
      |                 OIM
lowbw |       IRR    Std. Err.     z    P>|z|   [95% Conf. Interval]
------+-----------------------------------------------------------
smoke |  2.020686  .3260025    4.36   0.000    1.472897    2.772205
race2 |  1.969159  .4193723    3.18   0.001    1.29718     2.989244
race3 |  2.044699  .3655788    4.00   0.000    1.440257    2.90281
cases | (exposure)
------------------------------------------------------------------
```

Not surprisingly, the model is still overdispersed with a value of 4.43, which is similar to the binomial logistic model dispersion of 4.68. The primary difference due to reparameterization is in how the estimates are interpreted. The AIC and log-likelihood statistics are similar in each model, but the BIC is not. Its value indicates that the logistic model is preferable. However, both models are overdispersed, so it may be that the fit statistics are not distinct enough to be meaningful.

We know that the model is not finalized. Since the Poisson model is overdispersed, we must check for apparent versus real overdispersion, and take appropriate remedies based on each alternative. Given that the data are truly overdispersed, the data can be modeled as negative binomial to determine if overdispersion is accommodated.

Table 9.17 R: *Negative binomial (NB2) – lbwgrp data*

```
nb9_3 <- glm.nb(lowbw ~ smoke + race2 + race3 +
         offset(lncases), data=lbwgrp)
summary(nb9_3)
exp(coef(p9_3))
modelfit(nb9_3)  # see Table 5.3 for function
```

```
. glm lowbw smoke race2 race3, fam(nb ml) eform lnoff-
set(cases)

Generalized linear models              No. of obs      =        6
Optimization     : ML                  Residual df     =        2
                                       Scale parameter =        1
Deviance         =  5.898033751        (1/df) Deviance = 2.949017
Pearson          =  5.447387468        (1/df) Pearson  = 2.723694

Variance function: V(u) = u+(.0245)u^2 [Neg. Binomial]
Link function    : g(u) = ln(u)        [Log]
                                       AIC             = 7.824141
Log likelihood   = -19.47242342        BIC             = 2.314515
```

```
------------------------------------------------------------------------
      |                  OIM
lowbw |       IRR   Std. Err.    z    P>|z|  [95% Conf.  Interval]
------+-----------------------------------------------------------------
smoke |   2.035876  .4335387   3.34  0.001   1.341184    3.090397
race2 |   2.072214  .5540672   2.73  0.006    1.22699     3.49968
race3 |   2.063825  .5044597   2.96  0.003   1.278245    3.332204
cases | (exposure)
------------------------------------------------------------------------
Note: Negative binomial parameter estimated via ML and treated
      as fixed once estimated.

. abic
AIC Statistic    =   7.824141          AIC*n        =  46.944847
BIC Statistic    =     8.3392          BIC(Stata)   =  46.111885
```

Those using R's **glm.nb** should note that the value of α is not calculated. Instead θ is displayed, which is the inverse of α. R output shows the value of θ to be 40.8, which is indeed the inverse of 0.0245 shown in the Stata model.

The output indicates that the model is not statistically different from a Poisson model. The likelihood ratio test determining if α is statistically different from zero fails. In effect, the value of α is approximately zero, indicating a preference for the more parsimonious Poisson model.

```
Likelihood-ratio test of alpha=0:
chibar2(01) = 0.78 Prob>=chibar2 = 0.189
```

I tend to believe that, of the three choices, the model is best estimated using the negative binomial model, because it is the least overdispersed of the three. The logistic and Poisson models have dispersion statistics of 4.68 and 4.43, whereas the negative binomial has a dispersion statistic of 2.72. The fit statistics are not significantly different from one another to make a clear determination of better fit. This is particularly the case since there are only six observations in the grouped data.

If Poisson distributional assumptions are not a concern, calculating standard errors via bootstrap may be a viable modeling strategy. Recall that when standard errors are adjusted using a Huber–White sandwich, also called robust standard errors, or by bootstrap, the model is sometimes referred to as a pseudo-log-likelihood, or simply pseudo-likelihood, model. This appellation relies on the fact that under these conditions the standard errors do not derive directly from the underlying PDF or log-likelihood of the model.

Statisticians use a variety of specialized goodness-of-fit tests when evaluating a logistic model. These are discussed at length in texts such as Hilbe (2009),

Table 9.18 *Comparison of model-based and bootstraps SEs*

lowbw	Observed IRR	Bootstrap Std. Err.	z	P>|z|	Normal-based [95% Conf.	Interval]
smoke	2.020686	.7705485	1.84	0.065	.9569959	4.266656
race2	1.969159	.7648926	1.74	0.081	.9196939	4.216171
race3	2.044699	1.017267	1.44	0.151	.7711595	5.421438
cases	(exposure)					

Hardin and Hilbe (2007), Hosmer and Lemeshow (2000), and Collett (1989), but go beyond the scope of our discussion. Unfortunately no such fit statistics have been designed for count models.

An indicator of whether data should be modeled as a grouped logistic or as a rate parameterized count model relates to the ratio of successes to cases, or counts to exposure, respectively. This logic goes back to the derivation of the Poisson distribution from the binomial. Considered in this manner, Poisson models are rare binomial events.

If the ratio of the Poisson counts to the exposure is small, it is likely that a Poisson or negative binomial model will better fit the data. On the other hand, when the binomial numerator is close to the value of the denominator, it is likely that a logistic, probit, loglog, or complementary loglog model would be preferable. This topic is discussed at length in Hilbe (2009) and Hardin and Hilbe (2007).

Table 9.19 *Ratio of response values*

	lowbw	cases	%ratio
1	60	165	36.4
2	15	48	31.3
3	12	132	09.1
4	15	36	41.7
5	18	30	60.0
6	57	156	36.5

The percentage ratios of the variables *lowbw* and *cases* are provided in Table 9.19. The percentage ratio values do not indicate a clear preference for a binomial or count model. Given the rather large mean percentage of 36.5, the data appear to lean toward a logistic model rather than Poisson or NB.

However, this is not always the case. Given a binary response model, we may also have reason to need information concerning risk. When this occurs, the data must be converted to a count response. To do this, the binary format must first be converted to grouped, and the binomial denominator must be entered into the count model as an exposure. Example 3, in Section 9.5, demonstrates a successful use of this method.

Before concluding this section I shall provide sample code for constructing synthetic binary and binomial data, which can then be compared with synthetic count models. R generic synthetic binary logistic and probit functions are in the COUNT package: *Logitsyn.r; probitsyn.r.*

Table 9.20 R/ Stata: *Synthetic binary logistic regression*

```
# syn.logit.r
library(MASS)
nobs <- 50000
x1 <- runif(nobs)
x2 <- runif(nobs)
xb <- 2 + .75*x1 - 1.25*x2                    # linear predictor
exb <- 1/(1+exp(-xb))                         # fit; predicted prob
by  <- rbinom(nobs, size = 1, prob =exb)      # random logit variates
lry <- glm(by ~ x1 + x2, family=binomial(link="logit"))
summary(lry)
-------------------------------------------------------------------
set obs 50000
gen x1 = runiform()
gen x2 = runiform()
gen xb = 2 + 0.75*x1 - 1.25*x2
gen exb = 1/(1+exp(-xb))
gen by = rbinomial(1, exb)
glm by x1 x2, fam(bin 1)
```

Table 9.21 R/Stata: *Code changed for probit model*

```
exb <- pnorm(xb)
by  <- rbinom(nobs, size = 1, prob =exb)
pry <- glm(by ~ x1 + x2, family=binomial(link="probit"))
summary(pry)
-------------------------------------------------------------------
gen double exb = normprob(xb)
gen double py = rbinomial(1, exb)
glm py x1 x2, fam(bin 1) link(probit)
```

Table 9.22 R/Stata: *Synthetic binomial (grouped) logistic regression*

```
# syn.bin_logit.r
nobs <- 50000
x1 <- runif(nobs)
x2 <- runif(nobs)
d <- rep(1:5, each=10000, times=1)*100 # denominator
xb <- 2 + .75*x1 - 1.25*x2              # linear predictor; values
exb <- 1/(1+exp(-xb))                   # fit; predicted prob
by  <- rbinom(nobs, size = d, p = exb) # random binomial variate
dby <- d - by                           # denominator - numerator
gby <- glm(cbind(by,dby)~x1 + x2, family=binomial(link="logit"))
summary(gby)                            # displays model output
-------------------------------------------------------------------
. set obs 50000
. set seed 13579
. gen x1 = runiform()
. gen x2 = runiform()
*-  -  -  -  -  -  -  -  -  -  -  -  -  -  -  -
* Select User Specified or Random Denominator. Select Only One
. gen d = 100+100*int((_n-1)/10000) /// user specified
. gen d = ceil(10*runiform())       /// random
*-  -  -  -  -  -  -  -  -  -  -  -  -  -  -  -
. gen xb = 2 + 0.75*x1 - 1.25*x2
. gen exb = 1/(1+exp(-xb))
. gen by = rbinomial(d, exb)
. glm by x1 x2, nolog fam(bin d)
```

9.5 Examples: negative binomial regression

I shall present four examples demonstrating how negative binomial regression can be used to model count response data. These examples will also be used in later chapters when dealing with extensions to the basic form of negative binomial model.

Example 1: Modeling number of marital affairs

For the first example we shall evaluate data from Fair (1978). Although Fair used a tobit model with the data, the outcome measure can be modeled as a count. In fact, Greene (2003) modeled it as Poisson but, given the amount of overdispersion in the data, employing a negative binomial model is an appropriate strategy. The data are stored in the **affairs** data set.

Naffairs is the response variable, indicating the number of affairs reported by the participant in the past year. The classification of counts appears as Table 9.24.

Table 9.23 *Example 1: **affairs** data*

```
affairs.dta
---------------------------------------------------
obs:         601
---------------------------------------------------
naffairs     number of affairs within last year
kids         1 = have kids; 0 = no kids
vryunhap     ratemarr = = 1  very unhappily married
unhap        ratemarr = = 2  unhappily married
avgmarr      ratemarr = = 3  avg marriage
hapavg       ratemarr = = 4  happily married
vryhap       ratemarr = = 5  very happily married
antirel      relig = = 1     anti religious
notrel       relig = = 2     not religious
slghtrel     relig = = 3     slightly religious
smerel       relig = = 4     somewhat religious
vryrel       relig = = 5     very religious
yrsmarr1     yrsmarr = =     0.75 yrs
yrsmarr2     yrsmarr = =      1.5  yrs
yrsmarr3     yrsmarr = =      4.0  yrs
yrsmarr4     yrsmarr = =      7.0  yrs
yrsmarr5     yrsmarr = =     10.0 yrs
yrsmarr6     yrsmarr = =     15.0 yrs
```

Table 9.24 *Naffairs: frequency of counts*

Naffairs	Freq.	Percent	Cum.
0	451	75.04	75.04
1	34	5.66	80.70
2	17	2.83	83.53
3	19	3.16	86.69
7	42	6.99	93.68
12	38	6.32	100.00
Total	601	100.00	

The number of zeros in the data far exceeds the number reasonably expected by the distributional assumptions of both the Poisson and negative binomial. To observe differences between the observed and predicted number of zeros, as well as other values, we first model the data using Poisson regression.

Table 9.25 R: *Model of affairs data*

```
rm(list=ls())
data(affairs)
attach(affairs)
myTable(naffairs)
affmodel <-  glm(naffairs ~ kids + avgmarr + hapavg +  vryhap
  + notrel + slghtrel + smerel + vryrel + yrsmarr3 +  yrsmarr4
  + yrsmarr5 + yrsmarr6, family=poisson, data=affairs)
summary(affmodel)
confint(affmodel)
exp(coef(affmodel))
exp(confint(affmodel))
mu<-fitted.values(affmodel)
poi.obs.pred(len=15, model=affmodel)
```

```
  . glm naffairs kids avgmarr hapavg vryhap notrel slghtrel
  smerel vryrel yrsmarr3-yrsmarr6, fam(poi)

Generalized linear models                No. of obs      =       601
Optimization     : ML                    Residual df     =       588
                                         Scale parameter =         1
Deviance         =  2305.835984          (1/df) Deviance =   3.92149
Pearson          =  4088.616155          (1/df) Pearson  =  6.953429
                                  . . .
                                         AIC             =  4.701873
Log likelihood   = -1399.912931          BIC             = -1456.538
--------------------------------------------------------------------
        |              OIM
naffairs|    Coef.  Std. Err.     z    P>|z|  [95% Conf.  Interval]
--------+-----------------------------------------------------------
    kids| -.2226308 .1059723  -2.10  0.036   -.4303328  -.0149289
 avgmarr| -.8858196 .1050272  -8.43  0.000   -1.091669  -.6799701
  hapavg| -1.023898 .0859245 -11.92  0.000   -1.192307  -.8554889
  vryhap|  -1.38385 .1009577 -13.71  0.000   -1.581723  -1.185976
  notrel| -.6553382 .1111865  -5.89  0.000   -.8732597  -.4374166
slghtrel| -.5236987 .1113403  -4.70  0.000   -.7419218  -.3054756
  smerel| -1.370688 .1213036 -11.30  0.000   -1.608439  -1.132938
  vryrel| -1.363744 .1589703  -8.58  0.000    -1.67532  -1.052168
yrsmarr3|  .7578109 .1612081   4.70  0.000    .4418488   1.073773
yrsmarr4|  1.104536 .1698768   6.50  0.000    .7715832   1.437488
yrsmarr5|  1.480332 .1648648   8.98  0.000    1.157203   1.803461
yrsmarr6|  1.480467 .1555978   9.51  0.000    1.175501   1.785433
   _cons|  1.101651 .1648297   6.68  0.000    .7785906   1.424711
--------------------------------------------------------------------
```

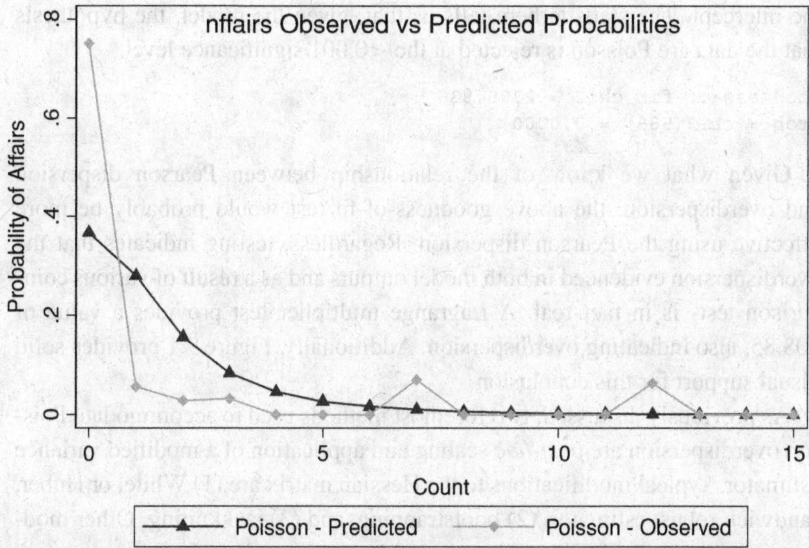

Figure 9.1 Poisson model for number of affairs: observed versus predicted probabilities

Exponentiating the parameter estimates results in them becoming incidence rate ratios.

```
. glm, eform nohead          [output not shown]
```

Observed counts from Table 9.24 can be graphed against predicted counts based on the fitted values, μ, from the above model.

Slightly above 75% of observed zero counts clearly differ from the approximate 38% zeros predicted on the basis of the distributional assumption of the Poisson model. From a count of 3 upwards, the empirical and predicted distributions are similar. This type of distributional violation typically results in substantial overdispersion, as indicated by the high value for the Pearson *chi*2-based dispersion statistic. In this case the dispersion is 6.95.

The code used to create Figure 9.1 is on the book's website, as discussed in the Preface. It can easily be encoded by adjusting the code for Figure 6.2, which is also on the book's website. The **poi.obs.pred.r** function in the COUNT package can be used to construct a table of observed and predicted values for creating the figure in R.

Several statistical packages provide a Poisson goodness-of-fit test. It is simply the deviance evaluated by a *chi*2 distribution with the degrees of freedom equal to the number of observations minus the number of predictors, including the intercept. The statistic here tells us that, given the model, the hypothesis that the data are Poisson is rejected at the <0.001 significance level.

```
Goodness-of-fit chi2 = 2305.836
Prob > chi2(588)     = 0.0000
```

Given what we know of the relationship between Pearson dispersion and overdispersion, the above goodness-of-fit test would probably be more effective using the Pearson dispersion. Regardless, testing indicates that the overdispersion evidenced in both model outputs and as a result of various comparison tests is in fact real. A Lagrange multiplier test provides a value of 508.85, also indicating overdispersion. Additionally, Figure 9.1 provides solid visual support for this conclusion.

As previously discussed, two foremost methods used to accommodate Poisson overdispersion are *post-hoc* scaling and application of a modified variance estimator. Typical modifications to the Hessian matrix are (1) White, or Huber, sandwich robust estimator, (2) bootstrapping, and (3) jackknifing. Other modifications are available as well in most of the major statistical packages.

The standard errors that result from an application of a robust variance estimator affects the parameter estimate *p*-values. Usually overdispersion serves to deflate *p*-values, perhaps misleading a researcher into believing that a predictor contributes to the model when in fact it does not. Applying a robust variance estimator, scaling of standard errors, or bootstrapping, usually inflates the *p*-values of an overdispersed count model. At times, however, given the interaction between predictors, such modifications appear to work in the opposite direction. But this situation is not the norm.

It is also important to mention that the source of overdispersion in Poisson models may be clearly identifiable. For example, overdispersion arises in longitudinal or clustered data when they are modeled without taking into consideration the extra correlation resulting from the similarities of the observations within groups. When an obvious source of overdispersion is identified, we may attempt to find the specifically appropriate remedy for it. For longitudinal and clustered data, a robust variance estimator may be applied by considering each group or cluster to be a single observation, with a summary variance statistic given to each respective cluster. Treated as individuals, each group is then assumed to be independent from one another. Care must be taken when applying this method though. Consider data taken from hospitals within a medium-sized city. If there are only, for instance, four hospitals in the city, and we apply a robust variance estimator on clusters, we effectively reduce the size of the data set to four independent components. The parameter estimates are not affected by this method though – only the Hessian matrix from which the algorithm abstracts model standard errors. Using clustering methods is extremely helpful when dealing with identified overdispersion, but the number of clusters must be sizeable.

Application of a robust variance estimator to the **affairs** data gives us the following output. Notice that *kids* and *yrsmarr3* no longer contribute to the model. However, no source of overdispersion has been identified other than the inflated zeros. We shall later return to this example when we discuss methods of dealing with excessive zeros in the data, e.g. zero-inflated Poisson, zero-inflated negative binomial, and hurdle models.

Table 9.26 R: ***Affairs*** *with robust SEs for IRR*

```
irr <- exp(coef(affmodel))
library("sandwich")
rse <- sqrt(diag(vcovHC(affmodel, type="HC0")))
ec <- c(irr)
rs <- c(rse)
ec * rs # vector of IRR robust SEs
```

```
. glm naffairs kids avgmarr hapavg vryhap notrel slghtrel
  smerel vryrel yrsmarr3-yrsmarr6, fam(poi) eform vce(robust)
  nohead
```

naffairs	IRR	Robust Std. Err.	z	P>|z|	[95% Conf.	Interval]
kids	.8004103	.2437552	-0.73	0.465	.4406457	1.453904
avgmarr	.412376	.0995915	-3.67	0.000	.2568755	.6620094
hapavg	.3591922	.0772231	-4.76	0.000	.2356818	.5474287
vryhap	.2506118	.0655191	-5.29	0.000	.1501296	.4183471
notrel	.5192664	.1333609	-2.55	0.011	.3138918	.8590145
slghtrel	.5923257	.1459787	-2.12	0.034	.3654112	.9601502
smerel	.2539321	.0731995	-4.75	0.000	.1443267	.4467745
vryrel	.2557017	.0953902	-3.66	0.000	.1230809	.5312229
yrsmarr3	2.1336	1.039479	1.56	0.120	.8211304	5.543882
yrsmarr4	3.017823	1.66825	2.00	0.046	1.021293	8.917379
yrsmarr5	4.394404	2.21657	2.93	0.003	1.635112	11.81007
yrsmarr6	4.394996	2.230482	2.92	0.004	1.625434	11.88359

Negative binomial regression employs an extra parameter, α, that directly addresses the overdispersion in Poisson models. Generally speaking, there is a direct relationship between the amount of overdispersion in a Poisson model and the value of α in a well-fitted negative binomial model. The relationship is clearly evident in the variance functions of the two models:

Poisson variance $= \mu$
Negative binomial variance $= \mu + \alpha\mu^2$
Alternative parameterizations of the negative binomial will be considered in later chapters.

Table 9.27 R: ***Affairs*** *– NB2 model*

```
affnb2 <-  glm.nb(naffairs~ kids + avgmarr + hapavg + vryhap +
    notrel + slghtrel + smerel + vryrel + yrsmarr3 + yrsmarr4
    + yrsmarr3 + yrsmarr4 + yrsmarr5 + yrsmarr6, data=affairs)
summary(affnb2)
confint(affnb2)
exp(coef(affnb2))
exp(confint(affnb2))
nb2.obx.pred(len=15, model=affnb2)    # table  obs  vs predicted
```

```
. glm naffairs kids avgmarr hapavg vryhap notrel slghtrel smerel
  vryrel yrsmarr3-yrsmarr6,  eform fam(nb ml)

Generalized linear models                 No. of obs      =       601
Optimization     : ML                     Residual df     =       588
                                          Scale parameter =         1
Deviance         =  339.9146951           (1/df) Deviance =  .5780862
Pearson          =  574.2568411           (1/df) Pearson  =  .9766273

Variance function: V(u) = u+(6.7601)u^2 [Neg. Binomial]
Link function    : g(u) = ln(u)            [Log]
                                          AIC             =  2.453377
Log likelihood   = -724.2398359           BIC             = -3422.459
------------------------------------------------------------------
         |                OIM
naffairs |      IRR   Std. Err.      z   P>|z|  [95% Conf.  Interval]
---------+--------------------------------------------------------
    kids | 1.091006   .3393095    0.28  0.779   .5930593    2.00704
 avgmarr | .3788406   .1626535   -2.26  0.024   .1633041   .8788527
  hapavg | .3754898   .1370964   -2.68  0.007   .1835748   .7680389
  vryhap | .2491712   .0936138   -3.70  0.000   .1193166   .5203494
  notrel |  .735144   .3482966   -0.65  0.516   .2904624   1.860608
slghtrel | .6610617   .3191906   -0.86  0.391   .2565921   1.703102
  smerel | .2307172   .1071631   -3.16  0.002   .0928358   .5733825
  vryrel | .2202639   .1199509   -2.78  0.005   .0757526   .6404555
yrsmarr3 |  1.95046   .7752988    1.68  0.093   .8949284   4.250947
yrsmarr4 | 3.801339   1.695028    2.99  0.003   1.586293   9.109401
yrsmarr5 | 3.283675   1.471676    2.65  0.008   1.364172   7.904079
yrsmarr6 | 4.165032    1.61167    3.69  0.000   1.950939   8.891866
------------------------------------------------------------------
```

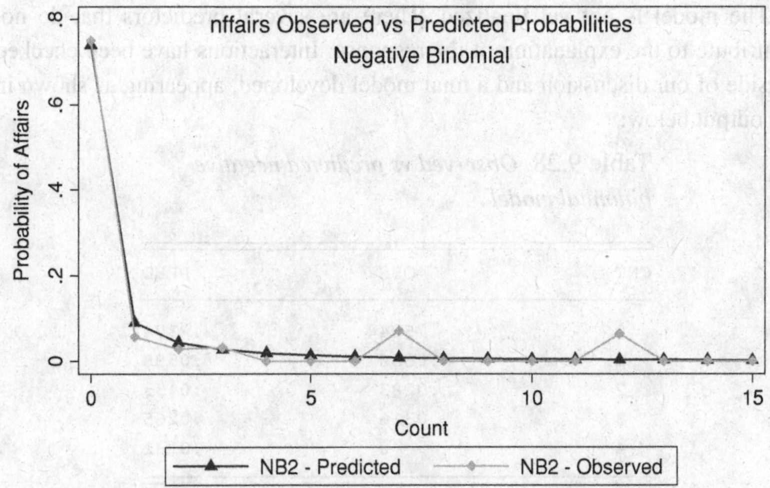

Figure 9.2 Negative binomial model for number of affairs: observed versus predicted probabilities

A boundary likelihood ratio test of the log-likelihood of the full model against the log-likelihood of the Poisson model ($\alpha = 0$) on the same data informs us whether the data are Poisson or non-Poisson. A significant *p*-value is usually taken to mean that the model is negative binomial. It may be, but it also may be some variant of the basic negative binomial model.

In any case, a likelihood ratio test of $\alpha = 0$ (the Poisson model) yields a *chi*2, with one degree of freedom, of 1351.35. The corresponding *p*-value is less than 0.000001, indicating that the negative binomial model with an α of 6.76 is significantly different from the Poisson. The negative binomial model produces AIC and BIC statistics of 2.45 and −3,422.46 respectively. These values compare with the Poisson model statistics of 4.70 and −1,456.54. The values for the negative binomial model are clearly and substantially less than those of the corresponding Poisson – indicating again that the data are better modeled as negative binomial rather than Poisson.

Visually comparing the observed and predicted counts for the negative binomial model may help in distinguishing it from the Poisson, as well as determining if it is a preferable model. The code for creating Figure 9.2 is on the book's website.

Figure 9.2 clearly shows the close association between the observed counts of *affairs* and the number of affairs predicted on the basis of the negative binomial model. The fit is superior to that shown for the Poisson model (Figure 9.1). Of particular interest is that the differences between predicted and observed zero counts are now minimal – only 0.01 (Table 9.28). On the other hand, the difference for the Poisson model is some 37.

The model is not yet finalized. There are several predictors that do not contribute to the explanation of the response. Interactions have been checked outside of our discussion and a final model developed, appearing as shown in the output below:

Table 9.28 *Observed vs predicted negative binomial model*

CNT	OBS	PRED
0	.7504	.7394
1	.0566	.0898
2	.0283	.0433
3	.0316	.0265
4	0	.0181

Table 9.29 R: ***Affairs*** *data; reduced NB2 model and graph*

```
affnb2r <-  glm.nb(naffairs~ avgmarr + hapavg + vryhap + smerel
    + vryrel + yrsmarr4 + yrsmarr5 + yrsmarr6, data=affairs)
summary(affnb2r)
confint(affnb2r)
exp(coef(affnb2r))
exp(confint(affnb2r))
deviance <- residuals(affnb2r, type="deviance")
dev <- sum(deviance*deviance)
pred <- predict(affnb2r, se.fit=TRUE, type="response")
mu <- pred$fit
stdp <- pred$se.fit                    # Std error of prediction
variance <- mu + (mu * mu)/pred$theta
h <- stdp * stdp * variance              # hat matrix diagonal
sdeviance <- rstandard(affnb2r)        # Std deviance
plot(mu, sdeviance)
```

```
. glm naffairs avgmarr-vryhap smerel vryrel yrsmarr4-
yrsmarr6, eform fam(nb ml)

Generalized linear models                No. of obs      =       601
Optimization     : ML                    Residual df     =       592
                                         Scale parameter =         1
Deviance         =  339.2113501          (1/df) Deviance =  .5729921
Pearson          =  517.6346425          (1/df) Pearson  =  .8743828
Variance function: V(u) = u+(6.9082)u^2 [Neg. Binomial]
Link function    : g(u) = ln(u)              [Log]
```

```
                                                 AIC          =  2.446937
Log likelihood    = -726.3044443                 BIC          = -3448.757
------------------------------------------------------------------------------
             |                 OIM
    naffairs |        IRR   Std. Err.      z    P>|z| [95% Conf. Interval]
-------------+----------------------------------------------------------------
     avgmarr |   .3720682   .1590264    -2.31   0.021    .1609937    .8598769
      hapavg |   .3681403   .1352423    -2.72   0.007    .1791888    .7563381
      vryhap |   .2514445   .0923022    -3.76   0.000    .1224549     .516307
      smerel |   .3047693    .088186    -4.11   0.000    .1728515    .5373648
      vryrel |    .279958   .1149339    -3.10   0.002    .1252105    .6259576
    yrsmarr4 |   2.824666   1.125855     2.61   0.009    1.293288    6.169344
    yrsmarr5 |   2.462933   .9828394     2.26   0.024    1.126623    5.384268
    yrsmarr6 |   3.173011   .9689469     3.78   0.000     1.74397    5.773035
------------------------------------------------------------------------------
```

All predictors significantly enter the model and both AIC and BIC statistics have been reduced, albeit not significantly. The model can be re-fitted using a robust variance estimator to determine if empirically based standard errors result in a difference in *p*-value significance. In this case there is minimal difference, with significance moving in the opposite direction, i.e. appearing more contributory rather than less.

```
. predict  sdev, deviance stan
. predict mu
. scatter sdev mu, title(Affairs: Negative Binomial)
```

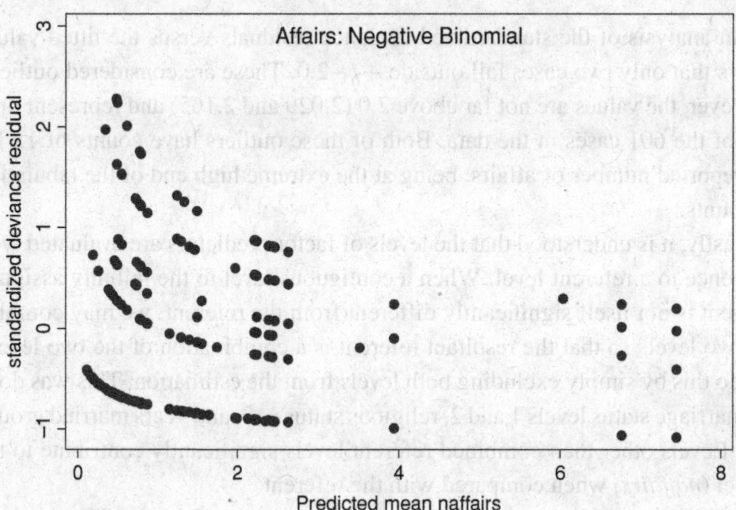

Figure 9.3 Negative binomial model for number of affairs: standardized deviance residuals versus fitted values

Table 9.30 R: ***Affairs** Reduced NB2, robust SE*

```
irr <- exp(coef(affnb2r))
library("sandwich")
rse <- sqrt(diag(vcovHC(affnb2r, type="HC0")))
ec <- c(irr)
rs <- c(rse)
ec * rs # vector of IRR robust SEs
```

```
. glm naffairs avgmarr-vryhap smerel vryrel yrsmarr4-yrsmarr6,
  eform fam(nb ml) vce(robust) nohead
```

naffairs	IRR	Robust Std. Err.	z	P>|z|	[95% Conf.	Interval]
avgmarr	.3720682	.1075804	-3.42	0.001	.2111081	.655753
hapavg	.3681403	.0885893	-4.15	0.000	.2297102	.5899925
vryhap	.2514445	.0716387	-4.85	0.000	.1438558	.439498
smerel	.3047693	.0710436	-5.10	0.000	.1929971	.4812731
vryrel	.279958	.0883555	-4.03	0.000	.1508174	.5196779
yrsmarr4	2.824666	1.026682	2.86	0.004	1.385417	5.759086
yrsmarr5	2.462933	.7091724	3.13	0.002	1.400746	4.33058
yrsmarr6	3.173011	.7823943	4.68	0.000	1.95697	5.144688

An analysis of the standardized deviance residuals versus the fitted values shows that only two cases fall outside $+/-2.0$. These are considered outliers. However, the values are not far above 2.0 (2.029 and 2.165) and represent only two of the 601 cases in the data. Both of these outliers have counts of 12 for the reported number of affairs, being at the extreme high end of the tabulation of counts.

Lastly, it is understood that the levels of factor predictors are evaluated with reference to a referent level. When a contiguous level to the initially assigned referent is not itself significantly different from the referent, we may combine the two levels so that the resultant referent is a combination of the two levels. We do this by simply excluding both levels from the estimation. This was done for marriage status levels 1 and 2, religious status 1–3, and years married groups 1–3. Levels other than combined referent levels significantly contribute to the model (*naffairs*) when compared with the referent.

It is also possible to evaluate the levels that are significantly different from the combined referent levels, now simply called the referent. They may not significantly differ from each other. If this turns out to be the case, then significant levels may themselves be combined. All major statistical packages allow levels

Table 9.31 *Testing interlevel predictor significance*

```
test avgmarr          =          vryhap
chi2(1)               =          1.19
Prob > chi2           =          0.2751

test smerel           =          vryrel
chi2(1)               =          0.04
Prob > chi2           =          0.8433

test yrsmarr4         =          yrsmarr6
chi2(1)               =          0.09
Prob > chi2           =          0.7623
```

to be tested for interlevel significance based on a *chi*2 or Wald statistic. Stata has a number of ways these can be evaluated. The simplest is to use the commands shown in part in Table 9.31.

Table 9.31 indicates that each nonreferent level in the model can be combined with one or more other significant levels. It is likely from an observation of the table that each model predictor can be dichotomized such that it is considered to be binary. For instance, with respect to marital status, a predictor called *marrstatus* can be defined as 1 = levels 3–5; 0 = levels 1–2 (referent). Religious status can be dichotomized as *relstatus* with 1 = levels 4–5; 0 = levels 1–3 (referent) and years married can likewise be dichotomized as *yrmarr* with 1 = levels 4–6; 0 = levels 1–3 (referent). Such a model may be preferable to the multileveled one. I leave it to the reader to determine.

Owing to the fact that the response, *naffairs*, consists of some 75% zero counts, we shall later return to these data when considering zero-inflated and hurdle models. Of interest will be a determination if either of these extended models fit the data better than the one we have developed here.

Example 2: Heart procedures

The second example relates to data taken from Arizona cardiovascular patient files in 1991. A subset of the fields was selected to model the differential length of stay for patients entering the hospital to receive one of two standard cardiovascular procedures: CABG and PTCA. CABG is an acronym for coronary artery bypass graft surgery; PTCA represents percutaneous transluminal coronary angioplasty. Angioplasty is performed by inserting a bulb through the artery to the place containing a blockage near the heart. The bulb is inflated or dilated, clearing the blockage in the affected part of artery. It is substantially safer than a CABG, and, as can be seen from Table 9.32, usually results in an earlier release from the hospital.

Length of stay values are found in the variable *los*; procedure data are found in *procedure*, with 1 = CABG and 0 = PTCA. CABG is considered to be the more difficult procedure of the two. Other controlling or confounding predictors include the following

```
sex          1 = Male; 0 = Female
admit        1 = Urgent/Emergency; 0 = Elective
age75        1 = age>75; 0= age <=75
hospital     encrypted facility code
```

The data, **azpro**, consist of 3,589 observations. The distribution of counts, together with a listing of the mean, median, and standard deviation, for each procedure is displayed in Table 9.33.

Table 9.32 R: ***azpro**: tabulation of procedure*

```
rm(list=ls())
data(azpro)
azpro$los <- unclass(azpro$los)
attach(azpro)
myTable(los)
tapply(los,procedure,myTable)
tapply(los,procedure,hist)
by(azpro["los"], procedure, myTable)
by( data.frame(los), procedure, myTable)
windows(record=TRUE)
by(azpro[,"los"], procedure, hist)
# OR
library("ggplot2")
qplot(los,geom="histogram",facets=procedure~.)
```

```
. use azpro
. tab los if procedure==0
. tab los if procedure==1
```

Table 9.33 *CABG/PTCA: upper frequencies andsummary statistics*

```
    |  PTCA             |  CABG
LOS |  Freq.   Percent  |  Freq.   Percent
----+-------------------+-----------------
  1 |   147       7.68  |
  2 |   399      20.86  |
  3 |   292      15.26  |      1      0.06
  4 |   233      12.18  |      1      0.06
```

```
        5 |     176      9.20 |     10      0.60
        6 |     149      7.79 |     48      2.86
          |       .       .     . |
       13 |      22      1.15 |     93      5.55
       14 |      14      0.73 |    114      6.80
       15 |      14      0.73 |     81      4.83
       16 |       7      0.37 |     52      3.10

               Mean     Median      SD
       ---------------------------------
       CABG   13.02       11       7.07
       PTCA    5.16        4       4.16
       =================================
```

A graphical representation of the differences in length of stay between the two procedures can be found in Figure 9.4.

It is evident that having a CABG results in a longer hospital stay. The question is whether the difference in stay is statistically significant between the two procedures, controlling for gender, type of admission, and age of patient. It is also desirable to determine the probable length of stay given patient profiles.

```
. label define procedure 0 "PTCA" 1 "CABG"
. label values procedure procedure
. hist los, by(procedure)
```

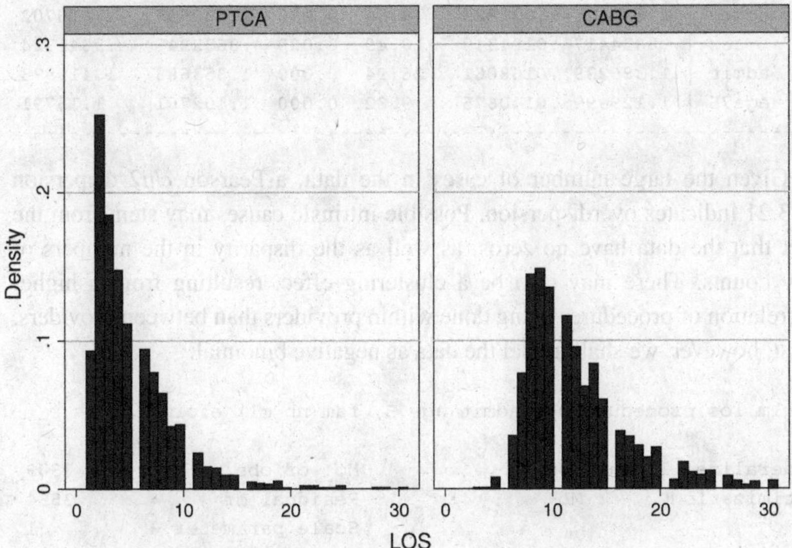

Figure 9.4 Length of stay (LOS) distribution: PTCA versus CABG

Modeling the data as Poisson, we have the following output:

Table 9.34 R: ***azpro*** *Poisson and NB2 model*

```
ex2poi <- glm(los ~ procedure + sex + admit + age75,
     family=poisson, data=azpro)
exp(coef(ex2poi))
exp(confint(ex2poi))
ex2nb <- glm.nb(los ~ procedure + sex + admit + age75, data=azpro)
exp(coef(ex2nb))
exp(confint(ex2nb))
```

```
. glm los procedure sex admit age75, fam(poi) eform

Generalized linear models                  No. of obs        =        3589
Optimization      : ML                     Residual df       =        3584
                                           Scale parameter   =           1
Deviance          =  8874.147204           (1/df) Deviance   =    2.476046
Pearson           =  11499.22422           (1/df) Pearson    =    3.208489

                                           AIC               =    6.238449
Log likelihood    = -11189.89758           BIC               =   -20463.15
---------------------------------------------------------------------------
          |                 OIM
      los |      IRR    Std. Err.      z    P>|z| [95% Conf.  Interval]
----------+----------------------------------------------------------------
procedure | 2.612576   .031825    78.84   0.000   2.550939    2.675702
      sex | .8834417  .0104349   -10.49   0.000   .8632245    .9041324
    admit | 1.386239  .0168061    26.94   0.000   1.353688    1.419573
    age75 | 1.129999  .0140675     9.82   0.000   1.102761     1.15791
---------------------------------------------------------------------------
```

Given the large number of cases in the data, a Pearson *chi*2 dispersion of 3.21 indicates overdispersion. Possible intrinsic causes may stem from the fact that the data have no zeros, as well as the disparity in the numbers of low counts. There may also be a clustering effect resulting from a higher correlation of procedures being done within providers than between providers. First, however, we shall model the data as negative binomial.

```
. glm los procedure sex admit age75, fam(nb ml) eform

Generalized linear models                 No. of obs        =        3589
Optimization      : ML                    Residual df       =        3584
                                          Scale parameter   =           1
Deviance          =  3525.650017          (1/df) Deviance   =    .9837193
Pearson           =  4947.825864          (1/df) Pearson    =    1.380532
Variance function: V(u) = u+(.1601022)u^2  [Neg. Binomial]
```

```
Link function     : g(u) = ln(u)              [Log]
                                              AIC            =  5.560626
Log likelihood    = -9973.543468              BIC            = -25811.64
------------------------------------------------------------------------------
         |                 OIM
     los |       IRR    Std. Err.      z    P>|z|  [95% Conf. Interval]
---------+--------------------------------------------------------------------
procedure | 2.667403 .0490528   53.35   0.000   2.572973    2.765298
      sex |  .881229 .0168211   -6.62   0.000   .8488693    .9148221
    admit | 1.448736 .0276089   19.45   0.000   1.395621    1.503871
    age75 | 1.127589 .0228369    5.93   0.000   1.083706    1.173249
------------------------------------------------------------------------------
```

The incidence rate ratios between the Poisson and negative binomial models are quite similar. This is not surprising given the proximity of α to zero. On the other hand, the AIC and BIC statistics for the negative binomial model are less than the Poisson – 12% and 26% respectively. A likelihood ratio *chi*2 value of 2,432.7, with 1 degree of freedom, indicates that the model value of α, at 0.16, is nevertheless significantly different from an α of zero.

The fact that hospital length-of-stay data exclude the possibility of having zero counts suggests that the data be modeled using a type of zero-truncated model, e.g. a zero-truncated negative binomial, or ZTNB. A negative binomial with endogenous stratification model is another possibility. Both of these models will be applied to the data in later chapters.

Example 3: Titanic survival data

These data come from the 1912 Titanic survival log. It consists of 1,316 passengers, as well as crew. The crew members have been excluded from the analysis. Only four variables are recorded, with each in binary format. The goal of the study is to assess the risk of surviving. These data have previously been examined using binary logistic regression. Here we shall demonstrate that modeling it as logistic is inferior to modeling the data as counts. Converting a binary model to grouped has been shown earlier in this chapter. The example data are defined in Table 9.35.

Modeled as a binary logistic model, and reporting the exponentiated parameter estimates as odds ratios, we have the following output (below Table 9.36).

Recall that goodness of fit is evaluated differently for binary binomial models than for count models. It makes no sense to talk about overdispersed binary response data, therefore the Pearson *chi*2 dispersion statistic has little value in

Table 9.35 *Titanic survivor dictionary*

```
n=1,316
---------------------------------------
survived       Survived           1/0
---------------------------------------
age            Child vs Adult     1/0
sex            Male vs Female     1/0
class1         class==1st class   1/0
class2         class==2nd class   1/0
class3         class==3rd class   1/0
```

Table 9.36 R: *Titanic logistic model*

```
rm(list=ls())
data(titanic)
attach(titanic)
ex3logit <- glm(survived ~ age + sex + factor(class),
    family=binomial, data=titanic)
exp(coef(ex3logit))
exp(confint(ex3logit))
```

```
. glm survived age sex class2 class3, fam(bin) eform

Generalized linear models                 No. of obs       =       1316
Optimization      : ML                    Residual df      =       1311
                                          Scale parameter  =          1
Deviance          =  1276.200769          (1/df) Deviance  =    .973456
Pearson           =  1356.674662          (1/df) Pearson   =    1.03484

                                          AIC              =   .9773562
Log likelihood    = -638.1003845          BIC              =  -8139.863
---------------------------------------------------------------------------
         |                 OIM
survived | Odds Ratio Std. Err.     z    P>|z|[95% Conf. Interval]
---------+-----------------------------------------------------------------
     age | .3479809  .0844397  -4.35  0.000  .2162749  .5598924
     sex | .0935308  .0135855 -16.31  0.000  .0703585  .1243347
  class2 | .3640159  .0709594  -5.18  0.000  .2484228  .5333952
  class3 | .1709522  .0291845 -10.35  0.000  .1223375  .2388853
---------------------------------------------------------------------------
```

assessing model worth. AIC and BIC statistics are important, but only when comparing models. Hosmer and Lemeshow (2000) and Hilbe (2009) provide extensive information regarding fit considerations for logistic models.

As seen earlier, an individual or observation level format may be converted to grouped format by using code similar to the following:

Table 9.37 *Titanic grouped data set*

	survive	cases	age	sex	class1	class2	class3
1	14	31	child	women	0	0	1
2	13	13	child	women	0	1	0
3	1	1	child	women	1	0	0
4	13	48	child	man	0	0	1
5	11	11	child	man	0	1	0
6	5	5	child	man	1	0	0
7	76	165	adults	women	0	0	1
8	80	93	adults	women	0	1	0
9	140	144	adults	women	1	0	0
10	75	462	adults	man	0	0	1
11	14	168	adults	man	0	1	0
12	57	175	adults	man	1	0	0

```
. egen grp = group(age sex class1 class2 class3)
. egen cases = count(grp), by(grp)
. egen survive = sum(survived), by(grp)
. sort grp
. by grp: keep if _n == 1
```

The above code groups the 1,316 cases into 12 covariate patterns. Table 9.37 consists of the entire re-formatted data set.

The data are modeled as a grouped logistic model. Parameter estimates, standard errors, and so forth are identical to the binary model.

As a grouped logistic model, we find that the data are indeed overdispersed with a Pearson *chi*2 based dispersion statistic of 14.41. The AIC and BIC statistics are 13.15 and 60.57 respectively.

It is easy to re-parameterize the model as a count response model. Converting the GLM family from binomial to Poisson is all that is required to make the change. Since the canonical link is the default (with most commercial software), it is not necessary to manually change link functions. On the

Table 9.38 R: ***Titanic*** *grouped logistic; Poisson, NB2 model*

```
rm(list=ls())
data(titgrp)
attach (titgrp)
y0 <- cases - survive
ex3gr <- glm( cbind(survive, y0) ~age + sex + factor(class),
 family=binomial, data=titanicgrp)
summary(ex3gr)
exp(coef(ex3gr))
exp(confint(ex3gr))
lncases <-log(cases)
ex3poi <- glm(survive ~ age + sex + factor(class) +
   offset(lncases), family=poisson, data=titgrp)
summary(ex3poi)
exp(coef(ex3poi))
exp(confint(ex3poi))
ex3nb <- glm.nb(survive ~ age + sex + factor(class) +
   offset(lncases), data=titgrp)
summary(ex3nb)
exp(coef(ex3nb))
exp(confint(ex3nb))
```

```
. glm survive age sex class2 class3, fam(bin cases) eform

Generalized linear models                  No. of obs      =         12
Optimization     : ML                      Residual df     =          7
                                           Scale parameter =          1
Deviance         =  110.8437538            (1/df) Deviance =   15.83482
Pearson          =  100.8828206            (1/df) Pearson  =   14.41183
                                  . . .
                                           AIC             =   13.14728
Log likelihood   = -73.88365169            BIC             =   60.56729
----------------------------------------------------------------------
        |                 OIM
survive | Odds Ratio Std. Err.    z    P>|z|  [95% Conf. Interval]
--------+-------------------------------------------------------------
    age |  .3479809  .0844397  -4.35  0.000  .2162749    .5598924
    sex |  .0935308  .0135855 -16.31  0.000  .0703585    .1243347
 class2 |  .3640159  .0709594  -5.18  0.000  .2484228    .5333952
 class3 |  .1709522  .0291845 -10.35  0.000  .1223375    .2388853
----------------------------------------------------------------------
```

other hand, the total number of observations per group, as indicated in the variable *cases*, is changed from the binomial denominator to the Poisson logged offset.

```
. glm survive age sex class2 class3, fam(poi) lnoffset(cases)
eform

Generalized linear models              No. of obs      =        12
Optimization     : ML                  Residual df     =         7
                                       Scale parameter =         1
Deviance         =  38.30402583        (1/df) Deviance =  5.472004
Pearson          =  39.06072697        (1/df) Pearson  =  5.580104
                                        . . .
                                       AIC             =  8.921621
Log likelihood   = -48.5297265         BIC             =  20.90968
----------------------------------------------------------------
         |              OIM
survive  |    IRR   Std. Err.    z     P>|z|[95% Conf.  Interval]
--------+-------------------------------------------------------
    age | .6169489  .0898438  -3.32  0.001  .4637587   .8207412
    sex | .3117178  .0296204 -12.27  0.000  .2587484   .3755308
 class2 | .6850337  .0805458  -3.22  0.001  .5440367   .8625725
 class3 |  .463439  .0496088  -7.18  0.000  .3757299   .5716227
  cases |(exposure)
----------------------------------------------------------------
```

The Poisson model produces an approximate threefold reduction in dispersion over the grouped logistic model. Additionally, the AIC and BIC statistics are reduced by some 50% and 300% respectively. The deviance statistic has also been reduced from 110.84 to 38.30. Clearly, the Poisson is the preferable model. On the other hand, with a dispersion statistic of 5.58, the model indicates that the data are overdispersed. Modeled as a negative binomial, we have:

```
. glm survive age sex class2 class3, fam(nb ml) lnoffset(cases)
eform

Generalized linear models                No. of obs      =        12
Optimization     : ML                    Residual df     =         7
                                         Scale parameter =         1
Deviance         =  12.47948809          (1/df) Deviance =  1.782784
Pearson          =  11.07146967          (1/df) Pearson  =  1.581639
Variance function: V(u) = u+(.104)u^2    [Neg. Binomial]
Link function    : g(u) = ln(u)          [Log]
                                         AIC             =  8.119471
```

```
Log likelihood      = -43.71682842      BIC              = -4.914858
------------------------------------------------------------------------
         |                OIM
survive  |       IRR    Std. Err.     z     P>|z|  [95% Conf. Interval]
---------+--------------------------------------------------------------
     age | . 5116907   .1289491   -2.66   0.008   .3122481    .8385238
     sex |  .3752549   .0887939   -4.14   0.000   .2360003    .5966781
  class2 |  .6875551   .2097046   -1.23   0.219   .3781732    1.250041
  class3 |  .4037074   .1157954   -3.16   0.002      .2301    .7082993
   cases | (exposure)
------------------------------------------------------------------------
Note: Negative binomial parameter estimated via ML and treated as
      fixed once estimated.
```

The AIC and BIC statistics are 8.12 (from 8.92) and −4.91 (from 20.91), with BIC alone indicating better fit. A likelihood ratio test indicates that the model is significantly different from Poisson ($\chi^2 = 9.63$; dof = 1; $P > \chi^2 = 0.001$).

To clean up the model it is necessary to combine class2 with class1 as the referent for *class*; α is slightly increased, with the deviance and AIC slightly decreased. More substantial change is found in the reduced model BIC statistic (−4.9 to −8.2). A final model is presented below.

```
. glm survive age sex  class3, fam(nb ml) lnoffset(cases) eform

Generalized linear models                  No. of obs      =        12
Optimization     : ML                      Residual df     =         8
                                           Scale parameter =         1
Deviance         =  11.71286566            (1/df) Deviance =  1.464108
Pearson          =  8.686216189            (1/df) Pearson  =  1.085777
Variance function: V(u) = u+(.1339)u^2  [Neg. Binomial]
Link function    : g(u) = ln(u)            [Log]
                                           AIC             =  8.061748
Log likelihood   = -44.37048795            BIC             = -8.166388
------------------------------------------------------------------------
         |                 OIM
survive  |        IRR    Std. Err.     z    P>|z|[95% Conf. Interval]
---------+--------------------------------------------------------------
     age |   .5410341   .1465874   -2.27  0.023    .318127   .9201289
     sex |   .3996861   .1034007   -3.54  0.000   .2407183   .6636345
  class3 |   .4819995   .1261313   -2.79  0.005   .2886033   .8049927
   cases | (exposure)
------------------------------------------------------------------------
Note: Negative binomial parameter estimated via ML and treated
      as fixed once estimated.
```

Example 4: Health reform data

A fourth example consists of data from a subset of the German Socio-Economic Panel (SOEP). The subset was created by Rabe-Hesketh and Skrondal (2005). Only working women are included in these data. Beginning in 1997, German health reform in part entailed a 200% increase in patient co-payment as well as limits in provider reimbursement. Patients were surveyed for the one-year panel (1996) before and the one-year panel (1998) after reform to assess whether the number of physician visits by patients declined – which was the goal of reform legislation.

The response, or variable to be explained by the model, is *numvisit,* which indicates the number of patient visits to a physician's office during a three month period. The data set, **mdvis**, consists of 2,227 cases.

Table 9.39 *German health reform data (**mdvis**)*

numvisit	visits to MD office 3mo prior - response
reform	1=interview yr post-reform: 1998;0=pre-reform:1996
badh	1=bad health; 0 = not bad health
age	Age(yrs 20-60)
agegrp	1=20-39; 2=40-49; 3=50-60
age1,2,3	age1=20-39; age2=40-49; age3=50-60
educ	education(1:7-10;2=10.5-12;3=HSgrad+)
educ1,2,3	educ1=7-10; educ2=10.5-12; educ3=HSgrad+
loginc	ln(household income DM)

A tabulation of *numvisit* provides an overview of the observed or empirical count distribution.

Table 9.40 R: *Table with Count, Freq, Prop, cumProp*

```
rm(list=ls())
data(mdvis)
attach(mdvis)
myTable(numvisit)
```

```
. use mdvis, clear
. tab numvisits
```

Table 9.41 *Tabulation of response: numvisits*

Visits MD	Freq.	Percent	Cum.	
0	665	29.86	29.86	<= large number of
1	447	20.07	49.93	0's
2	374	16.79	66.73	
3	256	11.50	78.22	
4	117	5.25	83.48	
	.	.	.	
36	1	0.04	99.82	
40	2	0.09	99.91	
50	1	0.04	99.96	
60	1	0.04	100.00	
Total	2,227	100.00		

Table 9.42 R: ***mdvis:*** *Poisson models*

```
ex4poi <- glm (numvisit ~ reform + badh + age +
      educ + loginc, family=poisson, data=mdvis)
summary(ex4poi)
exp(coef(ex4poi))
exp(confint(ex4poi))
library(sandwich)          # library for robust SE
vcovHC(poi)
exp(sqrt(diag(vcovHC(ex4poi))))
exp(sqrt(diag(vcovHC(ex4poi, type="HC0"))))
```

The data are first modeled using a Poisson regression. Results appear as:

```
. glm numvisit reform badh age educ loginc, fam(poi)

Generalized linear models                No. of obs        =       2227
Optimization       : ML                  Residual df       =       2221
                                         Scale parameter =            1
Deviance           =  7422.124433        (1/df) Deviance =   3.341794
Pearson            =   9681.69202    =>  (1/df) Pearson  =   4.359159
                                    .  .  .
                                     =>  AIC               =   5.343357
```

```
Log likelihood = -5943.828046            BIC       = -9698.256
----------------------------------------------------------------
         |             OIM
numvisit|     Coef. Std. Err.     z    P>|z|   [95% Conf. Interval]
--------+-------------------------------------------------------
  reform| -.1398842 .0265491   -5.27  0.000   -.1919195  -.0878489
    badh|  1.132628 .0302986   37.38  0.000    1.073244   1.192013
     age|  .0048853 .0012526    3.90  0.000    .0024302   .0073404
    educ| -.0118142 .0059588   -1.98  0.047   -.0234933  -.0001351
  loginc|  .1520247 .0359837    4.22  0.000     .081498   .2225515
   _cons|  -.421508  .268966   -1.57  0.117   -.9486718   .1056558
----------------------------------------------------------------
```

The model appears *prima facie* to fit well. Predictors appear significant. However, the dispersion statistic is over 4, indicating substantial overdispersion given the large number of observations. The AIC statistic equals 5.343, a value with which we shall later compare alternative models.

Exponentiated coefficients give us incidence rate ratios, and appropriately adjusted standard errors and confidence intervals.

```
. glm numvisit reform badh age educ loginc, fam(poi) eform nohead

----------------------------------------------------------------
         |             OIM
numvisit |      IRR   Std. Err.     z    P>|z|  [95% Conf. Interval]
---------+------------------------------------------------------
  reform | .8694589   .0230834   -5.27  0.000   .8253733   .9158993
    badh | 3.103804   .0940408   37.38  0.000   2.924853   3.293703
     age | 1.004897   .0012588    3.90  0.000   1.002433   1.007367
    educ | .9882553   .0058888   -1.98  0.047   .9767805   .9998649
  loginc | 1.164189   .0418918    4.22  0.000   1.084911    1.24926
----------------------------------------------------------------
```

Employing a robust sandwich variance estimator gives us the following table of results.

```
. glm numvisit reform badh age educ loginc, fam(poi) vce(robust)
  eform nohead

----------------------------------------------------------------
         |            Robust
numvisit |      IRR   Std. Err.     z    P>|z|  95% Conf. Interval]
---------+------------------------------------------------------
  reform | .8694589   .0515157   -2.36  0.018   .7741322   .9765241
    badh | 3.103804   .2590781   13.57  0.000   2.635382   3.655486
     age | 1.004897   .0029544    1.66  0.097   .9991234   1.010705
    educ | .9882553   .0112303   -1.04  0.299   .9664875   1.010513
  loginc | 1.164189   .0895704    1.98  0.048    1.00123   1.353671
----------------------------------------------------------------
```

There is an approximately 13% reduction in visits to a physician between 1996 and 1998, adjusted for health status, age, education, and the natural log of household income. Age and education do not contribute to the model. Recall that the model appears to indicate overdispersion.

The fact that the response consists of 30% zero counts is likely to be a cause of overdispersion. Table 9.43 displays the observed and predicted counts from 0 to 10. Predicted values are derived from the last Poisson model. Note that the robust standard errors are substantially greater than the corresponding model standard errors indicating that the data are overdispersed.

```
poi.obs.pred(len=11, model=ex4 poi)
```

Table 9.43 *Observed proportion and predicted probability for model visits from 0 to 10*

	Visits	%Visits obs.	%Visits pred.
1	0	0.298608	0.115397
2	1	0.200718	0.233520
3	2	0.167939	0.240503
4	3	0.114953	0.170382
5	4	0.052537	0.096959
6	5	0.045352	0.051449
7	6	0.034127	0.029828
8	7	0.009430	0.020123
9	8	0.012124	0.014646
10	9	0.004041	0.010482
11	10	0.027391	0.007053

Table 9.43 shows the observed proportions or empirical probabilities for each count up to 10 together with the predicted probability for each count based on the model (given the probability function $f(y;\, \mu) = e^{-\mu}\mu^{y}/y!$). Comparing each pair of percentages provides the analyst with values which may be analyzed using a *chi*2, or some other, test aimed at determining if both sets of counts come from the same population. No formal tests have appeared in the literature, but they may be constructed from the given data.

Note the near threefold higher value in the observed frequency for zero counts than for the predicted value. This is reflected in a graph of the table data, with additional counts through 20. The code for Figure 9.5 is on the book's website.

```
plot(0:20, avgp, type="b", xlim=c(0,20),
      main = "Observed vs Predicted Visits",
      xlab = "Number Visits", ylab = "Probability of Visits")
lines(0:20, propObs, type = "b", pch = 2)
legend("topright", legend = c("Predicted Visits","Observed
    Visits"), lty = c(1,1), pch = c(1,2))
```

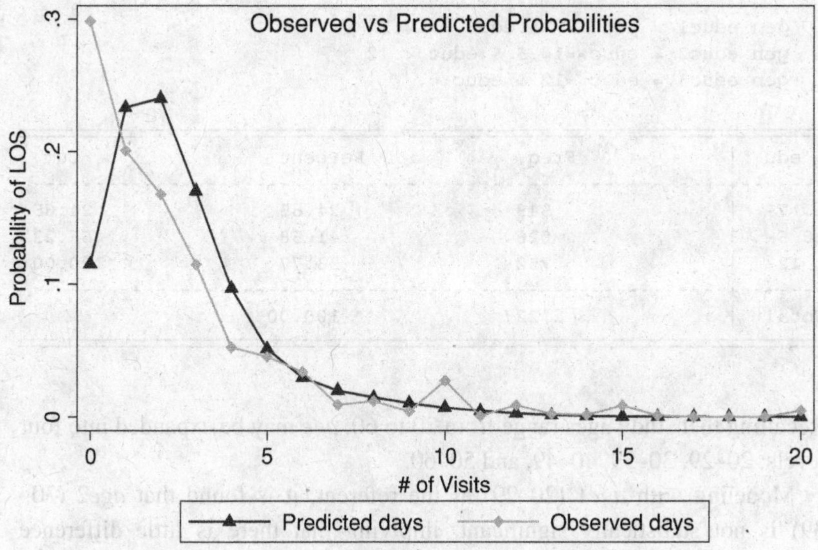

Figure 9.5 Poisson model for number of visits: observed versus predicted probabilities

The observed proportions approximate a geometric distribution, which is the discrete correlate of a negative exponential distribution. It is likely that if the model is estimated using a negative binomial regression, the value of α will approximate 1.0.

It is important, though, to first deal with *educ* and *age*. *Educ* does not appear to contribute to the model and *age* is questionable, particularly when subjected to scaling and adjusted by a robust sandwich variance estimator (not shown here). However, recall that both variables are discrete, with many levels. *Educ* has 16 levels and *age* has 41, one level for each year from 20 through 60. Both predictors may be considered as continuous; however, each age is measured as a unit, with no decimals. Categorizing each of these predictors may help understand the differential contribution of education levels and age groups, and also may help in dealing with model overdispersion.

Educ may be left as found, or separated into three separate binary predictors. Commercial statistical software generally prefers one type or another. In any event, one can have the software generate three dummies resulting in predictor levels *educ*1, *educ*2, and *educ*3. Levels are defined by the lowest number of years of education.

```
. gen educ1 = educ>=7 & educ < 10.5
. gen educ2 = educ>=10.5 & educ < 12
. gen educ3 = educ>=12 & educ<=.
```

edu	Freq.	Percent	Cum.
7-	549	24.65	24.65
10.5-	926	41.58	66.23
12-	752	33.77	100.00
Total	2,227	100.00	

Recalling that study ages range from 20 to 60, *age* may be expanded into four levels: 20–29, 30–39, 40–49, and 50–60.

Modeling with *age*1 (20–29) as the referent, it is found that *age*2 (30–39) is not statistically significant, implying that there is little difference between the two age divisions. In such a situation the two levels may be combined into an expanded reference group, 20–39. Levels can be labeled and tabulated.

age	Freq.	Percent	Cum.
20-39	1,352	60.71	60.71
40-49	515	23.13	83.83
50-60	360	16.17	100.00
Total	2,227	100.00	

Re-running the model with levels *educ*1–3 and *age*1–3, the model appears as in Table 9.44.

Table 9.44 R: ***mdvis*** *expanded data – Poisson model*

```
ex4poie <- glm(numvisit ~ reform + badh + factor(educ) +
  factor(agegrp) + loginc, family=poisson, data=mdvis)
exp(coef(ex4poie))
exp(confint(ex4poie))
library(haplo.ccs)        # load new library
sandcov(ex4poie, mdvis$id)
exp(sqrt(diag(sandcov(ex4poie, mdvis$id))))
```

```
. glm numvisit reform badh educ2 educ3 age2 age3 loginc, fam(poi) eform

Generalized linear models                 No. of obs      =      2227
Optimization     : ML                     Residual df     =      2219
                                          Scale parameter =         1
Deviance         =  7398.293267           (1/df) Deviance =  3.334066
Pearson          =  9518.948272           (1/df) Pearson  =  4.289747
                                 . . .
                                          AIC             =  5.334452
Log likelihood   = -5931.912464           BIC             =  -9706.67
----------------------------------------------------------------------
         |               OIM
numvisit |      IRR    Std. Err.     z    P>|z|    [95% Conf. Interval]
---------+------------------------------------------------------------
  reform | .8706594    .0231258   -5.21   0.000    .8264932   .9171858
    badh | 3.132308    .0938026   38.13   0.000     2.95375   3.321661
   educ2 | 1.085224    .0365147    2.43   0.015    1.015965   1.159204
   educ3 | .9221584    .0340499   -2.19   0.028    .8577794   .9913691
    age2 | 1.095001    .0352259    2.82   0.005    1.028092   1.166266
    age3 | 1.144818    .0410837    3.77   0.000    1.067062    1.22824
  loginc | 1.146791    .0410935    3.82   0.000    1.069012   1.230228
----------------------------------------------------------------------
```

Owing to overdispersion in the model, and the fact that there may be a clustering effect resulting from multiple visits by the same individual, it is wise to apply a robust cluster variance estimator to the model.

```
. glm numvisit reform badh educ2 educ3 age2 age3 loginc,
  fam(poi) eform cluster(id)

Generalized linear models                 No. of obs      =      2227
Optimization     : ML                     Residual df     =      2219
                                          Scale parameter =         1
Deviance         =  7398.293267           (1/df) Deviance =  3.334066
```

```
Pearson            =  9518.948272    (1/df) Pearson   =  4.289747
                                     .  .
                                     AIC              =  5.334452
Log pseudolikelihood = -5931.912464  BIC              =  -9706.67
                   (Std. Err. adjusted for 1518 clusters in id)
------------------------------------------------------------------
         |                Robust
numvisit |      IRR    Std. Err.     z    P>|z|  [95% Conf. Interval]
---------+--------------------------------------------------------
  reform | .8706594   .0488115   -2.47   0.013   .7800593   .9717823
    badh | 3.132308   .2628195   13.61   0.000   2.657318   3.692202
   educ2 | 1.085224   .0977545    0.91   0.364   .9095888   1.294773
   educ3 | .9221584   .0774285   -0.97   0.334   .7822307   1.087117
    age2 | 1.095001   .0877094    1.13   0.257    .935909   1.281138
    age3 | 1.144818   .1069266    1.45   0.148   .9533091   1.374798
  loginc | 1.146791   .0946715    1.66   0.097   .9754714   1.348198
------------------------------------------------------------------
```

Application of a negative binomial model is a reasonable approach to dealing with the excess dispersion in the Poisson model (4.29), particularly when a specific source of the overdispersion has not been identified.

Table 9.45 R: ***mdvis** expanded data – NB2 model*

```
ex4nbe <- glm.nb(numvisit ~ reform + badh + educ2 + educ3
   + age2 + age3 + loginc, data=mdvis)
exp(coef(ex4nbe))
exp(confint(ex4nbe))
library(haplo.ccs)
sandcov(ex4nbe, mdvis$id)
exp(sqrt(diag(sandcov(ex4nbe, mdvis$id))))
```

```
. glm numvisit reform badh educ2 educ3 age2 age3 loginc, fam(nb ml) eform

Generalized linear models                  No. of obs       =       2227
Optimization       : ML                    Residual df      =       2219
                                           Scale parameter  =          1
Deviance           =  2412.452905          (1/df) Deviance  =    1.08718
Pearson            =  2644.673466          (1/df) Pearson   =   1.191831
Variance function: V(u) = u+(.9982)u^2     [Neg. Binomial]
Link function      : g(u) = ln(u)          [Log]
                                           AIC              =   4.103197
```

```
Log likelihood    = -4560.909631              BIC              = -14692.51
------------------------------------------------------------------------------
             |                 OIM
    numvisit |        IRR   Std. Err.      z    P>|z|     [95% Conf. Interval]
-------------+----------------------------------------------------------------
      reform |    .871369   .0445206    -2.69   0.007     .7883371    .9631464
        badh |   3.134872   .2332147    15.36   0.000     2.709542    3.626969
       educ2 |   1.085687   .0716943     1.24   0.213      .953882    1.235704
       educ3 |    .970105   .0685734    -0.43   0.668     .8445984    1.114262
        age2 |   1.049625   .0666178     0.76   0.445     .9268511    1.188662
        age3 |   1.206782   .0867935     2.61   0.009     1.048115    1.389468
      loginc |   1.134513   .0798461     1.79   0.073     .9883313    1.302316
------------------------------------------------------------------------------
Note: Negative binomial parameter estimated via ML and treated as
      fixed once estimated.
```

Negative binomial regression has changed the model statistics. Note at first that the Pearson *chi*2 dispersion has been reduced from 4.29 to 1.19. AIC and BIC statistics deflate from 5.33 to 4.10 and from −9,706.67 to −14,692.51 respectively. The deviance itself has been substantially reduced from 7,398 to 2,412 – a threefold reduction. An interesting aside is that the value of α is very close to unity. Recall the discussion regarding Figure 9.5, which showed that the shape of the predicted counts from the Poisson model take the form of a geometric distribution. This speculation has been borne out with the value of $\alpha = 0.9982$ in the negative binomial model of the same data.

Adjusting the model by the clustering effect of *id*, and applying robust standard errors to the result, produces the following table of estimates. Notice that the cluster adjusted standard errors differ very little from base model standard errors. Moreover, a simple application of robust standard errors to the Hessian matrix without a clustering effect yields little difference to the model with clustering. This mutual lack of effect can be interpreted as indicating that there is little if any overdispersion in the resultant model data due to clustering and that any remaining variability in the data comes from a yet to be specified source. Fortunately, the unadjusted negative binomial model accounts for most of the Poisson overdispersion. This conclusion agrees with the observed value of the dispersion statistic.

```
. glm numvisit reform badh educ2 educ3 age2 age3 loginc, fam(nb ml)
  eform vce(robust) nohead
                               (Std. Err. adjusted for 1518 clusters in id)
```

```
------------------------------------------------------------------------------
             |               Robust
    numvisit |        IRR   Std. Err.      z    P>|z|     [95% Conf. Interval]
-------------+----------------------------------------------------------------
      reform |    .871369   .0446676    -2.69   0.007     .7880764     .963465
        badh |   3.134872   .2587493    13.84   0.000     2.666628    3.685337
       educ2 |   1.085687   .0880525     1.01   0.311     .9261247    1.272739
       educ3 |    .970105   .0786037    -0.37   0.708      .827655    1.137072
        age2 |   1.049625   .0772138     0.66   0.510     .9086928    1.212415
        age3 |   1.206782   .1161242     1.95   0.051     .9993571    1.457259
      loginc |   1.134513   .0952187     1.50   0.133     .9624292    1.337365
------------------------------------------------------------------------------
```

It appears that the negative binomial model – or geometric model – fits the data better than the Poisson model. However, it also appears that the cuts we made in *educ* and *age* have resulted in levels that do not significantly contribute to the model. For those who wish to continue model development, re-classification should be attempted. In addition, since the observed zeros in the data exceed the distributional assumption of the negative binomial, alternative models designed to deal with this specific type of problem should be investigated, such as zero-inflated and hurdle models. Both of these models will be discussed in later chapters and employed on these data.

When modeling a data situation in which two groups are being distinguished to determine differences of some sort between them, it is sometimes instructional to graph the residual data as a whole while allowing the intrinsic groupings to emerge. This may be done by graphing standardized deviance residuals by the predicted mean or μ. A correction value is used with the values of μ, calculated as 2*sqrt(μ).

Two groupings of residuals reflect the reform periods of 1996 and 1998. Standardized deviance residuals greater than 2 can be regarded as possible outliers. Pierce and Shafer (1986) recommend an adjusted standardized deviance using the following formula: $D_{\text{adj}} - d + 1/\{6^*\text{sqrt}(\text{mean}(y))\}$. However, I have not found this to be superior to traditionally defined standardized deviance residuals. On the other hand I have argued for the use of Anscombe residuals (Hilbe, 1994a; Hilbe and Turlach, 1995), which are theoretically more appropriate than standardized deviance residuals, although the two values are generally quite similar. When modeling, both should be attempted. If negative binomial Anscombe and standardized deviance residuals differ, then additional tests are required.

```
. qui glm numvisit reform badh educ2 educ3 age2 age3 loginc, fam(poi)
. predict  sdevnb, deviance stan
. predict mup
. replace mup = 2*sqrt(munp)
. scatter sdevnb mup, title(Poisson: Standardized Deviance vs 2*sqrt(mu))
```

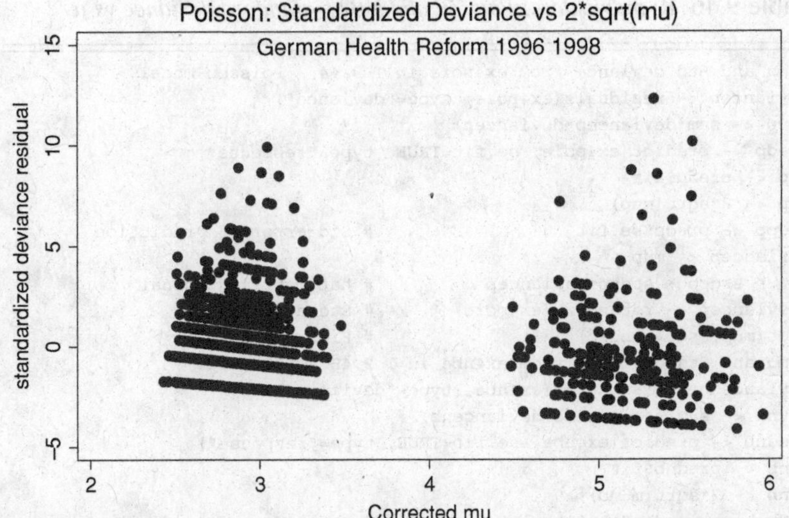

Figure 9.6 German health reform 1996 and 1998; standardized deviance residuals Versus corrected predicted mean number of visits. (A) Poisson model

```
. qui glm numvisit reform badh educ2 educ3 age2 age3 loginc, fam(nb ml)
. predict  sdevnb, deviance stan
. predict munb
. replace munb = 2*sqrt(munb)
. scatter sdevnb munb, title(NB2: Standardized Deviance vs 2*sqrt(mu))
```

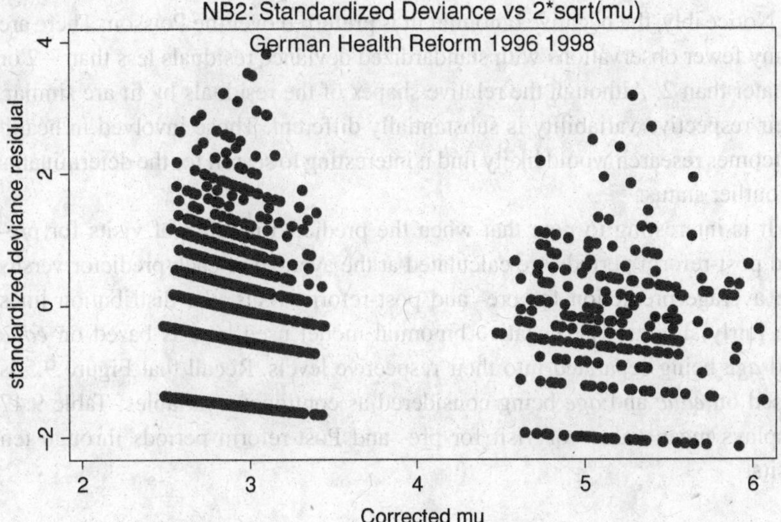

Figure 9.6 German health reform 1996 and 1998; standardized deviance residuals Versus corrected predicted mean number of visits. (B) negative binomial (NB2) model

Table 9.46 R: ***mdvis*** *data: Poisson and NB2 – standard deviance vs* μ

```
# mu and Std deviance from ex4poie in T 9.44 - Poisson model
deviancep <- residuals(ex4poie, type="deviance")
devp <- sum(deviancep*deviancep)
predp <- predict(ex4poie, se.fit=TRUE, type="response")
mup <- predp$fit
mup <- 2*sqrt(mup)
stdpp <- predp$se.fit                      # Std error of prediction
variancep <- mup
hp <- stdpp * stdpp*variancep              # hat matrix diagonal
sdeviancep <- rstandard(ex4poie)           # Std deviance
plot(mup, sdeviancep)                      # Figure 9.6A
# mu and Std deviance from ex4nbe in T 9.45 - NB2 model
deviancenb <- residuals(ex4nbe, type="deviance")
devnb <- sum(deviancenb*deviancenb)
prednb <- predict(ex4nbe, se.fit=TRUE, type="response")
munb <- prednb$fit
munb <- 2*sqrt(munb)
stdpnb <- prednb$se.fit                    # Std error of prediction
variancenb <- munb
hnb <- stdpnb * stdpnb*variancenb
sdeviancenb <- rstandard(ex4nbe)           # Std dev
plot(munb, sdeviancenb)                    # Figure 9.6B
```

Noticeably, the negative binomial fit is preferred over the Poisson. There are many fewer observations with standardized deviance residuals less than −2 or greater than 2. Although the relative shapes of the residuals by fit are similar, their respective variability is substantially different. Those involved in health outcomes research would likely find it interesting to search for the determinants of outlier status.

It is interesting to note that when the predicted number of visits for pre- and post-reform periods are calculated at the average of each predictor versus the average prediction for pre- and post-reform visits, the distribution lines are fairly similar. The negative binomial model used here is based on *educ* and *age* being separated into their respective levels. Recall that Figure 9.5 is based on *educ* and *age* being considered as continuous variables. Table 9.47 displays mean values of visit for pre- and Post-reform periods through ten visits.

```
. prvalue, x(reform=0)  maxcnt(11)
```

Table 9.47 *Pre (1996) and post (1998) mean predicted visits with CIs (confidence intervals) calculated by delta method*

```
PRE REFORM (1996)
-------------------------------------------------------------
                         95% Conf. Interval
-------------------------------------------------------------
 Rate:                 2.5115    [ 2.3345,     2.6885]
 Pr(y=0|x):            0.2845    [ 0.2701,     0.2989]
 Pr(y=1|x):            0.2037    [ 0.1976,     0.2099]
 Pr(y=2|x):            0.1458    [ 0.1443,     0.1473]
 Pr(y=3|x):            0.1043    [ 0.1032,     0.1053]
 Pr(y=4|x):            0.0746    [ 0.0723,     0.0768]
 Pr(y=5|x):            0.0533    [ 0.0507,     0.0560]
 Pr(y=6|x):            0.0381    [ 0.0355,     0.0408]
 Pr(y=7|x):            0.0273    [ 0.0248,     0.0297]
 Pr(y=8|x):            0.0195    [ 0.0173,     0.0216]
 Pr(y=9|x):            0.0139    [ 0.0121,     0.0158]
 Pr(y=10|x):           0.0100    [ 0.0085,     0.0115]

. prvalue, x(reform=1)  maxcnt(11)

POST REFORM (1998)
-------------------------------------------------------------
                         95% Conf. Interval
-------------------------------------------------------------
Rate:                 2.1884    [ 2.0329,      2.344]
Pr(y=0|x):            0.3134    [ 0.2981,     0.3287]
Pr(y=1|x):            0.2153    [ 0.2096,     0.2211]
Pr(y=2|x):            0.1479    [ 0.1472,     0.1485]
Pr(y=3|x):            0.1015    [ 0.0997,     0.1033]
Pr(y=4|x):            0.0697    [ 0.0668,     0.0725]
Pr(y=5|x):            0.0478    [ 0.0448,     0.0508]
Pr(y=6|x):            0.0328    [ 0.0300,     0.0356]
Pr(y=7|x):            0.0225    [ 0.0201,     0.0249]
Pr(y=8|x):            0.0154    [ 0.0134,     0.0174]
Pr(y=9|x):            0.0106    [ 0.0090,     0.0122]
Pr(y=10|x):           0.0073    [ 0.0060,     0.0085]

MEAN VALUES OF REMAINING PREDICTORS
------------------------------------------------------------
     badh      educ2      educ3        age2       age3     loginc
------------------------------------------------------------
.11360575  .41580602  .337674  .23125281 .16165245 7.7128263

Note: The graph in Figure 9.7 and values in Table 9.47 were
      calculated from a suite of programs created by Professor
      Scott Long of the University of Indiana for use with
      binary and count response models. Aimed at assisting
      researchers to assess model fit, the programs, termed
      SPOST, may be downloaded from http://www.indiana.edu/
      ~jslsoc/spost.htm.
```

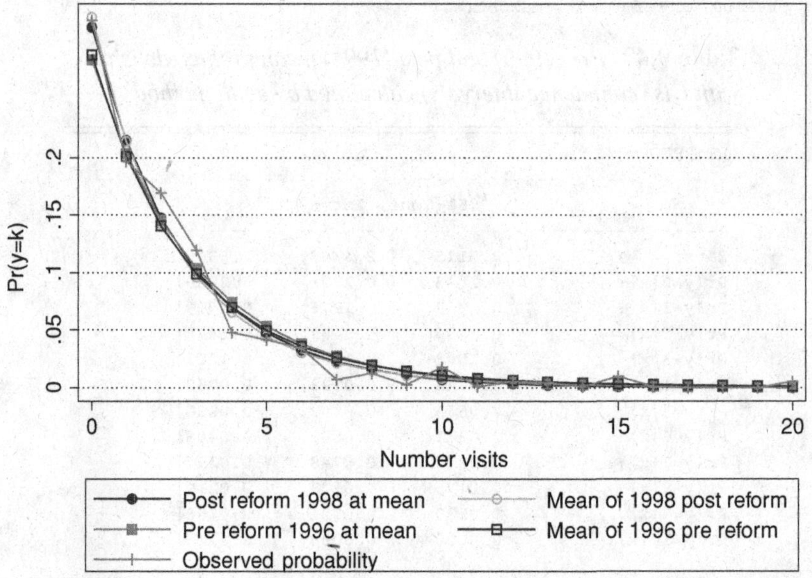

Figure 9.7 Hospital visitations: pre- and post-reform

Figure 9.7 visually displays the values from 0 through 10 given in Table 9.47. In addition, Figure 9.7 provides average prediction values for pre- and post-reform periods. These distributions are consistent with the previous distributions.

Summary

The examples used in this chapter demonstrate methods of handling overdispersed Poisson data without going beyond the basic negative binomial model. When identifying a count response model as overdispersed, it is first necessary to determine if the overdispersion is real or only apparent. If it is real, one next determines if it is significant, i.e. whether the variability significantly exceeds Poisson distributional assumptions.

Modeling data using a full maximum likelihood negative binomial allows the statistician to determine if the resulting value of α is statistically different from zero. If not, then the data is to be modeled as Poisson. If there is a statistically significant difference, then a model needs to be found that addresses the genesis of the extra correlation in the data. A basic negative binomial model may be sufficient to deal with the overdispersion, but if the data violate the distributional assumptions of both Poisson and negative binomial models, then adjustments need to be made to the negative binomial algorithm that directly deal with the source of overdispersion. Some data do not admit for an easy solution.

In the following chapters, the traditional negative binomial algorithm is enhanced to address count data that cannot be modeled using the traditional methods. First to be evaluated are data situations requiring an adjustment to the Poisson and/or negative binomial variance functions. Afterwards, data that violate Poisson or negative binomial distributional assumptions are discussed.

10
Alternative variance parameterizations

Negative binomial regression has traditionally been used to model otherwise overdispersed count or Poisson data. It is now considered to be the general catchall method used when Poisson data are found to be overdispersed, particularly when the source of overdispersion has not been identified. When we can identify that which gives rise to extra correlation, and hence overdispersion, the basic Poisson and negative binomial algorithms may themselves be further adjusted or enhanced to directly address the identified source of extra correlation. For example, when overdispersion results from an excess of zero counts in the response, an appropriate strategy is to model the data using either a zero-inflated Poisson (ZIP) or zero-inflated negative binomial (ZINB). Employing a hurdle model may also result in a better fit. On the other hand, if the response is structured such that zero counts are not possible, as in hospital length-of-stay data, a zero-truncated Poisson (ZTP) or zero-truncated negative binomial (ZTNB) model may be appropriate.

A variety of alternative models have been developed to address specific facts in the data that give rise to overdispersion. Models dealing with an excess or absence of zeros typically define a mixture that alters the distributional assumptions of the Poisson distribution. Other models are constructed to alter not the probability and log-likelihood distributions, but rather the Poisson and negative binomial variance functions. We discuss these types of models in this chapter.

Models that address overdispersion by making changes to the Poisson and negative binomial variance function are listed in Table 10.1. Note that the Poisson is regarded as the base count distribution as well as the base variance function. The negative binomial generically deals with overdispersion, and is itself modified given certain data situations.

Table 10.1 *Selected count model variance functions*

Model	Variance	
Poisson	: $V = \mu$	NB2: Poisson : $\alpha = 0$
QL Poisson	: $V = \mu\phi$	NB2: Geometric: $\alpha = 1$
Geometric	: $V = \mu(1 + \mu)$	
NB1	: $V = \mu(1 + \alpha)$	
NB2	: $V = \mu(1 + \alpha\mu)$	
NB-H	: $V = \mu(1 + (\alpha\nu)\mu)$	
NB-P	: $V = \mu + \alpha\mu^{\rho}$ <3 parameter>	
Sichel	: $V = \mu + h(\sigma,\nu)^2\mu^2$ <3 parameter>	
Delaporte	: $V = \mu + \sigma(1-\nu)^2\mu^2$ <3 parameter>	

10.1 Geometric regression: NB $\alpha = 1$

The geometric distribution is a special case of the negative binomial. Using the parameterization developed in Chapter 8, the geometric is a negative binomial with heterogeneity or overdispersion parameter, α, set to 1.0. GLM software that incorporates the negative binomial as a member family can be used to design geometric models by setting the value of α to a constant value of 1.0; the GLM algorithm should also be set with a log link with the negative binomial family. Maximum likelihood negative binomial algorithms, on the other hand, generally do not allow estimation of geometric models. Since α is estimated as an additional parameter, it cannot normally be constrained to a user-defined value unless the software allows constrained optimization. However, a geometric regression algorithm is simple to design with the appropriate programming language, e.g. SAS's IML, Stata's ML capabilities, or by programming in R.

10.1.1 Derivation of the geometric

This section describes the geometric model as a form of the NB2 negative binomial. Its derivation then is from a Poisson–gamma mixture, but without α. Since the geometric is NB2 with $\alpha = 1$, α is not required in the geometric probability or likelihood function. Otherwise it is identical to the negative binomial distribution we derived earlier. The derivation of the various statistics is also taken directly from the NB2 statistics, with α dropped from the equations.

Since it is this form of the geometric that is most commonly used in current statistical analysis, we begin with an overview of the geometric likelihood function and the GLM-based estimating algorithm. We shall wait until Section 10.1.4 to actually derive the geometric PDF.

The geometric log-likelihood function may be presented as:

$$\mathcal{L}(\mu; y) = \sum_{i=1}^{n} \left\{ y_i \ln\left(\frac{\mu_i}{1+\mu_i}\right) - \ln(1+\mu_i) \right\} \tag{10.1}$$

or

$$\mathcal{L}(\mu; y) = \sum_{i=1}^{n} \{ y_i \ln(\mu_i) - (y_i + 1) \ln(1+\mu_i) \} \tag{10.2}$$

Parameterized in terms of the log-link function $\ln(\mu) = \beta' x$ or $\ln(\mu) = x'b$

$$\mathcal{L}(\beta; y) = \sum_{i=1}^{n} \left\{ y_i \ln\left(\frac{\exp(x_i'\beta)}{1+\exp(x_i'\beta)}\right) - \ln\left(1+\exp(x_i'\beta)\right) \right\} \tag{10.3}$$

or

$$\mathcal{L}(\beta; y) = \sum_{i=1}^{n} \left\{ y_i x_i'\beta - (y_i + 1) \ln\left(1+\exp(x_i'\beta)\right) \right\} \tag{10.4}$$

As a special case of the negative binomial, the geometric mean, variance, and related functions take the same form. Therefore, we have

$$\begin{aligned} \text{MEAN} &= \mu_i \quad \text{or} \quad \exp(x_i'\beta) \\ \text{VARIANCE} &= \mu_i + \mu_i^2 \quad \text{or} \quad \mu_i(1+\mu_i) \\ &= \exp(x_i'\beta)\left(1+\exp(x_i'\beta)\right) \end{aligned}$$

$$\text{DERIVATIVE OF LINK} = \frac{1}{\mu_i(1+\mu_i)}$$

$$\text{DEVIANCE} = 2\sum_{i=1}^{n} \left\{ y_i \ln\left(\frac{y_i}{\mu_i}\right) - (1+y_i)\ln\left(\frac{1+y_i}{1+\mu_i}\right) \right\} \tag{10.5}$$

10.1.2 Synthetic geometric models

Synthetic geometric models may be created in the same manner as were the synthetic negative binomial models. Note that the only difference in the Table 10.2 code and that given in Table 9.4 is the pair of 1s for the **rgamma** function parameters. A maximum likelihood geometric command written by the author and available on the book's website is used to estimate the data created by the code.

Table 10.2 R: *Synthetic geometric model*

```
# geo_rng.r
library(MASS)
nobs <- 50000
x1 <- runif(nobs)
x2 <- runif(nobs)
xb <- 2 + .75*x1 - 1.25*x2    # parameter values
exb <- exp(xb)                # Poisson predicted value
xg <- rgamma(nobs, 1, 1, 1)   # gamma variate, param 1,1
xbg <-exb*xg                  # mix Poisson and gamma variates
gy <- rpois(nobs, xbg)        # generate NB2 variates
geo <-glm.nb(gy ~ x1 + x2)    # model geometric
summary(geo)
nbg <- glm(gy ~ x1 + x2, family=negative.binomial(1))
summary(nbg)                  # GLM NB2 with α=1
```

```
. set obs 50000
. gen x1 = runiform()
. gen x2 = runiform()
. gen xb = 2 + 0.75*x1 - 1.25*x2
. gen exb = exp(xb)
. gen xg = rgamma(1,1)
. gen xbg = exb * xg
. gen gy = rpoisson(xbg)
. georeg gy x1 x2
Geometric Estimates                      Number of obs    =    50000
                                         Model chi2(2)    = 7490.34
                                         Prob > chi2      =   0.0000
Log Likelihood =-141671.4615914          Pseudo R2        =   0.0258
-------------------------------------------------------------------
   gy |    Coef.    Std. Err.     z    P>|z|   [95% Conf. Interval]
------+------------------------------------------------------------
   x1 |  .7402257  .0169009   43.80  0.000    .7071006    .7733508
   x2 | -1.254667  .0169281  -74.12  0.000   -1.287845   -1.221488
_cons |  2.004008  .0128473  155.99  0.000    1.978828    2.029189
-------------------------------------------------------------------
        (LR test against Poisson, chi2(1) = 177583.7 P = 0.0000)
```

Note the likelihood ratio test of the geometric versus Poisson models. The high *chi*2 value clearly indicates that the model is not Poisson. Modeled as a GLM negative binomial with an α of 1, we have

```
. glm gy x1 x2, fam(nb 1)
Generalized linear models               No. of obs      =     50000
Optimization      : ML                  Residual df     =     49997
                                        Scale parameter =         1
```

```
Deviance          =  56888.15208     (1/df) Deviance =  1.137831
Pearson           =   50006.2332     (1/df) Pearson  =  1.000185
Variance function: V(u) = u+(1)u^2  [Neg. Binomial]
Link function    : g(u) = ln(u)     [Log]
                                     AIC             =  5.666978
Log likelihood    = -141671.4616     BIC             = -484068.3
-----------------------------------------------------------------
      |              OIM
   gy |      Coef.  Std. Err.      z    P>|z|   [95% Conf. Interval]
------+----------------------------------------------------------
   x1 |   .7402257  .0169009    43.80  0.000    .7071005   .7733508
   x2 |  -1.254667  .0169281   -74.12  0.000   -1.287845  -1.221488
_cons |   2.004008  .0128473   155.99  0.000    1.978828   2.029189
-----------------------------------------------------------------
```

Near identical results can be obtained by modeling the data as a GLM with a negative binomial family, where α is estimated, i.e. using the *fam(nb ml)* option.

Canonical geometric models are based on the geometric PDF, as will be discussed in Section 10.1.4. However, it is appropriate to provide the synthetic models in this section. The only canonical negative binomial regression function in R is **ml.nbc.r** in the COUNT package. A canonical geometric regression has not been implemented. However, we can model the data using the canonical negative binomial model with α approximating 1, as expected.

Table 10.3 R: *Synthetic canonical geometric model*

```
# syn.cgeo.r
nobs <- 50000
x2 <- runif(nobs)
x1 <- runif(nobs)
xb <- 1.25*x1 + .1*x2 - 1.5
mu <- 1/(exp(-xb)-1)
p <- 1/(1+mu)
r <- 1
gcy <- rnbinom(nobs, size=r, prob = p)
nbcmodel <- data.frame (gcy, x1, x2)
g2y <- ml.nbc(gcy ~ x1 + x2, data=nbcmodel)
g2y
```

```
. * geoc_rng.do
. * only observations with a p=(0-1) in model
. * x1=1.25  x2=0.1  _c=-1.5  alpha=1
. clear
. set obs 50000
```

```
. set seed 7531
. gen x1 = runiform()
. gen x2 = runiform()
. gen xb = 1.25*x1 + .1*x2 -1.5
. gen mu = 1/(exp(-xb)-1)
. gen p = 1/(1+mu)
. gen r = 1
. gen y = rnbinomial(r, p)
. cnbreg y x1 x2

Canonical Negative Binomial Regression  Number of obs  =    50000
                                        Wald chi2(2)   =  7154.23
Log likelihood = -67370.508             Prob > chi2    =   0.0000
---------------------------------------------------------------------
       y |    Coef. Std. Err.     z   P>|z|    [95% Conf. Interval]
---------+-----------------------------------------------------------
      x1 | 1.256393 .0149605   83.98 0.000     1.22707   1.285715
      x2 | .1116191 .0083594   13.35 0.000    .0952351   .1280032
   _cons |-1.507154 .0169326  -89.01 0.000   -1.540341  -1.473966
---------+-----------------------------------------------------------
/lnalpha | .0071225  .014837    0.48 0.631   -.0219575   .0362025
---------+-----------------------------------------------------------
   alpha | 1.007148 .0149431                  .9782818   1.036866
---------------------------------------------------------------------
AIC Statistic    =        2.695
```

If the canonical geometric data are modeled using the standard NB2 parameterization the results are quite different from the true parameters.

```
. glm y x1 x2, fam(nb ml)
Generalized linear models                  No. of obs      =     50000
Optimization     : ML                      Residual df     =     49997
                                           Scale parameter =         1
Deviance         =  46538.17393            (1/df) Deviance =  .9308193
Pearson          =  51458.45738            (1/df) Pearson  =  1.029231
Variance function: V(u) = u+(1.0377)u^2    [Neg. Binomial]
Link function    : g(u) = ln(u)            [Log]
                                           AIC             =  2.705337
Log likelihood   = -67630.43464            BIC             = -494418.3
---------------------------------------------------------------------
      |                OIM
    y |      Coef.  Std. Err.     z    P>|z|    [95% Conf. Interval]
------+--------------------------------------------------------------
   x1 |    2.63963  .0237984  110.92   0.000   2.592986     2.686274
   x2 |   .2876049  .0227987   12.61   0.000   .2429204     .3322895
_cons |  -1.604804  .0198132  -81.00   0.000  -1.643637     -1.56597
---------------------------------------------------------------------
Note: Negative binomial parameter estimated via ML and treated as
      fixed once estimated.
```

The parameter estimates of $x1$ and $x2$ are more than double the "true" parameter values, and the dispersion statistic and α parameter have been inflated by some 3–4%. However, if we were presented with these data, not knowing the true parameters, it is likely that the NB2 model would be declared to be well fitted. The dispersion is rather minimal and the predictors appear to contribute significantly to understanding y. Without the software to estimate a canonical model we have no way of knowing that a better fitted model might exist.

If GLM software allows for a canonical negative binomial, and allows the user to specify the value of the heterogeneity parameter, appropriate modeling can be obtained.

R

```
summary (glm(y~ x1 + x2,   family=negative.binomial (1)))
```

```
. glm y x1 x2, fam(nb 1) link(nb)
Generalized linear models                    No. of obs      =      50000
Optimization      : ML                       Residual df     =      49997
                                             Scale parameter =          1
Deviance          =  46682.18384             (1/df) Deviance =   .9336997
Pearson           =  50233.25344             (1/df) Pearson  =   1.004725
                                   .  .   .
                                             AIC             =   2.694945
Log likelihood    = -67370.62339             BIC             = -494274.3
------------------------------------------------------------------------------
      |                 OIM
    y |      Coef.   Std. Err.      z    P>|z|      [95% Conf. Interval]
------+-----------------------------------------------------------------------
   x1 |   1.261226   .0110899   113.73   0.000      1.23949     1.282962
   x2 |   .1121743   .0082997    13.52   0.000     .0959073     .1284413
_cons |   -1.51371   .0100473  -150.66   0.000    -1.533402    -1.494018
------------------------------------------------------------------------------
```

10.1.3 Using the geometric model

Refer to the various graphs presented in Chapter 8 representing the shape of negative binomial data given specified values of the mean and of α. The geometric distribution, with $\alpha = 1$, produces a shape that is the discrete correlate of the negative exponential distribution. Many types of data fit one of these shapes. If the counts of some item or event degenerate in a smooth decreasing manner, then a geometric model will likely fit the data. If the data are modeled using negative binomial regression, the value of α will approximate 1.0. We found this to be the case for Example 4 in the last chapter.

It may be instructional to observe the relationship between the Poisson and geometric models. After all, regardless of the shape of the geometric distribution, the geometric model represents an accommodation of Poisson overdispersion.

We shall first create a simulated geometric data set having the linear predictor composed of an intercept equal to -1, and two parameters with values of 2.0 and -0.5 respectively. Hence, with $\alpha = 1$, the linear predictor is constructed to be

$$x'\beta = b_0 + b_1x_1 + b_2x_2$$
$$-1 + 2x_1 - 0.5x_2$$

Plot of geometric distributions: five different means

```
. set obs 11      /* create Figure 10.1 */
. gen y = _n-1
. gen muh = .5
. gen byte mu1 = 1
. gen byte mu2 = 2
. gen byte mu5 = 5
. gen byte mu10 = 10
. gen ynbh =exp(y*ln(muh/(1+muh))   -ln(1+muh) +lngamma(y+1)
    -lngamma(y+1))
. gen ynb1 =exp(y*ln(mu1/(1+mu1))   -ln(1+mu1) +lngamma(y+1)
    -lngamma(y+1))
. gen ynb2 =exp(y*ln(mu2/(1+mu2))   -ln(1+mu2) +lngamma(y+1)
    -lngamma(y+1))
. gen ynb5 =exp(y*ln(mu5/(1+mu5))   -ln(1+mu5) +lngamma(y+1)
    -lngamma(y+1))
. gen ynb10=exp(y*ln(mu10/(1+mu10))-ln(1+mu10)+lngamma(y+1)
    -lngamma(y+1))
. lab var y      "count"
. lab var ynbh   "mean=.5"
. lab var ynb1   "mean= 1"
. lab var ynb2   "mean= 2"
. lab var ynb5   "mean= 5"
. lab var ynb10 "mean=10"
. graph twoway connected ynb10 ynb5 ynb2 ynb1 ynbh y,
    ms(s x d T) ///
  title("Negative Binomial Distributions: alpha=1")
```

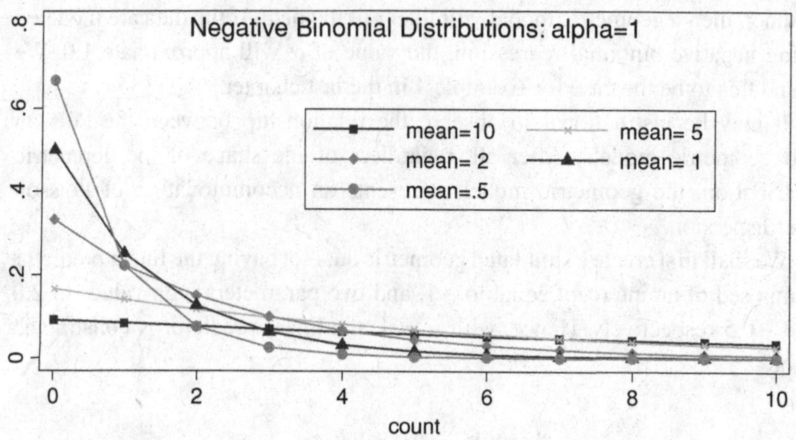

Figure 10.1 Negative binomial distributions: $\alpha = 1$

Modeled using a maximum likelihood negative binomial algorithm this produces the output shown in Table 10.4.

Table 10.4 R: *Synthetic geometric model*

```
# syn.geo.r
library(MASS)
nobs <- 50000
x2 <- runif(nobs)
x1 <- runif(nobs)
xb <- 2*x1 - .5*x2 + 1
exb <- exp(xb)
xg <- rgamma(nobs, 1, 1, 1)
xbg <-exb*xg
gy <- rpois(nobs, xbg)
gnb2 <-glm.nb(gy ~ x1 + x2)
summary(gnb2)
gpy <- glm(gy ~ x1 + x2, family=poisson)
summary(gpy)
```

```
*: geo_rng.do
. clear
. set obs 50000
. set seed 7331
. gen x1 = runiform()
. gen x2 = runiform()
. gen xb = 2*x1 - .5*x2 +1
. gen exb = exp(xb)
. gen xg = rgamma(1,1)
. gen xbg = exb * xg
. gen gy = rpoisson(xbg)
. glm gy x1 x2, fam(nb ml)
Generalized linear models                    No. of obs      =      50000
```

```
Optimization     : ML                Residual df      =     49997
                                     Scale parameter  =         1
Deviance         =  56603.37207      (1/df) Deviance  =  1.132135
Pearson          =   49902.9052      (1/df) Pearson   =   .998118
Variance function: V(u) = u+(.9964)u^2 [Neg.Binomial]
Link function    : g(u) = ln(u)            [Log]
                                     AIC              =   5.70072
Log likelihood   = -142515.0091      BIC              = -484353.1
-----------------------------------------------------------------
      |              OIM
   gy |    Coef.  Std. Err.      z   P>|z|  [95% Conf. Interval]
------+----------------------------------------------------------
   x1 |  2.007671  .0170817  117.53  0.000   1.974192   2.041151
   x2 | -.4970903  .0169087  -29.40  0.000  -.5302307  -.4639498
_cons |  1.002915  .0131404   76.32  0.000   .9771602   1.028669
-----------------------------------------------------------------
Note: Negative binomial parameter estimated via ML and treated
      as fixed once estimated.
```

The parameter estimates are very close to what was specified in the simulation set-up. In addition, the fitted value of α is approximately 1.0, as would be expected for a geometric model. Recall that unless a seed value is given to the random number generator, the values of the parameter estimates will differ slightly from run to run. Here the seed was set at 7,331 with the file, **geo_rng.do**.

We next model the geometric data using Poisson regression. The aim is to see the extent of overdispersion in the Poisson model, as indicated by the dispersion statistic. Notice that the parameter estimates are close to those defined, but the Pearson *chi*2 dispersion is at 7.83, compared with the value of 1.0 for the geometric model. This is a particularly large value considering the number of observations in the data. If the data were estimated using a maximum likelihood Poisson algorithm that fails to provide information concerning fit, a user might well believe that the model is in fact a good one. It is not.

```
. glm gy x1 x2, nolog fam(poi)
Generalized linear models            No. of obs       =     50000
Optimization     : ML                Residual df      =     49997
                                     Scale parameter  =         1
Deviance         =  341510.8742      (1/df) Deviance  =  6.830627
Pearson          =  391540.8482      (1/df) Pearson   =  7.831287
                                         . . .
                                     AIC              =  9.746325
Log likelihood   = -243655.1369      BIC              = -199445.6

-----------------------------------------------------------------
      |              OIM
   gy |    Coef.  Std. Err.      z    P>|z|  [95% Conf. Interval]
------+----------------------------------------------------------
   x1 | 2.011773  .0064839   310.27  0.000   1.999065   2.024482
   x2 | -.505516  .0059254   -85.31  0.000  -.5171296  -.4939024
_cons | 1.004517  .0053291   188.50  0.000   .9940723   1.014962
-----------------------------------------------------------------
```

The AIC statistic of the geometric model is calculated as 5.70 compared with the value of 9.75 for the Poisson.

10.1.4 The canonical geometric model

In a manner similar to the negative binomial, the geometric distribution can be derived from a series of Bernoulli trials, as well as from a Poisson–gamma mixture. The canonical geometric model is derived directly from the Bernoulli series distribution whereas the log-geometric is derived as a Poisson–gamma mixture via NB2. In either case the geometric model is a negative binomial with $\alpha = 1$, i.e. an NB-C with $\alpha = 1$ or NB2 with $\alpha = 1$.

The geometric distribution can be understood as the probability that there are y failures before the first success in a series of Bernoulli trials; p is defined as the probability of success for any given trial in the series. Given the above, the geometric distribution can be quantified as

$$f(y; p) = p(1 - p)^y, \quad y = 0, 1, \ldots, n \tag{10.6}$$

A second parameterization considers y as the trial of the first success. This means that the number of failures is $y - 1$. Both parameterizations can be formulated to result in the same model, but we shall employ the first parameterization since it is the more popular and it directly corresponds to the parameterization of the negative binomial given in the last chapter.

The derivation of the geometric from the above probability function is the basis of the canonical geometric model. The log-linked form which most statisticians are familiar with is based on the negative binomial (NB2). Both models may be estimated using GLM software, if it is possible to specify a value for the negative binomial heterogeneity parameter, α.

In exponential family form, the geometric PDF from equation 10.6 appears as

$$f(y; p) = \exp\{y_i \ln(1 - p_i) + \ln(p_i)\} \tag{10.7}$$

Following the same logic as in Chapter 8, the link and cumulant are given as

$$\text{LINK} = \ln(1 - p_i) \tag{10.8}$$

$$\text{CUMULANT} = -\ln(p_i) \tag{10.9}$$

Differentiating the cumulant with respect to $\ln(1 - p)$,

$$\text{MEAN} = (1 - p_i)/p_i = \mu_i \tag{10.10}$$

$$\text{VARIANCE} = (1 - p_i)/p_i{}^2 = V(\mu_i) = \mu_i(1 + \mu_i) \tag{10.11}$$

Parameterized in terms of μ, the geometric probability mass function is defined as

$$f(y;\mu) = \frac{1}{1+\mu_i}\left(\frac{\mu_i}{1-\mu_i}\right)^{y_i} \tag{10.12}$$

The log-likelihood, expressed in exponential family form, appears as

$$\mathcal{L}(\mu; y) = \sum_{i=1}^{n} y_i \ln\left(\frac{\mu_i}{1+\mu_i}\right) - \ln(1+\mu_i) \tag{10.13}$$

with $\theta = \ln(\mu/(1+\mu))$. We define the link and inverse link in terms of μ as

$$\text{LINK} = \eta_i = \ln(\mu_i/(1+\mu_i)) = -1\ln\left(\frac{1}{\mu_i}+1\right) \tag{10.14}$$

$$\text{INVERSE LINK} = \mu_i = \frac{1}{\exp(-\eta_i)-1} \tag{10.15}$$

The canonical form of the geometric log-likelihood function in terms of $x'b$ may be determined by substituting the value of the inverse link for every instance of μ.

CANONICAL LOG-LIKELIHOOD

$$\mathcal{L}(x'\beta; y) = \sum_{i=1}^{n}\left\{ y_i \ln\left(\frac{1}{\exp(-x_i'\beta)-1}\right) - (1+y_i)\ln\left(1+\frac{1}{\exp(-x_i'\beta)-1}\right)\right\} \tag{10.16}$$

Table 8.2 provides a sample GLM estimating algorithm for the canonical negative binomial. Using the same algorithm, but excluding the term α throughout the algorithm (or setting the value of α to 1.0), provides the appropriate canonical geometric parameter estimates.

It should be remembered that exponentiating the canonical geometric coefficients does not produce incidence rate ratios. The model is most appropriate, as is the canonical negative binomial, for estimating the counts based on significant predictors.

In the last chapter, Example 4 concerned itself with attempting to determine the difference between pre- and post-reform visits to a physician by participants in the German health system. It was noted that the negative binomial value of α was close to 1. It might be of interest to compare the (log) geometric model, or traditional negative binomial with $\alpha = 1$, with a canonically linked geometric model.

Table 10.5 R: *log-geometric model*

```
rm(list=ls())
data(mdvis)
attach(mdvis)
lgm <- glm(numvisit ~ reform + badh + educ2 + educ3
      + age2 + age3 + loginc, family = negative.binomial(1),
      data=mdvis)
summary(lgm)
```

LOG-GEOMETRIC MODEL

```
. glm numvisit reform badh educ2 educ3 age2 age3 loginc, fam(nb 1)

Generalized linear models                 No. of obs      =      2227
Optimization     : ML                     Residual df     =      2219
                                          Scale parameter =         1
Deviance          =  2410.140094          (1/df) Deviance =  1.086138
Pearson           =  2641.379859          (1/df) Pearson  =  1.190347
Variance function: V(u) = u+(1)u^2        [Neg. Binomial]
Link function    : g(u) = ln(u)           [Log]
                                          AIC             =  4.103197
Log likelihood    = -4560.910337          BIC             = -14694.82
------------------------------------------------------------------------
         |                OIM
numvisit |      Coef.  Std. Err.      z    P>|z|    [95% Conf. Interval]
---------+--------------------------------------------------------------
  reform | -.1376904   .0511247   -2.69   0.007    -.237893   -.0374878
    badh |  1.142591   .0744494   15.35   0.000    .9966731    1.288509
   educ2 |   .082214   .0660778    1.24   0.213    -.047296    .2117241
   educ3 | -.0303338   .0707307   -0.43   0.668   -.1689634    .1082959
    age2 |  .0484199   .0635086    0.76   0.446   -.0760547    .1728946
    age3 |  .1879734    .071968    2.61   0.009    .0469187     .329028
  loginc |  .1262021   .0704236    1.79   0.073   -.0118257    .2642299
   _cons | -.2478243   .5368377   -0.46   0.644   -1.300007    .8043583
------------------------------------------------------------------------
```

CANONICAL GEOMETRIC MODEL

```
. glm numvisit reform badh educ2 educ3 age2 age3 loginc, fam(nb 1)
      link(nb)
Generalized linear models                 No. of obs      =      2227
Optimization     : ML                     Residual df     =      2219
```

```
                                                  Scale parameter =          1
Deviance            =  2412.058271                (1/df) Deviance =   1.087002
Pearson             =  2654.882303                (1/df) Pearson  =   1.196432
Variance function: V(u) = u+(1)u^2                    [Neg. Binomial]
Link function     : g(u) = ln(u/(u+(1/1)))  [Neg. Binomial]
                                                  AIC             =   4.104059
Log likelihood      = -4561.869425                BIC             =   -14692.9
------------------------------------------------------------------------------
         |                 OIM
numvisit |      Coef.   Std. Err.      z    P>|z|    [95% Conf. Interval]
---------+--------------------------------------------------------------------
  reform | -.0313247   .0127882    -2.45   0.014    -.0563891   -.0062603
    badh |  .2442292   .0132114    18.49   0.000     .2183353    .2701232
   educ2 |   .017685   .0152966     1.16   0.248    -.0122959    .0476658
   educ3 | -.0199197   .0178609    -1.12   0.265    -.0549265    .0150871
    age2 |  .0227819   .0152921     1.49   0.136    -.0071901     .052754
    age3 |  .0289964   .0162879     1.78   0.075    -.0029272    .0609201
  loginc |  .0292594   .0167904     1.74   0.081    -.0036493    .0621681
   _cons | -.6165698   .1279526    -4.82   0.000    -.8673524   -.3657872
------------------------------------------------------------------------------
```

Although the parameter estimates and standard errors are different, the fit statistics are all extremely close. The log-likelihood function, AIC and BIC statistics, and dispersion statistics are each nearly identical with one another. Only one predictor, *age*3, has ambivalent results with respect to significance. Scaling standard errors and applying a robust variance estimator did not affect the difference. As it is, because the coefficients of *age*2 and *age*3 are similar, one might attempt to combine those two levels with *age*1. The three levels together would then serve as the referent. Education appears to be ruled out as contributory, and should probably be dropped. Moreover, assessing the possibility of interactions may lead to beneficial results. Modeling is far from complete.

A comparison of figures of standardized deviance residuals by fit can apprise the reader if there are major distributional differences between the models that somehow were not picked up by the fit statistics. Figure 9.6 displays the Poisson and negative binomial model residuals; Figure 10.2 shows the canonical geometric. However, since the negative binomial figure is based on an α of 1, it can be considered as (log) geometric.

```
. predict sdevg, deviance stan
. predict mug
. scatter sdevg mug, title(Canommnical Geometric Std Deviance
    vs Fit)
```

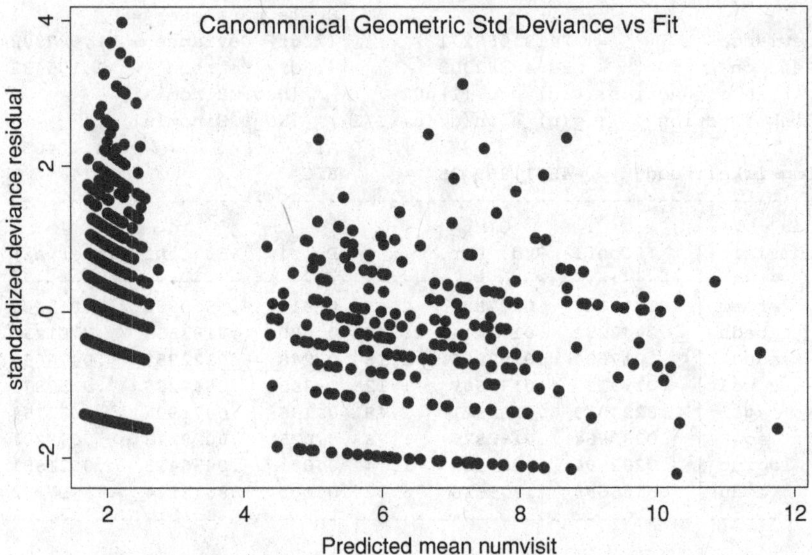

Figure 10.2 Canonical geometric model for predicted mean number of visits: standardized deviance residuals versus fitted values

10.2 NB1: The linear negative binomial model

10.2.1 NB1 as QL-Poisson

Cameron and Trivedi (1986) were the first to make the distinction between the linear (NB1) and quadratic (NB2) negative binomial models. The notion is based on the value of the exponent in the variance function. NB2, the traditional parameterization of the negative binomial, has a variance function appearing as $\mu + \alpha\mu^2$, or equivalently $\mu(1+ \alpha\mu)$. The first of these formulae is the more common representation of the variance. The square value of μ in the formula classifies the equation as quadratic. The NB1 model, on the other hand, is called the linear parameterization, owing to its form: $\mu + \alpha\mu$ or $\mu(1 + \alpha)$. The highest (implied) value of an exponent of μ in the formula is 1. A heterogeneous negative binomial variance function has been given as $\mu + \alpha\mu^p$, with p taking the value of 2 if quadratic, and 1 if linear. In Section 10.5 we shall address the situation where p is considered as another ancillary parameter – NB-P.

Table 10.1 displayed the range of multiplicative extensions that have been applied to the Poisson variance function. The top five in the list included the following:

Poisson:	$V = \mu$	NB1:	$V = \mu(1 + \alpha)$
QL Poisson:	$V = \mu(\phi)$	NB2:	$V = \mu(1 + \alpha\mu)$
Geometric:	$V = \mu(1 + \mu)$		

Expressed in the above manner, we see that the values within the parentheses are, taken together, multipliers of the basic Poisson variance. As such, there is little difference in the quasi-likelihood and NB1 parameterizations. If α is entered into the estimating equation as a constant, then the two formulae are identical – both are quasi-likelihood models. On the other hand, if α is estimated (maximum likelihood), then the models are clearly different. An example may help clarify these relationships. Parameters are defined as:

Constant	=	0.50
$x1$	=	1.25
$x2$	=	−1.50

Modeling the data as NB1, using maximum likelihood, we have the output shown in Table 10.6.

Table 10.6 R: *Creation of synthetic NB1 data*

```
library(COUNT)
require(MASS)
nb1syn  <- function(nobs = 50000, delta = 1,
                     xv = c(1, 0.75, -1.25))  {
  p <- length(xv) - 1
  X <- cbind(1, matrix(rnorm(nobs * p), ncol = p))
  xb <- X %*% xv
  d <- delta
  exb <- exp(xb)
  idelta <- (1/delta)*exb
  xg <- rgamma(nobs, idelta, idelta, 1/idelta)
  xbg <- exb*xg
  nb1y <- rpois(nobs, xbg)
  out <- data.frame(cbind(nb1y, X[,-1]))
  names(out) <- c("nb1y", paste("x", 1:p, sep=""))
  return(out)
}
# -----------------------------------------------------
sim.data <- nb1syn(nobs = 50000, delta=.5, xv = c(.5, 1.25, -1.5))
mynb <- ml.nb1(nb1y ~ ., data = sim.data)
mynb
library(gamlss)    # alternative method of modeling
jhgm <- gamlss(nb1y ~ x1 + x2,family=NBII, data=sim.data)
summary(jhgm)
```

```
. * nb1_rng.do
. * x1= 1.25;  x2= -1.5;  _c= .5
. * delta = .5  (1/.5 = 2.0)
. qui {
. clear
. set obs 50000
. gen x1= runiform()
. gen x2= runiform()
. gen xb = .5 + 1.25*x1 -- 1.5*x2
. gen exb=exp(xb)
. gen idelta = 2*exb
. gen xg=rgamma(idelta, 1/idelta)
. gen xbg = exb*xg
. gen nb1y = rpoisson(xbg)
. }

.  nbreg y1 x1 x2, dispersion(cons)

Negative binomial regression                    Number of obs    =      49770
                                                LR chi2(2)       =  108413.66
Dispersion      = constant                      Prob > chi2      =     0.0000
Log likelihood = -87238.249                     Pseudo R2        =     0.3832
------------------------------------------------------------------------------
        y1 |     Coef.  Std. Err.      z     P>|z|   [95% Conf. Interval]
-----------+------------------------------------------------------------------
        x1 |  1.250363 .0016506  757.50   0.000    1.247128   1.253599
        x2 | -1.500081 .0016055 -934.37   0.000   -1.503228  -1.496934
     _cons |  .5027491 .0035794  140.46   0.000    .4957337   .5097646
-----------+------------------------------------------------------------------
  /lndelta | -.6870266  .0222631                   -.7306615  -.6433917
-----------+------------------------------------------------------------------
     delta |  .5030697  .0111999                    .4815903    .525507
------------------------------------------------------------------------------
Likelihood-ratio test of delta=0:  chibar2(01) = 4039.17
Prob>=chibar2 = 0.000
```

We now model the data as a QL Poisson with $\phi = (1 + \delta) = 1.503$. The estimating algorithm uses built-in quasi-likelihood capability which is based on Fisher scoring. Fisher scoring in turn uses the expected rather than observed information matrix to calculate standard errors. This results in the comparative differences between the standard errors of the NB1 and QL-Poisson models.

```
. glm y1 x1 x2, fam(poi) disp(1.5030697) irls

Generalized linear models                        No. of obs      =      49770
Optimization      : MQL Fisher scoring           Residual df     =      49767
                    (IRLS EIM)                 Scale parameter =    1.50307
Deviance          =  43557.89747               (1/df) Deviance =   .8752366
```

```
Pearson            = 50154.52375     (1/df) Pearson  =  1.007787
Quasi-likelihood model with dispersion:  1.50307 BIC  = -494680.6
------------------------------------------------------------------
        |              EIM
     y1 |      Coef.  Std. Err.       z    P>|z|   [95% Conf. Interval]
--------+---------------------------------------------------------
     x1 |   1.250264  .0011009  1135.67  0.000    1.248106   1.252422
     x2 |  -1.499982  .0010706  -1401.10 0.000    -1.50208  -1.497884
  _cons |   .5030227  .0023855   210.87  0.000    .4983473   .5076982
------------------------------------------------------------------
AIC = 3.586933
```

R's **glm** function employs a quasi-Poisson family, which we earlier discovered is the same as the GLM Poisson model with scaled standard errors. The code in Table 10.6 produces the following results. Note the value of the dispersion parameter, $\delta + 1$. We interpret the NB1 heterogeneity or dispersion parameter as 0.5, which is the value we specified in the code. Keep in mind though that, unless the heterogeneity is entered into the algorithm as a constant, this relationship between the NB1 and quasi-Poisson breaks down.

```
> summary(jhgm)
Coefficients:
             Estimate   Std. Error   t value    Pr(>|t|)
(Intercept)  0.49452    0.01164       42.5      <2e-16 ***
x1           1.25182    0.01517       82.5      <2e-16 ***
x2          -1.49419    0.01540      -97.0      <2e-16 ***
---
Signif. codes:  0 '***' 0.001 '**' 0.01 '*' 0.05 '.' 0.1 ' ' 1
(Dispersion parameter for quasi-poisson family taken to
  be 1.502941)
```

The foremost reason to use a maximum likelihood NB1 model rather than a quasi-likelihood Poisson model with data rests with the fact that NB1 estimates ϕ as $(1 + \alpha)$, whereas ϕ, as a dispersion statistic, must be entered into the QL-Poisson model as a constant. On the down side, NB1 is rarely supported in commercial software. As of this writing, only Stata, R, SAS, and LIMDEP offer it as a capability. I have added an R NB1 function with this book, but otherwise it is available as a **gamlss** function.

10.2.2 Derivation of NB1

The NB1, or linear negative binomial, is derived as a Poisson–gamma mixture model; however, the manner of derivation differs from the traditional NB2 model (see Chapter 8). First derived by Evans (1953), the derivation begins, as does the mixture derivation of the NB2 distribution, with the Poisson distribution,

$$y_i \sim \text{Poisson}(\lambda_i) = f(y;\lambda) = \frac{e^{-\lambda_i}\lambda_i^{y_i}}{y_i!} \tag{10.17}$$

However, in this case the mean of the Poisson is itself a random variable such that

$$\lambda_i \sim \text{gamma}(\delta, \mu_i) \tag{10.18}$$

through which covariates are introduced via $\mu_i = \exp\left(x_i'\beta\right)$. If an offset is applied to the model, $\mu_i = \exp\left(x_i'\beta\right) + \text{offset}$. However, unless specifically addressed, I shall forego including the offset with the linear predictor for simplification purposes.

From the definition of the gamma distribution we know that the mean and variance are given by

$$E[\lambda] = \frac{\mu_i}{\delta} = \exp\left(x_i'\beta\right)/\delta \tag{10.19}$$

$$V[\lambda] = \frac{\mu_i}{\delta^2} = \exp\left(x_i'\beta\right)/\delta^2 \tag{10.20}$$

where δ is the gamma scale parameter. The resulting mixture is described as

$$f(y|x) = \int_0^\infty \frac{e^{-\lambda_i}\lambda_i^{y_i}}{y_i!}\frac{\delta^{\mu_i}\lambda^{\mu_i-1}}{\Gamma(\mu_i)}e^{-\lambda_i\delta}d\lambda \tag{10.21}$$

Solving to clear the integration, we have

$$f(y;\mu) = \int_0^\infty \frac{e^{-\mu_i}\mu_i^{y_i}}{y_i!}\frac{\delta^{\mu_i}}{\Gamma(\mu_i)}\mu_i^{\mu_i-1}e^{-\mu_i\delta}d\mu \tag{10.22}$$

$$= \frac{\delta^{\mu_i}}{\Gamma(y_i+1)\Gamma(\mu_i)}\int_0^\infty \left\{\mu_i^{(y_i+\mu_i)-1}e^{-\mu_i(\delta+1)}\right\}d\mu \tag{10.23}$$

$$= \frac{\delta^{\mu_i}}{\Gamma(y_i+1)\Gamma(\mu_i)}\frac{\Gamma(y_i+\mu_i)}{(\delta+1)^{y_i+\mu_i}} \times C \tag{10.24}$$

where C reduces to the value of 1. It appears as:

$$C = \int_0^\infty \frac{(\delta+1)^{y_i+\mu_i}}{\Gamma(y_i+\mu_i)}\lambda_i^{(y_i+\mu_i)-1}e^{-\mu_i(\delta+1)}d\lambda \tag{10.25}$$

Continuing from equation 10.23, less the value of C, we have

$$= \frac{\Gamma(y_i + \mu_i)}{\Gamma(y_i + 1)\Gamma(\mu_i)} \left(\frac{\delta}{1+\delta}\right)^{\mu_i} \left(\frac{1}{1+\delta}\right)^{y_i} \tag{10.26}$$

The mean and variance of equation 10.26, the NB1 distribution, are

$$\text{NB1 mean } = E[y_i] = \exp(x_i'\beta)/\delta \tag{10.27}$$

$$\text{NB1 variance } = V[y_i] = \exp(x_i'\beta)(1+\delta)/\delta^2 \tag{10.28}$$

The variance to mean ratio is $(1 + \delta)/\delta$, which is constant for all observations. This feature of the distribution results in constant overdispersion within the model, unlike NB2, in which δ is variable with a mean of 1. Defining $\alpha = 1/\delta$, the distribution may be re-expressed in more familiar terms as

$$\frac{\Gamma(y_i + \mu_i)}{\Gamma(y_i + 1)\Gamma(\mu_i)} \left(\frac{1}{1+\alpha}\right)^{\mu_i} \left(\frac{\alpha}{1+\alpha}\right)^{y_i} \tag{10.29}$$

As in the parameterization of NB2, specified in Chapter 8, $\alpha = 0$ is Poisson.

Standardizing the coefficients, β, by the addition of $-\ln(\alpha)$ to the linear predictor,

$$E(y) = \mu_i = \exp\left(x_i'\beta - \ln(\alpha)\right) \tag{10.30}$$

the NB1 distribution (equation 10.33.) may be expressed as the log-likelihood

$$\mathcal{L}(\mu, \alpha; y) = \sum_{i=1}^{n} \{\ln(\Gamma(y_i + \mu_i)) - \ln(\Gamma(y_i + 1)) - \ln(\Gamma(\mu_i)) + y_i \ln(\alpha) - (y_i + \mu_i)\ln(1 + \alpha)\} \tag{10.31}$$

NB1 score equations are derived as

$$\text{score}\left(x'\beta\right) = \frac{\exp\left(x_i'\beta\right)}{\alpha} \times \left\{\Psi\left(y_i + \frac{\exp\left(x_i'\beta\right)}{\alpha}\right) - \Psi\left(\frac{\exp\left(x_i'\beta\right)}{\alpha}\right) + \ln\left(\frac{1}{1+\alpha}\right)\right\} \tag{10.32}$$

$$\text{score}\left(\ln(\alpha)\right) = y_i - \left(y_i + \frac{\exp\left(x_i'\beta\right)}{\alpha}\right)\left(1 - \frac{1}{1 + \alpha\exp\left(x_i'\beta\right)}\right) - \text{score}\left(x_i'\beta\right) \tag{10.33}$$

Additional discussion regarding the derivation of the NB1, as well as of the logic of the standardization of the linear predictor, can be found in Hardin and Hilbe (2007).

10.2.3 Modeling with NB1

Using the same German health reform data as earlier in this chapter, the number of visits is modeled as NB1, as shown in Table 10.7.

Table 10.7 R: *NB1 estimation of reform data*

```
rm(list=ls())
data(mdvis)
attach(mdvis)
mdvis$educ <- factor(mdvis$educ)
mdvis$agegrp  <- factor(mdvis$agegrp)
med.nb1 <- ml.nb1(numvisit ~ reform + badh + educ + agegrp +
    loginc, data = mdvis) # Table 10.8
med.nb1
------------------------------------------------------------------
ALTERNATIVE:
library(gamlss)
med.nb1a<-gamlss(numvisit~reform+badh+factor(educ)
    +factor(agegrp) + loginc, family=NBII,data=mdvis)
summary(med.nb1a)
```

```
. nbreg numvisit reform badh educ2 educ3 age2 age3 loginc, irr
  disp(constant)
Negative Binomial Type 1 Regression         Number of obs  =      2227
                                            Wald chi2(7)   =    318.49
Log likelihood = -4600.3458                 Prob > chi2    =    0.0000
------------------------------------------------------------------
numvisit |       IRR   Std. Err.     z    P>|z|  [95% Conf. Interval]
---------+--------------------------------------------------------
  reform | .9019533   .0402234   -2.31   0.021   .8264641   .9843378
    badh | 2.607651    .147245   16.97   0.000   2.334453   2.912822
   educ2 | 1.050977   .0616253    0.85   0.396   .9368752   1.178974
   educ3 | 1.017186   .0629838    0.28   0.783   .9009369   1.148435
    age2 | .9936896   .0544585   -0.12   0.908   .8924856    1.10637
    age3 | 1.050923   .0655038    0.80   0.426   .9300703    1.18748
  loginc | 1.135741   .0686193    2.11   0.035   1.008907   1.278519
---------+--------------------------------------------------------
/lnalpha | .9881723     .05329   18.54   0.000   .8837258   1.092619
---------+--------------------------------------------------------
   alpha |  2.68632    .143154                    2.419899   2.982073
------------------------------------------------------------------
AIC Statistic =     4.139
```

The following table lists the differences in AIC and α for the models we have thus far discussed. Note that the NB2 and canonical linked model, NB-C, are nearly identical, even though NB-C does not use a log link in its algorithm. Both NB2 and NB1 employ the log link. On the other hand, parameter estimates for each of the models are similar, and indicate the same predictors as contributory to the model.

```
         AIC     alpha
NB2     4.104     .998
NB-H    4.104     .998
NB1     4.139    2.686
```

Of possible interest is a comparison of models based on the synthetic data created in Section 10.1. Using the same parameter specifications, synthetic data sets may be created for both NB2 and NB1 models. Modeling combinations of NB1 and 2 data with NB1 and 2 models gives us

PARAMETERS: $x1 = 2.0; x2 = -0.5;$ _cons $= -1.0; \delta/\alpha = 1.0$

```
# ALL: modification of parameters and delta/alpha
R    : nb1syn.r       # Table 10.6
Stata: do nb1_rng     # under Table 10.6
MODEL: NB1
DATA : NB1
---------------------------------------------------------------------
   y1 |     Coef.   Std. Err.      z    P>|z|   [95% Conf. Interval]
------+--------------------------------------------------------------
   x1 |  2.001738  .0035015   571.68 0.000    1.994876   2.008601
   x2 | -.5045447  .0034868  -144.70 0.000   -.5113788  -.4977106
_cons | -1.002748  .0081231  -123.44 0.000   -1.018668  -.9868266
------+--------------------------------------------------------------
delta |  1.015196  .0201667                    .9764294   1.055501
---------------------------------------------------------------------
Likelihood-ratio test of delta=0:   chibar2(01) = 9132.96
Prob>=chibar2 = 0.000 AIC Statistic =         2.292

R    : nb2_syn.r              # Table 9.4
     : ml.nb1(y2 ~ x1 + x2)  # Table 10.8
Stata: do nb2_rng
     : nbreg y2 x1 x2, disp(constant)
MODEL: NB1
DATA : NB2
---------------------------------------------------------------------
   y2 |      Coef.  Std. Err.      z     P>|z|  [95% Conf. Interval]
------+--------------------------------------------------------------
   x1 |   1.541805  .0078804  195.65   0.000   1.526359     1.55725
   x2 |  -.3880394  .0062148  -62.44   0.000  -.4002201   -.3758587
_cons |  -.0982971   .014988   -6.56   0.000  -.1276729   -.0689212
------+--------------------------------------------------------------
delta |   8.713806  .1342828                   8.454552    8.981011
---------------------------------------------------------------------
```

```
Likelihood-ratio test of delta=0:  chibar2(01) = 8.4e+04
Prob>=chibar2 = 0.000 AIC Statistic =          2.606

R     : nb1syn.r
        glm.nb(y1 ~ x1 + x2)
Stata: do nb1_rng
     : nbreg y1 x1 x2
MODEL: NB2
DATA : NB1
------------------------------------------------------------------
    y1 |      Coef.   Std. Err.      z    P>|z|   [95% Conf. Interval]
------+-----------------------------------------------------------
    x1 |   1.989207   .006577   302.45  0.000    1.976316    2.002098
    x2 |  -.4956426   .0051694  -95.88  0.000   -.5057745   -.4855108
 _cons |  -.9852668   .0088877 -110.86  0.000   -1.002686   -.9678472
------+-----------------------------------------------------------
alpha  |   .2370104   .0074614                    .2228283    .2520951
------------------------------------------------------------------
Likelihood-ratio test of alpha=0:  chibar2(01) = 3761.32
Prob>=chibar2 = 0.000 AIC Statistic =          2.408
```

A few observations regarding the above output: First, modeling NB2 data with an NB1 model substantially alters the specified parameter estimates, as well as the value of δ/α. Second, modeling NB1 data with an NB2 model does not substantially alter the specified parameter estimates, but the value of the ancillary parameter is changed. These relationships may be important in practical applications, but only if one knows *a priori* how the data are generated.

10.2.4 NB1: R maximum likelihood function

I am providing a maximum likelihood function for NB1 regression, **ml.nb1.r**. The function produces maximum likelihood estimates for model parameters, including the heterogeneity parameter. Any Poisson overdispersed count data may estimated. I have used the **medpar** data, which we used with NB2 models, as example data.

Table 10.8 R: *NB1 maximum likelihood function*

```
# NB1 maximum likelihood function
ml.nb1 <- function(formula, data, start = NULL,
                              verbose = FALSE) {
  mf <- model.frame(formula, data)
  mt <- attr(mf, "terms")
  y <- model.response(mf, "numeric")
```

```
  nb1X <- model.matrix(formula, data = data)
  nb1.reg.ml <- function(b.hat, X, y) {
    a.hat <- b.hat[1]
    xb.hat <- X %*% b.hat[-1]
    mu.hat <- exp(xb.hat)
    r.hat <- (1/a.hat) * mu.hat
    sum(dnbinom(y,
          size = r.hat,
          mu = mu.hat,
          log = TRUE))
    }
  if (is.null(start))
    start <- c(0.5, -1, rep(0, ncol(nb1X) - 1))
  fit <- optim(start,
          nb1.reg.ml,
          X = nb1X,
          y = y,
          control = list(
            fnscale = -1,
            maxit = 10000),
          hessian = TRUE
          )
  if (verbose | fit$convergence > 0) print(fit)
  beta.hat <- fit$par
  se.beta.hat <- sqrt(diag(solve(-fit$hessian)))
  results <- data.frame(Estimate = beta.hat,
                        SE = se.beta.hat,
                        Z = beta.hat / se.beta.hat,
                        LCL = beta.hat - 1.96 * se.beta.hat,
                        UCL = beta.hat + 1.96 * se.beta.hat)
  rownames(results) <- c("alpha", colnames(nb1X))
  results <- results[c(2:nrow(results), 1),]
  return(results)
}
# ------------------------------------------------------------------
-
library(COUNT)                  # ml.nbl.r in COUNT library
medpar$type <- factor(medpar$type)
med.nb1 <- ml.nb1(los ~ hmo + white + type, data = medpar)
med.nb1
```

```
> med.nb1
              Estimate         SE          Z         LCL         UCL
(Intercept)  2.34918407 0.06023641 38.9994022  2.23112070  2.46724744
hmo         -0.04533566 0.05004714 -0.9058592 -0.14342805  0.05275673
white       -0.12951295 0.06071130 -2.1332593 -0.24850711 -0.01051880
type2        0.16175471 0.04585569  3.5274735  0.07187757  0.25163186
type3        0.41879257 0.06553258  6.3906006  0.29034871  0.54723643
alpha        4.57898241 0.22015969 20.7984599  4.14746943  5.01049540
```

The output in Table 10.8 corresponds to the output provided by Stata's **nbreg** with the *dispersion(constant)* option, as well as with LIMDEP's NB1 output.

The data may also be estimated using the R **gamlss** library.

R: *NB1 using gamlss*

```
library(gamlss)
med.nb1a<-gamlss(los~hmo+white+factor(type),family=NBII,data=medpar)
summary(med.nb1a)
```

```
Mu link function: log
Mu Coefficients:
            Estimate Std. Error t value  Pr(>|t|)
(Intercept) 2.35027  0.06018    39.0524  2.827e-230
hmo         -0.04691 0.05008    -0.9368  3.490e-01
white       -0.12994 0.06067    -2.1419  3.237e-02
type2        0.16272 0.04582    3.5510   3.957e-04
type3        0.41933 0.06549    6.4029   2.039e-10

-----------------------------------------------------------------
Sigma link function: log
Sigma Coefficients:
  Estimate Std. Error   t value    Pr(>|t|)
 1.523e+00  4.814e-02 3.164e+01 1.847e-168
-----------------------------------------------------------------
No. of observations in the fit: 1495
Degrees of Freedom for the fit: 6
  Residual Deg. of Freedom: 1489
  at cycle: 5
Global Deviance: 9663.51
  AIC: 9675.51
  SBC: 9707.37
```

Note that **gamlss** employs t-values rather than z. We have argued that z is more appropriate, but only when there are many observations in the model. The fact that the *hmo* confidence intervals contain 0 indicates that it does not significantly contribute to the model. After shifting decimal places, this fact is evident in the **gamlss** output as well. Note also that **gamlss** displays the natural log of the heterogeneity parameter, 1.523, instead of its true value, 4.579. Keep this in mind when interpreting results.

10.3 NB-C: Canonical negative binomial regression

10.3.1 NB-C overview and formulae

Table 8.2 in Section 8.4.1 provides a schematic of the NB-C estimating algorithm. Section 10.1.4 provides an overview of the canonical geometric model, which is NB-C with a heterogeneity parameter equal to 1. Moreover, we earlier discovered that the NB-C model is directly based on the negative binomial PDF, whereas the traditional NB2 parameterization is not. The latter can be derived as a Poisson–gamma mixture, or as a non-canonical log-linked GLM negative binomial. The value, of course, of the mixture model parameterization – or log-negative binomial form of model – is that it allows us to model Poisson overdispersion. Both the Poisson and NB2 have a fitted probability expressed as $\exp(x'\beta)$, with $x'\beta$ as the model linear predictor. We therefore have prefaced this discussion with a background as to the nature of the NB-C model, and how it differs from the traditional negative binomial model.

In McCullagh and Nelder's (1989) foundational text on generalized linear models, the authors make reference to the canonical link of the negative binomial. They use it as an example of a model in which the variance function contains a second parameter to be estimated. Of course, the link and inverse link functions also contain the heterogeneity parameter, which we have designated as α. McCullagh and Nelder argue that the model we term NB-C is "problematic" in that the linear predictor is a function of α, a term of the variance function. In discussing this statement with Nelder, I found that the foremost problem they had in mind was the difficulty they experienced in attempting to estimate the model. They do not state it in their text, but the problem largely disappears when α is entered into the GLM estimating algorithm as a constant. As we have discussed, though, if the NB-C or NB2 models are to be considered as GLMs, which by definition are single-parameter estimation algorithms, α must be submitted as a constant. As such there should be only minimal additional difficulty in convergence, or estimation in general. When α is estimated as a second parameter using a full MLE algorithm, having it as a term in the link and variance can result in estimation difficulties. Typically more care needs to be given to starting values. However, contemporary MLE algorithms such as those employed in Stata and SAS are robust to starting values that are somewhat distant from the true values.

At present the only MLE commands or functions for NB-C are those designed by the author. The Stata command **cnbreg** may be downloaded from the SSC website and installed using the Stata command-line. Type

```
. ssc install cnbreg
```

The R command was written for this text, and will be given later. GLM NB-C models are available in Stata and SAS using the **glm** command and **GENMOD** procedure respectively.

It needs to be reiterated that the log-likelihood function, in terms of μ, of the NB-C model is identical to that of NB2. The NB1 log-likelihood differs from both NB-C and NB2. The difference in parameter estimates and model statistics between NB-C and NB2 comes from the differences in link and inverse link functions. When parameterized in terms of $x'\beta$, however, the log-likelihood functions of NB-C and NB2 differ. The NB-C log-likelihood appears as equation 8.38, repeated below.

$$\mathcal{L}(\beta; y, \alpha) = \sum_{i=1}^{n} \left\{ y_i \left(x_i'\beta\right) + \left(\frac{1}{\alpha}\right) \ln\left(1 - \exp\left(x_i'\beta\right)\right) + \ln\Gamma\left(y_i + \frac{1}{\alpha}\right) - \ln\Gamma\left(y_i + 1\right) - \ln\Gamma\left(\frac{1}{\alpha}\right) \right\} \quad (10.34)$$

Compare with the NB2 log-likelihood

$$\mathcal{L}(\beta; y, \alpha) = \sum_{i=1}^{n} \left\{ y_i \ln\left(\frac{\alpha \exp(x_i'\beta)}{1 + \alpha\exp(x_i'\beta)}\right) - \left(\frac{1}{\alpha}\right) \ln\left(1 + \alpha\exp(x_i'\beta)\right) + \ln\Gamma\left(y_i + \frac{1}{\alpha}\right) - \ln\Gamma\left(y_i + 1\right) - \ln\Gamma\left(\frac{1}{\alpha}\right) \right\} \quad (10.35)$$

The NB-C gradient, or first derivative of the log-likelihood function with respect to β, may be expressed as:

$$\frac{\partial \mathcal{L}}{\partial \beta_k} = \sum_{i=1}^{n} \left(y_i + \frac{e^{x_i'\beta}}{\alpha\left(e^{x_i'\beta} - 1\right)} \right) x_k \quad (10.36)$$

and with respect to α as

$$\frac{\partial \mathcal{L}}{\partial \alpha} = \sum_{i=1}^{n} - \frac{\ln\left(1 - e^{x_i'\beta}\right) + \Psi\left(y_i + \frac{1}{\alpha}\right) - \Psi\left(\frac{1}{\alpha}\right)}{\alpha^2} \quad (10.37)$$

where Ψ is the digamma function, the derivative of the log-gamma function. See equation 8.19.

The NB-C Hessian may be expressed with respect to $\beta;\alpha$ in matrix form as

$$\frac{\partial^2 \mathcal{L}}{\partial \beta_k; \partial \alpha} = \begin{bmatrix} \dfrac{2\alpha \ln\left(1 - e^{x_i'\beta}\right) + 2\alpha\Psi\left(y_i + \frac{1}{\alpha}\right) + \Psi'\left(y_i + \frac{1}{\alpha}\right) - 2\alpha\Psi\left(\frac{1}{\alpha}\right) - \Psi'\left(\frac{1}{\alpha}\right)}{\alpha^4} & -\dfrac{x_k e^{x_i'\beta}}{\alpha^2\left(e^{x_i'\beta} - 1\right)} \\ -\dfrac{x_k e^{x_i'\beta}}{\alpha^2\left(e^{x_i'\beta} - 1\right)} & -\dfrac{x_k^2 e^{x_i'\beta}}{\alpha\left(e^{x_i'\beta} - 1\right)^2} \end{bmatrix} \quad (10.38)$$

with Ψ' as the trigamma function (see equation 8.20). With respect to β_k and β_j, we have

$$\frac{\partial^2 \mathcal{L}}{\partial \beta_k \partial \beta_j} = \sum_{i=1}^{n} - \frac{e^{x_i'\beta} x_j x_k}{\alpha \left(e^{x_i'\beta} - 1\right)^2} \tag{10.39}$$

10.3.2 Synthetic NB-C models

An algorithm for generating synthetic NB-C data is straightforward to construct, in contrast to the NB1 and NB2 models that are developed as a Poisson–gamma mixture. The key terms are the inverse link function, $1/((\exp(-x'\beta)-1)*\alpha)$, which generates μ, the fitted value, and the fitted probability, $1/(1 + \alpha\mu)$. The value we assign to α is inverted before entering the negative binomial random number generator so that the model will reflect a variance with a direct relationship between μ and α.

Note that the negative binomial generator cannot be used to generate NB2 data. The fact that an NB-C model can be directly created in this manner emphasizes the relationship between the negative binomial PDF and the NB-C model.

Table 10.9 provides a generic function for creating synthetic NB-C data. The model based on it is given specified coefficients and a value for the heterogeneity parameter, alpha. Default values can be used, but here we provide values for *alpha*, the intercept and two predictors, creating a data set called *sim.data*. We could specify more, or fewer, predictors if we desired. The intercept is the first value indicated in *c*(). The data are modeled using the **ml.nbc** function found in the COUNT library. Note that the estimates are close to those we specified. Stata is then used to create synthetic NB-C data and model. The Stata data are modeled using the **cnbreg** command, found on the book's website, or on SCC's website.

Table 10.9 R: *Synthetic NB-C: generic function*

```
require(MASS)
   nbcsyn  <- function(nobs = 50000,
                       alpha = 1.15,
                       xv = c(-1.5, -1.25, -.1))  {
    q <- length(xv) - 1
    X <- cbind(1, matrix(runif(nobs * q), ncol = q))
    xb <- X %*% xv
    a <- alpha
    mu <- 1/((exp(-xb)-1)*a)
    p <- 1/(1+a*mu)
    r <- 1/a
    nbcy <- rnbinom(nobs, size=r, prob = p)
```

```
    out <- data.frame(cbind(nbcy, X[,-1]))
    names(out) <- c("nbcy", paste("x", 1:q, sep=""))
    return(out)
  }
sim.data <- nbcsyn(nobs = 50000, alpha=1.15, xv = c(-
1.5, 1.25, .1))
library(COUNT)
mynbc <- ml.nbc(nbcy ~ ., data = sim.data)
mynbc
```

NB-C R OUTPUT

```
               Estimate          SE         Z          LCL       UCL
(Intercept) -1.5022579 0.017722933 -84.76351 -1.53699488 -1.467521
x1           1.2486064 0.015767580  79.18821  1.21770197  1.279511
x2           0.1061846 0.009129308  11.63117  0.08829112  0.124078
alpha        1.1481083 0.017699703  64.86596  1.11341686  1.182800
```

STATA

```
. * nbc_rng.do
. * only observations with a p=(0-1) in model
. * x1=1.25  x2=0.1  _c=-1.5  alpha=1.15
. clear
. set obs 50000
. set seed 7787
. gen x1 = runiform()
. gen x2 = runiform()
. gen xb = 1.25*x1 + .1*x2 -1.5
. gen a =  1.15
. gen mu = 1/((exp(-xb)-1)*a)
. gen p = 1/(1+a*mu)
. gen r = 1/a
. gen y = rnbinomial(r, p)
. cnbreg y x1 x2
Canonical Negative Binomial Regression
                                          Number of obs   =      50000
                                          Wald chi2(2)    =    6386.70
Log likelihood = -62715.384               Prob > chi2     =     0.0000
------------------------------------------------------------------------
       y |    Coef.   Std. Err.     z    P>|z|   [95% Conf. Interval]
---------+--------------------------------------------------------------
      x1 | 1.252675   .015776    79.40   0.000    1.221754    1.283595
      x2 | .1009038  .0091313    11.05   0.000    .0830068    .1188008
   _cons |-1.504659   .0177159  -84.93   0.000   -1.539382   -1.469937
---------+--------------------------------------------------------------
/lnalpha |  .133643   .0153947    8.68   0.000    .1034699     .163816
---------+--------------------------------------------------------------
   alpha | 1.142985   .0175959                     1.109012    1.177998
------------------------------------------------------------------------
```

```
AIC Statistic     =       2.509

. abic
AIC Statistic     =   2.508775            AIC*n       = 125438.77
BIC Statistic     =   2.508837            BIC(Stata) = 125474.05
```

Modeling the data as a GLM yields

```
. glm y x1 x2, fam(nb 1.143) link(nb)
Generalized linear models                  No. of obs      =      50000
Optimization      : ML                     Residual df     =      49997
                                           Scale parameter =          1
Deviance          =  44474.65535           (1/df) Deviance =   .8895465
Pearson           =   49932.1741           (1/df) Pearson  =   .9987034
Variance function: V(u) = u+(1.143)u^2           [Neg. Binomial]
Link function     : g(u) = ln(u/(u+(1/1.143)))   [Neg. Binomial]
                                           AIC             =   2.508735
Log likelihood    = -62715.38385           BIC             = -496481.8
------------------------------------------------------------------------
       |                 OIM
     y |     Coef.   Std. Err.      z    P>|z|   [95% Conf.  Interval]
-------+----------------------------------------------------------------
    x1 |  1.252666   .0119357   104.95   0.000    1.229272    1.276059
    x2 |  .1009029   .0090694    11.13   0.000    .0831272    .1186786
 _cons | -1.504647   .0107897  -139.45   0.000   -1.525794   -1.483499
------------------------------------------------------------------------
```

Modeling the NB-C generated data using an NB2 model produces the following results

R:

```
nb2c <- glm.nb(nbcy~ x1 + x2, data = sim.data)
summary(nb2c)
modelfit(nb2c)        # Table 5.3
```

```
. nbreg y x1 x2
                      <header statistics excluded>
------------------------------------------------------------------------
         y|    Coef.    Std. Err.     z     P>|z|    [95% Conf. Interval]
----------+-------------------------------------------------------------
       x1 | 2.617604    .025509   102.61   0.000    2.567607    2.667601
       x2 | .2780534   .0244159    11.39   0.000    .2301992    .3259077
    _cons |-1.728094   .0211433   -81.73   0.000   -1.769534   -1.686654
----------+-------------------------------------------------------------
 /lnalpha |  .158443   .0153764                     .1283057    .1885803
----------+-------------------------------------------------------------
    alpha | 1.171685   .0180163                     1.136901    1.207534
------------------------------------------------------------------------
```

```
Likelihood-ratio test of alpha=0:  chibar2(01) = 1.9e+04
Prob>=chibar2 = 0.000

. abic
AIC Statistic    =    2.516108          AIC*n        = 125805.38
BIC Statistic    =    2.516169          BIC(Stata) = 125840.66
```

The AIC and BIC statistics in the left column do not appear to greatly differ, but the NB-C model does result in lower values. The statistics without division by *n* display the differences more clearly.

	NB-C	NB2
AIC	125,488.37	125,605.38
BIC	125,474.05	125,840.66

Given the criteria for determining if one model is a significantly better fit than another using the AIC statistic (Table 5.1), the NB-C model is clearly a superior fit. We would expect this to be the case. Without NB-C software, how could we determine that the NB2 model is not truly well fitted? If we modeled the data using an NB1 model the fit statistics appear as

```
. abic
AIC Statistic    =    2.552643             AIC*n        = 127632.16
BIC Statistic    =    2.552705             BIC(Stata) = 127667.44
```

which is substantially worse than the NB2 results. A Poisson model is considerably ill-fitted compared with NB-C and NB2. A Poisson model on the data results in a dispersion statistic of 2.2.

```
. abic
AIC Statistic    =   2.886441  AIC*n          =   144322.05
BIC Statistic    =   2.886453  BIC(Stata)    =    144348.5
```

A GLM NB2 model yields a dispersion of 1.022, which *prima facie* appears close to unity, indicating that the model is not NB2 overdispersed. But given 50,000 observations, the 2% increase in dispersion over unity is significant. A *seemingly unrelated estimation* test indicates that the model estimates are significantly different.

The problem remains though. If NB-C software is unavailable, it would be difficult to determine that the NB2 model is in fact not appropriate. The dispersion of 1.022 could be taken to be indicative of extra correlation in an

NB2 model instead of an indication of possible misspecification. A solution is to submit the NB2 model to a test of misspecification. When we do so with these data, the NB2 model fails the test. It fails the test if modeled as NB1 as well. NB-C is an alternative, but it is left untested. The data may not be negative binomial under any parameterization. We know the model is NB-C since we created it, but we need an NB-C function in order to verify it.

Table 10.10 R: *NB2 linktest*

```
sim.data <- nb2_syn(nobs = 50000, alpha = .5, xv = c(2, .75,
  -1.25))
nbtst <- glm.nb(nby ~ . , data = sim.data)
hat <-hatvalues(nbtst)
hat2 <-  hat*hat
nblink <- glm.nb(nby ~ hat + hat2, data = sim.data)
summary(nblink)
confint(nblink)
```

```
. qui nbreg y x1 x2
. linktest

Negative binomial regression                      Number of obs   =      50000
                                                  LR chi2(2)      =   10435.76
Dispersion        = mean                          Prob > chi2     =     0.0000
Log likelihood = -62735.412                       Pseudo R2       =     0.0768
------------------------------------------------------------------------------
           y |      Coef.   Std. Err.      z    P>|z|  [95% Conf. Interval]
-------------+----------------------------------------------------------------
        _hat |  1.076743   .0105706   101.86   0.000   1.056025    1.097461
      _hatsq |  .2486669   .0136734    18.19   0.000   .2218674    .2754663
       _cons | -.1327317   .0101315   -13.10   0.000   -.152589   -.1128744
-------------+----------------------------------------------------------------
    /lnalpha |   .136275   .0155248                    .1058468    .1667031
-------------+----------------------------------------------------------------
       alpha |  1.145997   .0177914                    1.111652    1.181403
------------------------------------------------------------------------------
Likelihood-ratio test of alpha=0:   chibar2(01) = 1.8e+04
Prob>=chibar2 = 0.000
```

10.3.3 NB-C models

Little work as been done with NB-C models. No research has been published using an NB-C model on data. However, we have observed that there are times when data can be better modeled using NB-C over other negative binomial parameterizations. The difficulty rests in the fact that NB-C models are appropriate, but not as interpretable, for modeling overdispersed Poisson data.

NB-C models do not utilize the log link employed by Poisson, NB1, NB2, and derived models. But they nevertheless account for overdispersion that exists in otherwise Poisson models.

Consider the **medpar** data we have used earlier in the text. First model the data using an NB-C, calculate AIC and BIC fit statistics, and then model the data using an NB2. We first display two tables containing R code for estimating the NB-C. Table 10.9 is similar in logic to the code in Table 10.7, which was used to estimate coefficients and α for synthetic NB-C data. Both are based on optimization.

Table 10.12 presents a general NB-C function, **ml.nbc()** that can be used to estimate any data appropriate for NB-C modeling. The algorithm searches for starting values to start the iteration process, but the NB-C model sometimes has difficulty converging owing to the inclusion of α in the link and variance functions. However, I have found that most data that are not too ill-formed can be modeled without difficulty.

Table 10.11 R: *NB-C solved using optimization*

```
See Table 10.11 in the COUNT scripts directory
```

Table 10.12 R: *NB-C function – ml.nbc()*

```
ml.nbc <- function(formula, data, start = NULL, verbose = FALSE) {
  mf <- model.frame(formula, data)
  mt <- attr(mf, "terms")
  y <- model.response(mf, "numeric")
  nbcX <- model.matrix(formula, data = data)
  nbc.reg.ml <- function(b.hat, X, y) {
    a.hat <- b.hat[1]
    xb.hat <- X %*% b.hat[-1]
    mu.hat <- 1 / ((exp(-xb.hat)-1)*a.hat)
    p.hat <- 1 / (1 + a.hat*mu.hat)
    r.hat <- 1 / a.hat
    sum(dnbinom(y,
                size = r.hat,
                prob = p.hat,
                log = TRUE))
  }
  if (is.null(start))
    start <- c(0.5, -1, rep(0, ncol(nbcX) - 1))
  fit <- optim(start,
               nbc.reg.ml,
               X = nbcX,
               y = y,
```

```
            control = list(
              fnscale = -1,
              maxit = 10000),
            hessian = TRUE
            )
  if (verbose | fit$convergence > 0) print(fit)
  beta.hat <- fit$par
  se.beta.hat <- sqrt(diag(solve(-fit$hessian)))
  results <- data.frame(Estimate = beta.hat,
                        SE = se.beta.hat,
                        Z = beta.hat / se.beta.hat,
                        LCL = beta.hat - 1.96 * se.beta.hat,
                        UCL = beta.hat + 1.96 * se.beta.hat)
  rownames(results) <- c("alpha", colnames(nbcX))
  results <- results[c(2:nrow(results), 1),]
  return(results)
}
-----------------------------------------------------------------
##  Load, factor, and run ml.nbc function
rm(list=ls())
data(medpar)
attach(medpar)
medpar$type <- factor(medpar$type)
med.nbc <- ml.nbc(los ~ hmo + white + type, data = medpar)
med.nbc
```

R: NB-C **ml.nbc()** *function output*

```
                Estimate          SE          Z         LCL          UCL
(Intercept) -0.20128564 0.013496886 -14.913488 -0.22773954 -0.174831744
hmo         -0.01399781 0.010729402  -1.304622 -0.03502744  0.007031817
white       -0.02491594 0.010756500  -2.316361 -0.04599868 -0.003833201
type2        0.04101961 0.008932639   4.592105  0.02351164  0.058527585
type3        0.10723427 0.010062008  10.657343  0.08751273  0.126955803
alpha        0.44516214 0.019775949  22.510280  0.40640128  0.483922999
```

The results are the same as **cnbreg** output. Type *help (ml.nbc)* to learn the usage of the function, then use it to obtain the table of estimates. The results using the same Stata maximum likelihood command as used for the synthetic data example earlier are given as:

NB-C MODEL

```
. use medpar
. cnbreg los white hmo type2 type3, nolog
Canonical Negative Binomial Regression
                                               Number of obs   =     1495
                                                 Wald chi2(4)  =   131.15
Log likelihood = -4796.6997                      Prob > chi2   =   0.0000
```

```
------------------------------------------------------------------------------
         los |      Coef.   Std. Err.      z    P>|z|     [95% Conf. Interval]
-------------+----------------------------------------------------------------
         hmo |-.0141484   .0107354    -1.32   0.188    -.0351894    .0068926
       white |-.0248636   .0107598    -2.31   0.021    -.0459524   -.0037749
       type2 | .0409448   .0089345     4.58   0.000     .0234335    .0584561
       type3 | .1072065   .0100628    10.65   0.000     .0874837    .1269293
       _cons |-.2012769   .0135036   -14.91   0.000    -.2277436   -.1748103
-------------+----------------------------------------------------------------
    /lnalpha |-.8091036   .0444641   -18.20   0.000    -.8962516   -.7219556
-------------+----------------------------------------------------------------
       alpha |  .445257   .0197979                      .4080965    .4858013
------------------------------------------------------------------------------
. abic
AIC Statistic  =   6.425016          AIC*n      = 9605.3994
BIC Statistic  =   6.431372          BIC(Stata) = 9637.2588
```

The values for the coefficient, α, and standard errors, are the same as those produced using the R optimization code. It is important to note that modeling identical data using NB2 and NB-C will result in different coefficients, but identical heterogeneity parameters; i.e. identical values for alpha.

NB2 MODEL

Table 10.13 R: *Model medpar using NB2*

```
nb2c <- glm.nb(los ~ hmo + white + type2 + type3, data = medpar)
summary(nb2c)
modelfit(nb2c)  # script from Table 5.3
```

```
. nbreg los white hmo type2 type3
Negative binomial regression                      Number of obs   =      1495
                                                  LR chi2(4)      =    118.03
Dispersion     = mean                             Prob > chi2     =    0.0000
Log likelihood = -4797.4766                       Pseudo R2       =    0.0122
------------------------------------------------------------------------------
         los |      Coef.   Std. Err.      z    P>|z|     [95% Conf. Interval]
-------------+----------------------------------------------------------------
         hmo |  -.0679552   .0532613    -1.28   0.202    -.1723455    .0364351
       white |  -.1290654   .0685418    -1.88   0.060    -.2634049     .005274
       type2 |    .221249   .0505925     4.37   0.000     .1220894    .3204085
       type3 |   .7061588   .0761311     9.28   0.000     .5569446    .8553731
       _cons |   2.310279   .0679474    34.00   0.000     2.177105    2.443453
-------------+----------------------------------------------------------------
    /lnalpha |   -.807982   .0444542                     -.8951107   -.7208533
-------------+----------------------------------------------------------------
       alpha |   .4457567   .0198158                      .4085624    .4863371
------------------------------------------------------------------------------
Likelihood-ratio test of alpha=0:  chibar2(01) = 4262.86
                                   Prob>=chibar2 = 0.000
. abic
```

```
AIC Statistic    =   6.426055        AIC*n       = 9606.9531
BIC Statistic    =   6.432411        BIC(Stata) = 9638.8125
```

The AIC and BIC statistics in the right columns above show lower values for the NB-C model compared with NB2. NB1 statistics are both some 30 points higher. The difference in both AIC and BIC statistics is 2.45. The criteria for a significant difference in AIC values is 2.5. The value calculated for the **medpar** models is at the lower boundary, with no clear result. Modeling the data using **glm** produces a dispersion value of 1.1278 for the NB-C model and 1.1335 for NB2. There does appear to be a preference for the NB-C model compared with NB2, but the significance of the difference is slight.

Based on what is attempted to be learned from the model, we may select either as preferable to the other. Predicted counts may be better fitted using NB-C, but interpretation in terms of risk ratios is possible only with the NB2 model. There are data situations in which the NB-C is clearly superior to other models, but, based on simulation studies, NB-C and NB2 generally fit count data equally well. More work needs to be done in exploring the NB-C model and determining the type of data which it best fits. Nevertheless, the problem of interpretation remains.

Cytel's **LogXact** software, Stata's **expoisson** command, and SAS's exact Poisson procedure employ exact statistical methods for determining Poisson parameter estimates and standard errors. ***Exact Poisson regression*** conditions on the number of events in each panel or stratum of counts, which is similar to fixed-effects Poisson regression. Exact statistical models are based on the canonical link of the distribution, therefore an exact negative binomial model would be based on the canonical link, not the traditional log link. No exact negative binomial model yet exists.

10.4 NB-H: Heterogeneous negative binomial regression

The heterogeneous negative binomial extends the negative binomial model by allowing observation-specific parameterization of the ancillary parameter, α. In other words, the value of α is partitioned by user specified predictors. α takes the form $\exp(z_i \nu)$, which, like α, is positive. The method is applicable to NB1, NB-C, and NB2 models.

There are two uses of the heterogeneous model. First, parameterization of α provides information regarding which predictors influence overdispersion. Second, it is possible to determine whether overdispersion varies over the significant predictors of α by observing the differential values of its standard errors. If the standard errors vary only a little between parameters, then the overdispersion in the model can be regarded as constant.

No R function currently exists for estimating a NB-H model.

HETEROGENEOUS NB2: ALPHA == 0.5

```
. gnbreg y1 x1 x2, lnalpha(x1 x2)
------------------------------------------------------------------------
        |      Coef.   Std. Err.      z    P>|z|    [95% Conf. Interval]
--------+---------------------------------------------------------------
y1      |
     x1 | -1.503948  .0049501   -303.82   0.000    -1.51365   -1.494246
     x2 |  .7491305   .004433    168.99   0.000     .740442    .757819
  _cons |  .9939296   .004998    198.86   0.000    .9841336   1.003725
--------+---------------------------------------------------------------
lnalpha |
     x1 |  .0041994  .0131544      0.32   0.750   -.0215828   .0299815
     x2 | -.0047799  .0110758     -0.43   0.666   -.0264881   .0169283
  _cons | -.6846148   .015551    -44.02   0.000   -.7150941  -.6541354
------------------------------------------------------------------------
AIC Statistic =          4.584
```

The above NB2 synthetic data set is created with parameters of $x1 = -1.5$, $x2 = 0.75$, and constant $= 1.0$, with α specified as 0.5. Parameterization of α produces estimates that have little variability, i.e. there is little difference in α parameter values as well as in standard errors. Of course, in setting up the data, $x1$ and $x2$ were created using the same formula. Synthetic data sets with different αs produce similar results.

Parameterization of α for a NB2 model of *numvisits* (in **mdvis**), as displayed in Example 4 in Chapter 9, is given as

```
. use mdvis
. gnbreg numvisit reform badh educ2 educ3 age2 age3 loginc,
  lnalpha(reform badh educ2 educ3 age2 age3 loginc)

Generalized negative binomial regression
                                          Number of obs   =       2227
                                          LR chi2(7)      =     259.25
                                          Prob > chi2     =     0.0000
Log likelihood = -4549.7517               Pseudo R2       =     0.0277
------------------------------------------------------------------------
           |      Coef.   Std. Err.     z    P>|z|    [95% Conf. Interval]
-----------+------------------------------------------------------------
numvisit   |
  reform   | -.1331439   .0507271   -2.62   0.009   -.2325672   -.0337205
    badh   |  1.127951   .0717968   15.71   0.000    .9872322     1.26867
   educ2   |  .0867164   .0686798    1.26   0.207   -.0478936    .2213264
   educ3   | -.0267123    .070227   -0.38   0.704   -.1643548    .1109302
    age2   |  .0429533   .0650913    0.66   0.509   -.0846232    .1705299
    age3   |  .1742675   .0751244    2.32   0.020    .0270264    .3215087
  loginc   |  .1132818    .070934    1.60   0.110   -.0257463    .2523099
   _cons   |  -.148793   .5421144   -0.27   0.784   -1.211318    .9137316
------------------------------------------------------------------------
```

```
------------------------------------------------------------------------------
lnalpha  |
  reform | -.009731   .0971176   -0.10   0.920   -.200078    .180616
    badh | -.1890283  .1238452   -1.53   0.127   -.4317604   .0537037
   educ2 |  .0521736  .1223241    0.43   0.670   -.1875772   .2919245
   educ3 | -.3076519  .1383376   -2.22   0.026   -.5787886  -.0365151
    age2 |  .2675544  .1180651    2.27   0.023    .036151    .4989577
    age3 |  .3246583  .1284251    2.53   0.011    .0729499   .5763668
  loginc | -.0967873  .1338098   -0.72   0.469   -.3590498   .1654751
   _cons |  .7203326  1.025238    0.70   0.482   -1.289097   2.729762
------------------------------------------------------------------------------

. abic
AIC Statistic   =   4.100361          AIC*n      = 9131.5029
BIC Statistic   =   4.125831          BIC(Stata) = 9222.8379
```

Having α parameterized tells us which predictors influence α. Here *educ*3, and *age* (*age*2,*age*3) influence the amount of overdispersion in the data. These two predictors also significantly contribute to the count aspect of a negative binomial-clog hurdle model on the same data, as will be observed in Chapter 11. The AIC value of 4.10 is also the same value as that of the hurdle model.

Heterogeneous negative binomial regression is a valuable tool for assessing the source of overdispersion. It can be used to differentiate sources influencing the model parameter estimates versus sources influencing overdispersion. A reduced model, indicating such influences, is given as

```
. gnbreg numvisit reform badh age3 loginc, lnalpha(age2 age3)

Generalized negative binomial regression
                                             Number of obs   =       2227
                                             LR chi2(4)      =     292.75
                                             Prob > chi2     =     0.0000
Log likelihood = -4558.3537                  Pseudo R2       =     0.0311
------------------------------------------------------------------------------
         |     Coef.   Std. Err.     z    P>|z|   [95% Conf. Interval]
---------+--------------------------------------------------------------------
numvisit |
  reform | -.1359128   .0508994   -2.67   0.008   -.2356739  -.0361517
    badh |  1.156625   .0746261   15.50   0.000    1.01036    1.302889
    age3 |  .1775751   .0723686    2.45   0.014    .0357352   .319415
  loginc |  .1302475   .0689939    1.89   0.059   -.0049781   .2654731
   _cons | -.2445701   .5329164   -0.46   0.646   -1.289067   .7999268
---------+--------------------------------------------------------------------
lnalpha  |
    age2 |  .2620468   .1147147    2.28   0.022    .0372102   .4868835
    age3 |  .3325014   .1244992    2.67   0.008    .0884876   .5765153
   _cons | -.1316017   .0657923   -2.00   0.045   -.2605521  -.0026512
------------------------------------------------------------------------------

. abic
AIC Statistic   =   4.100902          AIC*n      = 9132.707
BIC Statistic   =   4.108657          BIC(Stata) = 9178.375
```

All of the predictors, with the possible exception of loginc, significantly contribute to the model. Age adds extra correlation to the data. Note that the AIC and BIC statistics are the same in both the full and reduced models.

We next model the data using a heterogeneous NB1 model. The command used is an author-constructed model, and can be downloaded from the book's website.

```
. hnbreg1 numvisit reform badh educ2 educ3, lnalpha(age2 age3)

Heterogeneous Negative Binomial Type 1 Regression
                                            Number of obs    =        2227
                                            Wald chi2(4)     =      285.82
Log likelihood = -4600.9075                 Prob > chi2      =      0.0000
------------------------------------------------------------------------------
numvisit |      Coef.   Std. Err.      z    P>|z|    [95% Conf. Interval]
---------+--------------------------------------------------------------------
xb       |
  reform | -.1074906   .0446122    -2.41   0.016   -.1949289   -.0200524
    badh |  .9365314   .0565584    16.56   0.000    .8256791    1.047384
   educ2 |  .0868621   .0572634     1.52   0.129   -.025372     .1990963
   educ3 |  .0627662   .0598764     1.05   0.295   -.0545894    .1801217
   _cons | -1.194791   .1063683   -11.23   0.000   -1.403269   -.9863133
---------+--------------------------------------------------------------------
lnalpha  |
    age2 |  .0314284   .0298398     1.05   0.292   -.0270566    .0899134
    age3 |  .0713569   .0334843     2.13   0.033    .0057288     .136985
   _cons |  .9681074   .0544707    17.77   0.000    .8613468    1.074868
---------+--------------------------------------------------------------------
   alpha |  2.632957   .1434189                      2.366346    2.929606
------------------------------------------------------------------------------

. abic
AIC Statistic   =   4.139118            AIC*n       = 9217.8154
BIC Statistic   =   4.146873            BIC(Stata)  = 9263.4824
```

The AIC and BIC fit statistics indicate that the model is not as well fitted as its NB2 counterpart. Only *reform* and *badh* significantly contribute to the model. The highest level of *age* is an indicator of extra correlation in the data.

The data may also be evaluated using a heterogeneous canonical negative binomial model. Again, the command was written by the author. No R or commercial software exists for duplicating the heterogeneous NB1 or NB-C models given here.

```
. chnbreg numvisit reform badh educ2 educ3, lnalpha(age2 age3)

Heterogeneous Canonical Negative Binomial Regression
                                            Number of obs    =        2227
                                            Wald chi2(4)     =      271.70
Log likelihood =   -4564.776                Prob > chi2      =      0.0000
```

```
------------------------------------------------------------------------------
numvisit |    Coef.   Std. Err.    z    P>|z|    [95% Conf. Interval]
---------+--------------------------------------------------------------------
xb       |
  reform | -.030608  .0128913  -2.37  0.018  -.0558744  -.0053415
    badh | .2492683  .0157116  15.87  0.000   .2184741   .2800624
   educ2 | .0197366  .0149991   1.32  0.188  -.0096611   .0491343
   educ3 |-.0180732  .0174036  -1.04  0.299  -.0521837   .0160373
   _cons |-.3812757  .0213084 -17.89  0.000  -.4230394   -.339512
---------+--------------------------------------------------------------------
lnalpha  |
    age2 | .0212601  .0558061   0.38  0.703  -.0881178    .130638
    age3 |-.0739724  .0624411  -1.18  0.236  -.1963546   .0484099
   _cons | .0109156  .0509209   0.21  0.830  -.0888875   .1107188
---------+--------------------------------------------------------------------
   alpha | 1.010975  .0514798                  .9149485   1.117081
------------------------------------------------------------------------------

. abic
AIC Statistic   =   4.105724          AIC*n       = 9143.4473
BIC Statistic   =   4.109991          BIC(Stata) = 9177.6982
```

It is clear that this model does not indicate that age levels contribute to any extra dispersion in the model. In fact, only two primary predictors are significant at the 0.05 level, *reform* and *badh*. The fit, however, as indicated by the AIC and BIC statistics, is not significantly different from NB2, but is superior to the NB1 model.

The heterogeneous negative binomial, whether parameterized as NB2, NB1, or NB-C, is a valuable tool in the assessment of the source of overdispersion. By parameterizing α, we are able to determine which specific predictors contribute to the inflation of α, i.e. which are more highly correlated than allowed by the distributional assumptions of the model which is estimated.

10.5 The NB-P model: generalized negative binomial

Building on the prior work of Winkelmann and Zimmermann (1995) and Cameron and Trivedi (1998), who discussed what was termed a generalized event count [GEC(k)] model, Greene (2006a) created the NB-P model to allow more flexibility in the NB2 variance. Given the general negative binomial structure of $\mu + \alpha\mu^{1|2}$ where the exponent 1 is NB1 and exponent 2 is NB2, Greene's generalization takes the form:

$$\text{NB-P} \qquad \mu + \alpha\mu^{Q} \quad \text{or} \quad \mu(1 + \alpha\mu^{Q-1}) \tag{10.40}$$

where Q is a parameter to be estimated. This form of negative binomial is a three-parameter model, with μ, α, and Q as parameters. It is not a simple re-parameterization of the basic NB2 model, but rather an entirely separate model.

The model may be estimated using Limdep or by using the user-authored Stata command, *nbregp*, in Hardin and Hilbe (2011).

The NB2 model may be schematized as [see Greene, 2006a, (E24.3.4)]

$$\text{Prob}\,(Y = y) = \frac{\Gamma\,(y_i + \theta)}{\Gamma\,(\theta)\,\Gamma\,(y_i + 1)} u_i^{\theta}\,(1 - u_i)^{y_i} \tag{10.41}$$

with

$$u_i = \frac{\theta}{(\theta + \mu_i)} \tag{10.42}$$

and

$$\theta = \frac{1}{\alpha} \tag{10.43}$$

NB1 as

$$\text{Prob}\,(Y = y) = \frac{\Gamma\,(y_i + \mu_i\theta)}{\Gamma\,(\mu_i\theta)\,\Gamma\,(y_i + 1)} u_i^{\mu_i\theta}\,(1 - u_i)^{y_i} \tag{10.44}$$

with

$$u = \frac{\theta}{(\theta + 1)} \tag{10.45}$$

The NB-P distribution takes the same form as NB2, as expressed in equations 10.41 through 10.43. However, for each value of θ in equations 10.41 and 10.42, we substitute the value $\theta\mu^{2-P}$. For ease of interpretation, however, Greene substitutes parameter $2-P$ as Q, with NB2 having $Q = 0$ and NB1 as $Q = 1$. Parameterized in this manner, the NB-P distribution replaces NB2 values of θ with the value $\theta\mu^{Q}$.

The NB-P probability mass function is given by Greene (2006a) as

$$\text{Prob}(Y = y|x) = \frac{\Gamma(\theta\lambda_i^Q + y_i)}{\Gamma(\theta\lambda_i^Q)\Gamma(y_i + 1)} \left(\frac{\theta\lambda_i^Q}{\theta\lambda_i^Q + \lambda_i}\right)^{\theta\lambda_i^Q} \left(\frac{\lambda_i}{\theta\lambda_i^Q + \lambda_i}\right)^{y_i} \tag{10.46}$$

The NB-P log-likelihood is generated from equation 10.46 as

$$\mathcal{L}\,(\lambda; y, \theta, Q) = \sum_{i=1}^{n} \left\{ \theta\lambda_i^Q \ln\left(\frac{\theta\lambda_i^Q}{\theta\lambda_i^Q + \lambda_i}\right) + y_i \ln\left(1 - \frac{\theta\lambda_i^Q}{\theta\lambda_i^Q + \lambda_i}\right) \right.$$
$$\left. + \ln\Gamma\left(y_i + \theta\lambda_i^Q\right) - \ln\Gamma\left(\theta\lambda_i^Q\right) - \ln\Gamma\,(y_i + 1) \right\} \tag{10.47}$$

The gradient functions are obtained as the derivative of the log-likelihood by λ, θ, and Q respectively.

$$\frac{\partial\mathcal{L}(\mu; y, \theta, Q)}{\partial\lambda_i} = \sum_{i=1}^{n}\left\{\left[\Psi\left(y_i + \theta\lambda_i^Q\right) - \Psi\left(\theta\lambda_i^Q\right) + \ln\left(\frac{\theta\lambda_i^Q}{\theta\lambda_i^Q + \lambda_i}\right)\right]\theta Q\lambda_i^{Q-1} + \left[\left(\theta\lambda_i^Q + \lambda_i\right) + \frac{y_i\left(\theta\lambda_i^Q - \lambda_i\right)}{\theta\lambda_i^Q}\right]\left(\frac{(Q-1)\theta\lambda_i^Q}{\theta\lambda_i^Q + \lambda_i}\right)\right\} \tag{10.48}$$

$$\frac{\partial\mathcal{L}(\mu; y, \theta, Q)}{\partial\theta} = \sum_{i=1}^{n}\left\{\left[\Psi\left(y_i + \theta\lambda_i^Q\right) - \Psi\left(\theta\lambda_i^Q\right) + \ln\left(\frac{\theta\lambda_i^Q}{\theta\lambda_i^Q + \lambda_i}\right)\right]\lambda_i^Q + \left[\left(\theta\lambda_i^Q + \lambda_i\right) + \frac{y_i\left(\theta\lambda_i^Q - \lambda_i\right)}{\theta\lambda_i^Q}\right]\left(\frac{\lambda_i^{Q+1}}{\left(\theta\lambda_i^Q + \lambda_i\right)^2}\right)\right\} \tag{10.49}$$

$$\frac{\partial\mathcal{L}(\mu; y, \theta, Q)}{\partial Q} = \sum_{i=1}^{n}\left\{\left[\Psi\left(y_i + \theta\lambda_i^Q\right) - \Psi\left(\theta\lambda_i^Q\right) + \ln\left(\frac{\theta\lambda_i^Q}{\theta\lambda_i^Q + \lambda_i}\right)\right]\theta\lambda_i^Q\ln(\lambda_i) + \left[\left(\theta\lambda_i^Q + \lambda_i\right) + \frac{y_i\left(\theta\lambda_i^Q - \lambda_i\right)}{\theta\lambda_i^Q}\right]\frac{\theta\lambda_i^{Q+1}}{\left(\theta\lambda_i^Q + \lambda_i\right)^2}\right\} \tag{10.50}$$

An example will show the value of the NB-P model. We use the same German health data, called **rwm**, as used by Greene. It is taken from earlier years than the **mdvis** data we have used in previous examples.

The response is *docvis,* number of visits to the doctor, with three predictors:

age: age from 25 through 64
hh: monthly net household income in German marks/1,000. Converted from *hhninc* by *hh* = *hhninc*/10
educ: years of schooling, ranging from 7 through 18(+).

First, we use an NB2 model to estimate parameters and the value of α.

Table 10.14 R: *NB2 rwm data*

```
rm(list=ls())
data(rwm)
attach(rwm)
hh <- hhninc/10
nb2mp <- glm.nb(docvis ~ age + hh + educ, data=rwm)
summary(nb2mp)
```

```
. use rwm
. gen hh = hhninc/10
. nbreg docvis age hh educ

Negative binomial regression                 Number of obs   =      27326
                                             LR chi2(3)      =    1027.40
Dispersion       = mean                      Prob > chi2     =     0.0000
Log likelihood = -60322.021                  Pseudo R2       =     0.0084
---------------------------------------------------------------------------
   docvis |     Coef.   Std. Err.   z     P>|z|  [95% Conf. Interval]
---------+-----------------------------------------------------------------
      age | .0204292  .0008006  25.52  0.000  .0188601    .0219984
       hh |-.4768144  .0522786  -9.12  0.000 -.5792785   -.3743503
     educ |-.0459575  .0042257 -10.88  0.000 -.0542398   -.0376752
    _cons | .9132608  .0633758  14.41  0.000  .7890465    1.037475
---------+-----------------------------------------------------------------
/lnalpha | .6608039  .0115374                   .638191    .6834168
---------+-----------------------------------------------------------------
    alpha | 1.936348  .0223404                  1.893053    1.980634
---------------------------------------------------------------------------
Likelihood-ratio test of alpha=0:  chibar2(01) = 8.9e+04
Prob>=chibar2 = 0.000
AIC Statistic =         4.415
```

Next we model the data as NB-P, where both α as well as Q are estimated. LIMDEP 9.0 is used to model the data.

```
---------------------------------------------------------------------------
Negative Binomial (P) Model
Maximum Likelihood Estimates
Model estimated: Mar 31, 2006 at 09:13:20PM.
Dependent variable                       DOCVIS
Weighting variable                         None
Number of observations                    27326
Iterations completed                         15
Log likelihood function               -60258.97
Number of parameters                          6
Info. Criterion: AIC =                  4.41082
  Finite Sample: AIC =                  4.41082
Info. Criterion: BIC =                  4.41262
```

```
Info. Criterion:HQIC =          4.41140
Restricted log likelihood    -104814.1
McFadden Pseudo R-squared      .4250871
Chi squared                   89110.23
Degrees of freedom                    1
Prob[ChiSqd > value] =         .0000000
-----------------------------------------------------------------
Variable Coefficient Standard Error b/St.Er P[|Z|>z] Mean of X
-----------------------------------------------------------------
 Constant | .77029290 .05940343  12.967  .0000
 AGE      |   .02177762   .00074029    29.418    .0000  43.5256898
 HHNINC   | -.38749687    .05121714    -7.566    .0000    .35208362
 EDUC     | -.04127636    .00412037   -10.018    .0000  11.3206310
----------+------------------------------------------------------
          | Dispersion parameter for count data model
----------+------------------------------------------------------
 Alpha    | 3.27757050    .14132403    23.192    .0000
          | Negative Binomial. General form, NegBin P
 P        | 2.45563168    .03595933    68.289    .0000
-----------------------------------------------------------------
```

Table 10.15 displays a comparison between NB2 and NB-P estimates.

Table 10.15 *Comparison: NB2 and NB-P results*

	NB2	NB-P
age	.0204	.0218
hh	-.4768	-.3875
educ	-.0460	-.0413
constant	.9133	.7703
alpha	1.936	3.275
power	2.000	2.456
AIC	4.415	4.411

A likelihood ratio test between the NB2 and NB-P models results in:

$$-2\{\mathcal{L}(\text{NB2}) - \mathcal{L}(\text{NB-P})\}$$

$$-2\{(-60322.021) - (-60268.97)\} = 106.10$$

A reverse cumulative upper tail χ^2 distribution with 1 degree of freedom gives us a probability value: `chi2tail(1, 106.10) = 7.011e-25`. The traditional cutoff point for a 0.05 significance level is `chiprob(1, 3.84) = 0.05004352`. Therefore, any χ^2 having a value exceeding 3.84 will be significant, i.e. the models differ significantly from one another. Greene presents a Wald t-test of significance for comparing the two values of power: (NB-P – NB2)/$SE_{NB\text{-}P}$ or (2.456 – 2)/0.036 = 12.65. Using a reverse cumulative upper tail Students's T distribution with 1 degree of freedom, we have: `ttail(1,12.65) = 0.02511062`. A test of power gives additional evidence that the two models are significantly different. Given the lower value of the NB2 log-likelihood function, and slightly lower value of the NB-P AIC statistic, it appears that the NB-P model may be slightly preferred over the NB2.

NB-P models generalize the basic negative binomial model, providing more flexibility in fitting data. The usefulness of the model has yet to be explored, but I suspect that this will soon be remedied.

10.6 Generalized Waring regression

The generalized Waring (GW) distribution has a rather old and interesting history – one that has not generally been noticed by many statisticians. The distribution has its origins in work of Edward Waring (1736–1798), Lucasian professor of mathematics at Cambridge University and originator of the well-known *Waring's Problem* in mathematics. The distribution Waring first described was advanced by Newbold (1927) and later by Irwin (1968), who gave it its present form. Irwin was interested in accident theory, and was able to determine that a generalized Waring distribution could be used to better understand the proneness certain individuals had toward getting into accidents.

Inherent to our understanding of traffic accidents, for example, is the unpredictability of the event. The basic approach to understanding and predicting accidents of this sort is to regard them as random chance events per unit time or area. As such, accidents, or rather counts of accidents, tend to follow a Poisson distribution with the mean given as $\lambda^y e^{-\lambda}/y!$. Y is the random variable describing the number of accidents sustained by an individual person. If accidents are counted over various time periods or by area, the above equation is amended to incorporate such an adjustment, t. t is multiplied by mean λ, and entered into a regression model based on the distribution as an offset. We discussed the nature of this relationship in Chapter 6.

Another more popular approach, particularly among actuaries, is to understand how individuals differ from one another in their proneness to having accidents. Individuals having one accident are more likely to have another than is an individual who has never had an accident. This is especially the case if emphasis is placed on individuals who are responsible for the accident. Regardless, more sophisticated methods using this approach assume that accidents consist of two components, random and non-random. Non-random components to an accident entail understanding the individual's psychology.

Some individuals tend to be more careless, or text while driving, or have a history of taking narcotic medications, and so forth. This proneness factor adds heterogeneity to otherwise equidispersed Poisson count data. As far back as Greenwood and Woods (1919) it was understood that a negative binomial model can be used to deal with accident data where one's proneness to having accidents can be considered.

Irwin parameterized the negative binomial PDF, with parameters k and $1/\nu$, as

$$\binom{y+k-1}{y} \nu^{y} (1+\nu)^{-(y+k)}, \quad y = 0, 1, \ldots \tag{10.51}$$

Note the different parameterization from what we derived in equations 8.8 and 8.28. k in equation 10.51 is the same as r in 8.21, and ν is the equivalent of p. The exponent of the final term in equation 10.51 adjusts for the singular y in the choose function, resulting in a negative binomial having the same interpretation as described Section 8.2.2.

Irwin's "proneness-liability" model represented an advance over the simple proneness view of heterogeneity. He proposed that the non-random component can be further partitioned into psychological and other external factors, e.g. the time of the day and type of roadway driven by the individual. Together, both components provide a better explanation as to why some people have more accidents than others. Irwin coined the terms *accident proneness*, ν, to indicate an individual's predisposition to accidents, and *accident liability*, $\lambda|\nu$, to indicate an individual's exposure to the external risk of having an accident.

The key to Irwin's generalized Waring distribution is that ν itself is given to follow a betaII distribution of the second type (Johnson and Kotz, 1969) with parameters a and ρ. The distribution is displayed as:

GENERALIZED WARING PDF

$$\frac{\Gamma(a+\rho)}{\Gamma(a)\,\Gamma(\rho)} \nu^{(a-1)} (1+\nu)^{-(a+\rho)} \tag{10.52}$$

with $a, \rho \geq 0$.

The distribution therefore has three parameters, a , k and ρ, reflecting the observed variability, which can be partitioned into three additive components: randomness; proneness or the internal differences between individuals; and liability, or the presence of other external factors that are not included as model predictors. Irwin did not separately estimate proneness and liability. Xekalaki (1983) solved that problem. Over the years since 1983 she has fully explicated the various moments and ancillary statistics of the distribution, and developed a working regression model for the distribution.

Following on the work of Irwin and Xekalaki, Rodríguez-Avi *et al.* (2009) have recently developed an R package, **GWRM**, for the estimation of the generalized Waring model. It will be employed to show an example of model output.

We shall use the German health data, **rwm5yr,** with number of visits to the doctor as the response, and *age*, annual household income divided by 10, and education, each entered into the model as continuous predictors. The **GWRM** package must be installed and loaded into memory. We first model the data using a standard NB2 negative binomial model, followed by a generalized Waring model. Note that the negative bimomial heterogeneity parameter, α, or which is referred to as θ in the software, is partitioned into two betaII parameters in the generalized Waring output. This provides us information on two separate sources of overdispersion in the model.

The GW mean and variance are defined as follows, indicating the added beta variation into the algorithm.

$$\mu = \frac{ak}{\rho - 1} \tag{10.53}$$

$$V(\mu) = \mu + \mu\left(\frac{k+1}{\rho - 2}\right) + \mu^2\left(\frac{k+\rho-1}{k(\rho-2)}\right) \tag{10.54}$$

The log-likelihood is expressed as (with $:\alpha_i = \mu_i k/(\rho - 1)$):

$$\begin{aligned}\mathcal{L}(k, \rho, \beta; \mathrm{y}) = \sum_{i=1}^{\mathrm{n}} \{ & \ln\Gamma(a_i + y_i) - \ln\Gamma(a_i) + \ln\Gamma(k + y_i) - \ln\Gamma(k) \\ & - \ln\Gamma(a_i + k + \rho + y_i) + \ln\Gamma(a_i + \rho) + \ln\Gamma(k + \rho) \\ & - \ln\Gamma(\rho) - \ln\Gamma(y_i + 1)\}\end{aligned} \tag{10.55}$$

Table 10.16 R: *Comparison of NB2 and generalized Waring model*

```
rm(list=ls())
library(GWRM)
data(rwm5yr)
attach(rwm5yr)
hh <- hhninc/10
nb2mp <- glm.nb(docvis ~ age + hh + educ, data=rwm5yr)
summary(nb2mp)
gw2mp <- GWRM.fit(docvis ~ age + hh + educ, data=rwm5yr)
GWRM.display(gw2mp)
```

NEGATIVE BINOMIAL NB2 MODEL (partial output)

```
Coefficients:
            Estimate Std. Error z value Pr(>|z|)
(Intercept) 0.895511  0.078618  11.391  <2e-16 ***
age         0.021224  0.001004  21.139  <2e-16 ***
hh         -0.667788  0.075566  -8.837  <2e-16 ***
educ       -0.044406  0.005196  -8.547  <2e-16 ***
---
AIC: 85891

              Theta:  0.48274
          Std. Err.:  0.00657
 2 x log-likelihood:  -85881.22200
```

GENERALIZED WARING MODEL

```
$Table
       covars   estimates          se          z            p
1 (Intercept)  0.71331480 0.083105983   8.583194  9.227429e-18
2         age  0.02388392 0.001044956  22.856390 1.262469e-115
3          hh -0.60903300 0.077922213  -7.815910  5.456737e-15
4        educ -0.04042529 0.005498805  -7.351651  1.957737e-13
$betaII
  par      coef     value
1   k -0.470137 0.6249166
2  ro  1.226077 4.4078340
$Fit
  loglikelihood      AIC      BIC    df
1     -42788.81 85589.62 85636.92 19603
```

The difference in coefficient and standard errors between the NB2 and GW models is minimal, indicating that there appears to little if any non-NB overdispersion. The AIC statistic is 300 lower in the Waring model than in the NB2, indicating a better fit. However, the betaII parameters k and ρ (rho) tell us the following:

K proneness: A measure of one's likelihood to visit the doctor again given an earlier visit. A value of 0.62 indicates that the model explained more heterogeneity than the NB2 α parameter. There was definite evidence of proneness in the data.

P liability: A measure of external factors in the environment not observed or considered in the model. There is no one-to-one relationship between this value and external sources of heterogeneity; it is best regarded as a comparative statistic.

When there is a binary or categorical predictor in the model, the variability can be broken into components reflective of each level. The same may be done for continuous predictors, but the output is unwieldly and the interpretation difficult. Here we take the same data, but for ease of interpretation employ a single binary predictor, *outwork*. We shall first observe if the Waring model fits the data better than the NB2, and then we shall partition the three dispersion statistics across levels. Comparisons can also be made with the above model with three continuous predictors.

Table 10.17 R: GWRM *single binary predictor model*

```
nb2 <- glm.nb(docvis ~ outwork, data=rwm5yr)
summary(nb2)
gwr <- GWRM.fit(docvis ~ outwork, data=rwm5yr, method=1)
GWRM.display(gwr)
cat <-factor(c("W", "NW"), levels=c("W", "NW"))
levels <- data.frame(outwork=cat)
bvalues <- GWRM.stats(gwr, covars=levels)
bvalues
```

NEGATIVE BINOMIAL (Partial display of output)

```
Coefficients:
             Estimate Std. Error z value Pr(>|z|)
(Intercept)   0.98046    0.01397   70.19   <2e-16 ***
outwork       0.44266    0.02340   18.92   <2e-16 ***
---
AIC: 86266
           Theta:  0.46844
           Std. Err.:  0.00632
2 x log-likelihood:  -86259.68500
```

GENERALIZED WARING MODEL

```
> GWRM.display(gwr)
$Table
       covars estimates         se        z            p
1 (Intercept) 0.9635455 0.01556121 61.91970 0.000000e+00
2     outwork 0.4864380 0.02451766 19.84031 1.336439e-87
$betaII
  par      coef     value
1   k -0.5056023 0.6031422
2  ro  1.2040453 4.3335750
```

```
$Fit
  loglikelihood      AIC    BIC    df
1      -42974.73 85957.46 85989 19605
$params
  outwork          a          k        ro
1       W 14.48615 0.6031422 4.333575
2      NW 23.56190 0.6031422 4.333575
$stats
  outwork Mean.est Low.bound Upp.bound  Var.est
1       W 2.620973  2.541035  2.700911 23.63555
2      NW 4.263044  4.091637  4.434451 58.02315
$partvar
  outwork Randomness  Liability Proneness
1       W 0.11089112 0.07618107 0.8129278
2      NW 0.07347144 0.05047413 0.8760544
```

Note the substantially lower AIC value for the Waring model compared to the NB2 – 85957 to 85266, nearly 600 difference. The important statistics are the ones in the final block, displaying the partitioned levels of *outwork*, with W:*outwork* = 1 and NW:*outwork* = 0. Here we see that proneness is at 81% for patients who are out of work compared with 87% for those working. Working patients are more apt to repeat visits than those out of work. Considerably more heterogeneity is explained by this model compared with a Poisson model (randomness). There is only 7.6% and 5% liability respectively in the model, indicating that there are not many external factors unaccounted for by the model.

More work needs to be done on the interpretation of the three components of variability provided in the Generalized Waring model. The model has the potential of providing a better fitted model to count data than other simpler parameterizations of the negative binomial. Generalized Waring regression is a type of negative binomial model, one with the standard heterogeneity parameter, α, partitioned into two components. It may well become a well-used statistical model in coming years.

10.7 Bivariate negative binomial

Bivariate models are models with two related response variables. The degree of relation between the response variables is given as a correlation value, ρ. Typically the responses consist of long vectors of numbers that give separate characteristics for a subject. For example, in education a bivariate count model can be constructed with ACT and SAT scores recorded for students who took both tests during an academic year. The number of days spent in the hospital following two different types of surgery is another example of a bivariate relationship. We shall employ an example of the number of times per month a teen smokes and drinks, from among students admitting both activities. Other similar relationships can be assessed, with responses normally being characteristics of a given object or subject.

Bivariate probit models are perhaps the most well-known bivariate models. Software such as Stata, R, SAS, LIMDEP, Matlab, and others have a bivariate probit as a basic offering. Bivariate Poisson and negative binomial models, however, have not enjoyed such software support. Given their potential value in understanding the relationship of two related count variables, though, it is somewhat surprising that they have been generally ignored by the research community.

The equation for the bivariate negative binomial probability function was first constructed by Marshall and Olkin (1990), and is displayed as:

BIVARIATE NEGATIVE BINOMIAL

$$f\left(y_1, y_2|x_1, x_{2;}\right) = \frac{\Gamma\left(y_1 + y_2 + \alpha^{-1}\right)}{y_1!y_2! + 1}\left[\frac{\mu_1}{\mu_1 + \mu_2 + 1}\right]^{y_1}\left[\frac{\mu_2}{\mu_1 + \mu_2 + 1}\right]^{y_2} \times \left[\frac{1}{\mu_1 + \mu_2 + 1}\right]^{\alpha^{-1}} \quad (10.56)$$

The corresponding correlation function that exists between the two vectors of counts is given as follows:

$$C\left[y_1; y_2\right] = \frac{\lambda_1\lambda_2}{\sqrt{\left(\lambda_1^2 + \alpha\lambda_1\right)\left(\lambda_2^2 + \alpha\lambda_2\right)}} \quad (10.57)$$

A number of other parameterizations and generalizations of the basic bivariate negative binomial have been proposed, but have not found widespread use. Perhaps the most well known of this group are the *bivariate Poisson-lognormal mixture model*. The model employs the log-normal mixing distribution rather than the gamma distribution, which results in the NB2 negative binomial. The drawback, however, is that there is no closed form solution of the model, and some variety of quadrature or simulation needs to be used for estimation.

Aside from the standard coefficients and standard errors given in standard regression output, the bivariate negative binomial model produces several other statistics of interest. First, since there are two correlated response variables and predictors for each response, estimation provides two sets of coefficients and standard errors, together with a correlation value relating the two. Output also includes statistics symbolized as ϕ, α, λ, and ρ. Statistics ϕ and α are moments of gamma variance in the Poisson–gamma mixture, which constitute the derived bivariate negative binomial distribution. Based on the above bivariate negative binomial PDF, they are defined as:

$$\phi^{k+1} = \frac{\sum_{i=1}^{n}\left\{\left(y_{1i}-\mu_{1i}^{k}\right)^2+\left(y_{2i}-\mu_{2i}^{k}\right)^2-\left(\mu_{1i}^{k}+\mu_{2i}^{k}\right)\right\}}{\sum_{i=1}^{n}\left\{\left(\mu_{1i}^{k}\right)^2+\left(\mu_{2i}^{k}\right)^2\right\}} \tag{10.58}$$

$$\alpha^{k+1} = \frac{\sum_{i=1}^{n}\left\{\left(y_{1i}-\mu_{1i}^{k}\right)\left(y_{2i}-\mu_{2i}^{k}\right)\right\}}{\phi^{k+1}\sum_{i=1}^{n}\left\{\mu_{1i}^{k}\mu_{2i}^{k}\right\}} \tag{10.59}$$

Without superscripts, we have

$$\lambda = \frac{(1+\alpha)}{\phi\alpha} \qquad \rho = \frac{\alpha}{(\alpha+1)} \tag{10.60}$$

where ρ is the correlation between y_1 and y_2 expressed in terms of α.

Using an R bivariate negative binomial function, **binegbin.glm**, kindly created for this text by Masakazu Iwasaki, Clinical Statistics Group Manager for Research & Development at Schering-Plough K.K., Tokyo, we provide an example using the data found in a comma delimited text file, **smokedrink.dat**, which is available on this book's website as well as in the R COUNT package. The parameter estimates, standard errors and correlations between the two response levels are given as (assume that the text file and binegbin.glm files are in the source directory):

Table 10.18 R: *Bivariate negative binomial*

```
smdr <- read.table("c:\\source\\smokedrink.raw", sep=",")
y1=smdr[,1]
y2=smdr[,2]
age=smdr[,3]
gender=smdr[,4]
x1=cbind(1,age,gender)
x2=cbind(1,age,gender)
source("c://source/binegbin.glm")
smdr.out=binegbin.glm(y1,y2,x1,x2)
```

OUTPUT

```
[1] 6        Total number of parameters
smdr.out  intercept         age       gender
[[1]][1] 1.49257599 0.05030777  0.01045346  <- Estimates of beta1
[[2]][1] 0.78098900 0.04748482 -0.09142176  <- Estimates of beta2
[[3]]
# Asymptotic Standard Errors for Betas
                  age      gender
0.45825810 0.03024894 0.07983801        <--- Beta 1
0.49042631 0.03235953 0.08531008       <--- Beta 2
[[4]][1] 0.8159685     <--- Estimate of Phi
[[5]][1] 0.2208678     <--- Estimate of Alpha
[[6]][1] 6.774275      <--- Estimate of Lamda
[[7]][1] 0.1809105     <--- Estimate of Rho
[[8]]                  Estimate of Asymptotic Covariance Matrix
                                       age        gender
        0.2100004844 -1.375867e-02 -4.205794e-03  0.0409078396 -2.681761e-03
age    -0.0137586670  9.149982e-04  7.261959e-05 -0.0026817825  1.784637e-04
gender -0.0042057940  7.261959e-05  6.374107e-03 -0.0008229687  1.424420e-05
        0.0409078396 -2.681782e-03 -8.229687e-04  0.2405179641 -1.575356e-02
       -0.0026817615  1.784637e-04  1.424420e-05 -0.0157535624  1.047139e-03
       -0.0008236196  1.428740e-05  1.246959e-03 -0.0047335704  8.097279e-05

       -8.236196e-04
age     1.428740e-05
gender  1.246959e-03
       -4.733570e-03
        8.097279e-05
        7.277810e-03

[[9]][1]   -0.001387864      <--- Mean of y1-mu1
[[10]][1] -2.890762e-05     <--- Mean of y2-mu2

[[11]][1] 82.42534     <--- Mean Square Error of y1
[[12]][1] 20.8426      <--- Mean Square Error of y2

[[13]][1] 9.524704     <--- Mean of mu1
[[14]][1] 4.267732     <--- Mean of mu2
```

For a detailed examination of the derivation of this form of bivariate negative binomial, together with examples, see Iwasaki and Tsubaki (2005a, 2005b, 2006).

As an aside, it may prove helpful to have information related to the *bivariate Poisson*. The logic is the same as the bivariate negative binomial, where the probability of observing y is the sum of the joint probabilities over y_1 and y_2 where each y is a function of specific explanatory predictors.

The probability distribution function of the multiplicative approach can be given as:

$$f(y_1, y_2|x_1, x_2) = \frac{y_2!}{u!(y_2-u)!}\left(\frac{\gamma}{\lambda_2+\gamma}\right)^u\left(\frac{\lambda_2}{\lambda_2+\gamma}\right)^{y_2-u} \times \exp(-\lambda_1)\frac{\lambda_1^{y_1-u}}{(y_1-u)!} \tag{10.61}$$

y_1 is Poisson distributed with parameter λ_1; $u \mid y_2$ is binomially distributed with the probability of 1 equal to $\gamma/(\lambda_2+\gamma)$ and binomial denominator m equal to y_2; γ is defined as the variance of u, and the correlation between y_1 and y_2 is expressed as:

$$\text{Corr}(y_1; y_2) = \frac{\gamma}{\sqrt{(\lambda_1+\gamma)(\lambda_2+\gamma)}} \tag{10.62}$$

As with the bivariate negative binomial, the key to understanding the bivariate Poisson model is to realize that it is in effect a two-way regression between y_1 and y_2. Both count responses are understood on the basis of each other, and are related by equation 10.62.

The only software currently existing that estimates a bivariate Poisson is included in R's **bivpois** library. The data to be modeled are required to be members of a data frame. We use the **rwm5yr** data, which are in data frame format, to display example code.

```
library(bivpois)
data(rwm5yr)
attach(rwm5yr)
bvp <- lm.bp(docvis~reform+outwork+hhkids, hospvis~reform+
    outwork+hhkids, data=rwm5yr)
```

Post-estimation model statistics are not obtained using the **summary** function. Rather each component of the statistical output is accessed using individual functions. To obtain model coefficients, as well as the constant or intercept, type `bvp$coef`. The log-likelihood is obtained by typing `mean(bvp$loglikelihood)`. Other model statistics can be determined by reviewing the documentation for **bivpois**.

10.8 Generalized Poisson regression

A generalization to the basic Poisson model was developed by Consul and Jain (1973), which they aptly termed *generalized Poisson regression*. It has since undergone various modifications, with models created with names such as the *restricted generalized Poisson*, three parameterizations of a *hybrid generalized Poisson*, and so forth. Refer to Consul and Famoye (1992) for a good overview of the base generalized Poisson model and its derivation.

Generalized Poisson is similar to the negative binomial in that it incorporates an extra heterogeneity or dispersion parameter. However, whereas the negative binomial heterogeneity parameter, α, is based on the single parameter gamma distribution with a mean of 1, the heterogeneity parameter employed by the generalized Poisson is based on the lognormal distribution. This allows modeling of both underdispersed as well as overdispersed data. This type of Poisson-lognormal mixture model was first used in species abundance literature (Bulmer, 1974).

There are several different models that are referred to as generalized Poisson models. A commonly used parameterization comes from Consul (1989), and is used in R's **VGAM** package. The distribution is described in detail in Consul and Famoye (1992). The probability function is given as:

$$f(y;\lambda,\theta) = \theta\,(\theta + \lambda_i y_i)^{y_i - 1}\,\frac{\exp(-\theta - \lambda_i y_i)}{y_i\,!} \tag{10.63}$$

Equation 10.59 reduces to the traditional standard Poisson when $\theta = 0$. The first two moments of the distribution are:

MEAN $$E(Y) = \frac{\theta}{(1-\lambda_i)}$$

VARIANCE $$V(Y) = \frac{\theta}{(1-\lambda_i)^3}$$

The value of θ serves as a second parameter, analogous to the negative binomial model. It reflects the amount of extra Poisson overdispersion in the data. We use the **azpro** data as an example of how it can be used. The value of θ, the heterogeneity parameter, is displayed as (Intercept):2 in the statistical output. In this instance, $\theta = 1.02352$.

R:

```
rm(list=ls())   # caution - function drops all objects from memory
library(COUNT); library(VGAM)
data(azpro)
attach(azpro)
vglm(los~ procedure+sex+admit+age75,genpoisson(zero=1),data=azpro)
```

```
                     Min       1Q    Median        3Q     Max
elogit(lambda,  -0.66905 -0.61874  -0.46808 -0.065816 35.5485
log(theta)     -10.72271 -0.60294   0.11635  0.643891  1.2008

Coefficients:
                   Value Std. Error t value
(Intercept):1    0.76482   0.019314 39.5984
(Intercept):2    1.02352   0.025664 39.8817
procedure        0.94721   0.018462 51.3067
sex             -0.11318   0.018064 -6.2658
admit            0.30120   0.018377 16.3906
age75            0.11505   0.019052  6.0384

Log-likelihood: -9909.837 on 7172 degrees of freedom
```

We employ a user-created Stata implementation found in Hardin and Hilbe (2007), and in an unpublished manuscript authored by Yang *et. al.* (2006) to model the same data.

```
. gpoisson los procedure sex admit age75
Generalized Poisson regression                    Number of obs =      3589
                                                  LR chi2(4)    =   2378.52
Dispersion      =  1.574262                       Prob > chi2   =    0.0000
Log likelihood = -9909.8371                       Pseudo R2     =    0.1071
-----------------------------------------------------------------------------
         los |     Coef.   Std. Err.      z   P>|z|   [95% Conf. Interval]
-------------+---------------------------------------------------------------
   procedure |  .9472089   .0183127   51.72 0.000   .9113167    .983101
         sex | -.1131816   .0179749   -6.30 0.000  -.1484118   -.0779514
       admit |   .301198   .0181977   16.55 0.000   .2655312    .3368649
       age75 |   .115044   .0189695    6.06 0.000   .0778645    .1522234
       _cons |   1.47734   .0237253   62.27 0.000   1.430839    1.523841
-------------+---------------------------------------------------------------
  /tanhdelta |  .3823905   .0089607                 .3648278    .3999533
-------------+---------------------------------------------------------------
       delta |  .3647817   .0077684                 .3494594    .379909
-----------------------------------------------------------------------------
Likelihood-ratio test of delta=0:  chibar2(1) = 2560.12
                                   Prob>=chibar2 = 0.0000
```

The estimates are near identical, but not the heterogeneity parameter. The Stata parameter is calculated in terms of a hyperbolic tangent; the R parameter is in terms of an extended logit, defined in R as [exp(*lambda*) − 1]/[exp(*lambda*) + 1].

Given the R output, *lambda* is found as the value of intercept:1, which is 0.76482. Applying the formula for extended logit, we have a calculated value of .3648, which is the same as *delta* in the Stata model.

```
. (exp(.76482)-1) / (exp(.76482)+1)
.36479858
```

.3648 is the estimated value of the Stata dispersion parameter.

Another well-known parameterization of the generalized Poisson model is given by the parameterization found in Famoye and Singh (2006) and Winkelmann (2008).

$$f(y;\mu,\alpha) = \left(\frac{\mu_i}{1+\alpha\mu_i}\right)^{y_i} \frac{(1+\alpha y_i)^{y_i-1}}{y_i!} \exp\left[\frac{-\mu_i(1+\alpha y_i)}{1+\alpha\mu_i}\right] \tag{10.64}$$

with α specifying the heterogeneity parameter, and the mean and variance of y defined as

$$E(y) = \mu_i, \qquad V(y) = \mu_i(1+\alpha\mu_i)^2$$

The log-likelihood function can be given as

$$\mathcal{L}(\mu,\alpha;y) = \sum_{i=1}^{n}\left\{y_i \ln\left(\frac{\mu_i}{1+\alpha\mu_i}\right) + (y_i-1)\ln(1+\alpha y_i) - \left[\frac{-\mu_i(1+\alpha y_i)}{1+\alpha\mu_i}\right] - \ln\Gamma(y_i+1)\right\} \tag{10.65}$$

or in terms of $x\beta$ as

$$\mathcal{L}(\beta,\alpha;y) = \sum_{i=1}^{n}\left\{y_i \ln\left(\frac{\exp(x_i'\beta)}{1+\alpha\exp(x_i'\beta)}\right) + (y_i-1)\ln(1+\alpha y_i) - \left[\frac{-\exp(x_i'\beta)(1+\alpha y_i)}{1+\alpha\exp(x_i'\beta)}\right] - \ln\Gamma(y_i+1)\right\} \tag{10.66}$$

Like the R and Stata model with respect to θ, as α approaches zero, this parameterization of generalized Poisson also reduces to the basic Poisson. In most real data situations equidispersion rarely occurs. However, since the α of the generalized Poisson model can be both positive and negative, any value close to zero should be considered as Poisson. Likelihood ratio tests, as we discussed earlier, can assess if the generalized model is statistically different from a Poisson model. We do not show any examples based on this parameterization, but it is employed in LIMDEP and yields values consistent with R and Stata. New *gpoisson* and zero-inflated generalized Poisson (*zigp*) Stata commands can be found with Hardin and Hilbe (2012).

Mention should be made of the *double Poisson regression* model, which is also designed to estimate both over- and underdispersion. Double Poisson models were originated by Efron (1986) based on his work with the double exponential distibution. The following characterization of the double Poisson probability distribution comes from Efron, as discussed in Cameron and Trivedi (1998).

$$f(y;\mu,\phi) = K(\mu_i,\phi)\,\phi^{.5}\exp(-\phi\mu_i)\exp(-y_i)\,y_i^{y_i}\left(\frac{e\mu_i}{y_i}\right)^{\phi y_i}$$

with
$$\frac{1}{K(\mu_i,\phi)} \cong 1 + \frac{1-\phi}{12\phi\mu_i}\left(1+\frac{1}{\phi\mu_i}\right) \tag{10.67}$$

The double Poisson mean is approximately μ with an approximate variance of μ/ϕ. If $\phi = 1$, the double Poisson reduces to the standard distribution. Therefore, the Poisson is nested within the double Poisson, allowing a likelihood ratio test to determine the preferred model.

Estimation is accomplished by parameterizing the value of μ to $\exp(x'\beta)$, as is standard for count models. As noted in Cameron and Trivedi (1998), the maximum likelihood estimate of ϕ is given as the mean value of the following deviance statistic:

$$\phi = \frac{1}{n}\sum_{i=1}^{n} y_i \frac{\ln(y_i)}{\mu_i} - (y_i - \mu_i) \tag{10.68}$$

R software exists to estimate double Poisson models. It is an option within the bivariate Poisson package, **bivpois,** using the *lm.bp()* function. One need only add the term *zeroL3=TRUE* to the otherwise bivariate equation to estimate a double Poisson. However, it has been demonstrated that the model estimates are not exact, nor reliable, since the normalization term has no closed form solution (Winkelmann, 2008). Note how $K(\mu,\phi)$ is defined in equation 10.67. As such, the mean and variance functions do not have a closed form solution either. The consequence has been that statisticians use other models to assess underdispersion. We will discuss it no further.

10.9 Poisson inverse Gaussian regression (PIG)

A brief mention should be made of the Poisson inverse Gaussian model, which is a mixture of the Poisson and inverse Gaussian distributions. It is typically referred to as a PIG model. The PIG probability distribution is itself a two-parameter version of the *three-parameter Sichel distribution*, first developed by Sichel (1971) as a mixture of the Poisson and generalized inverse Gaussian distributions. The Sichel distribution has three shape parameters, $\alpha > 0$, $0 < \theta < 1$, and $-\infty < \gamma < \infty$; the PIG is Sichel with $\gamma = -0.5$. The PIG has also been referred to as a *univariate Sichel distribution*.

The PIG probability distribution entails a *Bessel function of the second kind*, which is typically not found in statistical software applications. Without the function, the PIG PDF cannot be converted to a log-likelihood function or its derivatives, and therefore cannot be estimated. R's **gamlss** library includes both a PIG model and a Sichel model. We shall use the former on the same data that we employed for the generalized Poisson model in the previous section. Note that we shall use the **gamlss** function for estimating finite mixture models later in Chapter 13. In addition, although we will primarily use the **glm.nb** function for estimating negative binomial models using R, the **gamlss** function can be used as well. It can also be used to incorporate predictor smoothing into the model. I recommend the **gamlss** library to anyone interested in tackling some of these more flexible count models.

Poisson inverse Gaussian models are appropriate when modeling correlated counts with long sparse extended tails. The **azpro** data is skewed to the right, but not overly so. Moreover it is not highly correlated. As a consequence, we should expect that the generalized Poisson model of the data is a better fit as it is evaluated using an AIC statistic.

Table 10.19 R: *PIG model of azpro data*

```
library(gamlss)       # azpro still in memory
proc <- as.numeric(procedure)
pivg <- gamlss(los~ proc + gender + admit + age75,
  data=azpro, family = PIG)
summary(pivg)
```

```
Mu Coefficients:
               Estimate        Std. Error     t value      Pr(>|t|)
(Intercept)      0.1679         0.05630         2.982     2.879e-03
proc             1.0006         0.01818        55.022     0.000e+00
gender          -0.1243         0.01868        -6.653     3.302e-11
admit            0.3658         0.01964        18.630     5.136e-74
age75            0.1205         0.02007         6.004     2.117e-09
-------------------------------------------------------------------
Sigma link function:  log
Sigma Coefficients:
               Estimate        Std. Error     t value      Pr(>|t|)
(Intercept)      -1.749         0.02923        -59.84             0
-------------------------------------------------------------------
No. of observations in the fit:  3589
Degrees of Freedom for the fit:  6
      Residual Deg. of Freedom:  3583
                      at cycle:  4
```

```
Global Deviance:     19731.06
            AIC:     19743.06
            SBC:     19780.18
*******************************************************************
```

The exponentiated coefficients, or incidence rate ratios, are

```
exp(coef(pivg))
(Intercept)        proc      gender       admit       age75
  1.1828194   2.7197886   0.8831422   1.4416995   1.1280616
```

and the value of the heterogeneity parameter, which is the exponentiated value of sigma, is

```
exp(-1.749)
[1] 0.1739478
```

Comparing AIC statistics with the generalized Poisson clearly indicates that the PIG model is not appropriate for these data. However, if the data were more highly skewed to the right, and had strong evidence of correlation, the PIG model would likely be preferred.

It should be noted that **gamlss** software provides the capability to model *zero-inflated Poisson-inverse Gaussian data* (ZIPIG). Refer to the **gamlss** manual for details, and to the next chapter for an overview of zero-inflated models. User-authored PIG (*pigreg*) and ZIPIG (*zipig*) Stata commands are available with Hardin & Hilbe (2012).

10.10 Other count models

A number of extended count models have been developed in order to accommodate both under- and overdispersion. For clustered models, panels of observations may be overdispersed with respect to one another, but underdispersed within the clusters. In fact, this is the usual situation, and is rarely addressed. Models such as the *Malcolm Faddy birth processes* models and the *Conway–Maxwell Poisson* model being emphasized by Kimberly Sellers et al. estimate three parameters, one of which can be used to measure and adjust for intracluster underdispersion. I have written materials on these models that can be downloaded from the book's website, together with references to code for their estimation.

Gamlss may be used to model a variety of mixtures models. Hinde's noteworthy *Poisson-normal mixture* model and a *NB2-normal mixture* model are easily estimated using the **gamlssNP** function code that appears below. The same logic of estimation obtains for a Poisson-normal model as it did above for a PIG model. *Gamlss* provides other distributions for such count mixtures as well, including the normal and lognormal mixtures with *Poisson*, *NB2*, *NB1*, *PIG*, *Sichel*, and *Delaporte*. The function for a *Poisson-normal* model is:

```
gamlssNP(y~x,family=PO,K=20, mixture="gq")
```

Summary

There are two major varieties of enhancements to the negative binomial model. One type relates to adjustments made to the basic NB2 model in light of distributional abnormalities with the data. That is, the count data to be modeled do not always match the distributional assumptions of the negative binomial model. Likewise, count data do not always accord with the distributional assumptions of the Poisson. Although NB2 is used to adjust for Poisson overdispersion, it does so without a knowledge of the possible reasons for the overdispersion. However, certain distributional properties of the data violate both Poisson and NB2 assumptions, and we can identify the source of overdispersion. For instance, both Poisson and NB2 require that there at least be the possibility of zero counts in the data, and, if there are zero counts, that there is not an excessive number compared with the assumptions of the respective distributions. When there are no possibilities of zero counts, or there are far more than what accords with the respective model assumptions, then one may construct zero-truncated or zero-inflated Poisson and negative binomial models. We address these models in the next chapter.

The second type of enhancement adjusts the Poisson variance function, thereby creating new statistical models. Alterations in variance functions are shown in Table 10.1. An overview of each of these models was presented in this chapter. The following is a brief summary of how each of these models relates to the basic NB2 model.

Geometric:	NB2 with $\alpha{=}{=}1$
NB1:	NB2 with no exponent
NB-H:	NB2 with α parameterized
NB-P:	NB2 with the exponent parameterized

Unfortunately most commercial software packages do not offer these models. Only LIMDEP offers them all. Stata's **nbreg** command can be used to model both NB2 and NB1 by using the *dispersion(mean)* and *dispersion(constant)* options respectively. *Dispersion(mean)* is the default. Stata also has the NB-H model, but calls it *generalized negative binomial*. However, since there is a previous tradition in the literature regarding generalized negative binomial models that differs considerably from NB-H, Stata's usage is a misnomer. Except for the *generalized Waring negative binomial* and NB-P, which was first developed for LIMDEP by that name in 2006, I do not discuss what have been termed generalized negative binomial models in this text since the models that have been developed have thus far been proven to result in biased estimates. A model based on a three-parameter distribution developed by Faddy, and enhanced by Faddy and Smith in 2010, may become a viable statistical software function in R. However, as of the time of this writing it is still in the review stage prior to journal publication. The only software thus far developed for the Faddy model is for specific applications, not global usage.

11
Problems with zero counts

I previously indicated that extended Poisson and negative binomial models have generally been developed to solve either a distributional or variance problem arising in the base Poisson and NB2 models. We shall discover later that some extended negative binomial models rely on the NB1 parameterization rather than on the NB2, e.g. fixed-effects negative binomial. But, except for a few models, most negative binomial models used in actual research are based on the NB2 model.

Changes to the negative binomial variance function were considered in the last chapter. In this chapter, we address the difficulties arising when the zero counts in Poisson and NB2 models differ considerably from model distributional assumptions. In the first section we address models that cannot have zero counts.

11.1 Zero-truncated count models

Many times we are asked to model count data that structurally exclude zero counts. Hospital length-of-stay data are an example of this type of data. When a patient first enters the hospital, the count begins upon registration, with the length of stay given as 1. There can be no 0 days – unless we are describing patients who do not enter the hospital. This latter situation describes a different type of model where there may be two generating processes – one, for example, for patients who may or may not enter the hospital, and another for patients who are admitted. This type of model will be discussed later.

The Poisson and negative binomial distributions both include zeros. When data structurally exclude zero counts, the underlying probability distribution must be amended to preclude this outcome if the data are to be properly modeled. This is not to say that Poisson and negative binomial regression routines are not commonly used to model such data. The point here is that they should not be. The Poisson and negative binomial probability functions, and their respective log-likelihood functions, need to be amended to exclude zeros, and at the same time provide for all probabilities in the distribution to sum to 1. Models based on this type of amendment are commonly termed *zero-truncated count models*. A common feature of zero-truncated models is that the mean of the distribution is shifted left, resulting in underdispersion.

With respect to the Poisson distribution, the probability of a zero count, based on the log-likelihood function given in equation 4.46, is $\exp(-\mu)$. This value needs to be subtracted from 1 with the remaining probabilities rescaled on this difference. The resulting log-likelihood function, with $\mu = \exp(x\beta)$, is

$$\mathcal{L}(\mu; y|y>0) = \sum_{i=1}^{n} \{y_i(x_i'\beta) - \exp(x_i'\beta) - \ln\Gamma(y_i+1) - \ln[1-\exp(-\exp(x_i'\beta))]\} \tag{11.1}$$

The logic of the zero-truncated negative binomial is the same. The probability of a zero count is

$$(1+\alpha\mu_i)^{-1/\alpha} \tag{11.2}$$

Subtracted from 1 and together conditioned out of the negative binomial log-likelihood by rescaling, we have

$$\mathcal{L}(\mu; y|y>0) = \sum_{i=1}^{n} \left\{\mathcal{L}_{\text{NB2}} - \ln\left[1 - \{1+\alpha\exp(x_i'\beta)\}^{-1/\alpha}\right]\right\} \tag{11.3}$$

where $\mathcal{L}_{\text{NB2}}$ is the negative binomial log-likelihood as given in equation 8.39.

For completeness, the log-likelihood for the zero-truncated NB-C is:

$$\mathcal{L}(\mu; y|y>0) = \sum_{i=1}^{n} \left\{\mathcal{L}_{\text{NB-C}} - \ln\left[1 - \left\{1 + \frac{1}{\alpha(\exp(x_i'\beta)-1)}\right\}^{-1/\alpha}\right]\right\} \tag{11.4}$$

It is clear that an IRLS-type estimating algorithm is not appropriate for estimating zero-truncated models. IRLS or GLM-type models are based on a likelihood derived from a probability function that is a member of the exponential family of distributions. GLM methodology allows the canonical link function to be changed for a given distributional family, but this does not affect the underlying probability function. In the case of zero-truncated models, the likelihood itself has changed. The amended PDF and log-likelihood is not a simple reparameterization, but rather an altogether new model. As with all models based on distributional violations of the base Poisson and negative binomial, estimation is by maximum likelihood, or some other estimation method, e.g. quadrature or simulation.

The effect of truncation on a negative binomial model can be substantial. Much depends on the shape of the observed distribution of counts. If the mean count is high, then the theoretical probability of obtaining a zero count is less than if the mean count is low. The graphs given in Section 8.3 provide clear evidence of the relationship of mean count to the probability of a zero count. Given an observed distribution with a low mean count and no zeros, the difference between using a negative binomial model and a zero-truncated model can be substantial. In such a case a truncated model should be used with the data.

I'll again derive synthetic data to demonstrate the difference between models. First create a 50,000-observation negative binomial data set with pre-established parameters associated with following predictors

$$x1 = 0.75, \quad x2 = -1.25, \quad \text{constant} = 0.5$$

and

$$\alpha = 0.75$$

Next, the data are modeled using a negative binomial algorithm. Code generating a synthetic model with these specific parameters is given in Table 11.1, together with model results.

Table 11.1 R: *Synthetic NB2 model*

```
library(COUNT)
nobs <- 50000
x1 <- runif(nobs)
x2 <- runif(nobs)
xb <- .5 + .75*x1 - 1.25*x2
a <- .75
ia <- 1/a
exb <- exp(xb)
xg <- rgamma(nobs, a, a, ia)
xbg <-exb*xg
y1 <- rpois(nobs, xbg)
summary(glm.nb(y1 ~ x1 + x2))
myTable(y1)
```

```
. clear
. set obs 50000
. set seed 4744
. gen a= .75
. gen x1 = runiform()
. gen x2 = runiform()
. gen xb = .5 + 0.75*x1 - 1.25*x2
. gen ia = 1/a
. gen exb = exp(xb)
```

```
. gen xg = rgamma(ia, a)
. gen xbg = exb * xg
. gen y1 = rpoisson(xbg)

. glm y1 x1 x2, fam(nb ml)

Generalized linear models              No. of obs      =      50000
Optimization     : ML                  Residual df     =      49997
                                       Scale parameter =          1
Deviance         =  52186.45814        (1/df) Deviance =   1.043792
Pearson          =  49879.32727        (1/df) Pearson  =   .9976464
Variance function: V(u) = u+(.7544)u^2  [Neg. Binomial]
Link function    : g(u) = ln(u)         [Log]
                                       AIC             =   3.149111
Log likelihood   = -78724.76798        BIC             =    -488770
--------------------------------------------------------------------
     |              OIM
  y1 |     Coef.  Std. Err.    z    P>|z|  [95% Conf.   Interval]
-----+--------------------------------------------------------------
  x1 |  .7555417  .0193011  39.15   0.000  .7177123     .7933711
  x2 | -1.269299  .0194346 -65.31   0.000  -1.30739    -1.231208
_cons |  .5024642  .0145073  34.64   0.000  .4740305     .5308979
--------------------------------------------------------------------
Note: Negative binomial parameter estimated via ML and treated
      as fixed once estimated.
```

A tabulation of the first 11 counts, from 0 to 10, is given in Table 11.2. The mean count of *y*1 is 1.4, quite low considering the large range of counts. We would expect that there would be a marked difference in the estimates, α, and the remaining goodness-of-fit statistics when the data are modeled without zeros.

Table 11.2 *Synthetic data distribution: counts 0–10*

y1	Freq.	Percentage	Cum.
0	20,596	41.19	41.19
1	12,657	25.31	66.51
2	7,126	14.25	80.76
3	4,012	8.02	88.78
4	2,270	4.54	93.32
5	1,335	2.67	95.99
6	781	1.56	97.55
7	479	0.96	98.51
8	278	0.56	99.07
9	175	0.35	99.42
10	106	0.21	99.63

I should mention that using random uniform in place of random normal values for $x1$ and $x2$ reduces the value of the mean considerably, and tightens the distribution. If we had employed random normal covariates in place of uniform, the mean of $y1$ would have been 4.8 and the cumulative percentage for $y1 = 10$ would have been 89.90%. Instead, the defined distribution produced 99.63% of the counts at or equal to $y1 = 10$. There are also 2,000 more zero counts with the uniform distribution compared with the random normal.

Specifying α as 0.75 results in excessive negative binomial zeros, although it is possible to check the observed versus theoretical negative binomial distributions as we previously did to determine the extent to which the two distributions vary. Next we drop all zero counts from the data, then re-estimate the model to obtain new parameter estimates and goodness-of-fit statistics.

R

```
y1 <- ifelse(y1==0, NA, y1)
summary (glm.nb(y1 ~ x1 + x2))
```

```
. drop if y1==0
. glm y1 x1 x2, fam(nb ml)

Generalized linear models                 No. of obs      =      29404
Optimization     : ML                     Residual df     =      29401
                                          Scale parameter =          1
Deviance         =  24192.64223           (1/df) Deviance =    .822851
Pearson          =  30100.39271           (1/df) Pearson  =   1.023788
Variance function: V(u) = u+(.0924)u^2  [Neg. Binomial]
Link function    : g(u) = ln(u)           [Log]
                                          AIC             =   3.591551
Log likelihood   = -52799.98778           BIC             = -278310.9
----------------------------------------------------------------------
      |              OIM
   y1 |     Coef.  Std. Err.      z    P>|z| [95% Conf.  Interval]
------+---------------------------------------------------------------
   x1 | .4331122   .0147154    29.43   0.000  .4042707    .4619538
   x2 | -.722878   .0150439   -48.05   0.000 -.7523634   -.6933926
_cons | .9375719   .0109945    85.28   0.000  .9160232    .9591206
----------------------------------------------------------------------
Note: Negative binomial parameter estimated via ML and treated
      as fixed once estimated.
```

Note the changes that have occurred by deleting zero counts from a full negative binomial distribution having those parameters.

Modeling the data without zero counts with a zero-truncated negative binomial results in a model appearing very much like the model we had when zero counts were included. Recall that the zero-truncated model is determining parameter estimates based on data without zero counts.

R's **gamlss.tr** library is used to model right- and left-truncated count models in R. Family = PO indicates the Poisson, NBI is our NB2 model and NBII is what we have referred to as NB1. **gamlss** can be used for other truncated distributions as well. In line 3 below, the $y > 0$ indicates a zero-truncation, which is again indicated at the beginning of the **gen.trun** function with another 0. A left-truncated at 3 model would have 3 s in place of the 0s in the code.

R: *Zero-truncated NB2 on synthetic data*

```
library(gamlss.tr)
sdata <-data.frame(y1, x1, x2)
lsdata<-subset(sdata, sdata$y>0)
lmpo <- gamlss(y1~x1+x2,data=lsdata,family=NBI)
gen.trun(0, "NBI", type="left", name = "lefttr")
lt0nb <- gamlss(y1~x1+x2, data=lsdata, family=NBIlefttr)
summary(lt0nb)
```

```
. ztnb y1 x1 x2

Zero-truncated negative binomial regression
                                          Number of obs =      29404
                                          LR chi2(2)    =    2888.74
Dispersion     = mean                     Prob > chi2   =     0.0000
Log likelihood = -46162.174               Pseudo R2     =     0.0303
------------------------------------------------------------------------
      y1 |     Coef.  Std. Err.     z   P>|z|  [95% Conf. Interval]
---------+--------------------------------------------------------------
      x1 |  .7387867 .0259546  28.46 0.000    .6879166   .7896567
      x2 | -1.257216 .0268201 -46.88 0.000   -1.309783   -1.20465
   _cons |  .5051103 .0216343  23.35 0.000    .4627077   .5475128
---------+--------------------------------------------------------------
/lnalpha | -.2751051 .0351903                -.3440767  -.2061334
---------+--------------------------------------------------------------
   alpha |  .7594923 .0267267                 .7088746   .8137245
------------------------------------------------------------------------
Likelihood-ratio test of alpha=0:  chibar2(01) = 5927.17
                                   Prob>=chibar2 = 0.000
```

We shall next use the **medpar** data set which we have used before to model length-of-stay (*los*) data. Noting that the mean value of *los* is 9.9, we should expect that zero counts would not have as great an impact on the model as with the synthetic data with a mean count value of 1.4.

Table 11.3 R: *NB2 and zero truncated negative binomial model*

```
rm(list=ls())
data(medpar)
attach(medpar)
mnb <- gamlss(los~ white+died+type2+type3,data=medpar,family=NBI)
summary(mnb)
library(gamlss.tr)
lmedpar<-subset(medpar, los>0)
lmnb <- gamlss(los~white+died+type2+type3,data=lmedpar,family=NBI)
summary(lmnb)
gen.trun(0, "NBI", type="left", name = "lefttr")
ltm0nb <- gamlss(los~white+died+type2+type3, data=lmedpar,
   family=NBIlefttr)
summary(ltm0nb)
modelfit(ltm0nb)
```

NEGATIVE BINOMIAL

```
. nbreg los white died type2 type3

Log likelihood = -4781.7268
------------------------------------------------------------------
   los |      Coef.   Std. Err.      z    P>|z|  [95% Conf. Interval]
------+-----------------------------------------------------------
white | -.1258405   .0677911   -1.86   0.063  -.2587085    .0070276
 died | -.2359093   .0404752   -5.83   0.000  -.3152393   -.1565792
type2 |  .2453806   .0501704    4.89   0.000   .1470485    .3437128
type3 |  .7388372   .0750077    9.85   0.000   .5918248    .8858496
_cons |  2.365268   .0679452   34.81   0.000   2.232097    2.498438
------+-----------------------------------------------------------
alpha |   .434539    .01947                     .398006    .4744253
------------------------------------------------------------------
. abic
AIC Statistic    =   6.404986          AIC*n      = 9575.4541
BIC Statistic    =   6.411341          BIC(Stata) = 9607.3125
```

ZERO-TRUNCATED NEGATIVE BINOMIAL

```
. ztnb los white died type2 type3

Log likelihood = -4736.7725
------------------------------------------------------------------
   los |      Coef.   Std. Err.      z    P>|z|  [95% Conf. Interval]
------+-----------------------------------------------------------
white |   -.131835  .0746933   -1.77   0.078  -.2782312    .0145612
 died | -.2511928   .0446812   -5.62   0.000  -.3387663   -.1636193
type2 |  .2601118   .0552939    4.70   0.000   .1517378    .3684858
type3 |  .7691718   .0825861    9.31   0.000    .607306    .9310376
_cons |  2.333412   .0749931   31.11   0.000   2.186428    2.480395
------+-----------------------------------------------------------
alpha |  .5315121   .0292239                    .4772126      .59199
------------------------------------------------------------------
```

```
. abic
AIC Statistic   =   6.344846         AIC*n        = 9485.5449
BIC Statistic   =   6.351202         BIC(Stata) = 9517.4043
```

As expected, the two models are similar. Moreover, the AIC and BIC statistics tend to favor the zero-truncated model, as they should in such a situation.

We may compare the AIC and BIC statistics for zero-truncated NB1 and NB-C models.

NB1

```
. ztnb los white died type2 type3, disp(constant)

Zero-truncated negative binomial regression
------------------------------------------------------------------
     los |     Coef. Std. Err.      z   P>|z| [95% Conf. Interval]
---------+--------------------------------------------------------
   white | -.1131934 .0697439   -1.62  0.105 -.2498889   .0235021
    died | -.4314338 .0517247   -8.34  0.000 -.5328124  -.3300552
   type2 |  .2236895 .0532101    4.20  0.000  .1193996   .3279793
   type3 |  .5617603 .0711944    7.89  0.000  .4222219   .7012987
   _cons |  2.388953 .0686497   34.80  0.000  2.254402   2.523504
---------+--------------------------------------------------------
/lndelta |  1.659652   .05495                1.551952   1.767352
---------+--------------------------------------------------------
   delta |  5.257479 .2888983                4.720675   5.855326
------------------------------------------------------------------
Likelihood-ratio test of delta=0:  chibar2(01) = 4198.19
                                   Prob>=chibar2 = 0.000

. abic
AIC Statistic   =   6.349041         AIC*n        = 9491.8164
BIC Statistic   =   6.355397         BIC(Stata) = 9523.6758
```

NB-C

```
 . ztcnb los white died type2 type3 // user authored command

Zero-Truncated Canonical Negative Binomial Regression
------------------------------------------------------------------
     los |    Coef.   Std. Err.     z  P>|z|   [95% Conf. Interval]
---------+--------------------------------------------------------
   white | -.0205202  .0101728  -2.02 0.044  -.0404585 -.0005819
    died | -.0415983  .0078251  -5.32 0.000  -.0569352 -.0262614
   type2 |  .041943   .0086101   4.87 0.000   .0250675  .0588185
   type3 |  .104109   .0095818  10.87 0.000   .0853291  .1228888
   _cons | -.1704912  .0130334 -13.08 0.000  -.1960363 -.1449461
---------+--------------------------------------------------------
/lnalpha | -.6345178   .054989 -11.54 0.000  -.7422942 -.5267414
---------+--------------------------------------------------------
   alpha | .5301911   .0291547              .4760206   .5905261
------------------------------------------------------------------
. abic
AIC Statistic   =   6.343721         AIC*n        = 9483.8623
BIC Statistic   =   6.350076         BIC(Stata) = 9515.7217
```

The zero-truncated NB-C model appears to fit the data better than either NB2 or NB1. The AICn value is 9,483.86 whereas NB2 is 9,485.54 and NB1 is 9491.82. From the table of AIC differences, however, the NB-C and NB2 models estimates are not statistically different. Given the rate ratio interpretation that can be given to the NB2 model, it is to be preferred.

Marginal effects and discrete change calculations are identical to those employed for Poisson and NB2, although the NB-C model does not interpret coefficients as log-counts.

Zero-truncated models are subsets of the more general truncated count models we shall discuss in the next chapter. However, given the frequency of overdispersion in count data, as well as the frequency with which count models exclude the possibility of zero counts, zero-truncated negative binomial models are important to modeling counts and should therefore be part of the standard statistical capabilities of commercial statistical packages. Note that new user-authored Stata commands for zero-inflated generalized Poisson (*zigp*) and ZI Poisson-inverse Gaussian (*zipig*) are available with Hardin and Hilbe (2012).

11.2 Hurdle models

We have emphasized the necessity of creating models that remedy variance and distributional problems with the Poisson and negative binomial NB2 models. Zero-truncated models attempt to accommodate the data when zeros are structurally excluded from the model. But what happens when there are far more zero counts in the data than are allowed on the basis of Poisson and negative binomial distributional assumptions? This topic was discussed earlier. But we did not attempt to address ways to deal with the situation.

As two-part models, hurdle and zero-inflated count models are the two foremost methods used to deal with count data having excessive zero counts. But whereas zero-inflated models require excessive zero counts in the data, hurdle models can adjust for models with too few zero counts. However, some zero counts are necessary.

It is theoretically possible for a hurdle model to have multiple hurdles. One type of double hurdle model could have censoring at one and at another higher count. Of course, truncation would have to correspond to censoring. A realistic double hurdle model would have both truncation at 0 and truncation at 1, with associated levels for censoring. The problem with actually applying such a model rests in its sensitivity to misspecification. If the data do not closely follow the hurdle points, it will not converge.

Currently Stata, LIMDEP, and R support estimating hurdle models, but they do not all provide the same models. Check the software for what is available.

Note that, although LIMDEP supports hurdle models, they are not part of the menu system and must be estimated using command-line code. The R **pscl** package contains excellent support for hurdle models; Stata's hurdle commands were written by the author and have been published at the Boston School of Economics statistical software center [http://ideas.repec.org/s/boc/bocode.html]. Most software assumes that both parts of the model consist of the same predictors, although this need not be the case.

Hurdle models are sometimes referred to as zero-altered models (Heilbron, 1989). Zero-altered Poisson and negative binomial models are thus referred to, respectively, as ZAP and ZANB. They have also been termed overlapping models, or zero-inflated models.

Hurdle models are based on the assumption that zero counts are generated from a different process than are the positive counts in a given data situation. A hurdle model handles this by partitioning the model into two parts: a binary process generating positive counts (1) versus zero counts (0); and a process generating only positive counts. The binary process is generally estimated using a binary model; the positive count process is estimated using a zero-truncated count model. The binary process can also be estimated using a 'right censored at one' count model; Poisson, geometric, or negative binomial. Censoring takes place at a count value of one such that, for example, a Poisson count of zero is given the binary process value of zero and counts of one or greater are given the value of 1. We shall discuss censoring in the following chapter, but it should be understood that I am referring to econometric or cut-point censoring here, not a survival parameterization of censoring. What these designations mean is addressed in Chapter 12.

The following list provides the type of binary and count components that are used to estimate hurdle models.

BINARY: binomial, Poisson, geometric, NB2
COUNT: Poisson, geometric, negative binomial

Binomial links that have been provided for the binomial component include: logit, probit, complementary loglog, Cauchit, and log.

The first hurdle models were designed by Mullahy (1986) and were later popularized by Cameron and Trivedi (1986, 1998). Mullahy used a Poisson distribution to govern the probability of a positive versus zero count, and a zero-truncated Poisson distribution – also called a positive-Poisson distribution – to model the positive counts. We now call this a Poisson–Poisson hurdle model. Mullahy also employed a geometric distribution for the binary portion of the hurdle, which was found to provide identical results to having a binomial-logit distribution.

The reason for this is that a binomial-logit and 'right censored at one' geometric model are based on the same log-likelihood function. Consequently, using either distribution for the hurdle component of a model produces identical coefficients.

As long as a software package has the capability of modeling zero-truncated count models and standard binary response models (e.g. logit and probit), most common hurdle models can be estimated in parts. I show how this is done in this section. In order to estimate hurdle models with the binary component based on a count model, software must have commands or functions for estimating censored count models. Only LIMDEP and authored-written Stata commands are available at present; however, one may estimate econometric censored models by abstracting the estimated binary component of a hurdle model. We discuss this in greater length when addressing censored models. For now it is enough to know that it is possible to use a right-censored (econometric) Poisson model and a zero-truncated negative binomial to construct a viable NB2–Poisson hurdle model. I shall use this model on data later in this section.

11.2.1 Theory and formulae for hurdle models

The notion of hurdle comes from considering the data as being generated by a process that commences generating positive counts only after crossing a zero barrier, or hurdle. Until the hurdle is crossed, the process generates a binary response (1/0). The nature of the hurdle is left unspecified, but numerically may simply be considered as the data having a positive count. In this sense, the hurdle is crossed if a count is greater than zero. In any case, the two processes are conjoined using the following log-likelihood:

$$\mathcal{L} = \ln(f(0)) + \{\ln[1 - f(0)] + \ln P(t)\} \tag{11.5}$$

where $f(0)$ represents the probability of the binary part of the model and $P(t)$ represents the probability of a positive count.

In the case of a logit model, the probability of zero is

$$f(0) = P(y = 0; x) = \frac{1}{1 + \exp(x_i' \beta_b)} \tag{11.6}$$

with β_b indicating the parameter estimates from the binary component of the hurdle model. It follows that $1 - f(0)$ is given by

$$\frac{\exp(x_i' \beta_b)}{1 + \exp(x_i' \beta_b)} \tag{11.7}$$

From equation 11.1 we know that the zero-truncated negative binomial NB2 log-likelihood is

$$P(y|x>0) = y_i \ln\left(\frac{\exp(x_i'\beta)}{1+\exp(x_i'\beta)}\right) - \frac{\ln(1+\exp(x_i'\beta))}{\alpha} + \ln\Gamma\left(y_i+\frac{1}{\alpha}\right) - \ln\Gamma(y_i+1) - \ln\Gamma\left(\frac{1}{\alpha}\right) - \ln\left(1-(1+\exp(x_i'\beta))\right)^{-1/\alpha} \tag{11.8}$$

Putting the above together, we have a two-part NB-logit hurdle model likelihood:

$$\text{if } (y=0): \quad \frac{1}{1+\exp(x_i'\beta_b)}$$

$$\text{if } (y>0): \quad y_i \ln\left(\frac{\exp(x_i'\beta)}{1+\exp(x_i'\beta)}\right) - \frac{\ln(1+\exp(x_i'\beta))}{\alpha} + \ln\Gamma\left(y_i+\frac{1}{\alpha}\right) - \ln\Gamma(y_i+1) - \ln\Gamma\left(\frac{1}{\alpha}\right) - \ln\left(1-(1+\exp(x_i'\beta))\right)^{-1/\alpha} \tag{11.9}$$

The log-likelihood for a negative binomial – complementary loglog hurdle model is given as

$$\text{if } (y=0): \quad -\exp(x_i'\beta_b)$$

$$\text{if } (y>0): \quad \ln(1-\exp(-\exp(x_i'\beta_b))) + y_i \ln\left(\frac{\exp(x_i'\beta)}{1+\exp(x_i'\beta)}\right) - \frac{\ln(1+\exp(x_i'\beta))}{\alpha} + \ln\Gamma\left(y_i+\frac{1}{\alpha}\right) - \ln\Gamma(y_i+1) - \ln\Gamma\left(\frac{1}{\alpha}\right) - \ln\left(1-(1+\exp(x_i'\beta))\right)^{-1/\alpha} \tag{11.10}$$

11.2.2 Synthetic hurdle models

Although is it possible to create synthetic hurdle models for any of the described combinations, perhaps the simplest is the Poisson–logit hurdle model. However, it is nearly always the case that a NB2–logit or NB2–complementary loglog model is a better fitted model than the alternatives. Econometricians tend to favor a NB2–probit hurdle model. Biostatisticians, however, rarely employ hurdle models in their research, although they would likely prove useful in explaining various types of health related situations.

I show below a synthetic NB2–logit model with the following parameters associated with both predictors $x1$ and $x2$, and values for the intercepts and α.

```
NB2 intercept = 1 ;     x1 = 0.75; x2 = -1.25; α = 0.5
logit intercept = -2 ; x1 = -0.90 ; x2 = - 0.10
```

Table 11.4 R: *Synthetic NB2–logit hurdle model*

```
library(MASS)     # syn.logitnb2.hurdle.r
library(pscl)
nobs <- 50000
x1 <- runif(nobs)
x2 <- runif(nobs)
xb <- 2 + .75*x1 - 1.25*x2
a <- .5
ia <- 1/.5
exb <- exp(xb)
xg <- rgamma(nobs, a, a, ia)
xbg <- exb*xg
nby <- rpois(nobs, xbg)
nbdata <- data.frame(nby, x1, x2)
pi <- 1/(1+exp(-(.9*x1 + .1*x2 + .2)))  # filter
nbdata$bern <- runif(nobs) > pi
nbdata <- subset(nbdata, nby > 0) # Remove response zeros
nbdata$nby[nbdata$bern] <- 0      # Add structural zeros
hlnb2 <- hurdle(nby ~ x1 + x2, dist = "negbin",
  zero.dist = "binomial", link = "logit", data = nbdata)
summary(hlnb2)
```

Table 11.5 *Synthetic NB2–logit hurdle model*

```
* nb2logit_hurdle.do
* LOGIT: x1=-.9, x2=-.1, _c=-.2
* NB2   : x1=.75, n2=-1.25, _c=2, alpha=.5
clear
set obs 50000
set seed 1000
gen x1 = invnorm(runiform())
gen x2 = invnorm(runiform())
* NEGATIVE BINOMIAL- NB2
gen xb = 2 + 0.75*x1 - 1.25*x2
gen a = .5
gen ia = 1/a
gen exb = exp(xb)
gen xg = rgamma(ia, a)
gen xbg = exb * xg
```

```
gen nby = rpoisson(xbg)
* BERNOULLI
drop if nby==0
gen pi =1/(1+exp(-(.9*x1 + .1*x2+.2)))
gen bernoulli = runiform()>pi
replace nby=0 if bernoulli==0
rename nby y
* logit bernoulli x1 x2, nolog /// test
* ztnb y x1 x2 if y>0, nolog   /// test
* NB2-LOGIT HURDLE
hnblogit y x1 x2, nolog
```

```
. hnblogit y x1 x2, nolog
Negative Binomial-Logit Hurdle Regression
                                                  Number of obs =    43443
                                                  Wald chi2(2)  =  5374.14
Log likelihood = -84654.938                       Prob > chi2   =   0.0000
------------------------------------------------------------------------------
             |      Coef.   Std. Err.      z    P>|z|     [95% Conf. Interval]
-------------+----------------------------------------------------------------
logit        |
          x1 | -.8987393   .0124338   -72.28   0.000    -.9231091    -.8743695
          x2 | -.0904395    .011286    -8.01   0.000    -.1125597    -.0683194
       _cons | -.2096805   .0106156   -19.75   0.000    -.2304867    -.1888742
-------------+----------------------------------------------------------------
negbinomial  |
          x1 |   .743936   .0069378   107.23   0.000     .7303381     .7575339
          x2 | -1.252363   .0071147  -176.02   0.000    -1.266307    -1.238418
       _cons |  2.003677   .0070987   282.26   0.000     1.989764      2.01759
-------------+----------------------------------------------------------------
/lnalpha     | -.6758358   .0155149   -43.56   0.000    -.7062443    -.6454272
------------------------------------------------------------------------------
AIC Statistic =    3.897
```

11.2.3 Applications

Using the German health reform data, **mdvis**, we model *numvisit* using several types of hurdle models; *numvisit* has far more zero counts than expected, as shown in the histogram in Figure 11.1.

R

```
rm (list=ls ())
data (mdvis)
hist (mdvis$numvisit)
```

```
. use mdvisits
. histogram numvisit, bin(20) percent
```

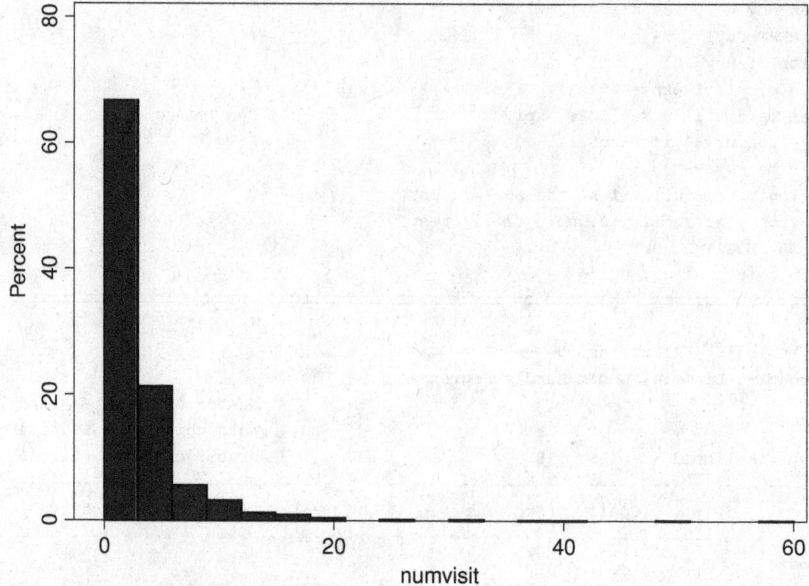

Figure 11.1 Histogram of *numvisit*

numvisit provides a count of the number of visits a patient makes to a physician in a year. Since the number of zero counts is high, we employ a hurdle model. Keep in mind that zero-inflated models also estimate counts with a high percentage of zeros. We begin with an NB2 complementary loglog model.

NEGATIVE BINOMIAL – COMPLEMENTARY LOGLOG HURDLE

Table 11.6 R: *NB2 complementary loglog hurdle model*

```
library(pscl)
attach(mdvis)
hnbc <- hurdle(numvisit ~ reform + badh + educ3 + age3, data=mdvis,
    dist = "negbin", zero.dist="binomial", link="cloglog")
summary(hnbc)
alpha <-1/hnbc$theta
alpha
```

```
. hnbclg numvisit reform badh educ3 age3

Log Likelihood = -1331.9847
------------------------------------------------------------------
            |  Coef.  Std. Err.    z    P>|z|[95% Conf.   Interval]
------------+-----------------------------------------------------
cloglog     |
     reform | -.1073391 .0543833 -1.97  0.048 -.2139284  -.0007497
       badh |  .5771608 .0853904  6.76  0.000  .4097988   .7445228
      educ3 |  .1123926 .0573353  1.96  0.050  .0000175   .2247678
       age3 |  .0180807 .0742734  0.24  0.808 -.1274926    .163654
      _cons |  .1445252 .0454318  3.18  0.001  .0554804  .2335699
------------+-----------------------------------------------------
negbinomial |
     reform | -.1182993 .0639946 -1.85  0.065 -.2437265   .0071278
       badh |  1.159176 .0862206 13.44  0.000  .9901863   1.328165
      educ3 | -.1960328 .0682512 -2.87  0.004 -.3298028  -.0622629
       age3 |  .2372101  .084559  2.81  0.005  .0714776   .4029426
      _cons |  .7395257 .0671674 11.01  0.000  .6078801   .8711714
------------+-----------------------------------------------------
      alpha | 1.1753772
------------------------------------------------------------------
AIC Statistic =      4.096
```

The model output provides parameter estimates for both the binary complementary loglog model and the negative binomial. The joint model can be separated into partitioned models by first creating a second response variable, which we call *visit*, using the following logic

visit = 1 if numvisit > 0; i.e. is a positive count
visit = 0 if numvisit == 0.

Table 11.7 R: *Components to hurdle models*

```
visit <- ifelse(mdvis$numvisit >0, 1, 0)
table(visit)
clog <- glm(visit ~ reform + badh + educ3 + age3, data=mdvis,
    family=binomial(link="cloglog"))
summary(clog)
library(pscl)
hnbc2 <- hurdle(numvisit ~ reform + badh + educ3 + age3, data=mdvis,
   dist = "negbin", zero.dist="binomial", link="cloglog")
summary(hnbc2)
logit <- glm(visit ~ reform + badh + educ3 + age3, data=mdvis,
    family=binomial(link="logit"))
summary(logit)
```

```
. gen int visit = numvisit>0
. tab visit    |
     visit     |    Freq.      Percent        Cum.
---------------+--------------------------------------------------------
             0 |      665        29.86        29.86
             1 |    1,562        70.14       100.00
---------------+--------------------------------------------------------
     Total        2,227        100.00
```

The binary component, a complementary loglog regression, is modeled as

COMPLEMENTARY LOGLOG MODEL

```
. cloglog visit reform badh educ2 age3
Complementary log-log regression            Number of obs      =      2227
                                            Zero outcomes      =       665
                                            Nonzero outcomes   =      1562
                                            LR chi2(4)         =     51.55
Log likelihood = -1331.9847                 Prob > chi2        =    0.0000
--------------------------------------------------------------------------
 visit |     Coef.  Std. Err.      z   P>|z|   [95% Conf.  Interval]
-------+------------------------------------------------------------------
reform | -.1073391  .0543833   -1.97   0.048   -.2139284   -.0007497
  badh |  .5771608  .0853904    6.76   0.000    .4097988    .7445228
 educ3 |  .1123926  .0573353    1.96   0.050    .0000175    .2247678
  age3 |  .0180807  .0742734    0.24   0.808   -.1274926     .163654
 _cons |  .1445252  .0454318    3.18   0.001    .0554805    .2335699
--------------------------------------------------------------------------
AIC Statistic =          1.202
```

The parameter estimates are identical to that of the hurdle model.

We next model the data using a zero-truncated negative binomial, making certain to exclude zero counts from the modeling process.

ZERO-TRUNCATED NEGATIVE BINOMIAL MODEL

```
. ztnb numvisit reform badh educ3 age3 if numvisit>0
Zero-truncated negative binomial regression
                                              Number of obs    =      1562
                                              LR chi2(4)       =    233.36
Dispersion      = mean                        Prob > chi2      =    0.0000
Log likelihood = -3223.6195                   Pseudo R2        =    0.0349
--------------------------------------------------------------------------
numvisit |     Coef.   Std. Err.      z    P>|z|    [95% Conf.  Interval]
---------+----------------------------------------------------------------
  reform | -.1182993   .0639946    -1.85   0.065   -.2437265    .0071278
    badh |  1.159176   .0862206    13.44   0.000    .9901863    1.328165
   educ3 | -.1960328   .0682512    -2.87   0.004   -.3298028   -.0622629
```

```
    age3 |  .2372101   .084559   2.81  0.005   .0714776   .4029426
   _cons |  .7395257  .0671674  11.01  0.000   .6078801   .8711714
---------+------------------------------------------------------------
/lnalpha |  .1615891  .1106842                 -.055348   .3785262
---------+------------------------------------------------------------
   alpha |  1.175377  .1300957                 .9461558   1.460131
-----------------------------------------------------------------------
Likelihood-ratio test of alpha=0:  chibar2(01) = 1677.79
                                   Prob>=chibar2 = 0.000
AIC Statistic   =        4.134          BIC Statistic = -10615.008
```

Again, the parameter estimates are identical to the hurdle model. Notice, however, that the AIC statistic is lower for the conjoined hurdle model than for the zero-truncated model, indicating a significantly better fit [(1562*4.134) – (1562*4.096) = 6457.308 – 6397.952 = 59.356] given the criteria recommended from Table 5.1.

The same relationship maintains for the NB–logit model. The logit model appears as:

LOGISTIC MODEL

```
. logit visit reform badh educ3 age3
Logistic regression                          Number of obs   =      2227
                                             LR chi2(4)      =     51.96
                                             Prob > chi2     =    0.0000
Log likelihood = -1331.7768                  Pseudo R2       =    0.0191
-----------------------------------------------------------------------
 visit |      Coef.  Std. Err.      z  P>|z|   [95% Conf. Interval]
-------+---------------------------------------------------------------
reform | -.1879245   .0939389   -2.00  0.045  -.3720413   -.0038076
  badh |  1.144087   .1940181    5.90  0.000   .7638189    1.524356
 educ3 |  .2018225   .1003517    2.01  0.044   .0051367    .3985082
  age3 |  .0238393   .1301832    0.18  0.855  -.2313152    .2789937
 _cons |  .7795456    .078196    9.97  0.000   .6262844    .9328069
-----------------------------------------------------------------------
AIC Statistic =      1.201
```

The zero-truncated negative binomial is the same as before. Submitting the data of a negative binomial-logit model produces the following output.

NEGATIVE BINOMIAL–LOGIT

The logit models are the same. All hurdle models can be partitioned or broken apart in this fashion. Assuming a significant *p*-value is associated with a

Table 11.8 R: *NB2–logit hurdle model*

```
hnblog <- hurdle(numvisit ~ reform + badh + educ3 + age3,
      data=mdvis, dist = "negbin")  # logit link is default
summary(hnblog)
aich <- 4.13*(1- 665/2227) + 1.20
exp(coef(hnblog))
predhnbl <- hnblog$fitted.values
```

```
. hnblogit numvisit reform badh educ3 age3

Negative Binomial-Logit  Hurdle Regression
                                          Number of obs   =       2227
                                          Wald chi2(4)    =      42.65
Log likelihood = -4555.3963               Prob > chi2     =     0.0000
------------------------------------------------------------------------
            |     Coef.  Std. Err.     z  P>|z|  [95% Conf.Interval]
------------+-----------------------------------------------------------
logit       |
     reform | -.1879245 .0939389 -2.00 0.045 -.3720413 -.0038076
       badh |  1.144088 .1940181  5.90 0.000  .7638189  1.524356
      educ3 |  .2018225 .1003517  2.01 0.044  .0051367  .3985082
       age3 |  .0238393 .1301832  0.18 0.855 -.2313152  .2789937
      _cons |  .7795456  .078196  9.97 0.000  .6262844  .9328069
------------+-----------------------------------------------------------
negbinomial |
     reform | -.1182993 .0639946 -1.85 0.065 -.2437264  .0071278
       badh |  1.159175 .0862205 13.44 0.000  .9901862  1.328164
      educ3 | -.1960328 .0682512 -2.87 0.004 -.3298027 -.0622629
       age3 |    .23721 .0845589  2.81 0.005  .0714776  .4029424
      _cons |  .7395271 .0671672 11.01 0.000  .6078818  .8711725
------------+-----------------------------------------------------------
   /lnalpha |  .1615866  .110684  1.46 0.144 -.0553499  .3785232
------------+-----------------------------------------------------------
      alpha |  1.175374 .1300951                .946154  1.460127
------------------------------------------------------------------------
AIC Statistic =      4.096
```

predictor, a positive signed coefficient for the binary portion of a hurdle model generally means that the corresponding predictor increases the probability of a non-zero count. A positive signed coefficient that is part of the zero-truncated portion of the model indicates that the corresponding predictor increases the value of the count. Higher values generally indicate a greater change in value.

The AIC statistic for the hurdle model can be calculated from a knowledge of the AIC statistics of both constituent components, and the percentage of response values greater than zero. Although the hurdle model algorithm does not use this approach to calculate an AIC statistic, it can be calculated by hand as:

$$\text{AIC}_{\text{hurdle}} = \left(\text{AIC}_{\text{ZT}} * \left(1 - \frac{n_{>0}}{n}\right)\right) + \text{AIC}_{\text{binary}} \tag{11.11}$$

Calculating the AIC statistic for the hurdle model based on the above formula provides:

```
. di 4.13* (1- 665/2227) + 1.20
4.096749
```

which is nearly the same as the AIC observed for the negative binomial–logit model (4.096). Rounding errors will at times produce minor discrepancies between the hand-calculated value and the model-calculated value. Regardless, we see that the AIC statistic for the hurdle model is proportioned between both constituent model AIC statistics, with the count model adjusted by the percentage of non-zero counts in the response.

The log-likelihood value of the hurdle model is the sum of component model log-likelihoods. Other model statistics can likewise be calculated by combining the statistical values of the constituent parts of the hurdle model.

Interpretation is now considered. Each predictor is evaluated in terms of the contribution it makes to each respective model. For example, with respect to the NB–logit model, a positive significant coefficient in the negative binomial frame indicates that the predictor increases the rate of physician visits, in the same manner as any negative binomial model is interpreted. A positive coefficient in the logit frame is interpreted in such a manner that a one-unit change in a coefficient decreases the odds of no visits to the doctor by exp(β). If a logistic coefficient is 0.2018, then the odds of zero visits is decreased by exp(0.2018) = 1.2236033, or about 22%.

Parameterizing the estimates in exponential form so that the counts can be interpreted as incidence rate ratios, and the logistic model as odds ratios, we have

```
. hnblogit numvisit reform badh educ3 age3, nolog eform

Negative Binomial-Logit  Hurdle Regression
                                                   Number of obs   =      2227
                                                   Wald chi2(4)    =     42.65
Log likelihood = -4555.3963                        Prob > chi2     =    0.0000
------------------------------------------------------------------------------
             | exp(b)   Std. Err.    z    P>|z| [95% Conf.  Interval]
-------------+----------------------------------------------------------------
logit        |
      reform | .8286773   .077845 -2.00   0.045   .6893258   .9961997
        badh | 3.139575 .6091346  5.90   0.000   2.146458   4.592186
       educ3 | 1.223631 .1227934  2.01   0.044    1.00515   1.489601
```

```
        age3 | 1.024126  .133324  0.18  0.855  .7934893  1.321799
-------------+------------------------------------------------------
negbinomial  |
      reform | .8884301 .0568547 -1.85  0.065   .783702  1.007153
        badh | 3.187304  .274811 13.44  0.000  2.691735  3.774109
       educ3 | .8219853 .0561015 -2.87  0.004  .7190656  .9396358
        age3 | 1.267707 .1071959  2.81  0.005  1.074094  1.496221
-------------+------------------------------------------------------
    /lnalpha | .1615866  .110684  1.46  0.144 -.0553499  .3785232
-------------+------------------------------------------------------
       alpha | 1.175374 .1300951                .946154  1.460127
--------------------------------------------------------------------
AIC Statistic =      4.096
```

Predicted values, μ, may also be calculated for the count model.

```
. hnblogit_p mu, eq(#2) irr
. l mu numvisit reform badh educ3 age3 in 1/5

     +------------------------------------------------+
     |       mu   numvisit  reform  badh  educ3  age3 |
     |------------------------------------------------|
  1. | 2.359472         30       1     0      0     1 |
  2. | 2.094944         25       0     0      0     0 |
  3. | 2.183009         25       0     0      1     1 |
  4. | 2.094944         25       0     0      0     0 |
  5. | 2.359472         20       1     0      0     1 |
     +------------------------------------------------+
```

To check the first fitted value:

```
. di exp(-.1182993 + .23721 + .7395271)
2.3594718
```

which is consistent to the value shown on line 1 of the above table.

Finally we shall model a negative binomial–Poisson hurdle model, consisting of a zero-truncated negative binomial and a 'right censored at one' Poisson model. Only R has software to estimate the model directly, but it is not difficult to construct such a model using Stata or any other software with a pliable programming language or a censored Poisson function or model. The model can easily be estimated by its component parts, as observed below.

NEGATIVE BINOMIAL – POISSON HURDLE MODEL

R

```
hnbp <- hurdle(numvisit ~ reform + badh + educ3 + age3,
      data=mdvis, dist = "negbin", zero="poisson")
summary(hnbp)
alpha <- 1/hnbp$theta
alpha
```

```
Count model coefficients (truncated negbin with log link):
                Estimate Std. Error z value Pr(>|z|)
(Intercept)      0.73953    0.06717  11.010  < 2e-16 ***
reform1998      -0.11830    0.06399  -1.849  0.06452 .
badhbad health   1.15918    0.08622  13.444  < 2e-16 ***
educ3           -0.19603    0.06825  -2.872  0.00408 **
age3             0.23721    0.08456   2.805  0.00503 **
Log(theta)      -0.16158    0.11068  -1.460  0.14433
Zero hurdle model coefficients (censored poisson with log link):
                Estimate Std. Error z value Pr(>|z|)
(Intercept)      0.14452    0.04543   3.181  0.00147 **
reform1998      -0.10734    0.05438  -1.974  0.04841 *
badhbad health   0.57716    0.08539   6.759 1.39e-11 ***
educ3            0.11239    0.05734   1.960  0.04996 *
age3             0.01808    0.07427   0.243  0.80766
---
Signif. codes: 0 '***' 0.001 '**' 0.01 '*' 0.05 '.' 0.1 ' ' 1

Theta: count = 0.8508
Number of iterations in BFGS optimization: 20
Log-likelihood: -4556 on 11 Df

count
1.175371  <- alpha
```

Next we estimate the zero-truncated negative binomial component of the model, followed by a right-censored Poisson model. Note that R's α statistic of 1.175 ($\alpha = 1/\theta$) is identical to the zero-truncated negative binomial NB2. It is not an α statistic for the hurdle model as a whole.

```
. ztnb numvisit reform badh educ3 age3 if numvis>0, nolog

Zero-truncated negative binomial regression
                                            Number of obs   =      1562
                                            LR chi2(4)      =    233.36
Dispersion     = mean                       Prob > chi2     =    0.0000
Log likelihood = -3223.6195                 Pseudo R2       =    0.0349
------------------------------------------------------------------------
numvisit |     Coef. Std. Err.      z    P>|z|  [95% Conf. Interval]
---------+--------------------------------------------------------------
  reform | -.1182993 .0639946   -1.85   0.065 -.2437265    .0071278
    badh |  1.159176 .0862206   13.44   0.000  .9901863    1.328165
   educ3 | -.1960328 .0682512   -2.87   0.004 -.3298028   -.0622629
    age3 |  .2372101  .084559    2.81   0.005  .0714776    .4029426
   _cons |  .7395257 .0671674   11.01   0.000  .6078801    .8711714
---------+--------------------------------------------------------------
/lnalpha |  .1615891 .1106842                  -.055348    .3785262
---------+--------------------------------------------------------------
   alpha |  1.175377 .1300957                  .9461558    1.460131
---------+--------------------------------------------------------------
Likelihood-ratio test of alpha=0:  chibar2(01) = 1677.79
                                   Prob>=chibar2 = 0.000
```

```
. abic

AIC Statistic    =   4.135236       AIC*n        = 6459.2388
BIC Statistic    =   4.141319       BIC(Stata) = 6491.3613

. gen rcen=1
. replace rcen=-1 if numvis>=1
. cpoissone numvisit reform badh educ3 age3, censor(rcen)
  cright(1)

Censored Poisson Regression             Number of obs    =       2227
                                        Wald chi2(4)     =      54.12
Log likelihood = -1331.9847             Prob > chi2      =     0.0000
-------------------------------------------------------------------
numvisit |      Coef. Std. Err.      z   P>|z|  [95% Conf. Interval]
---------+---------------------------------------------------------
  reform | -.1073391 .0543833    -1.97  0.048 -.2139284   -.0007497
    badh |  .5771609 .0853904     6.76  0.000  .4097988    .7445229
   educ3 |  .1123926 .0573353     1.96  0.050  .0000175    .2247678
    age3 |  .0180807 .0742734     0.24  0.808 -.1274926     .163654
   _cons |  .1445252 .0454318     3.18  0.001  .0554804    .2335699
-------------------------------------------------------------------
AIC Statistic =      1.201
```

The models are the same, indicating the underlying nature of the double count models. The overall hurdle AIC statistics can be calculated as shown earlier as:

```
. di 4.135236 * (1-1562/2227) + 1.201
2.4358145
```

which is a substantial drop in value from a straightforward NB2 model of 4.104091. The NB2–Poisson hurdle model is therefore a much superior fitted model than estimating the data using NB2. Note that the **epoissone** command can be downloaded from the SSC website, or from the book's website.

A variant version of the likelihood ratio test can be constructed to determine if the negative binomial hurdle model is statistically different from an NB2 model. Gerdtham (1997) proposed a test statistic based on the following:

$$\mathrm{LR}_{\mathrm{hurdle}} = 2(L_{\mathrm{binary}} + L_{\mathrm{TruncCount}} - L_{\mathrm{count}})$$

The test is evaluated as other likelihood ratio tests.

It is possible to reduce the number of predictors in the binary component without affecting the values of the count component. Simply create the binary component by hand and model with the desired predictors. Remember to obtain the new AIC statistic, and amend the overall hurdle AIC, or $\mathrm{LR}_{\mathrm{hurdle}}$ as defined above. I know of no software that provides the capability to reduce predictors in the binary component, but it is not difficult to calculate.

The results of the hurdle models will be compared with a zero-inflated model of the same data in Section 11.3. Remember, we do not evaluate a model as true, but rather as a comparatively better fit than alternative models of the data.

11.2.4 Marginal effects

The marginal effects of hurdle models can be obtained separately for each respective component. Since we can partition the model into either a right-censored binary model, or simply a binary model, and a zero-truncated count model, marginal effects can be calculated for each of these procedures. Sofware will differ in how this is to be effected.

In Stata, marginal effects are calculated using the **margins** command. Following the discussion of marginal effects and discrete change presented in Section 6.6, and continued for negative binomial models in Section 9.3, the **margins** command produces the following results for the NB–logit hurdle model discussed earlier. Note, in order to obtain the correct discrete change values the non-continuous predictors must be indicated as factor variables.

ZERO-TRUNCATED NEGATIVE BINOMIAL

```
. qui ztnb numvisit i.reform i.badh i.educ3 i.age3 if numvisit>0
. margins, dydx(*) atmeans noatlegend

Conditional marginal effects                  Number of obs   =  1562
Model VCE     : OIM
Expression    : Predicted number of events, predict()
dy/dx w.r.t. : 1.reform 1.badh 1.educ3 1.age3
------------------------------------------------------------------------
           |          Delta-method
           |      dy/dx  Std. Err.      z    P>|z|  [95% Conf.  Interval]
-----------+------------------------------------------------------------
  1.reform | -.2672941   .1448893   -1.84   0.065  -.5512718    .0166837
    1.badh |  4.199705   .4883787    8.60   0.000    3.2425      5.15691
   1.educ3 | -.4308447   .1469813   -2.93   0.003  -.7189228   -.1427666
    1.age3 |  .5819139   .2251402    2.58   0.010   .1406471    1.023181
------------------------------------------------------------------------
Note: dy/dx for factor levels is the discrete change from the
      base level.
```

LOGIT

```
. gen bincnt=numvisit>0
. qui logit bincnt i.reform i.badh i.educ3 i.age3
. margins, dydx(*) atmeans noatlegend
```

```
Conditional marginal effects               Number of obs   =  2227
Model VCE      : OIM
Expression     : Pr(bincnt), predict()
dy/dx w.r.t. : 1.reform 1.badh 1.educ3 1.age3
------------------------------------------------------------------
           |          Delta-method
           |     dy/dx  Std. Err.      z    P>|z|   [95% Conf.  Interval]
-----------+------------------------------------------------------
  1.reform | -.0388068  .0193626   -2.00   0.045  -.0767568   -.0008568
    1.badh |  .1893701  .0237363    7.98   0.000   .1428477    .2358924
   1.educ3 |  .0411216  .0201298    2.04   0.041    .001668    .0805752
    1.age3 |  .0049103  .0267235    0.18   0.854  -.0474668    .0572875
------------------------------------------------------------------
Note: dy/dx for factor levels is the discrete change from the
      base level.
```

The '*' indicates that marginal effects or discrete changes are calculated for all predictors used in the model. This need not be the case. Note that the software recognizes if a predictor is continuous or binary, and calculates the appropriate statistic. Here all predictors are binary, and therefore only discrete changes are calculated.

Sometimes the marginal effects of a predictor give conflicting information when comparing the two components of the model. When this occurs it is probable that the hurdle model is misspecified.

11.3 Zero-inflated negative binomial models

11.3.1 Overview of ZIP/ZINB models

Zero-inflated count models were first introduced by Lambert (1992) to provide another method of accounting for excessive zero counts. Like hurdle models, they are two-part models, consisting of both binary and count model sections. Unlike hurdle models, though, zero-inflated models provide for the modeling of zero counts using both binary and count processes. The hurdle model separates the modeling of zeros from the modeling of counts, entailing that only one process generates zeros. This mixture of modeling zeros is reflected in the log-likelihood function. In difference to hurdle models, which estimate zero counts using different distributions, zero-inflated models incorporate zero counts into both the binary and count processes. Moreover, whereas the binary component of hurdle models estimates non-zero counts, the binary portion of zero-inflated models estimates zero counts. That is, if one (1) is considered as a binary success, then for zero-inflated models, one (1) is the estimation of zero counts. Care must be taken, then, when comparing the binary component of hurdle versus zero-inflated models. The two components of a zero-inflated model cannot be severed as they can be for hurdle models. On the other hand, an intercept-only hurdle model produces identical results to an intercept-only zero-inflated model.

In a similar manner to how the Poisson model is nested in NB2, the zero-inflated Poisson, or ZIP, model is nested within the zero-inflated negative binomial, or ZINB. This allows for comparison of the two models using a boundary likelihood ratio test as described in Section 7.4.2. A Wald test can also be used to assess model worth. In addition, use of the AIC and BIC fit statistics can be used to determine the best fitted zero-inflated model among the various types of model. There have been several alternative parameterizations to the zero-inflated models.

For instance, a zero-deflated model has been created to model negative counts, but the mean of the binary component cannot then be regarded as a probability, as it is in zero-inflated models. A zero-and-two inflated model has also been designed to assess if the count of two occurs more frequently than allowed by the distributional assumptions of the count model (Melkersson and Roth, 2000). Researchers have explored a variety of alternative two-part models, but typically revert to the more standard hurdle and zero-inflated models discussed in this text.

11.3.2 ZINB algorithms

Commercial software implementations typically allow the zero-inflated binary process to be either probit or logit. Counts, including zeros, are estimated using either a Poisson or negative binomial regression. The following types of zero-inflated models are available in R. Stata, SAS, and LIMDEP also provide for most of the listed options.

Table 11.9 *Zero-inflated components*

Binary		Count
logit	probit	Poisson
cauchit	log	negative binomial (NB2)
cloglog		geometric

The log-likelihood functions of the NB–logit and NB–probit models are listed below. Note that β_1 signifies the binary component linear predictor; β signifies the count component.

ZERO-INFLATED NEGATIVE BINOMIAL–LOGIT

$$\text{if } (y == 0): \quad \sum_{i=1}^{n} \left\{ \ln\left(\frac{1}{1+\exp(-x_i'\beta_1)}\right) + \frac{1}{1+\exp(x_i'\beta_1)}\left(\frac{1}{1+\alpha\exp(x_i'\beta)}\right)^{1/\alpha} \right\} \tag{11.12}$$

$$\text{if } (y > 0): \quad \sum_{i=1}^{n} \left\{ \ln\left(\frac{1}{1+\exp(-x_i'\beta_1)}\right) + \ln\Gamma\left(\frac{1}{\alpha}+y_i\right) - \ln\Gamma(y_i+1) - \ln\Gamma\left(\frac{1}{\alpha}\right) + \left(\frac{1}{\alpha}\right)\ln\left(\frac{1}{1+\alpha\exp(x_i'\beta)}\right) + y_i\ln\left[1-\frac{1}{1+\alpha\exp(x_i'\beta)}\right] \right\} \tag{11.13}$$

ZERO-INFLATED NEGATIVE BINOMIAL–PROBIT

$$\text{if } (y == 0): \quad \sum_{i=1}^{n} \left\{ \ln(\Phi x'_i\beta_1) + (1-\Phi(x'_i\beta_1))\left(\frac{1}{1+\alpha\exp(x_i'\beta)}\right)^{1/\alpha} \right\} \tag{11.14}$$

$$\text{if } (y > 0): \quad \sum_{i=1}^{n} \left\{ \ln(1-\Phi(x'_i\beta_1)) + \ln\Gamma\left(\frac{1}{\alpha}+y_i\right) - \ln\Gamma(y_i+1) - \ln\Gamma\left(\frac{1}{\alpha}\right) + \left(\frac{1}{\alpha}\right)\ln\left(\frac{1}{1+\alpha\exp(x_i'\beta)}\right) + y_i\ln\left[1-\frac{1}{1+\alpha\exp(x_i'\beta)}\right] \right\} \tag{11.15}$$

where $\exp(x\beta_1)$ is the fit, or μ, from the binary process, and $\exp(x\beta)$ is the same with respect to the count process. Φ represents the normal or Gaussian cumulative distribution function. These terms are sometimes given entirely different symbols, e.g. $x\beta_1$ is commonly represented as z for the binary covariate and γ for their associated coefficients. Given these terms the logistic cumulative probability function is expressed as

$$\Lambda = \frac{\exp(z_i\gamma)}{1 + \exp(z_i\gamma)}$$

The Gaussian cumulative function is given as

$$\Phi(z_i\gamma) = \int_0^{z_i\gamma} \frac{1}{\sqrt{2\pi}} \exp \frac{-\mu^2}{2} d\mu$$

Inflation refers to the binary process. Unlike hurdle models, the binary process typically have different predictors than in the count process. The important point is for the statistician to use the model to determine which variables or items in the data have a direct bearing on zero counts. This is why the zero-inflated model, unlike hurdle models, has its count process predict zeros. Note in the first equation of the zero-inflated NB–logit model, the three terms predicting zero counts are (1) logistic inverse link i.e. μ, the prediction that $y == 0$, (2) $1 - \mu$, and (3) the negative binomial prediction of a zero count. If the last formula is unfamiliar, recall that the formula has been expressed in a variety of ways, e.g. $\{\alpha^{-1}/(\alpha^{-1} + \mu)\}\alpha^{-1}$.

Score and gradient NB2–logit ZINB functions can be given as below.

$$\frac{\partial \mathcal{L}}{\partial \beta} = \sum_{i:y_i=0}^{n} \left[\frac{-\exp(x_i'\beta)\left(1 + \alpha\exp(x_i'\beta)\right)^{-\alpha^{-1}-1}}{\exp(x_i'\beta_1) + \left(1 + \alpha\exp(x_i'\beta)\right)^{-\alpha^{-1}}} \right] x_i$$

$$+ \sum_{i:y_i>0}^{n} \left[\frac{y_i - \exp(x_i'\beta)}{1 + \alpha\exp(x_i'\beta)} \right] x_i$$

$$+ \sum_{i:y_i=0} \left[\frac{\exp(z_i'\gamma)}{\exp(z_i'\gamma) + (1 + \alpha\exp(x_i'\beta))^{-\alpha^{-1}}} \right] z_i + \sum_{i=1}^{N} \left[\frac{\exp(z_i'\gamma)}{1 + \exp(z_i'\gamma)} \right] z_i$$

$$\frac{\partial \mathcal{L}}{\partial \alpha} = \sum_{\{i:y_i=0\}} \left[\frac{\alpha^{-2}\left[(1 + \alpha\exp(x_i'\beta))\ln(1 + \alpha\exp(x_i'\beta)) - \alpha\exp(x_i'\beta)\right]}{\exp(z_i'\gamma)(1 + \alpha\exp(x_i'\beta))^{(1+\alpha)/\alpha} + (1 + \alpha\exp(x_i'\beta))} \right]$$

$$+ \sum_{\{i:y_i=0\}} \left\{ -\alpha^{-2} \sum_{j=0}^{y_i-1} \frac{1}{(j + \alpha^{-1})} + \alpha^{-2}\ln(1+\alpha\exp(x_i'\beta)) + \frac{y_i - \exp(x_i'\beta)}{\alpha(1 + \alpha\exp(x_i'\beta))} \right\}$$

$$\delta = \frac{1}{\alpha}$$

$$\mu_i = \exp(x_i'\beta)$$
$$\xi = \frac{1}{(1+\alpha\exp(x_i'\beta))}$$
$$\tau_1 = \frac{1}{(1+\alpha\exp(x_i'\beta_1))}$$
$$\tau_2 = 1 - \tau_1(1-\xi^\delta)$$
$$\tau_3 = \exp(x_i'\beta_1)(\tau_1)^2$$

$$\text{Score 1} \quad (y == 0): -\frac{(\tau_1 * \mu_i * \xi^{\delta+1})}{\tau_2}$$
$$(y > 0): \frac{\xi(y_i - \mu_i)}{\tau_2}$$
$$\text{Score 2} \quad (y == 0): \frac{\tau_3(1-\xi^\delta)}{\tau_2}$$
$$(y > 0): \frac{\tau_3}{\tau_1}$$
$$\text{Score 3} \quad (y == 0): \frac{-\alpha\tau_1\xi^\delta\{\delta^2\ln(\xi) + \delta\mu_i\xi\}}{\tau_2}$$
$$(y > 0): -\frac{\{\alpha\xi(\mu_i - y_i) + \ln(\xi) + \Psi(y_i+\mu_i) - \Psi(\delta)\}}{\alpha}$$

11.3.3 Applications

I shall again use the German health reform data used in the previous section. Modeled as a NB–logit, we have the output shown in Table 11.10.

Table 11.10 R: *Zero-inflated NB–logit model*

```
library(pscl)
# mdvis assumed to be in memory and attached
model1 <- zeroinfl(numvisit ~ reform + badh + educ3 + age3 |
   reform + badh + educ3 + age3, data=mdvis, dist = "negbin")
summary(model1)
alpha <- 1/model1$theta
alpha
```

R: ZINB *alternative using* **gamlss** *function*

```
library(gamlss)
m1zinb<-gamlss(numvisit ~ reform + badh + educ3 + age3,
     sigma.fo =~ educ3, family=ZINBI, data=mdvis)
```

```
# using all predictors in binary component not stable; use of educ3 optimal
summary(mlzinb)
```

```
. global xvars reform badh educ3 age3
. zinb numvisit $xvars, inflate($xvars) vuong zip

Zero-inflated negative binomial regression
                                          Number of obs =    2227
                                          Nonzero obs   =    1562
                                          Zero obs      =     665
Inflation model = logit                   LR chi2(4)    =  297.60
Log likelihood  = -4561.673               Prob > chi2   =  0.0000
------------------------------------------------------------------
numvisit |   Coef.   Std. Err.    z   P>|z|  [95% Conf.  Interval]
---------+--------------------------------------------------------
numvisit |
  reform | -.1216958   .056136  -2.17  0.030  -.2317204  -.0116713
    badh |  1.102749  .0754847  14.61  0.000   .9548017   1.250696
   educ3 | -.1241022  .0616922  -2.01  0.044  -.2450167  -.0031877
    age3 |  .2020721  .0739543   2.73  0.006   .0571244   .3470198
   _cons |  .8310162  .0554129  15.00  0.000   .7224089   .9396235
---------+--------------------------------------------------------
Inflate  |
  reform |  .4725348  .9826678   0.48  0.631  -1.453459   2.398528
    badh | -2.109341  4.397292  -0.48  0.631  -10.72788   6.509193
   educ3 | -18.66752  6121.076  -0.00  0.998  -12015.76   11978.42
    age3 |  .7502029   .829699   0.90  0.366  -.8759772   2.376383
   _cons | -3.304505  1.121312  -2.95  0.003  -5.502236  -1.106775
---------+--------------------------------------------------------
/lnalpha | -.0753906  .0731984  -1.03  0.303  -.2188569   .0680756
---------+--------------------------------------------------------
   alpha |  .9273811  .0678828                 .8034367   1.070446
------------------------------------------------------------------
Likelihood-ratio test of alpha=0: chibar2(01) =  1666.19
                                    Pr>=chibar2 =  0.0000
Vuong test of zinb vs. standard negative binomial:
                                    z =  1.06  Pr>z = 0.1451
. abic
AIC Statistic  =    4.106576    AIC*n      = 9145.3457
BIC Statistic  =    4.120386    BIC(Stata) = 9208.1387
```

Post-reform, bad health, age 50–60, and patients with a post-high-school education are all predictors of positive counts, i.e. the number of visits to the physician. Therefore, patients made [exp(−0.1216958) = 0.88541767] about 11% fewer visits to the doctor following reform, which was a marked goal of reform legislation. The Vuong test and likelihood-ratio test results will be addressed in the following subsection.

During the post-reform period (1998), there is a decrease in the expected rate of visits to the doctor by a factor of 0.885, holding all other predictors constant. Patients having the opportunity to visit their doctor, and who are in bad health, increased visits to their doctor some threefold [exp(1.103) = 3.01]. The binary equation section describes the change in odds for always having zero visits versus not always having zero visits. As such, we can interpret the output as showing that the 50–60 year old patient group has some [exp(0.75) = 2.12] two times greater odds of not visiting a doctor than does the 20–39 year age group. This interpretation may sound rather counterintuitive, but it is possible that the older patient group takes better care of their health than the more reckless younger group, or it is likely that limited incomes in old age cause self-rationing of medical care. Still, it is not surprising that the predictor is not significant. That is, we may not rely on it for a meaningful conclusion regarding visits to the doctor. Additional discussion of the logic of this interpretation and related probabilities can be found in Long and Freese (2006).

Those using zero-inflated models on their data must take special care to correctly interpret the model. It is somewhat trickier than interpreting the hurdle model, which is fairly straightforward. A source of mistakes relates to the fact that the model predicts zero counts in two quite different ways. First, in zero-inflated models zero counts are predicted as usual on the basis of a standard Poisson or negative binomial model. Secondly, within the framework of the binary process, a prediction of success is a prediction that the response has a zero count. A response of zero indicates a positive count – or rather, a non-zero count. Unfolding the relationships can be a source of confusion.

11.3.4 Zero-altered negative binomial

Mention should be made of zero-altered negative binomial (ZANB) models. Many statisticians have defined ZANB as the same model as the ZINB. However, there is an interpretation in which the models clearly differ. R's **gamlss** library contains the only software implementation of the model.

Zero-inflated models are two-part models with the binary component being an estimation of Prob($y = 0$), and the count component an estimation of the full range of counts, including zero. The zero-adjusted negative binomial model also estimates Prob($y = 0$), but the negative binomial component is truncated at zero. The zero-altered model is therefore similar to a hurdle model, but with zero estimated as the binary component instead of 1.

R: *ZANB alternative using* **gamlss** *function*

```
library(gamlss)
data (mdvis)
m1zanb<-gamlss(numvisit ~ reform + badh + educ3 + age3,
          nu.fo =~ reform + badh + age3 + educ3,
          family=ZANBI, data=mdvis)
summary(m1zanb)
```

Recall that with the **gamlss** function, the now traditional NB2 is given the symbol NBI, and the NB1 linear negative binomial is represented as NBII. The results are interpreted as a zero-inflated model, except that binary component predictor values are interpreted as the binary component predictors of NB–logit hurdle models. The R source code is provided on the book's website. It is also provided when downloading the **gamlss** software.

It perhaps should be mentioned here that the **gamlss** library provides several more distributions that can be parameterized into regression models. Aside from standard count models thus far discussed, models such as Poisson inverse Gaussian (PIG), zero-inflated PIG (ZIPIG), the three-parameter Delaporte (DEL), Sichel (SICHEL), and Sichel 2 (SI) models are available for count modeling purposes. The Delaporte model entails a shifted-gamma Poisson distribution; and the Sichel distribution takes two parameterizations. Note that the PIG distribution is an SI distribution with a constant shape parameter value of -0.5. See Rigby and Stasinopoulos (2008) for details.

11.3.5 Tests of comparative fit

The standard fit test for ZINB models is the Vuong test. This test is a comparison of ZINB and ZIP. Essentially, the Vuong test is a comparison of predicted fit values of ZINB and ZIP, assessing if there is a significant difference between the two. Given that $P_{\text{ZIP}}(y|x)$ is the probability of observing y on the basis of x in a ZIP model, and $P_{\text{ZINB}}(y|x)$ is the probability of observing the same y on the basis of the same x using a ZINB model, the Vuong test formula may be expressed as:

$$V = \frac{\sqrt{n}\bar{u}}{\text{SD}\,(u_i)} \tag{11.16}$$

where u is the log ratio of the sum of probabilities, given as

$$u_i = \ln\left(\frac{\sum_i P_{\text{ZIP}}(y_i|\text{x}_i)}{\sum_i P_{\text{ZINB}}(y_i|x_i)}\right)$$

Note that $\bar{u}$ is the mean of u and SD(u) is its standard deviation. The Vuong test uses a normal distribution to assess comparative worth. At a 95% confidence level, values of V greater than $+1.96$ indicate that ZIP is the preferred model. Values lower than -1.96 indicate that ZINB is the preferred model. Values between these to critical points indicate that neither model is preferred over the other.

The Vuong test compares the probabilities of the numerator and denominator, with values greater than 1.96 favoring the probabilities from the numerator. If the ZIP and ZINB models are reversed in the above formula, the interpretation will be reversed as well. Most tests use the formula expressed in equation 11.13. Be certain that you know the which formula is being used with your software prior to reporting a conclusion.

The Vuong test may be calculated by hand for the ZINB model above using the following code:

Table 11.11 R: *Vuong test – ZIP and ZINB*

```
modelp <- zeroinfl(numvisit ~ reform + badh + educ3 + age3 |
    reform + badh + educ3 + age3, data=mdvis, dist = "poisson")
pp <- predict(modelp)
pnb <- predict(model1)
u <- log(pp/pnb)
umean <- mean(u)
nobs <- dim(mdvis)[1]
stdu <- sqrt(var(u))
v <- (sqrt(nobs) * umean)/stdu
dnorm(v)
```

```
. qui zip numvisit $xvars, inflate($xvars)
. predict cp
. qui zinb numvisit $xvars, inflate($xvars)
. predict cnb
. gen u = ln(cp/cnb)
. sum u
. gen v = (sqrt(r(N)) * _result(3))/sqrt(_result(4))
. sum v
    Variable |     Obs        Mean    Std. Dev.       Min        Max
-------------+-----------------------------------------------------------
           v |    2227     1.06432           0    1.06432    1.06432
. 1-normprob(v)
.14359178
```

which is the value (1.064) displayed in the model for the Vuong statistic. The p-value is also the same, given rounding error. The `_result(#)` and `r(N)` code are values saved by the command for use in subsequent statistical calculations. Many of the commands have the same coding values, but not all. You need to check the software and command to determine the correct coding.

A border likelihood ratio test may also be used to compare a ZINB with a ZIP. The method and formulae are identical to those that were used to compare NB2 with Poisson models. Recall that the logic of the comparison is based on the distance that the value of α is from 0. Is α significantly greater than 0 such that the model is NB2 rather than Poisson (α=0)? Again, the test is given as:

$$\text{LR} = -2(\mathcal{L}_P - \mathcal{L}_{\text{NB}})$$

A modified *chi*2 test is used to evaluate the significance, with one degree of freedom and with the test statistic divided by 2. For the models used in this section, we have

R

```
LLp <- loglik(modelp)
LLnb <- loglik(model1)
LRtest <- -2*(LLp -- LLnb)
dchisq(1, LRtest)/2
```

```
. di -2*(-5394.77 - (-4561.673))
1666.194
. di chiprob(1, 1666.194)/2
0
```

These values are consistent with model output.

11.3.6 ZINB marginal effects

In a similar manner to hurdle models, marginal effects or discrete change calculations are broken into the two components of the model. Discrete change from 0 to 1 is calculated for each component for *reform*, *badh*, *age*3, *and educ*3. There are no continuous predictors in the model, but if there were, the displayed effects would be interpreted as average marginal effects.

We can use **listcoef** or **margins** to calculate marginal effects or discrete change.

The Vuong test appears to tell us that the ZINB model does not fit the data as well as a standard NB2 model does . That is, the extra zero counts were not a significant factor to the model fit. Adding other predictors, or dropping some, may lead to an alternative conclusion. Given the fact that none of the binary

Table 11.12 R: *ZINB marginal effects*

```
# calculate discrete change for reform in count portion;
# use same method for all other predictors
meanreform_c <- mean(mdvis$reform)
meanbad_c <- mean(mdvis$badh)
meaneduc_c <- mean(mdvis$educ3)
meanrage_c <- mean(mdvis$age3)
dcref1 <- exp(.8310162  -.1216958*meanreform_c +
   1.102749*meanbad_c -.1241022*meaneduc_c + .2020721*meanage_c)
dcref0 <- exp(.8310162 + 1.102749*meanbad_c -.1241022*meaneduc_c
   + .2020721*meanage_c)
dc_reform <- dcref1 -- dcref0
mean(dc_reform)
# use same method for badh, educ3, and age3
```

```
. listcoef
zinb (N=2227): Factor Change in Expected Count
 Observed SD: 4.0161991
Count Equation: Factor Change in Expected Count for
                  Those Not Always 0
------------------------------------------------------------------
numvisit |      b         z     P>|z|     e^b   e^bStdX    SDofX
---------+--------------------------------------------------------
  reform |  -0.12170   -2.168    0.030    0.8854    0.9410    0.5001
    badh |   1.10275   14.609    0.000    3.0124    1.4191    0.3174
   educ3 |  -0.12410   -2.012    0.044    0.8833    0.9430    0.4730
    age3 |   0.20207    2.732    0.006    1.2239    1.0772    0.3682
---------+--------------------------------------------------------
ln alpha | -0.07539
   alpha |  0.92738 SE(alpha) = 0.06788
------------------------------------------------------------------
Binary Equation: Factor Change in Odds of Always 0
------------------------------------------------------------------
 Always0 |      b         z     P>|z|    e^b   e^bStdX    SDofX
---------+--------------------------------------------------------
  reform |   0.47253    0.481    0.631    1.6041    1.2666    0.5001
    badh |  -2.10934   -0.480    0.631    0.1213    0.5120    0.3174
   educ3 | -18.66752   -0.003    0.998    0.0000    0.0001    0.4730
    age3 |   0.75020    0.904    0.366    2.1174    1.3182    0.3682
------------------------------------------------------------------
 Vuong Test =  1.06  (p=0.145)
```

component predictors were significant, it is not surprising at all to conclude that the ZINB model is a poor fit. We remodel the data so that average discrete changes are produced for both the count and binary (inflate) components of the zero-inflated negative binomial.

```
. qui zinb numvisit i.reform i.badh i.educ3 i.age3, inflate(i.reform
   i.badh i.educ3 i.age3)
```

AVERAGE DISCRETE CHANGE – COUNT (NB2) COMPONENT

```
. margins, dydx(*) predict(equation(numvisit))

Average marginal effects                          Number of obs   = 2227
Model VCE      : OIM
Expression     : Linear prediction, predict(equation(numvisit))
dy/dx w.r.t. : 1.reform 1.badh 1.educ3 1.age3
------------------------------------------------------------------------
         |            Delta-method
         |      dy/dx   Std. Err.      z    P>|z|    [95% Conf. Interval]
---------+--------------------------------------------------------------
1.reform | -.1216958    .056136    -2.17   0.030   -.2317204   -.0116713
  1.badh |  1.102749    .0754847   14.61   0.000    .9548017    1.250696
 1.educ3 | -.1241022    .0616922   -2.01   0.044   -.2450167   -.0031877
  1.age3 |  .2020721    .0739543    2.73   0.006    .0571244    .3470198
------------------------------------------------------------------------
Note: dy/dx for factor levels is the discrete change from the
      base level.
```

AVERAGE DISCRETE CHANGE – BINARY (Logit) COMPONENT

```
. margins, dydx(*) predict(equation(inflate))

Average marginal effects                          Number of obs   = 2227
Model VCE      : OIM
Expression     : Linear prediction, predict(equation(inflate))
dy/dx w.r.t. : 1.reform 1.badh 1.educ3 1.age3
------------------------------------------------------------------------
         |            Delta-method
         |      dy/dx   Std. Err.      z    P>|z|    [95% Conf. Interval]
---------+--------------------------------------------------------------
1.reform |  .4725348    .9826678    0.48   0.631   -1.453459    2.398528
  1.badh | -2.109341    4.397292   -0.48   0.631   -10.72788    6.509193
 1.educ3 | -18.66752    6121.076   -0.00   0.998   -12015.76    11978.42
  1.age3 |  .7502029     .829699    0.90   0.366   -.8759772    2.376383
------------------------------------------------------------------------
Note: dy/dx for factor levels is the discrete change from the
      base level.
```

The interpretation of the predictor average discrete changes for each component is the same as described in Sections 6.6.2 and 9.3. Keep in mind that probabilities related to the binary component are aimed at explaining predictors giving zero counts.

The AIC statistic for the NB–logit hurdle model (9,133) compared with the AIC statistic of the ZINB model (9,145) indicates that the hurdle model is a superior fit. The difference between the two is 12, which according to the criteria given in Table 5.1 indicates that the difference is significant. The NB–logit model is preferred.

11.4 Comparison of models

Figure 11.2 consists of lines representing the differences between the observed and predicted probabilites for the NB2, ZIP, and ZINB models. It uses the **prcounts** command authored by Long and Freese (2006) to develop the requisite statistics. Parts of the lines that are above 0 on the *y*-axis indicate underprediction of counts; parts of the line below indicated overprediction. Zero counts are highly overpredicted, with 1-counts underpredicted. Counts of 4 and over appear to be equipredicted, i.e. the predicted values closely approximate the observed values. Given the fact that there are excessive number of zeros based on the distributional mean, the results should be of no surprise.

Table 11.13 *Graphing predicted vs observed probabilities*

```
* Modified from code in Long and Freese (2006), Page 406
use mdvis, clear    /// nbr2_f11.11.do
global xvars reform badh educ3 age3

* NB2
qui nbreg numvisit reform badh educ3 age3 if numvis>0, nolog
prcounts pnb2, plot max(9)
lab var pnb2preq "Predicted NB2"
lab var pnb2obeq "Observed NB2"

* ZIP
qui zip numvisit $xvars, inflate($xvars)
prcounts pzip, plot max(9)
lab var pzippreq "Predicted ZIP

* ZINB
qui zinb numvisit $xvars, inflate($xvars)
prcounts pzinb, plot max(9)
lab var pzinbpreq "Predicted ZINB
```

```
*  Differences
gen obs = pnb2obeq
gen dnb2 = obs - pnb2preq
lab var dnb2 "NB2 Diff"

gen dzip = obs - pzippreq
lab var dzip "ZIP Diff"

gen dzinb = obs - pzinbpreq
lab var dzinb "ZINB Diff"

graph twoway connected dnb2 dzip dzinb pnb2val, ///
  ytitle(Observed-Predicted) ylab(-.10(.05).10) ///
  xlab(0(1)9) msymbol(Oh Sh O S)
```

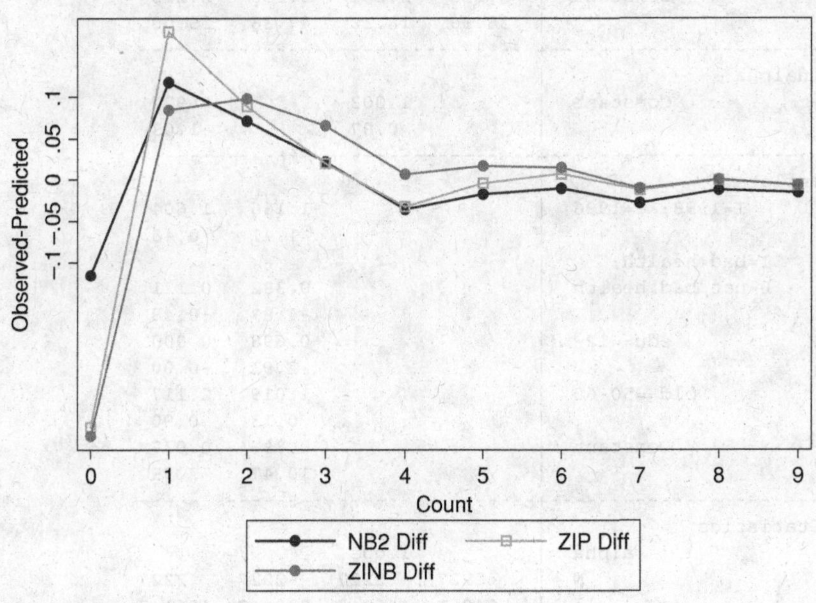

Figure 11.2 Difference in observed vs predicted

Finally, a batch program authored by Long and Freese (2006) called **countfit** can be used to compare summary statistics across models. The command below provides statistics for a Poisson, NB2, ZIP, and ZINB models. For space reasons I show only the results which bear most heavily on the data we have been considering in this chapter. The **countfit**, **prcounts**, and **prvalues** commands used in this chapter may be downloaded from http://www.indiana.edu/~jslsoc/Stata/.

```
. countfit numvisit reform badh educ3 age3,

  inf( reform badh educ3 age3) prm nb zip zinb max(10)
-------------------------------------------------------
            Variable |     PRM     NBRM      ZIP    ZINB
---------------------+---------------------------------
numvisit             |
    1=1998; 0=1996   |   0.875    0.873    0.904   0.885
                     |   -5.03    -2.65    -3.58   -2.17
    1=bad health;    |
    0=not bad health |   3.132    3.126    2.586   3.012
                     |   38.80    15.51    31.25   14.61
            edu==12- |   0.886    0.928    0.824   0.883
                     |   -4.17    -1.37    -6.15   -2.01
          old==50-60 |   1.117    1.194    1.111   1.224
                     |    3.31     2.60     3.00    2.73
            Constant |   2.258    2.201    3.156   2.296
                     |   35.96    18.36    47.36   15.00
---------------------+---------------------------------
lnalpha              |
            Constant |            1.003            0.927
                     |             0.07            -1.03
---------------------+---------------------------------
inflate              |
      1=1998; 0=1996 |                     1.166   1.604
                     |                      1.41    0.48
    1=bad health;    |
    0=not bad health |                     0.382   0.121
                     |                     -4.87   -0.48
            edu==12- |                     0.698   0.000
                     |                     -2.91   -0.00
          old==50-60 |                     1.019   2.117
                     |                      0.13    0.90
            Constant |                     0.396   0.037
                     |                    -10.47   -2.95
---------------------+---------------------------------
Statistics           |
               alpha |            1.003
                   N |    2227     2227     2227    2227
                  ll | -5948.1  -4563.9  -5394.8 -4561.7
                 bic | 11934.7   9174.1  10866.6  9208.1
                 aic | 11906.1   9139.8  10809.5  9145.3
-------------------------------------------------------

Comparison of Mean Observed and Predicted Count
            Maximum        At      Mean
Model       Difference     Value   |Diff|
--------------------------------------------
PRM         0.184          0       0.040
NBRM        0.026          2       0.011
ZIP         0.093          1       0.025
ZINB        0.027          2       0.011
```

```
ZINB: Predicted and actual probabilities
Count    Actual      Predicted      |Diff|      Pearson
-----------------------------------------------------
 0       0.299       0.310         0.011        0.927
 1       0.201       0.201         0.001        0.005
 2       0.168       0.141         0.027       11.436
 3       0.115       0.098         0.017        6.416
 4       0.053       0.068         0.016        8.156
 5       0.045       0.048         0.002        0.290
 6       0.034       0.034         0.000        0.010
 7       0.009       0.024         0.015       19.742
 8       0.012       0.017         0.005        3.453
 9       0.004       0.013         0.009       13.005
10       0.027       0.009         0.018       77.480
-----------------------------------------------------
Sum      0.967       0.964         0.121      140.920
Tests and Fit Statistics
NBRM        BIC= -7992.569  AIC=     4.104  PreferOver  Evidence
-------------------------------------------------------------------
  vs ZIP   BIC= -6300.006  dif= -1692.563  NBRM   ZIP   Very strong
           AIC=     4.854  dif=    -0.750  NBRM   ZIP
-------------------------------------------------------------------
  vs ZINB  BIC= -7958.492  dif=   -34.077  NBRM   ZINB  Very strong
           AIC=     4.107  dif=    -0.002  NBRM   ZINB
           Vuong=   1.057  prob=    0.145  ZINB   NBRM  p=0.145
-------------------------------------------------------------------
ZIP         BIC= -6300.006  AIC=     4.854  PreferOver  Evidence
-------------------------------------------------------------------
  vs ZINB  BIC= -7958.492  dif=  1658.486  ZINB   ZIP   Very strong
           AIC=     4.107  dif=     0.747  ZINB   ZIP
           LRX2= 1666.194  prob=    0.000  ZINB   ZIP   p=0.000
```

It appears that NB2 is the preferred model for the German health reform data over Poisson and zero-inflated models. This is consistent with what we found earlier. Hurdle models are not provided statistics since they are not part of the official Stata application. However, it is possible to make the appropriate comparisons by hand using the methods demonstrated in this section.

Summary

We discussed two data situations which we know give rise to overdispersion on both Poisson and negative binomial (NB2) models. The first relates to when there is no possibility of a zero count in the data. Hospital length of stay is a good example of this type of data. Zero-truncated models adjust the probability functions of the Poisson and NB2 models so that respective zero counts are excluded, but the sum of probabilities is still 1.

The situation of having an excess of zero counts is also frequently found in research data. Hurdle and zero-inflated models are the two most commonly used types of model that have been developed to adjust for excessive zeros in count data. Both are two-part models, but they are each based on different reasoning. They differ in accounting for the origin or generation of the extra zeros. This accounting is then reflected in the respective estimating algorithms.

We next turn to a discussion of models involving censoring and truncation.

12

Censored and truncated count models

There are many times when certain data elements are lost, discarded, ignored, or are otherwise excluded from analysis. Truncated and censored models have been developed to deal with these types of data. Both models take two forms: truncation or censoring from below, and truncation or censoring from above. Count model forms take their basic logic from truncated and censored continuous response data, in particular, from Tobit (Amemiya, 1984) and censored normal regression (Goldberger, 1983) respectively.

The traditional parameterization used for truncated and censored count data can be called the econometric or cut-point parameterization. This is the form of model discussed in standard econometric texts and is the form found in current econometric software implementations. I distinguish this from what I term a survival parameterization, the form of which is derived from standard survival models. This parameterization only relates to censored Poisson and censored negative binomial models. I shall first address the more traditional econometric parameterization. Note, however, that recently developed censored Poisson and negative binomial models have been developed for which the likelihood functions allow both parameterizations to be estimated using the same model. Interval censoring is also provided. Refer to Hardin and Hilbe (2012).

12.1 Censored and truncated models – econometric parameterization

Censored and truncated count models are related, with only a relatively minor algorithmic difference between the two. The essential difference relates to how response values beyond a user-defined cut-point are handled. Truncated models eliminate the values altogether; censored models revalue them to the value of the cut-point. In both cases the probability function and log-likelihood functions must be adjusted to account for the change in distribution of the response. We begin by considering truncation.

12.1.1 Truncation

In order to understand the logic of truncation, we begin with the basic Poisson probability mass function, defined as

$$\text{Prob}(Y = y) = \frac{e^{-\mu_i} \mu_i^{y_i}}{y_i!}, \quad y = 0, 1, \ldots \tag{12.1}$$

Recall that when we discussed zero-truncated Poisson in the last chapter we adjusted the basic Poisson PDF to account for the structural absence of zeros. Given the probability of a zero count as $e^{-\mu}$, or $\exp(-\mu)$, it is subtracted from 1 to obtain the probability of a non-zero positive count. The Poisson probability mass function is then rescaled by the resultant value, $1 - \exp(-\mu)$, to obtain the zero-truncated Poisson PDF. The same logic maintains for zero-truncated negative binomial regression. The probability of a negative binomial count of zero is $(1 - \alpha\mu)^{-1/\alpha}$. Subtracting this value from 1 gives the negative binomial formula of a non-zero positive count. The negative binomial PDF is then rescaled by $1 - (1 - \alpha\mu)^{-1/\alpha}$ to obtain the zero-truncated negative binomial.

In the more general case, zero-truncated count models can be considered as left- or lower-truncated count models. The lower cut-point, C, is at 1. If we wish to extend C to any higher value in the observed distribution, the value to be divided from the basic PDF must reflect the total probability of counts up to the cut. The smallest response value in the observed distribution is $C + 1$. For example, if C is specified as 1, then both the probability of zero counts and counts of 1 need to be calculated, summed, subtracted from 1, and used to rescale the resulting basic count PDF. In the case of Poisson,

$$\text{Prob}(Y = (y = 0)) = e^{-\mu_i} \tag{12.2}$$

and

$$\text{Prob}(Y = (y = 1)) = \mu_i e^{-\mu_i} \tag{12.3}$$

These values are then summed and subtracted from 1.

$$\text{Prob}(Y = (y = 0, 1)) = 1 - \left(e^{-\mu_i} + \mu_i e^{-\mu_i}\right) \tag{12.4}$$

This value is then divided from the Poisson PDF to obtain a 1-truncated Poisson, or more accurately a left-truncated at 1 Poisson, PDF. Remaining values in the distribution have a minimum at $C + 1$, or in this case, $1 + 1 = 2$.

The same logic applies with respect to repeatedly greater left values, depending on the value of the defined cut-point. A left-truncated cut at $C = 3$ specifies that the response values in the model have values starting at $C + 1$, or 4. This does not mean that the distribution must have a value of 4, only that no values have non-zero probability for the distribution less than 4. The left-truncated negative binomial follows the same reasoning.

To demonstrate the relationships involved when left-truncating to a specified cut-point, consider a Poisson mean 0.6 and a $C = 4$

$$\begin{aligned}
\text{Prob}(Y = (y = 0)) &= [\mu^y e^{-\mu}]/1 = [0.6^0 e^{-0.6}]/1 \\
&= (1)(0.54881164) \qquad = 0.5488 \ : 5488 \\
\text{Prob}(Y = (y = 1)) &= [\mu^y e^{-\mu}]/1 = [0.6^1 e^{-0.6}]/1 \\
&= (0.6)(0.54881164)/1 \qquad = 0.3293 \ : 0.8781 \\
\text{Prob}(Y = (y = 2)) &= [\mu^y e^{-\mu}]/1 = [0.6^2 e^{-0.6}]/2 \\
&= (0.36)(0.54881164)/2 \qquad = 0.0988 \ : 0.9769 \\
\text{Prob}(Y = (y = 3)) &= [\mu^y e^{-\mu}]/1 = [0.6^3 e^{-0.6}]/6 \\
&= (0.216)(0.54881164)/6 \qquad = 0.0196 \ : 0.9966 \\
\text{Prob}(Y = (y = 4)) &= [\mu^y e^{-\mu}]/1 = [0.6^4 e^{-0.6}]/24 \\
&= (0.1296)(0.54881164)/24 \qquad = 0.00296 : 0.9996
\end{aligned}$$

The same may be calculated using the following code from the command line. With an amendment of the mean and cut-point, the denominator of the truncated Poisson may be calculated for any mean and left-truncated cut-point. The same logic can be used for the right cut.

```
. set obs 10
. local j=0
. local mu = .6
. local cut=4
. forvalues i =0/'cut' {
  2. gen x'i' = ('mu'^'i' * exp(-'mu'))/
       round(exp(lnfactorial('i')))
  3. local j = x'i' + 'j'
  4. di 'j'
  5. }
.54881161
.87809861
.9768847
.99664192
.9996055

. di 1 - .9996055
.0003945
```

We can formalize the left-truncated Poisson PDF as:

$$\text{Prob}(Y = y|Y > C) = \frac{e^{-\mu_i}\mu_i^{y_i}/y_i!}{\text{Prob}\,(y_i > C)}$$
$$= \frac{e^{-\mu_i}\mu_i^{y_i}/y_i!}{1 - \sum_{j=0}^{C} e^{-\mu_i}\mu_i^{j_i}/j_i!}, \quad \text{for } y = 0, 1, \ldots, C-1 \tag{12.5}$$

where C is the user-defined cut-point and *j* is the running index in the summations. See Greene (2006a) for details of derivation as well as formulae for gradients and marginal effects.

A little algebraic manipulation allows formulation of the left-truncated Poisson log-likelihood function as

$$\mathcal{L}(\beta; y) = \sum_{i=1}^{n} \left\{ y_i \ln(\mu_i) - \mu_i - \ln\Gamma(y_i + 1) - \left[1 - \sum_{j=0}^{C} (j_i \ln(-\mu_i) - \mu_i - \ln\Gamma\,(j_i + 1)) \right] \right\} \tag{12.6}$$

An example may help show the differences in parameter estimates and associated model statistics for data in which the left side has been dropped up to a point and a left-truncated model with a cut defined at the same point. The first set of models will come from the **mdvis** data. The left truncation is defined with a cut of 3, meaning that the sample response, *numvisit,* starts with the count value of 4. A tabulation of counts is provided in Table 12.1.

Table 12.1 ***mdvis*** *data, truncated at 3*

	numvisit	Freq.
	0	665
	1	447
	2	374
	3	256
CUT		
	4	117
	5	101
	.	.
	50	1
	60	1
	Total	485

POISSON: ALL DATA

```
. use mdvis
. glm numvisit reform badh educ3, fam(poi) nohead
------------------------------------------------------------------------
         |                OIM
numvisit |      Coef.  Std. Err.      z   P>|z|   [95% Conf. Interval]
---------+--------------------------------------------------------------
  reform | -.129116   .0264999  -4.87   0.000   -.1810549   -.0771772
    badh | 1.157007   .0290431  39.84   0.000    1.100084    1.213931
   educ3 |-.1299779    .028922  -4.49   0.000   -.1866639   -.073292
   _cons | .8319684   .0219431  37.91   0.000    .7889607    .874976
------------------------------------------------------------------------
```

POISSON: DROPPED VALUES 1–3

```
. glm numvisit reform badh educ3 if numvisit>3, fam(poi) nohead

------------------------------------------------------------------------
         |                OIM
numvisit |      Coef.  Std. Err.      z   P>|z|   [95% Conf. Interval]
---------+--------------------------------------------------------------
  reform |-.0589452   .0327728  -1.80   0.072   -.1231787    .0052882
    badh | .3997378   .0331004  12.08   0.000    .3348622    .4646133
   educ3 |-.2199466   .0368481  -5.97   0.000   -.2921676   -.1477256
   _cons | 2.005811   .0277107  72.38   0.000    1.951499    2.060123
------------------------------------------------------------------------

. abic
AIC Statistic    =   6.341249          AIC*n       = 3075.5059
BIC Statistic    =   6.347621          BIC(Stata) = 3092.2424
```

LEFT-TRUNCATED POISSON: CUT = 3

Compare the difference with a left-truncated at 3 model. Using Stata's **tpoisson** command (new with version 11.1), we have

```
. tpoisson numvisit reform badh educ3 if numvisit>3, ll(3)

Truncated Poisson regression                Number of obs =      485
Truncation point: 3                         LR chi2(3)    =   240.51
                                            Prob > chi2   =   0.0000
Log likelihood = -1486.0242                 Pseudo R2     =   0.0749
```

```
------------------------------------------------------------------------
numvisit3 |     Coef. Std. Err.     z   P>|z|   [95% Conf. Interval]
----------+-------------------------------------------------------------
   reform |-.0765541 .0372434  -2.06  0.040  -.1495498  -.0035584
     badh | .4831879 .0367094  13.16  0.000   .4112388    .555137
    educ3 |-.2988265 .0446251  -6.70  0.000  -.3862902  -.2113629
    _cons | 1.938093 .0318925  60.77  0.000   1.875585   2.000601
------------------------------------------------------------------------

. abic
AIC Statistic    =     6.14443              AIC*n       = 2980.0486
BIC Statistic    =    6.150802              BIC(Stata) = 2996.7852
```

Observe the sizeable difference in AIC*n and BIC(Stata) values between the model in which the counts 0–3 were excluded, compared with the left truncated at 3 model. Given the criteria we have been using to compare AIC values, the left-truncated model is clearly a better fitted model.

Stata's **tpoisson** command only provides left truncation at present. For a user-developed truncated Poisson allowing both sides of truncation, see **trpoisson** (Hilbe, 2009), which is published on SSC.

The code in Table 12.2 can be used to compare the two models using R.

Table 12.2 R: *Left-truncated Poisson*

```
data(mdvis)
attach(mdvis)
mpo <- gamlss(numvisit~reform+badh+educ3,data=mdvis,family=PO)
summary(mpo)
library(gamlss)
library(gamlss.tr)
lmdvis<-subset(mdvis, mdvis$numvisit>3)
lmpo <-
gamlss(numvisit~reform+badh+educ3, family=trun(3, "PO", "left"),
   data=lmdvis)
summary(lmpo)
-----------------
lmdvis<-subset(mdvis, mdvis$numvisit>3)  # alternative method
gen.trun(3, "PO", "left")                   # saved globally
                                              for session
lt3po <-
gamlss(numvisit~reform+badh+educ3, family=POleft, data=lmdvis)
summary(lt3po)
```

```
Mu Coefficients:

              Estimate  Std. Error  t value   Pr(>|t|)
(Intercept)   1.93812    0.03187    60.815  5.803e-228
reform       -0.07654    0.03723    -2.056   4.033e-02
```

```
badh          0.48314     0.03670    13.166   5.005e-34
educ3        -0.29875     0.04461    -6.697   5.943e-11
-------------------------------------------------------
No. of observations in the fit:  485
Degrees of Freedom for the fit:  4
      Residual Deg. of Freedom:  481
                      at cycle:  2

Global Deviance:     2972.048
            AIC:     2980.048
            SBC:     2996.785
```

Notice that the model for which counts 0–3 were simply dropped results in a p-value for *reform* that is not significant at the 0.05 level, whereas it is significant in the left-truncated model. The reason for the difference is due to the fact that the truncated model adjusts the log-likelihood.

The left-truncated Poisson, and right-truncated negative binomial below, were estimated using R's **gamlss.tr** function, one of the **gamlss** suite of functions authored by Robert Rigby and Mikis Stasinopoulos. Rigby wrote the functions used here specifically for this text. The functions rely on **gamlss** and have since been incorporated into the **gamlss** package on CRAN. Note that a wide range of truncated models can be developed from the **gamlss.tr** library. For modeling counts, Poisson, NB2, and NB1 are the most important. Note that Rigby refers to NB1 as NBII and NB2 as NBI. This may result in confusion unless kept in mind.

Right-truncated models have a cut on the upper or right side of the distribution. The right-truncated Poisson PDF may be specified as

$$\begin{aligned}\text{Prob}(Y = y|Y < C) &= \frac{e^{-\mu_i}\mu_i^{y_i}/y_i!}{\text{Prob}\ (y_i < C)} \\ &= \frac{e^{-\mu_i}\mu_i^{y_i}/y_i!}{\sum_{j=0}^{C-1} e^{-\mu_i}\mu_i^{j_i}/j_i!}, \quad \text{for } y = 0, 1, \ldots, C-1\end{aligned} \tag{12.7}$$

A right cut at 10 provides that values up to and including 9 will have non-zero probabilities in the truncated model i.e. $C - 1$.

The left-truncated negative binomial PDF may be expressed as

$$H = \frac{\Gamma\left(y_i + \alpha^{-1}\right)}{\Gamma(\alpha^{-1})\Gamma(y_i + 1)}(\alpha\mu_i)^{y_i}(1 + \alpha\mu_i)^{-(y_i+\alpha^{-1})} \tag{12.8}$$

$$I_j = \frac{\Gamma\left(j_i + \alpha^{-1}\right)}{\Gamma\left(\alpha^{-1}\right)\Gamma\left(j_i + 1\right)}(\alpha\mu_i)^{j_i}\left(1 + \alpha\mu_i\right)^{-(j_i+\alpha^{-1})} \tag{12.9}$$

$$\text{Prob}\,(Y = y|Y > C) = \frac{H}{1 - \sum_{j=0}^{C} I_j} \tag{12.10}$$

The right-truncated negative binomial PDF is formulated in the same manner as is the right-truncated Poisson

$$\text{Prob}(Y = y|Y < C) = \frac{H}{\sum_{j=0}^{C-1} I_j} \tag{12.11}$$

We shall use the same **mdvis** data for an example of a right-truncated negative binomial model. Using a cut of 15, only values of *numvisit* up to 14 will be included in the model. The Stata **tnbreg** command (new with version 11.1) can be used to model only left-truncated negative binomial models. However, R will be used here to display example results. R code is given in Table 12.3, and assumes that the code in Table 12.2 has previously been run. R **gamlss** and LIMDEP ouput are displayed.

Table 12.3 R: *Right-truncated negative binomial: cut = 15*

```
tmdvis<-subset(mdvis, mdvis$numvisit<15)
rtold <- gamlss(numvisit~reform+badh+educ3, data=tmdvis,
   family=trun(14, "NBI", type="right"))
summary(rtold)
```

```
Mu Coefficients:
             Estimate  Std. Error  t value  Pr(>|t|)
(Intercept)   0.73389     0.04030  18.2124  4.239e-69
reform       -0.13721     0.04973  -2.7590  5.845e-03
badh          1.18689     0.07275  16.3140  1.533e-56
educ3        -0.01416     0.05254  -0.2695  7.876e-01
------------------------------------------------------
Sigma link function:  log
Sigma Coefficients:
             Estimate  Std. Error  t value  Pr(>|t|)
(Intercept)   -0.1334  0.05172     -2.579  0.009969
------------------------------------------------------
No. of observations in the fit:  2183
Degrees of Freedom for the fit:  5
      Residual Deg. of Freedom:  2178
Global Deviance:     8474.596
            AIC:     8484.596
            SBC:     8513.038
  exp(-0.1334)
[1] 0.875115
```

RIGHT-TRUNCATED NEGATIVE BINOMIAL: CUT = 15

Variable	Coefficient	St Error	b/St.Er.	P[|Z|>z]	Mean of X
Constant	.73384757	.04065421	18.051	.0000	
REFORM	-.13716032	.05221156	-2.627	.0086	.50664224
BADH	1.18445765	.12792755	9.259	.0000	.10352726
EDUC3	-.01416990	.05690943	-.249	.8034	.34127348
	Dispersion parameter for count data model				
Alpha	.87479181	.04757499	18.388	.0000	

12.1.2 Censored models

Censored models have a similar form to the truncated, but there are important differences. For left-censored models, a cut of C indicates that C is the smallest recordable response value of the censored model. As such, all values of the original response which are actually less than C are measured and recorded as the value of C; thus this value in the data actually means "less than or equal to C." If $C = 3$, the lowest measurable response is a 3, and all values of the original response less than 3 now have a value recorded as 3. On the other hand, a $C = 3$ value for truncated models specifies that 4 is the lowest possible value of the truncated response and that there are no response values under 4; if there are, then they are dropped from a truncated analysis.

There is a similar difference of interpretation for right censoring. A right-censored cut at C indicates that the largest recordable value of the response is C and that all values greater than C are recorded as C; thus this value in the data actually means "greater than or equal to C." If C = 15, then all response values greater than 15 are revalued to 15. No values are dropped from the model analysis.

To summarize this point, censored cut-points differ from truncated cuts in two ways.

Truncated Left: If $C = 3$, only values greater than 3 are supported by the underlying distribution; lower values, if they exist, are dropped. Right: If $C = 15$, only values less than 15 are supported by the underlying distribution; higher values, if they exist, are dropped.

Censored Left: If $C = 3$, 3 is the smallest observable value in model; this value is inexact and means only that the observation is less than or equal to 3. Any response in the data that is less than 3 is also considered to be less than or equal to 3. Right: If $C = 15$, 15 is highest observable value in model; this value is inexact and

> means only that the observation is greater than or equal to 15. Any response in the data that is greater than 15 is also considered to be greater than or equal to 15. Left: P(Y<=C), Right: P(Y>=C).

The disparity of meanings for what a cut value indicates may give rise to considerable confusion. One must keep the difference clearly in mind when engaging truncated and censored count models.

The econometric Poisson and negative binomial censored log-likelihoods may be expressed for the purpose of estimation using the following Stata code. Corresponding code in R or other software packages may be used as well. We use the R **gamlss.cens** function to model left- and right-censored example data. It was authored by the same two statisticians who wrote **gamlss.tr**, which was used for estimating truncated count models. Note that censoring can be done on a number of alternative count distributions. Check the **gamlass.cens** software for possibilities.

CENSORED POISSON

```
LEFT:    ln(poisson(exp(xb)),C) or ln(1-gammap(C+1,exp(xb)))
RIGHT:   ln(1-poissontail(exp(xb)),C)) or ln(1-gammaptail(C,
         exp(xb)))
```

CENSORED NEGATIVE BINOMIAL

```
LEFT:    ln(ibeta(C, 1/alpha, 1/(1+exp(-xb-ln(alpha)))))
RIGHT:   ln(ibetatail(C-1, 1/alpha, 1/(1+exp(-xb-ln(alpha)))))
```

We shall use the **mdvis** data used for truncated models to compare with censored value output. We begin with respective examples of left-truncated and left-censored negative binomial regression with a cut of 3. For ease of comparison, I display results of a left-truncated negative binomial using LIMDEP, and censored negative binomial, both cut at 3, using the R code provided in Table 12.4.

Table 12.4 R: *Left-censored NB2: cut = 3*

```
library(gamlss.cens)
library(survival)
lcmdvis <- mdvis
cy <- with(lcmdvis, ifelse(numvisit<3, 3, numvisit))
ci <- with(lcmdvis, ifelse(numvisit<=3, 0, 1))
Surv(cy,ci, type="left")[1:100]
cbind(Surv(cy,ci, type="left")[1:50], mdvis$numvisit[1:50])
```

```
lcmdvis <- data.frame(lcmdvis, cy, ci )
rm(cy,ci)
gen.cens("NBI",type="left")
lcat30<-gamlss(Surv(cy, ci, type="left") ~ reform+badh+educ3,
  data=lcmdvis, family=NBIlc)
summary(lcat30)
```

LEFT-TRUNCATED NEGATIVE BINOMIAL: CUT = 3

Variable	Coefficient	St Error	b/St.Er.	P[|Z|>z]	Mean of X
Constant	-.34794507	1.76257038	-.197	.8435	
REFORM	-.11831607	.13485136	-.877	.3803	.45154639
BADH	.90189762	.16928103	5.328	.0000	.31134021
EDUC3	-.48182675	.14780265	-3.260	.0011	.31340206
	Dispersion parameter for count data model				
Alpha	7.44973611	14.2858346	.521	.6020	

LEFT-CENSORED NEGATIVE BINOMIAL: CUT = 3

Variable	Coefficient	St Error	b/St.Er.	P[|Z|>z]	Mean of X
Constant	.73734541	.05285178	13.951	.0000	
REFORM	-.14399988	.05908096	-2.437	.0148	.50606197
BADH	1.23620067	.09635204	12.830	.0000	.11360575
EDUC3	-.15383921	.06588073	-2.335	.0195	.33767400
	Dispersion parameter for count data model				
Alpha	1.47560302	.11474405	12.860	.0000	

The parameter estimates have the same signs, but quite different values. Moreover, *reform* is not contributory to the truncated model whereas it is for the censored model. The values of α are substantially different: 7.45 to 1.48. It is likely that the deletion of values less than 4, which consist of 1,742 out of the original 2,227 patients, or 78% of the cases, result in considerable overdispersion in the remaining data. This situation does not exist for censored models. All observations are kept; censored values are just revalued.

Modeling a right-truncated negative binomial with a cut of 15 may be compared with the truncated model at the end of the previous section. R code is presented using the **gamlss.cens** function. I give the relevant output below Table 12.5.

Table 12.5 R: *Right-censored NB2: cut = 15*

```
data(mdvis)
library(gamlss.cens)
library(survival)
rcmdvis <- mdvis
cy <- with(rcmdvis, ifelse(numvisit>=15, 14, numvisit))
ci <- with(rcmdvis, ifelse(numvisit>=15, 0, 1))
rcmdvis <- data.frame(rcmdvis, cy, ci )
rm(cy,ci)
gen.cens("NBI",type="right")
rcat30<-gamlss(Surv(cy, ci) ~ reform+badh+educ3,
  data=rcmdvis, family=NBIrc, n.cyc=100)
summary(rcat30)
```

RIGHT-CENSORED NEGATIVE BINOMIAL: CUT = 15

```
Family:  c("NBIrc", "right censored Negative Binomial type I")

-------------------------------------------------------------------
Mu link function:  log
Mu Coefficients:
              Estimate  Std. Error  t value   Pr(>|t|)
(Intercept)    0.79320     0.04104   19.327  4.738e-77
reform        -0.10697     0.05050   -2.118  3.424e-02
badh           1.13474     0.07224   15.708  8.242e-53
educ3         -0.07049     0.05368   -1.313  1.893e-01
-------------------------------------------------------------------
Sigma link function:  log
Sigma Coefficients:
              Estimate  Std. Error  t value  Pr(>|t|)
(Intercept)   -0.02892     0.04791  -0.6037    0.5461
-------------------------------------------------------------------
No. of observations in the fit:  2227
Degrees of Freedom for the fit:  5
     Residual Deg. of Freedom:  2222
                     at cycle:  3

Global Deviance:    8868.304
           AIC:     8878.304
           SBC:     8906.846

. di exp(-0.02892)
.97149418           <= value of alpha
```

Unlike the left-truncated model, the right deletes only 44 out of 2,227 observations, or only 2%. The differences between the right-censored and right-truncated models, as expected, do not substantially differ. In fact, the models display quite similar output.

For completeness, mention should be made about the probability and log-likelihood functions. The logic is quite simple. All four of the relevant censored models – right-censored Poisson, right-censored negative binomial, left-censored Poisson, left-censored negative binomial – take the same form as the truncated models we discussed in the last section. The difference is, however, in how the cuts are managed, as well as how values beyond the cut-points are handled. The values at the cut, *C*, are included in the model, unlike the case with truncation, and values beyond cut-points are revalued to the value of *C*. With truncation, values of *C* and beyond are dropped. The only alteration this causes in the PDF and log-likelihood functions is at the cut.

Tests for overdispersion based on score functions for both left- and right-truncated Poisson and negative binomial models are discussed by Gurmu and Trivedi (1992). Greene (2006a) provides additional discussion of both truncated and econometric censored Poisson and negative binomial models.

Gamlss has the capability of incorporating censoring and truncation to the following count models, shown with R model code: Poisson (PO), NB2 (NBI), NB1 (NBII), Poisson inverse Gaussian (PIG), Sichel (SICHEL), Sichel-2 (SI), Delaporte (DEL). Censoring may be left, right, or interval. Interval censoring indicates observations that are censored between two specified counts levels.

12.2 Censored Poisson and NB2 models – survival parameterization

A prime motivating feature of survival models is the capability to censor specific observations. Censoring in this sense generally relates to the time when information about the observation is part of the model. For example, suppose we are following two groups of cancer patients for a 10-year period. Patients are registered into the study upon diagnosis. One group – the *treatment group* – is given a new type of treatment. A *control group* is given the standard treatment. Patients are followed until either they die or the study closes. What happens, though, if a study patient withdraws from the study following eight years of participation? The patient has not died, but rather has moved away from the study area. They have contributed a substantial amount of information to the study, and we know that the patient has survived through eight years. Patients who withdraw in such a fashion are said to be *right-censored*. On the other hand, if another patient has had the event of study interest occur prior to actual entry into the study, they are said to be *left-censored.* Potential contributing information is lost to the study results.

Admittedly, left censoring rarely occurs with this type of data, and if it does, typically only a few cases are censored. Regardless, in a single study, patients can be both right- and left-censored, as well as not censored at all. Moreover, they can be censored at a variety of times. This situation is substantially different from the econometric notion of censoring in which a single cut-point defines censoring for all affected cases.

The majority of survival models have a continuous time response, as would be the case in the example above. However, there is a separate type of model called a *discrete response survival model.* Rather than having the response defined in terms of time, the response can be construed as counts. These types of models have traditionally been modeled as piecewise models, but can in fact be modeled using a count model that has the capability of accounting for censored observations. Censored Poisson and negative binomial models can be devised for such count response data. The models may be used for any type of count response, regardless of whether it is being used in a survival context. The only difference is that censoring – in the survival sense – is allowed as a capability.

The essential difference between the two approaches to censoring is that the survival parameterization considers censor points as observation-defined whereas the econometric parameterization considers them as dataset-defined. The econometric parameterization uses cut-points at specified values in the data set, with all values above or below the cuts censored. The survival parameterization is more general in that cut-points may be defined by observations above or below an assigned or specified value, but may also be defined for individual observations within the cuts. The econometric parameterization defines ranges of truncated and censored values whereas the survival parameterization defines only individual observations. Values are censored by virtue of their place in the data set, or values are censored by virtue of external reasons, e.g. lost to study, late entry, and so forth.

First employed by Hilbe and published in SAS by Hilbe and Johnston (1995), later published by Hilbe and Judson in Stata (1998) and revised by Hilbe (2005a), the survival parameterized censored Poisson log-likelihood appears as:

CENSORED POISSON LOG-LIKELIHOOD FUNCTION

$$\begin{aligned}\mathcal{L}(\beta; y) = \delta\left\{-\exp\left(x'_i\beta\right) + y_i\left(x'_i\beta\right) - \ln\Gamma\left(y_i+1\right)\right\} \\ +\xi\left\{\ln\Gamma_I\left(y_i, \exp\left(x'_i\beta\right)\right)\right\} \\ +\tau\left\{\ln\left(1 - \ln\Gamma_I\left(y_i+1, \exp\left(x'_i\beta\right)\right)\right)\right\}\end{aligned} \tag{12.12}$$

where δ: 1 if observation not censored; 0 otherwise
ζ: 1 if observation is left censored; 0 otherwise
τ: 1 if observation is right censored; 0 otherwise

and ln Γ_{I} is the 2-parameter incomplete gamma function. Note that the final term in equation 12.12 is a modification of the Poisson survival function with $\Gamma(y+1,\mu)$ being the numerator of the Poisson cumulative distribution function (CDF). The censored Poisson log-likelihood function can also be expressed in the same manner as given for the econometric or cut point parameterization above, with C indexed by observations, C_i. In this manner both parameterizations may be estimated using the same algorithm or function.

First derived and published in Hilbe (2005a), the survival parameterized censored negative binomial log-likelihood function is given as:

CENSORED NEGATIVE BINOMIAL LOG-LIKELIHOOD FUNCTION

$$\begin{aligned}\mathcal{L}(x\beta; y, \alpha) = {} & \delta\{y_i \ln(\exp(x_i\beta)/(1+\exp(x_i\beta))) - \ln(1+\exp(x_i\beta))/\alpha \\ & + \ln\Gamma(y_i + 1/\alpha) - \ln\Gamma(y_i+1) - \ln\Gamma(1/\alpha)\} \\ & + \zeta\{\ln(\beta_{\text{I}}(C_i, 1/\alpha, 1/(1+\exp(x_i'\beta - \ln(\alpha)))))\} \\ & + \tau\{\ln(\beta_{\text{I}}(C_i - 1, 1/\alpha, 1/(1+\exp(x_i'\beta - \ln(\alpha)))))\}\end{aligned} \tag{12.13}$$

Other terms are α = heterogeneity parameter and β_{I} = incomplete beta function. The 3 parameter β_{I} function returns the cumulative Beta distribution, or incomplete beta function, for censored responses. The function is identical to the econometric parameterization given in the last section, but with C indexed.

Using the **medpar** data used earlier in the text, we model length of hospital stay (*los*) on *white*, being a member of an HMO (*hmo*), and whether the patient is over 80 years of age (*age*80). A right censor indicator, *die*, was created from the variable, *died*, such that 1 specifies that the patient has died, 0 that the patient is alive, and −1 that the patient is lost from the study. Patients lost from the study after participating are, therefore, right censored. Values of −1 are randomly assigned from among patients who were alive.

Table 12.6 ***Medpar** censor variable*

die	Freq.	Percentage	Cum.
-1	58	3.88	3.88
0	924	61.81	65.69
1	513	34.31	100.00
	Total	1,495	100.00

A censored Poisson model (Hilbe, 2005a) on the data gives the following output:

```
. cpoisson los white hmo age90, cen(died)
Censored Poisson Regression                  Number of obs   =     1495
                                             Wald chi2(3)    =   193.88
Log likelihood = -4623.6239                  Prob > chi2     =   0.0000
------------------------------------------------------------------------
  los |     Coef.   Std. Err.       z   P>|z|   [95% Conf. Interval]
------+-----------------------------------------------------------------
white |-.2879192  .0326395   -8.82  0.000  -.3518915 -.2239469
  hmo | -.1753515  .027617   -6.35  0.000  -.2294799   -.121223
age80 | -.1865871  .0228685  -8.16  0.000  -.2314086  -.1417656
_cons |  2.901129  .0312814  92.74  0.000   2.839819   2.962439
------------------------------------------------------------------------
AIC Statistic =          6.191
```

Comparing the model with a standard Poisson model on the same data, we note that the parameter estimates and standard errors are somewhat similar. However, the Pearson *chi*2 dispersion is extremely high at 7.71.

```
. glm los white hmo ago80, fam(poi)
Generalized linear models              No. of obs        =       1495
Optimization      : ML                 Residual df       =       1491
                                       Scale parameter   =          1
Deviance          =  8800.483496       (1/df) Deviance   =   5.902403
Pearson           =  11490.80115       (1/df) Pearson    =   7.706775

                                       AIC               =   9.714805
Log likelihood    = -7257.816534       BIC               =   -2098.55
------------------------------------------------------------------------
      |           OIM
  los |    Coef.  Std. Err.      z    P>|z|   [95% Conf. Interval]
------+-----------------------------------------------------------------
white |-.1858699  .0273143   -6.80   0.000  -.2394049  -.1323349
  hmo |-.1448544   .023748   -6.10   0.000  -.1913997  -.0983091
age80 |-.0712421  .0203222   -3.51   0.000  -.1110729  -.0314113
_cons | 2.493478  .0260726   95.64   0.000   2.442377   2.544579
------------------------------------------------------------------------
```

Generally speaking, negative binomial models tend to dampen any overdispersion that may reside in a Poisson model. However, in so doing, the significance of various model predictors may be affected. Typically the negative binomial inflates the standard errors of overdispersed Poisson parameter estimates. This results in one or more predictors showing a non-contributory relationship to the model, whereas they appeared significant in the Poisson model. The same is the case with censored models.

```
. censornb los white hmo age80, cen(died)
Censored Negative Binomial Regression      Number of obs =      1495
                                           Wald chi2(3)  =    128.83
Log likelihood = -1981.6047                Prob > chi2   =    0.0000
----------------------------------------------------------------
  los   |      Coef.  Std. Err.      z   P>|z|  [95% Conf. Interval]
--------+-------------------------------------------------------
xb      |
white   |  .0026736  .1153047    0.02  0.982  -.2233194   .2286667
hmo     |-.4839241  .1019323   -4.75  0.000  -.6837077  -.2841404
age80   |-.7338938  .0814828   -9.01  0.000  -.8935971  -.5741904
_cons   | 4.027752  .1104592   36.46  0.000   3.811256   4.244248
--------+-------------------------------------------------------
lnalpha |
_cons   |  .7238855   .058532   12.37  0.000   .6091649   .8386061
--------+-------------------------------------------------------
alpha   | 2.062431  .1207182                   1.838895    2.31314
----------------------------------------------------------------
AIC Statistic =          2.656
```

Note also the markedly reduced AIC statistic. The censored negative binomial appears to be the preferred model of the two.

It may be of interest to compare the censored negative binomial, parameterized as an econometric model, with the survival parameterization. Since there is a substantial drop in values after *los* = 24, we shall set a cut at 24. For the survival model we shall create a right censor variable that has a code of −1 for *los* values of 24 or greater.

Table 12.7 R: *Right-censored negative binomial: cut = 24*

```
rm(list=ls())
data (medpar)
attach (medpar)
library(gamlss.cens)
library(survival)
rcmed <- medpar
cy <- with(rcmed, ifelse(los>=24, 23, los))
ci <- with(rcmed, ifelse(los>=24, 0, 1))
rcmed <- data.frame(rcmed, cy, ci )
rm(cy,ci)
gen.cens("NBI",type="right")
rcat24<-
gamlss(Surv(cy,ci)~white+hmo+age80,data=rcmed, family=NBIrc,n.cyc=100)
summary(rcat24)
```

CENSORED NB2: ECONOMETRIC

```
*******************************************************************
Family:  c("NBIrc", "right censored Negative Binomial type I")
-------------------------------------------------------------------
Mu link function:  log
Mu Coefficients:
              Estimate  Std. Error  t value    Pr(>|t|)
(Intercept)    2.40571    0.06421   37.466  4.678e-217
white         -0.13858    0.06676   -2.076   3.808e-02
hmo           -0.11229    0.05179   -2.168   3.031e-02
age80         -0.05183    0.04551   -1.139   2.549e-01
-------------------------------------------------------------------
Sigma link function:  log
Sigma Coefficients:
              Estimate  Std. Error  t value   Pr(>|t|)
(Intercept)    -0.8636    0.05645    -15.3  3.355e-49
-------------------------------------------------------------------
No. of observations in the fit:  1495
Degrees of Freedom for the fit:  5
      Residual Deg. of Freedom:  1490
                      at cycle:  3
Global Deviance:       9111.633
            AIC:       9121.633
            SBC:       9148.182
*******************************************************************
. di exp(-0.8636)
.42171203          <= alpha
```

CENSORED NB2: SURVIVAL

```
. gen rgtc= -1 if los>=24
. replace rgtc = 1 if rgtc==.
. tab rgtc

       rgtc |      Freq.     Percent        Cum.
------------+-----------------------------------
         -1 |         74        4.95        4.95
          1 |      1,421       95.05      100.00
------------+-----------------------------------
      Total |      1,495      100.00

. censornb los white hmo age80, cen(rgtc)

Censored Negative Binomial Regression      Number of obs =   1495
                                           Wald chi2(3)  =   5.13
Log likelihood = -4299.2669                Prob > chi2   = 0.1627
```

```
------------------------------------------------------------------------------
  los  |    Coef.  Std. Err.     z    P>|z|  [95% Conf. Interval]
-------+----------------------------------------------------------------------
xb     |
 white | -.1162715  .0635172  -1.83   0.067  -.2407629   .0082199
   hmo | -.0340281  .0480665  -0.71   0.479  -.1282368   .0601805
 age80 |  .0459967  .0421791   1.09   0.275  -.0366727   .1286661
 _cons |  2.238545  .0612939  36.52   0.000   2.118411   2.358679
-------+----------------------------------------------------------------------
lnalpha|
 _cons | -1.116643  .0524391 -21.29   0.000  -1.219422  -1.013865
-------+----------------------------------------------------------------------
 alpha |  .3273768  .0171673                  .2954008   .3628141
------------------------------------------------------------------------------
AIC Statistic =            5.757
```

The parameter estimates, standard errors, and αs are similar between the two parameterizations. The difference, of course, is that the survival parameterization has a substantially greater scope of censoring capabilities. Censoring can be anywhere in the data, not only at the tails.

The econometric and survival parameterizations of the right-censored Poisson show a similar list of parameter estimates and standard errors, with the exception of the binary predictor *age*80, for which the effects are reversed. However, since *age*80 is not a significant predictor in the econometric model, its effect should not be taken seriously. We use the censored Poisson commands published by the author for the analysis. Note the differences in AIC statistic between the two. The survival parameterization is a significantly better fit.

CENSORED POISSON: ECONOMETRIC

```
. gen cvar=1
. replace cvar= -1 if los>=24
. cpoissone los white hmo age80, censor(cvar) cright(24)

Censored Poisson Regression                   Number of obs   =      1495
                                              Wald chi2(3)    =     39.34
Log likelihood = -5980.2892                   Prob > chi2     =    0.0000
------------------------------------------------------------------------------
  los |      Coef.  Std. Err.      z   P>|z|   [95% Conf. Interval]
------+-----------------------------------------------------------------------
white | -.1282919     .02898   -4.43  0.000  -.1850916   -.0714922
  hmo | -.0947459   .0240634  -3.94  0.000  -.1419094   -.0475824
age80 | -.0286865   .0207154  -1.38  0.166   -.069288     .011915
_cons |  2.364869   .0277503  85.22  0.000   2.310479    2.419259
------------------------------------------------------------------------------
AIC Statistic = 8.006
```

CENSORED POISSON: SURVIVAL

```
. cpoisson los white hmo age80, censor(cvar)
Censored Poisson Regression                    Number of obs   =     1495
                                               Wald chi2(3)    =    20.65
Log likelihood = -5188.4206                    Prob > chi2     =   0.0001
--------------------------------------------------------------------------
  los |     Coef.  Std. Err.     z    P>|z|   [95% Conf. Interval]
------+-------------------------------------------------------------------
white |   -.11738    .031577  -3.72   0.000  -.1792698    -.0554903
  hmo |-.0346937   .0249728  -1.39   0.165  -.0836395     .0142521
age80 | .0477128   .0215142   2.22   0.027   .0055458     .0898799
_cons | 2.239266   .0302947  73.92   0.000   2.179889     2.298642
--------------------------------------------------------------------------
AIC Statistic =            6.946
```

Summary

Truncation and censoring primarily deal with how data is gathered. When there is missing data due to late entry into a study, or because data elements at a given point have been excluded from a study, we have truncation and censoring. Truncation occurs when we do not have knowledge of counts at either a point near the beginning or at the end of the counting process. In either case, the truncated data is actually excluded from the model. The probability function of either the Poisson or NB2 model is adjusted to account for the missing counts. Left truncated Poisson and negative binomial may be estimated with Stata's **tpoisson** and **tnbreg** commands respectively. Limdep, R's **gamlss** function, and the author's **trpoisson** Stata command provide both left and right truncation.

Censoring may be parameterized in either an econometric or a survival sense. In the former, censoring is similar to truncation, but the censored data to the outside of the cut-point(s) are revalued to that nearest value included in the model. For instance, if the cut-point is at 3, counts of 0, 1, and 2 are not dropped as in truncation, but are set to 3. Censored data sets can be identified with numerous values at either cut-point.

The survival parameterization of censoring allows censoring to take place anywhere in the data, not only at cut points, although such may be accommodated. This type of censoring is based on survival models such as Cox regression and the various parametric survival model. The survival parameterization of censored Poisson and negative binomial regression are currently available only as user authored Stata commands; refer to **cpoisson** and **censornb** respectively. Econometric parameterized censored models may be estimated using Limdep and **gamlss**, as well as with the authors's Stata command, **cpoissone**.

13

Handling endogeneity and latent class models

This chapter takes us into areas that have been primarily addressed in the econometric literature. However, the methods discussed are just as appropriate for research in biostatistics and medical health, and in social and political analysis. Recall that hurdle, zero-inflated, and selection models all partition the response distribution into two components. Zero-inflated models explain excessive zero counts using a binary generating process, which is mixed with a Poisson generating process, including zero. Hurdle models, on the other hand, are also partitioned into binary and count components, but the counts are generated by two separate processes with no overlapping mixture. Zero versus positive counts are formatted into a binary model and a zero-truncated count model generating the positive counts. We found that zero-inflated models employ traditional binary models such as logit and probit regression for understanding the zero counts, whereas the binary component of the hurdle model may be a right censored at 1 count model or a traditional binary model. The finite mixture models discussed in Section 11.1 extend hurdle models in such a manner that processes generating disparate types of counts may be identified.

We follow an examination of finite mixture models with an overview of the methods used to deal with the problem of endogenous predictors. Actually, endogeneity touches on finite mixture models, as well as on count models with unexplained heterogeneity in general, e.g. generalized Waring regression. I have found, however, that the concepts involved are more easily understandable following an evaluation of hurdle, zero-inflated and finite mixture models. Moreover, the notion of endogeneity is important to the understanding of random- and mixed-effects models, which we address in the following chapter.

We shall address quantile count models following our overview of finite mixture models and of models with endogeneity in general. Quantile count models relate to our discussion of finite mixture models, but also apply to counts in general, not only to specific exponential family distributions. However, since quantile models for count data are growing in use, it is worthwhile to present an overview of the method, comparing results with a negative binomial distribution.

13.1 Finite mixture models

13.1.1 Basics of finite mixture modeling

Finite mixture models assume that the response is not a single count distribution, but rather a mixture of two or more latent distributions. Studies have been performed proving that the optimal finite mixture models have either two or three component distributions. There are four standard types of finite mixture models – Poisson–Poisson, NB2–NB2, NB1–NB1, and geometric–geometric – however, mixtures of different distributions may also be constructed.

Perhaps the best way to envision a finite mixture model is to consider the count response as having been generated from $K = 1$ (or more) separate processes. Each process has a different mean. For a Poisson–Poisson finite mixture, counts can be generated from a Poisson distribution having a mean of $\lambda(1.25)$ and another Poisson distribution with a mean of $\lambda(3.75)$. Each component distribution consists of a given percentage or proportion of counts a_k, with the percentage of both sets of counts equaling $\sum_{k=1}^{K} a_k = 1$. The total mixture may be expressed as:

$$f(x) = \sum_{k=1}^{K} a_k f_k(x) \tag{13.1}$$

where $f_k(x)$ is the kth class distribution function and a_k indicates the proportion of total observations that is represented by the kth distribution. If there are two latent classes or distributions, $K = 2$. Each class has its own defined distribution.

Stata's user-authored **fmm** (Deb, 2007) and R's **gamlss** (Rigby) and **flexmix** (Leisch, Gruen) provide finite mixture modeling capability. Each function allows the user to specify distributions, how many mixture components are estimated, and the means and proportions of each. The software algorithm can select means and proportions if desired, and, unless one has previous knowledge of the distributions, it is wise to let the software estimate these values. **fmm** does not allow the user to mix distributions, only to mix the same distribution, but with different mean values. **Flexmix** uses an EM algorithm for estimation and provides substantial diagnostic output, but does not provide the ability to estimate negative binomial models. Type `?flexmix::flexmix` in the R editor to determine the supported distributions. **gamlss** allows the user to engage a broad variety of distributions, and is the software we shall employ for R examples.

Because of the differences in the algorithms used for estimation, and how mean values for component distributions are calculated, there is considerable variability in the estimated coefficients between functions. This should be kept in mind when comparing results.

We shall use the **azdrg112** data set that relates the hospital length of stay for patients having a CABG or PTCA (*typel*) heart procedure. The data come from the 1995 Arizona Medicare data for DRG (Diagnostic Related Group) 112. Other predictors include *gender* (1 = female) and *age*75 (1–age 75+). *Type* is labeled as 1 = emergency or urgent admission; 0 = elective. Length of stay (*los*) ranges from 1 to 53 days.

R: *NB2–NB2 finite mixture model using gamlss*

```
rm(list=ls())
data(azdrg112)
attach(azdrag112)
library(gamlss.mx)
fmNB <- gamlssNP(los~gender+type1+age75 data=azdrg112,
   random=~1, family=NBI, K=2)
summary(fmNB)
```

```
. fmm los gender typel age75, components(2)  mixtureof(negbin2)

2 component Negative Binomial-2 regression  Number of obs =    1798
                                            Wald chi2(6)   = 404.11
Log likelihood = -4211.0078                 Prob > chi2    = 0.0000
-----------------------------------------------------------------------
        los |     Coef.   Std. Err.      z    P>|z|  [95% Conf. Interval]
-----------------------------------------------------------------------
component1  |
     gender | .0953431   .0331238    2.88   0.004    .0304216   .1602646
      typel | .6394801   .0372654   17.16   0.000    .5664413   .7125188
      age75 | .1387465   .0345786    4.01   0.000    .0709736   .2065194
      _cons | .8853407   .0372932   23.74   0.000    .8122474    .958434
-----------+-----------------------------------------------------------
component2  |
     gender | .2681655   .0922572    2.91   0.004    .0873448   .4489862
      typel | .5213427   .1135142    4.59   0.000     .298859   .7438264
      age75 | .0735899   .1012519    0.73   0.467   -.1248602     .27204
      _cons | 1.675311   .1702868    9.84   0.000    1.341555   2.009067
-----------+-----------------------------------------------------------
```

```
/imlogitpi1| 1.724653  .3768203   4.58  0.000   .9860983 2.463207
 /lnalpha1 |-3.263588  .4674468  -6.98  0.000  -4.179767 -2.347409
 /lnalpha2 |-1.460399  .2595833  -5.63  0.000  -1.969173 -.9516248
-----------+-----------------------------------------------------
    alpha1 | .0382509  .0178803                 .0153021  .0956165
    alpha2 | .2321437  .0602606                 .1395723  .3861132
       pi1 | .8487271  .0483797                 .7283166  .9215219
       pi2 | .1512729 .0483797                  .0784781  .2716834
-----------------------------------------------------------------

. abic
AIC Statistic    =    4.696338           AIC*n       = 8444.0156
BIC Statistic    =    4.713442           BIC(Stata)  = 8504.4541
```

The AIC and BIC statistics displayed in the right column above are the same values as using **estat ic** after estimation. I prefer to have all four values, therefore use the *abic* command in its place. Those using R may obtain the same values using the **modelfit** function.

Interpretation of coefficients is the same as for standard NB2 models, except that consideration for the component must be taken. Exponentiate to obtain incidence rate ratios. The two NB2 component distributions consist of 85% (*pi*1) and 15% (*pi*2) of the total observations respectively. The mean predicted values of the two components are calculated as:

```
. predict mean1, equation(component1)
. predict mean2, equation(component2)
. su mean*

Variable |    Obs       Mean      Std. Dev.       Min       Max
---------+-----------------------------------------------------
   mean1 |    1798    4.138442    1.196006     2.42381   5.806108
   mean2 |    1798    8.771076    2.420452    5.340455   12.65951
```

The second component distribution has a mean value over twice that of the first. However, since it is only 15% of the combined distribution, the overall distributional mean value is not appreciably different (4.64) from the mean value of the first component.

```
. di .8487*4.1384 + .15127*8.771
4.8390493
```

Average marginal effects and discrete change values may also be calculated following the estimation of a finite model. It is important to partition the effects by their respective components. Interpretation is the same as usual, except the effects relate to component values. Note that there is a considerable difference in gender effects between the two components. The type of procedure also differs. Older patients do not contribute to the smaller distribution.

MARGINAL EFFECTS/DISCRETE CHANGE

```
. margins, dydx(*) predict(equation(component1))

Average marginal effects                      Number of obs   =   1798
Model VCE     : OIM
Expression    : predicted mean: component1,predict(equation
                (component1))
dy/dx w.r.t. : gender type1 age75
-----------------------------------------------------------------
       |          Delta-method
       |     dy/dx   Std. Err.      z    P>|z|   [95% Conf. Interval]
-------+---------------------------------------------------------
gender | .3945719    .1387753    2.84   0.004    .1225774   .6665665
 type1 | 2.646451    .1892651   13.98   0.000    2.275499   3.017404
 age75 | .5741945    .1428388    4.02   0.000    .2942355   .8541535
-----------------------------------------------------------------

. margins, dydx(*) predict(equation(component2))
Average marginal effects                      Number of obs   =   1798
Model VCE     : OIM
Expression    : predicted mean: component2, predict(equation
                (component2))
dy/dx w.r.t. : gender type1 age75
-----------------------------------------------------------------
       |          Delta-method
       |     dy/dx   Std. Err.      z    P>|z|   [95% Conf. Interval]
-------+---------------------------------------------------------
gender |    2.3521   .8755292    2.69   0.007    .6360942   4.068106
 type1 | 4.572736    .9794754    4.67   0.000       2.653   6.492473
 age75 | .6454625    .8945142    0.72   0.471   -1.107753   2.398678
-----------------------------------------------------------------
```

It may be of interest to use R's **flexmix** command to construct a Gaussian–Poisson mixture distribution. Notably the distributions will not be easy to pull apart. The output is not displayed here, but is simple to recreate.

Table 13.1 R: *Gaussian–Poisson finite mixture model*

```
rm(list=ls())     # deletes all objects; use only with care
library(COUNT)
library(flexmix)
```

```
data(rwm5yr)
attach(rwm5yr)
factor(edlevel)
fmm_pg <- flexmix(docvis~outwork+edlevel+age, data=rwm5yr, k=2,
        model=list(FLXMRglm(docvis~., family="gaussian"),
                   FLXMRglm(docvis~., family="poisson")))
parameters(fmm_pg, component=1, model=1)
parameters(fmm_pg, component=2, model=1)
summary(fmm_pg)
```

13.1.2 Synthetic finite mixture models

Synthetic finite mixture models may be constructed to provide a better understanding of the relationship between the component distributions. We construct both Poisson–Poisson and NB2–NB2 synthetic finite mixture models with specified coefficient values for each distribution. The algorithms also allow the proportion of each component to be given.

For the Poisson–Poisson synthetic finite mixture model we define each component distribution as:

```
COMPONENT 1: intercept= 1   x1= .25   x2= -0.75  proportion= 0.1
COMPONENT 2: intercept= 2   x1= .75   x2= -1.25  proportion= 0.9
```

Table 13.2 R: *Synthetic Poisson–Poisson finite mixture model*

```
library(flexmix)
nobs <- 50000
x1 <- runif(nobs)
x2 <- runif(nobs)
xb1 <- 1 + 0.25*x1 - 0.75*x2
xb2 <- 2 + 0.75*x1 - 1.25*x2
exb1 <- exp(xb1)
exb2 <- exp(xb2)
py1 <-  rpois(nobs, exb2)
py2 <-  rpois(nobs, exb1)
poixpoi <- py2
poixpoi <- ifelse(runif(nobs) > .9, py1, poixpoi)
pxp <- flexmix(poixpoi ~ x1 + x2, k=2,
   model=FLXMRglm(family="poisson"))
summary(pxp)
parameters(pxp, component=1, model=1)
parameters(pxp, component=2, model=1)
```

```
. clear
. set obs 50000
. gen x1 = runiform()
. gen x2 = runiform()
. gen xb1 = 1 + 0.25*x1 - .75*x2
. gen xb2 = 2 + 0.75*x1 - 1.25*x2
. gen exb1=exp(xb2)
. gen exb2=exp(xb1)
. gen py1 = rpoisson(exb2)
. gen py2 = rpoisson(exb1)
. su py*
-------------------------------------------------------------------
Variable |  Obs        Mean   Std. Dev.    Min     Max
---------+---------------------------------------------------------
     py1 |50000    2.18006    1.56282       0      12
     py2 |50000    6.27174   3.659017       0      31
-------------------------------------------------------------------

. rename py2 poixpoi
. replace poixpoi = py1 if runiform() > .9

. fmm poixpoi x1 x2, components(2) mixtureof(poisson)

2 component Poisson regression                Number of obs   =     50000
                                              Wald chi2(4)    = 40415.33
Log likelihood = -117724.82                   Prob > chi2     =    0.0000
-------------------------------------------------------------------
    poixpoi   |    Coef.  Std. Err.      z  P>|z|  [95% Conf. Interval]
-------------+-----------------------------------------------------
component1   |
         x1  | .2532785   .056633     4.65 0.000  .1522798   .3742772
         x2  |-.7486015  .0621454   -11.56 0.000 -.8404044  -.5967987
      _cons  |  1.00564  .0450579    22.32 0.000  .9173277   1.093952
-------------+-----------------------------------------------------
component2   |
         x1  | .7514944  .0070839   104.43 0.000  .7259101   .7536786
         x2  |-1.252912  .0073356  -169.70 0.000  -1.25919  -1.230435
      _cons  | 2.001304  .0054573   366.91 0.000  1.991608   2.013001
-------------+-----------------------------------------------------
 /imlogitpi1 |-2.174919  .0395665   -54.97 0.000 -2.252468   -2.09737
-------------+-----------------------------------------------------
        pi1  | .1020255  .0036249                .0951368   .1093527
        pi2  | .8979745  .0036249                .8906473   .9048632
-------------------------------------------------------------------
```

Developing an NB2–NB2 finite mixture model is a bit more complicated, but the logic is the same as for the Poisson–Poisson mixture model

Table 13.3 R: *Synthetic NB2–NB2 finite mixture model*

```
library(gamlss.mx)
nobs <- 50000
x1 <- (runif(nobs))
x2 <- qnorm(runif(nobs))
xb1 <- 1 + .25*x1 - .75*x2
xb2 <- 2 + .75*x1 - 1.25*x2
a1 <- .5
a2 <- 1.5
ia1 <- 1/a1
ia2 <- 1/a2
exb1 <- exp(xb1)
exb2 <- exp(xb2)
xg1 <- rgamma(nobs, a1, a1, ia1)
xg2 <- rgamma(nobs, a2, a2, ia2)
xbg1 <-exb1*xg1
xbg2 <-exb2*xg2
nby1 <- rpois(nobs, xbg1)
nby2 <- rpois(nnobs, xbg2)
nbxnb <- nby2
nbxnb <- ifelse(runif(nobs) > .9, nby1, nbxnb)
fmmnb <- data.frame(nbxnb, x1, x2)
nxn <- gamlssNP(nbxnb~x1+x2,random=~1,family=NBI,K=2,data=fmmnb)
summary(nxn)
```

Table 13.4 *NB2–NB2 Finite mixture model*

```
. clear
. set obs 50000
. set seed 7455
. gen x1 = runiform()
. gen x2 = runiform()
. gen xb1 = 2 + 0.75*x1 - 1.25*x2
. gen xb2 = 1 + 0.25*x1 - .75*x2
. gen a1 = .5
. gen a2 = 1.5
. gen ia1=1/a1
. gen ia2=1/a2
. gen exb1 = exp(xb1)
. gen exb2 = exp(xb2)
. gen xg1 = rgamma(ia1, a1)
. gen xg2 = rgamma(ia2, a2)
```

```
. gen xbg1 = exb1*xg1
. gen xbg2 = exb2*xg2
. gen nby1 = rpoisson(xbg1)
. gen nby2 = rpoisson(xbg2)
. rename nby1 nbnb
. replace nbnb = nby2 if runiform() > .8  ///2nd comp=.8
```

```
. su xbg* nby*
----------------------------------------------------------------
Variable |   Obs      Mean     Std. Dev.       Min         Max
---------+------------------------------------------------------
    xbg1 |  50000  6.289644   5.510383   .0232447    64.74759
    xbg2 |  50000  2.170705   2.774493   4.57e-07    35.69936
    nby1 |  50000   6.28498   6.074539          0          80
    nby2 |  49997   2.17023   3.156802          0          41
----------------------------------------------------------------

. fmm nbnb x1 x2, components(2) mixtureof(negbin2)

2 component Negative Binomial-2 regression    Number of obs =   49999
                                              Wald chi2(4)  = 3143.43
Log likelihood = -134864.74                   Prob > chi2   =  0.0000
----------------------------------------------------------------------
    nbnb |    Coef.   Std. Err.      z    P>|z|   [95% Conf. Interval]
---------+------------------------------------------------------------
component1 |
      x1 |  .2593626  .1864768    1.39  0.164   -.1061253   .6248504
      x2 | -.8829247  .1445243   -6.11  0.000   -1.166187  -.5996622
   _cons |  .9267742  .4281033    2.16  0.030    .0877071   1.765841
---------+------------------------------------------------------------
component2|
      x1 |  .7611433  .0282369   26.96  0.000      .7058    .8164867
      x2 | -1.258077  .0258545  -48.66  0.000   -1.308751  -1.207403
   _cons |  2.008653  .0286862   70.02  0.000    1.952429   2.064877
---------+------------------------------------------------------------
/imlogitpi1 |-1.380332 .4230962  -3.26  0.001   -2.209585  -.5510784
 /lnalpha1 |  .3334062 .1101137    3.03  0.002    .1175874    .549225
 /lnalpha2 | -.719585  .0506254  -14.21  0.000   -.818809  -.6203611
---------+------------------------------------------------------------
  alpha1 |  1.395714  .1536872                     1.12478    1.73191
  alpha2 |  .4869543  .0246523                    .4409565   .5377502
     pi1 |  .2009557  .0679376                     .098893   .3656143
     pi2 |  .7990443  .0679376                    .6343857    .901107
```

13.2 Dealing with endogeneity and latent class models

13.2.1 Problems related to endogeneity

Discussions of endogeneity seem to occur primarily in econometric literature, including research related to health outcomes. In fact, much of the current discussion involving endogeneity, finite mixture models, and hurdle models has focused on health outcomes research. These methods have yet to find a solid place in biostatistics, traditional social statistics, or other areas of statistical specialization. However, this is not to say that these methods cannot be used effectively in research outside econometrics. In fact, negative binomial regression itself appears to have been used in econometric research long before its use in areas such as biostatistics and medical statistics, where it has now become a standard model.

Authors such as Cameron and Trivedi (2005, 2009), Winkelmann (2008), and Jones *et al.* (2007) have provided the statistical community with excellent analyses of endogeny and of the nature of endogenous regressors. This discussion relates to the field of *latent class models*, which will be touched on here and in the following chapter. I will not attempt to duplicate what has already been presented on the subject, but will instead provide an overview of the subject and show how the concepts involved can be employed outside econometric research.

In econometric literature, the observed explanatory predictors in a model are commonly referred to as *exogenous variables*. We earlier referred to the explanatory predictors of foremost interest as risk factors, and the remaining predictors as confounders. Corresponding to exogenous variables, econometricians typically refer to the model response or dependent variable(s) as endogenous. The endogenous variable(s) of a statistical model, given the set of explanatory predictors, are determined on the basis of an underlying probability distribution, or mixture of distributions. In the case of quasi-likelihood models, which do not necessarily have a PDF substructure, a created log-likelihood function takes the role of a PDF-based log-likelihood and, for the purposes of defining endogeny and exogeny, can be regarded as the same.

In Chapter 7 we showed the consequence of omitting a variable that is essential to explaining a given model response. Exclusion of such an unobserved variable leads to model overdispersion – also called heterogeneity. Usually the necessary variable is not collected, or is simply unknown. We are aware of it by virtue of otherwise unexplained model overdispersion. It may also be the case that more than one unobserved variable is involved; however, the unexplained heterogeneity may be collapsed to a single unobserved variable, also called an endogenous variable. It is called this because of its conjoined relationship with the response, and its independence from other observed or exogenous predictors.

The terms exogenous and endogenous have unique meanings when applied to statistical models. The standard or classical types of explanatory predictors that one employs in a count model are termed *exogenous*, meaning that they are independent of the regression error term. *Endogenous* predictors, on the other hand, are correlated with the error. Consequently they are jointly estimated with the response and result in biased estimates; this problem is sometimes referred to as *endogeny*. Endogenous predictors are related to instrumental predictors, usually called *instrumental variables* (IVs). Instrumental variables are used to adjust for the bias caused by inclusion of endogenous predictors into a model. They are not themselves independent of the response.

To reiterate, the problem of endogeny occurs when independent predictor(s) are correlated with the model error term and therefore with the response. When this occurs, model coefficients are biased. Essentially anytime a predictor may not be fully or correctly specified, it is endogenous. Statisticians have devised a number of ways of handling endogeny, including the use of instrumental variable regression, generalized method of moments, structural models, and maximum likelihood simulated estimation, to name a few. Since there are many excellent texts on econometrics that deal with these types of model, we shall not address them all here, or in detail. In this section I shall simply provide several examples of how endogeny can be accommodated or adjusted in a simple count model context. Keep in mind that, owing to the complexity of these types of models, and the myriad variations and difficulties that relate to each approach, it is likely best to refer to specialty texts when wanting to engage in this type of modeling (Cameron and Trivedi, 1998, 2009; Winkelmann, 2007; Jones *et al.*, 2007). We shall begin with the basic approach of including instrumental variables in two-stage regression. The instruments are aimed at adjusting the model to include the endogenous predictor.

13.2.2 Two-stage instrumental variables approach

We use the **rwm** data set, which is a more complete version of the German health reform data we have used thus far in the book. We model visits to the doctor per year as the response, with a number of predictors. We select *outwork*, i.e. the patient is out of work, as the endogenous variable, and assume that being out of work will bear on the decision to seek medical help; *outwork* is therefore closely related to the number of visits made, and is independent of other predictors such as *age*, educational level (*edlevel*), whether health *reform* has been passed, and marital status (*married*).

Table 13.5 *Expanded German pre- and post-reform health data*

```
Response : docvis
Exogenous: reform [1=post health care reform (>1990)
                   0=pre health care reform (1984-1988)]
           age [ages 25-64]
           married [1;0]
           edlevel [educational level 1-4 [1=not HS grad;
                   2=HS grad; 3=Coll/Univ; 4 Grad school]
Endogenous: outwork [1;0] patient is unemployed
Instrumental variables: bluec [1;0] patient a blue collar
                   worker female [1;0] patient is female
```

Endogeneity refers to the situation when there is an independent predictor in the model that is correlated with an unobservable predictor or predictors that are latently included in the regression error term. Again, we suppose that *outwork* is endogenous. The instrumental variables, *bluec* (blue collar) and *female,* are correlated with *outwork*, but are assumed uncorrelated with the error term and do not affect the number of patient visits to the doctor, as adjusted by the model predictors.

For this example, *outwork* is correlated *bluec* (0.35) and with *female* (0.39).

Stage 1: Model binary endogenous predictor

```
. global exog reform age married edlevel2-edlevel4
. regress outwork $exog bluec female, vce(robust)

----------------------------------------------------------------------
             |               Robust
     outwork |      Coef.   Std. Err.      t    P>|t|   [95% Conf. Interval]
-------------+--------------------------------------------------------
      reform |  -.0640675   .0053406  -12.00   0.000   -.0745353   -.0535996
         age |   .0071279   .0002301   30.98   0.000    .0066769    .0075788
     married |  -.0000639   .0059944   -0.01   0.991   -.0118133    .0116855
    edlevel2 |   -.107084   .0107629   -9.95   0.000   -.1281798   -.0859881
```

```
edlevel3 | -.0330511  .0098303  -3.36  0.001   -.0523189 -.0137832
edlevel4 | -.2257406   .008775 -25.73  0.000   -.2429399 -.2085412
   bluec | -.3212469  .0048546 -66.17  0.000   -.3307623 -.3117316
  female |  .2745116  .0052845  51.95  0.000    .2641537  .2848694
   _cons |  .0031371  .0117838   0.27  0.790   -.0199596  .0262339
------------------------------------------------------------------
```

The two instruments are both significant predictors of the endogenous predictor, *outwork*. Blue collar workers were more likely to be working and females more likely not.

Next obtain the residuals, or errors, from an OLS regression. They will be used as the adjustor for endogeneity in the count model.

```
. predict endores, residual
```

Stage 2: Negative binomial with endogenous regressor

At the second stage, we employ a count regression model of our choice, including the endogenous predictor, but excluding the two instruments. Standard errors should be based on a robust or sandwich variance estimator.

NB2: ADJUSTED STANDARD ERRORS

```
. nbreg docvis outwork $exog endores, vce(robust)

Negative binomial regression                 Number of obs   =    27326
Dispersion          = mean                   Wald chi2(8)    = 1030.58
Log pseudolikelihood = -60193.249            Prob > chi2     =   0.0000
------------------------------------------------------------------
         |            Robust
  docvis |     Coef.  Std. Err.     z    P>|z|   [95% Conf. Interval]
---------+--------------------------------------------------------
 outwork |  .6343118  .0495773  12.79  0.000    .5371421  .7314816
  reform |   .077695  .0225299   3.45  0.001    .0335371  .1218528
     age |  .0154656  .0010119  15.28  0.000    .0134823  .0174489
 married | -.1043179  .0260301  -4.01  0.000    -.155336 -.0532999
edlevel2 |   -.05766  .0380269  -1.52  0.129   -.1321914  .0168714
edlevel3 | -.2175772  .0352558  -6.17  0.000   -.2866773 -.1484771
edlevel4 | -.3702709  .0434191  -8.53  0.000   -.4553707  -.285171
 endores | -.4203726  .0555943  -7.56  0.000   -.5293355 -.3114097
   _cons |   .332742  .0478728   6.95  0.000     .238913  .4265709
---------+--------------------------------------------------------
/lnalpha |  .6462406  .0136388                  .6195091  .6729721
---------+--------------------------------------------------------
   alpha |  1.908353  .0260276                  1.858016  1.960054
```

The residual term, *endores*, is a significant but negatively related predictor of doctor visits. If it were not a significant predictor, then it would be advisable to bootstrap the standard errors instead of applying a robust variance estimator. Such a choice is executed by substituting *vce(boot)* for *vce(robust)* on the command line; *vce(boot)* also is available for use with the **glm** command.

The fact that *endores*, the residual or error term, is negative indicates that the underlying latent factor that increases the probability of being out of work has a negative relationship on the number of yearly visits made to the doctor.

The table of NB2 estimates, without adjustment by the two instruments, is given as

NB2: MODEL STANDARD ERRORS

```
. glm docvis outwork $exog, fam(nb ml) nohead
------------------------------------------------------------------------
         |                 OIM
  docvis |      Coef.   Std. Err.      z    P>|z|     [95% Conf. Interval]
---------+--------------------------------------------------------------
 outwork |   .3017821   .0200451    15.06   0.000     .2624944    .3410698
  reform |   .0570794   .0203072     2.81   0.005      .017278    .0968808
     age |   .0184986   .0008441    21.92   0.000     .0168442     .020153
 married |  -.1155529   .0218417    -5.29   0.000    -.1583617    -.072744
edlevel2 |  -.0552856   .0379256    -1.46   0.145    -.1296184    .0190471
edlevel3 |  -.2075875   .0323301    -6.42   0.000    -.2709534   -.1442216
edlevel4 |  -.4253496    .037918   -11.22   0.000    -.4996675   -.3510318
   _cons |   .3284203   .0397335     8.27   0.000     .2505441    .4062966
------------------------------------------------------------------------
```

Note the difference between the coefficients and standard errors of *outwork*. For the adjusted model they are 0.6343 (0.0496) and for the unadjusted model 0.3018 (0.0200). The endogenous model coefficient is 2.1 times greater than the unadjusted model. Model standard errors for both are nearly identical, but the robust variance inflates the value nearly 2 and a half times.

We expect that the value of the endogenous predictor coefficient will be inflated. This tends to occur when the instruments are only moderately correlated with the endogenous predictor(s). Recall that *bluec* and *female* were correlated with *outwork* at 0.35 and 0.39 respectively. In any case, *outwork* as endogenous has a much larger effect on the model. Being out of work leads to a 100 *(exp(0.6343118) − 1) = 88.57% increase in doctor visits per year compared with a 35.23% increase in doctor visits for the unadjusted model.

Finally, although α is not displayed in the above output for the unadjusted model, its value is higher than the value of α in the endogenous model, albeit not significantly higher. Nevertheless, the statistics given in the example model are consistent with what we would expect in a latent class endogenous model.

For additional information regarding this method of controlling for endogeneity, see Cameron and Trivedi (2009).

13.2.3 Generalized method of moments (GMM)

The generalized method of moments approach is commonly known by its acronym, GMM. GMM and the structural equation approach discussed in the previous section are the two foremost methods of handling endogeneity through the use of instrumental variables. The primary difference between the two methods with respect to count models is that, whereas the endogenous term is linearly defined in the above structural approach, it may be non-linearly characterized using GMM.

We shall model the same data with GMM as we illustrated using the structural equation approach.

```
. use rwm
. global exog reform age married edlevel2-edlevel4
. gmm (docvis - exp({xb: outwork $exog} + {b0})),
  instruments(bluec female $exog) twostep nolog

Final GMM criterion Q(b) =  .0019724

GMM estimation
Number of parameters =  8
Number of moments    =  9
Initial weight matrix: Unadjusted                 Number of obs = 27326
GMM weight matrix:     Robust
------------------------------------------------------------------------------
             |               Robust
             |      Coef. Std. Err.      z    P>|z|    [95% Conf. Interval]
-------------+----------------------------------------------------------------
 /xb_outwork |   .5301813   .042979   12.34   0.000    .4459439    .6144187
  /xb_reform |   .0660372  .0228462    2.89   0.004    .0212595     .110815
     /xb_age |   .0163065  .0010342   15.77   0.000    .0142794    .0183335
 /xb_married |  -.1208362  .0257775   -4.69   0.000   -.1713592   -.0703131
/xb_edlevel2 |  -.0698137  .0380924   -1.83   0.067   -.1444734    .0048459
/xb_edlevel3 |  -.2279121  .0350233   -6.51   0.000   -.2965565   -.1592676
/xb_edlevel4 |  -.3797723  .0456862   -8.31   0.000   -.4693157   -.2902289
         /b0 |   .3259051  .0483991    6.73   0.000    .2310446    .4207657
------------------------------------------------------------------------------
Instruments fo equation 1: bluec female reform age married
  edlevel2 edlevel3 edlevel4 _cons
```

The test for overidentifying restrictions is also referred to *Hansen's test*, *Sargan's test*, and the *Hansen–Sargan test*. *Overidentification* with respect to GMM models relates to the situation when there is one more instrument than there are predictors in the model. The test statistic is distributed as *chi*2 with one degree of freedom, given the single added instrument. Rejection of the test indicates that at least one instrument is invalid. A *p*-value of under 0.05 specifies that the null hypothesis is not rejected. Note that the weight matrix can be retrieved using *e*(*W*).

```
. estat overid

  Test of overidentifying restriction:
  Hansen's J chi2(1) = 53.8989 (p = 0.0000)

Final GMM criterion Q(b) =  .0682142

GMM estimation
Number of parameters =    8
Number of moments    =    9
Initial weight matrix: Unadjusted                   Number of obs = 27326
-----------------------------------------------------------------------------
             |              Robust
             |      Coef.  Std. Err.      z    P>|z|  [95% Conf. Interval]
-------------+---------------------------------------------------------------
  /xb_outwork |  .5361135 .0427501   12.54   0.000  .4523248    .6199021
   /xb_reform |  .0649832 .0227588    2.86   0.004  .0203768    .1095896
      /xb_age |  .0156943 .0010307   15.23   0.000  .0136742    .0177143
  /xb_married | -.1046109 .0259549   -4.03   0.000 -.1554815   -.0537403
 /xb_edlevel2 |  -.061699 .0376805   -1.64   0.102 -.1355515    .0121534
 /xb_edlevel3 | -.2211948 .0347378   -6.37   0.000 -.2892797   -.1531098
 /xb_edlevel4 | -.3577878 .0448471   -7.98   0.000 -.4456865    -.269889
          /b0 |  .3421576 .0481893    7.10   0.000  .2477083     .436607
-----------------------------------------------------------------------------
Instruments for equation 1: bluec female reform age married
   edlevel2 edlevel3 edlevel4 _cons
```

13.2.4 NB2 with an endogenous multinomial treatment variable

It often occurs in research that the presumed endogenous predictor is not a binary selection variable, but is rather distributed as multinomial. A variable may have order to the various levels and still be considered as if it were orderless. Of course information is lost, but if this is kept in mind we may nevertheless model it as multinomial.

Deb and Trivedi (2006) developed a negative binomial model with a multinomial endogenous treatment. Instruments which influence the treatment are incorporated into the algorithm. The authors express the model as:

$$E\left(y_i|d_i, x_i, l_i\right) = x_i'\beta + \sum_{j=1}^{J} \gamma_j d_{ij} + \sum_{j=1}^{J} \lambda_j l_{ij} \tag{13.2}$$

where x is the set of exogenous predictors, β the associated parameters, γ the endogenous treatment predictor, d the binary treatment choices, λ the negative binomial mean, and l latent factors that include unobserved information regarding the treatment that is not explicitly available for analysis.

The algorithm, written in Stata and C, models the data using a simulated log-likelihood function. To speed up the process, the authors use quasi-random draws from Halton sequences (Deb and Trivedi).

We continue to use data from the German health reform data (1984–1993) as an example of how to employ this type of model into your own work. Two hundred Halton draws are declared, and a robust variance estimator is used for standard errors. Robust, or sandwich, standard errors should normally be used for models such as this. The variables used in the model are:

Response: *docvis*
Endogenous: *edlevel* (educational level – 4 levels; level 1 the reference)
Instruments: *outwork*, *public* (public insurance)
Exogenous: *reform*, *age*, *female*, *bluec*, *hhinc* (income)

```
. mtreatnb docvis reform age female married bluec hhninc,
  mtreat(edlevel age female married bluec hhninc outwork public)
  sim(200) basecat(1) robust

Multinomial treatment-effects NB regression  Number of obs=    27326
                                             Wald chi2(30)= 6026.11
Log pseudolikelihood = -77803.013            Prob > chi2   =  0.0000
---------------------------------------------------------------------
            |               Robust
            |    Coef.   Std. Err.    z    P>|z|  [95% Conf. Interval]
------------+--------------------------------------------------------
_outcome_2  |
        age |-.0370231   .0027425  -13.50  0.000  -.0423983  -.031648
     female | .2794096   .0600008    4.66  0.000   .1618102   .397009
    married | .0068414   .0697608    0.10  0.922  -.1298872  .1435701
      bluec |-1.961017   .1045334  -18.76  0.000  -2.165898 -1.756135
     hhninc | .1939782   .0164152   11.82  0.000    .161805  .2261514
```

```
     outwork | -.3903107  .0679958   -5.74  0.000    -.52358 -.2570415
      public | -.3598509  .0828719   -4.34  0.000  -.5222768 -.1974249
       _cons | -1.413144  .1468447   -9.62  0.000  -1.700954 -1.125333
-------------+----------------------------------------------------------
_outcome_3   |
         age |  -.052142  .0025012  -20.85  0.000  -.0570441 -.0472398
      female | -.9527259  .0552585  -17.24  0.000   -1.06103 -.8444213
     married | -.8925594  .0537059  -16.62  0.000   -.997821 -.7872977
       bluec |  -2.51009  .0970663  -25.86  0.000  -2.700337 -2.319844
      hhninc |  .2277653  .0149329   15.25  0.000   .1984973  .2570333
     outwork |  .0907913   .060826    1.49  0.136  -.0284255  .2100081
      public | -.5566288  .0704043   -7.91  0.000  -.6946187  -.418639
       _cons |  .7900667    .12816    6.16  0.000   .5388777  1.041256
-------------+----------------------------------------------------------
_outcome_4   |
         age |  -.027876  .0028511   -9.78  0.000   -.033464  -.022288
      female | -.8497098  .0644874  -13.18  0.000  -.9761028 -.7233167
     married | -.8621573  .0698411  -12.34  0.000  -.9990434 -.7252712
       bluec | -4.021489  .2494866  -16.12  0.000  -4.510474 -3.532505
      hhninc |   .328774  .0176885   18.59  0.000   .2941051  .3634428
     outwork | -.9330238  .0844161  -11.05  0.000  -1.098476 -.7675712
      public | -1.941322  .0651195  -29.81  0.000  -2.068954  -1.81369
       _cons |  .1871402  .1426241    1.31  0.189  -.0923979  .4666783
-------------+----------------------------------------------------------
docvis
I_outcome_2 | -.0981386  .1054356   -0.93  0.352  -.3047885  .1085114
I_outcome_3 | -.1334312  .0712231   -1.87  0.061  -.2730259  .0061634
I_outcome_4 | -.8512805  .0461607  -18.44  0.000  -.9417537 -.7608072
     reform |  .1361852  .0214147    6.36  0.000   .0942132  .1781572
        age |  .0224237  .0008883   25.24  0.000   .0206827  .0241646
     female |  .3610168  .0207591   17.39  0.000   .3203297  .4017038
    married | -.0644162  .0243882   -2.64  0.008  -.1122163 -.0166161
      bluec | -.0901635  .0267307   -3.37  0.001  -.1425547 -.0377722
     hhninc | -.0335631  .0060921   -5.51  0.000  -.0455035 -.0216228
      _cons | -.0379665  .0563887   -0.67  0.501  -.1484864  .0725534
-------------+----------------------------------------------------------
    /lnalpha |  .2887962  .0271657   10.63  0.000   .2355524  .3420401
/lambda__o~2|  .0888935  .1083747    0.82  0.412   -.123517   .301304
/lambda__o~3|  .0368572  .0692882    0.53  0.595  -.0989452  .1726596
/lambda__o~4|  .6575081  .0270845   24.28  0.000   .6044234  .7105928
------------|-----------------------------------------------------------
      alpha| |  1.33482  .0362613                         1.265608  1.407817
-------------------------------------------------------------------------
Notes:
1. _outcome_1 is the base outcome
2. 200 Halton sequence-based quasi-random draws per observation
```

The instrumental variables, *outwork* and *public* are significant predictors in all but the third level where *outcome* is not a significant predictor. Those on public insurance, i.e. in the national health care system, have lower levels of education. Those who are unemployed likewise have lower levels of education than those who are working. Both of these conclusions make intuitive sense.

The negative binomial model is displayed in the *docvis* frame. Higher levels of education are inversely proportional to visits to the doctor. That is, those who are higher educated tend to visit the doctor less frequently. However, educational levels 1 and 2 are statistically the same, and level 3 is questionable ($p = 0.061$). Valid conclusions are based primarily on those with graduate-level education. The other predictors are interpreted in the normal manner for a negative binomial regression.

The bottom frame provides information regarding the latent factors implicit in the treatment levels. Only treatment level 4, graduate-level education, is a significant predictor of doctor visits. Patients with graduate degrees are more likely to visit the doctor than those who are not secondary school graduates. This gives a contrary relationship to doctor visits than was indicated in the negative binomial model. The conclusion is that the levels need to be collapsed, then re-estimated.

Deb and Trivedi (2006) suggest a constrained log-likelihood test for the model. The *constrained log-likelihood* is defined as twice the difference between the multinomial treatment-effects negative binomial model and a standard negative binomial regression on the data. The *p*-value is based on 1 – *chi*2 (dof = 3), given 4–1 educational levels as the treatment variable. In Stata the commands are:

```
. scalar L = 2*(e(ll) - e(ll_exog))    // constrained likelihood ratio test
. di L
. di 1 - chi2(3,L)                     // p-value
```

If *p* is less than 0.05, the null hypothesis of exogeneity is statistically rejected.

Those desiring to implement this type of model to their research are advised to refer to Deb and Trivedi (2006).

13.2.5 Endogeneity resulting from measurement error

Our final example relates to incorporating predictors into a count model that have known measurement error. It is assumed that errors are missing at random (MAR), and that the mean of the variables with measurement error has not changed. Models do exist for *congeneric measurement error* where the mean and variance of the variables with measurement error are not assumed to be constant. We focus here on the basic method detailed in Carroll *et al.* (1995) and in Hardin *et al.* (2003), which emphasize the use of instrumental variables as a method of accommodating additive measurement error in models that are members of the family of generalized linear models. Poisson, NB2, and NB-C models are therefore addressed by this method.

Software for this method of handling measurement error has been written in Stata with C subroutines to assist the speed of convergence. The Stata command named **qvf**, is built on Stata's **glm** command (Hardin and Hilbe, 2001), but employs its own command line structure.

qvf y exogenous, endogenous (instruments), family( ), link () other-options

The *ml* option for negative binomial models is not available. Therefore it is advised to model the data without endogeneity, note the value of α, and insert it as the second term in the *family()* option, e.g. *family(nb #).* Try various values of α over and under the initial one used to determine the model having a Pearson-dispersion value closest to 1.0. That model will be the optimal model for the data, assuming it is negative binomial.

We shall use the same variables for an example as used in the previous section, except that *outwork* and *public* will now be endogenous, with assumed measurement error. Instruments bearing on how such errors occur are *income* and being a blue collar worker, *bluec*. Again, recall that instrumental variables are uncorrelated with the model error term; however, they are correlated with the endogenous predictors, which in this case are those predictors which were measured with error. The amount of error is not specified, as it is in some models that deal with adjustments for measurement error. To summarize, we have:

Response:	*docvis*
Exogenous:	*reform, age, female, married*
Endogenous:	*outwork, public*
Instruments:	*hhinc, bluec*

Deb and Trivedi (2002) recommend that models with instrumental variables should employ *Murphy–Topel* standard errors, as defined by Murphy and Topel (1985). See Hardin and Carroll (2003) for complete details of the variance estimator.

```
. nbreg docvis  reform age female married outwork public

------------------------------------------------------------------------------
         |               Robust
  docvis |      Coef.   Std. Err.      z    P>|z|     [95% Conf. Interval]
---------+--------------------------------------------------------------------
  reform |   .0437702   .0224815     1.95   0.052    -.0002928    .0878332
     age |   .0196857   .0009551    20.61   0.000     .0178137    .0215576
  female |   .2714386   .0233322    11.63   0.000     .2257082     .317169
 married |  -.0763476   .0262629    -2.91   0.004    -.1278218   -.0248733
 outwork |   .2146703   .0257537     8.34   0.000      .164194    .2651466
  public |   .3083967   .0329233     9.37   0.000     .2438682    .3729252
   _cons |  -.1838371    .054843    -3.35   0.001    -.2913274   -.0763468
---------+--------------------------------------------------------------------
/lnalpha |   .6410648   .0137668                      .6140823    .6680473
---------+--------------------------------------------------------------------
   alpha |   1.898501   .0261363                       1.84796    1.950425
------------------------------------------------------------------------------
Likelihood-ratio test of alpha=0:  chibar2(01) = 8.7e+04
                                   Prob>=chibar2 = 0.000
```

The value of α is 1.8985, which is used for the negative binomial endogenous model.

```
. qvf docvis reform age female married outwork public
  (reform age female married hhninc bluec), fam(nb 1.898501)mtopel

IV Generalized linear models                   No. of obs      =      27326
Optimization       : MQL Fisher scoring        Residual df     =      27319
                     (IRLS EIM)                Scale param     =          1
Deviance           = -320084.1567              (1/df) Deviance= -11.71654
Pearson            =  38331.67137              (1/df) Pearson =  1.403114

Variance Function: V(u) = u+(1.898501)u^2      [Neg. Binomial]
Link Function    : g(u) = ln(u)                [Log]
Standard Errors  : Murphy-Topel
------------------------------------------------------------------------------
        |            Murphy-Topel
 docvis |      Coef.   Std. Err.      z    P>|z|     [95% Conf. Interval]
--------+---------------------------------------------------------------------
 reform |   .0729464   .0214012     3.41   0.001     .0310009     .114892
    age |    .017226   .0009459    18.21   0.000      .015372    .0190799
 female |    .145454   .0243657     5.97   0.000      .097698      .19321
married |  -.0770793   .0231641    -3.33   0.001    -.1224801   -.0316786
outwork |    .443848   .0482553     9.20   0.000     .3492693    .5384267
 public |   1.106616   .0867013    12.76   0.000     .9366842    1.276547
  _cons |  -.8020014   .0855888    -9.37   0.000    -.9697525   -.6342504
------------------------------------------------------------------------------
```

The two endogenous predictors, *outwork* and *public*, have coefficients that differ considerably from the traditional NB2 model. The predictor *outwork* doubles its effect and *public* is some three-and-a-half times larger. The difference is due to the impact of the instruments. In addition, *reform* and *female* have a larger effect in the IV model, reflecting the fact that females in post-health-reform years have a bearing on measurement error. We would rather not see such an effect, but it is fairly minimal. Regardless, being out of work and on public health insurance – Medicare – in the United States leads to visiting a doctor more frequently during the year.

Exponentiating the instrumental variables negative binomial model provides the usual incidence rate ratios, and assists in the interpretation of the model predictors.

```
------------------------------------------------------------------------------
        |             Murphy-Topel
 docvis |     e^coef   Std. Err.      z    P>|z|   [95% Conf. Interval]
--------+---------------------------------------------------------------------
 reform | 1.075673   .0230207     3.41   0.001   1.031486    1.121752
    age | 1.017375   .0009623    18.21   0.000   1.015491    1.019263
 female | 1.156565   .0281805     5.97   0.000    1.10263    1.213137
married | .9258164   .0214457    -3.33   0.001   .8847235    .9688179
outwork | 1.558694   .0752153     9.20   0.000   1.418031    1.713309
 public | 3.024106    .262194    12.76   0.000   2.551507    3.584242
------------------------------------------------------------------------------
```

13.3 Sample selection and stratification

There are a variety of sample selection methods. As with the other models discussed in this chapter, sample selection methods generally assume that there is an endogenous predictor influencing a count response. Selection models employ a binary predictor as the primary endogenous variable. The binary predictor is typically itself the response of a probit regression of explanatory predictors. In the paradigm sample selection case the binary predictor is used to select which observations of the count response are going to be observed. Hence the name, sample selection.

As mentioned, there are variations on this theme. The first such variation is a negative binomial model with endogenous stratification. A full information maximum likelihood (FIML) method is used to estimate coefficients, unlike most other selection techniques. We then address traditional sample selection models, followed by a popular method in econometrics called an *endogenous switching model*. We shall find that sample selection and endogenous switching models primarily differ in how the endogenous binary term is conceived.

13.3.1 Negative binomial with endogenous stratification

Negative binomial with endogenous stratification is a model that is perhaps most noted for its application in the area of recreation research (Shaw, 1988; Englin and Shonkwiler, 1995) and survey sampling of shopping mall visitors. The model simultaneously accommodates three features of on-site samples dealing with count data. The first accommodation is an overdispersion relative to the Poisson model; the second is truncation of zero counts; the third is endogenous stratification due to oversampling.

Endogenous stratification occurs when the likelihood of sampling observations is dependent on the choice made by a subject of study, which is in itself the response variable. For example, in the field of recreational demand analysis, one is more likely to interview subjects who visit a particular site more frequently than those who rarely visit it. This implies oversampling of those who visit more frequently, and reports of more visits than are likely to occur. This is an instance of endogeneity. Likewise, patients who visit a doctor more frequently are more likely to be sampled if the survey about number of visitations is conducted at the clinic.

If the data are in fact Poisson, and therefore equidispersed, but nevertheless are truncated and endogenously stratified (ES), the Poisson PDF may be amended to appear as

$$f_{\mathrm{ES}}(y;\lambda) = \frac{e^{-\lambda_i}\lambda_i^{y_i-1}}{(y_i-1)!} \tag{13.3}$$

The log-likelihood can be expressed in terms of $x'\beta$ as:

$$\mathcal{L}_{\mathrm{ES}}(\beta;y) = \sum_{i=1}^{n}(y_i-1)(\lambda_i) - \exp(x_i'\beta) - \ln\Gamma(y_i) \tag{13.4}$$

which is equivalent to a Poisson with the response subtracted by 1, i.e. $y-1$. The model has also been called a *size-biased Poisson* and *shifted Poisson* aside from a *Poisson with endogenous stratification*. The rescaled probability function and log-likelihood for the negative binomial with endogenous stratification can be given as:

$$f_{\mathrm{NB-ES}}(y;\lambda,\alpha) = \frac{y_i\Gamma(y_i+\alpha)}{\Gamma(y_i+1)\Gamma(\alpha)}\alpha^{-y_i}\lambda_i^{y_i-1}\left[\frac{\lambda_i}{\alpha}+1\right]^{-(y_i+\alpha)} \tag{13.5}$$

$$\mathcal{L}_{\text{NB-ES}}(\beta; y, \alpha) = \sum_{i=1}^{n} \Big\{ y_i \ln(\alpha) + (y_i - 1)\ln\left(\exp\left(\mathrm{x}_i'\beta\right)\right) - \left(y_i + \frac{1}{\alpha}\right)\ln\left(1 + \alpha \exp\left(x_i'\beta\right)\right) - \ln\Gamma(y_i + 1) - \ln\Gamma\left(\frac{1}{\alpha}\right) + \ln\Gamma\left(y_i + \frac{1}{\alpha}\right) + \ln(y_i) \Big\};\ y_i > 0 \tag{13.6}$$

The following data are taken from Loomis (2003). The study relates to a survey taken on reported frequency of visits to national parks during the year. The survey was taken at park sites, thus incurring possible effects of endogenous stratification. I shall model a subset of the data, with *anvisits*, annual count of reported visits to parks, as the response. Predictors include *gender*, distance traveled to the closest park, and annual income. *Travel* and *income* are factored as shown in Table 13.6.

Table 13.6 *Recreation model: predictors*

	income	Freq.	Percent	Cum.
income1	<=25000	53	13.91	13.91
income2	25001-<=55000	97	25.46	39.37
income3	55001-<=95000	87	22.83	62.20
income4	>95000	144	37.80	100.00
	Total	381	100.00	
	travel	**Freq.**	**Percent**	**Cum.**
travel1	<.25 hrs	95	25.89	25.89
travel2	.25-<4 hrs	142	38.69	64.58
travel3	>=4 hrs	130	35.42	100.00
	Total	367	100.00	
	gender	**Freq.**	**Percent**	**Cum.**
female	0	155	38.75	38.75
male	1	245	61.25	100.00
	Total	400	100.00	

Modeling the data, **loomis**, with a negative binomial algorithm with endogenous stratification (Hilbe and Martinez-Espineira, 2005) produces the following output. Note that there are no R functions for this model. The only software available is in Stata, developed by the author; **nbstrat** is posted to the SSC site.

```
. nbstrat anvisits gender travel2 travel3 income2 income3 income4

Negative Binomial with Endogenous Stratification
                                          Number of obs =        342
                                          Wald chi2(6)  =     519.24
Log likelihood = -1232.0184              Prob > chi2   =     0.0000
-----------------------------------------------------------------
anvisits |     Coef.  Std. Err.     z   P>|z|[95% Conf.  Interval]
---------+-------------------------------------------------------
  gender | -.6011335  .1272006  -4.73  0.000  -.8504421  -.3518249
 travel2 | -.5569893  .1453207  -3.83  0.000  -.8418127  -.2721658
 travel3 | -3.080732  .1580607 -19.49  0.000  -3.390525  -2.770939
 income2 |  .4045486  .1919525   2.11  0.035   .0283287   .7807686
 income3 | -.7505286  .1953772  -3.84  0.000  -1.133461  -.3675962
 income4 |  -.599445  .1827182  -3.28  0.001  -.9575661   -.241324
   _cons | -12.10614  124.4169  -0.10  0.922  -255.9588   231.7465
---------+-------------------------------------------------------
/lnalpha |  16.60685  124.4169   0.13  0.894  -227.2457   260.4594
---------+-------------------------------------------------------
   alpha |  1.63e+07  2.03e+09                 2.03e-99   1.3e+113
-----------------------------------------------------------------
AIC Statistic =          7.252
```

Modeling the same data using a zero-truncated negative binomial gives us the output shown in Table 13.7.

Table 13.7 R: *Zero-truncated component of endogenous stratification*

```
rm(list=ls())
data(100mis)
attach(loomis)
library(gamlss)
library(gamlss.tr)
lloomis <- subset(anvisits, loomis$anvisits>0)
estnb <- gamlss(anvisits~ gender + travel2 + travel3 + income2 +
      income3 + income4, data=lloomis, family=NBI)
gen.trun(0,, "NBI", type= "left", name = "lefttr")
estnb0 <- gamlss(anvisits~ gender + travel2 + travel3 + income2 +
     income3 + income4, data=lloomis, family=NBIlefttr)
summary(estnb0)
modelfit(estnb0)  # function from Table 5.3
```

```
. ztnb anvisits gender travel2 travel3 income2 income3 income4
Zero Truncated Negative Binomial Regression
                                              Number of obs   =      342
                                              Wald chi2(6)    =   264.39
Log likelihood = -1188.9244                   Prob > chi2     =   0.0000
----------------------------------------------------------------------
anvisits |     Coef.  Std. Err.      z   P>|z|   [95% Conf. Interval]
---------+------------------------------------------------------------
  gender | -.7376444   .208367   -3.54  0.000  -1.146036  -.3292526
 travel2 | -.5795156  .2498082   -2.32  0.020  -1.069131  -.0899005
 travel3 | -3.646618  .2709248  -13.46  0.000  -4.177621  -3.115615
 income2 |  .7553192  .3404561    2.22  0.027   .0880375   1.422601
 income3 | -.8828776  .3153164   -2.80  0.005  -1.500886  -.2648688
 income4 |  -.510249  .3077044   -1.66  0.097  -1.113339   .0928406
   _cons |  4.233715  .3241964   13.06  0.000   3.598301   4.869128
---------+------------------------------------------------------------
/lnalpha |  1.339343  .2412675    5.55  0.000   .8664672   1.812218
---------+------------------------------------------------------------
   alpha |  3.816534  .9208055                  2.378493   6.124017
----------------------------------------------------------------------
AIC Statistic    =        6.994           BIC Statistic =   -1749.631
```

The AIC statistics of the two models indicate that the zero-truncated model without adjustment for endogenous stratification is preferred. However, the value of α for the latter model is far too high. This indicates that the model does not fit well with the data. The first 10 counts are given in Table 13.8.

Table 13.8 *Loomis – frequency of anvisits: 1–10*

anvisits	Freq.	Percentage	Cum.
1	133	32.44	32.44
2	30	7.32	39.76
3	15	3.66	43.41
4	8	1.95	45.37
5	5	1.22	46.59
6	15	3.66	50.24
7	3	0.73	50.98
8	3	0.73	51.71
9	2	0.49	52.20
10	10	2.44	54.63

The extreme number of 1s in the data is the likely cause of the model not fitting.

The negative binomial with endogenous stratification model has seen relatively little use and is apparently fragile in the presence of ill-structured data. However, if the data are appropriate for the model it performs better than its strictly zero-truncated counterpart. This is particularly the case when there is evidence of substantial oversampling.

13.3.2 Sample selection models

A variety of sample selection models can be found in statistical literature. The most common usage of sample selection has been within the domain of continuous response models. Heckman selection models, bivariate probit, and normal models with censoring have been most commonly used in research. However, as discussed in Cameron and Trivedi (1998) these models are not appropriate for count response models.

Selection models occur when the response is observed to occur only if a specified selection criterion is met. Generally the selection variable is binary. If it has the value of 1, the response is observed; if it has the value of 0, the response is not observed. In effect, sample selection is a type of truncation where the response is not observed unless a criterion has been met. There is a similarity of this type of situation with the survival parameterized censored Poisson and negative binomial discussed in Chapter 12. Greene calls this type of model as "incidental truncation;" Winkelmann prefers the label "endogenous selectivity." Regardless, this type of model corresponds to many life situations, but is rarely used in models outside of econometrics. However, it will be clear that these models have applicability for data situations in nearly every discipline.

Sample selection models for count data may be represented as:

$$\ln(\mu_i) = x_i'\beta + \gamma S + \varepsilon_i \tag{13.7}$$

where S is the endogenous binary sample selection term, γ is its coefficient, and ε is the Gaussian distributed error distributed normally with a mean of 0 and variance of 1, i.e. ε = N[0,1]; β are, as standard, the coefficients of the exogenous count model, with x as the corresponding data element. It is assumed that there is an endogenous relationship between the selection criterion and the count response.

The variance of ε determines the amount of overdispersion in the data. In fact, the Poisson sample selection model provides for overdispersion in a similar manner to the negative binomial. A drawback of sample selection negative binomial models is that overdispersion is obtained from two sources, α and ε, which may be combined into a single overdispersion term, often referred to as ρ, or *rho*. Care must be taken when using negative binomial sample selection models for an overcorrection of data overdispersion.

We use the German health data for an example of sample selection, with the number of doctor visits per year as the count response. Exogenous predictors include the patient being married, their age in years, and education level. Patients have neither public nor private insurance, but pay for medical care from personal funds. Given that situation, we regard being out of work as a selection of the likelihood of visiting the doctor. Variables that influence being out of work are being female and having children. This example is for demonstration of the method only, but can still be regarded as a sensible model for the data. Adaptive quadrature is the method used for estimation. Poisson is the family, and a robust variance estimate is used for standard errors.

```
. global exovars married age edlevel2 edlevel3 edlevel4
. ssm docvis outwork $exovars, s(outwork= $exovars female hhkids)
   adapt q(16) fam(poiss) link(log) sel robust

Sample Selection Poisson Regression
(Adaptive quadrature -- 16 points)
                                          Number of obs   =     27326
                                          Wald chi2(13)   =   7821.82
Log likelihood = -74397.715               Prob > chi2     =    0.0000
------------------------------------------------------------------------
  docvis  |     Coef.   Std. Err.      z    P>|z|   [95% Conf. Interval]
----------+-------------------------------------------------------------
docvis    |
 outwork  | 1.080154   .0538615   20.05   0.000    .9745873     1.18572
 married  |-.0841867   .0241366   -3.49   0.000   -.1314936   -.0368797
     age  | .0157214   .0010309   15.25   0.000    .0137009    .0177419
edlevel2  | .0552977   .0398279    1.39   0.165   -.0227635    .1333589
edlevel3  |-.1746433   .0360294   -4.85   0.000   -.2452596   -.1040271
edlevel4  |-.2725468   .0430507   -6.33   0.000   -.3569247   -.1881689
   _cons  |-.6556359   .0449902  -14.57   0.000   -.7438151   -.5674567
----------+-------------------------------------------------------------
selection |
 married  |-.0523007   .0224148   -2.33   0.020    -.096233   -.0083684
     age  | .0305701   .0009444   32.37   0.000    .0287191    .0324211
edlevel2  |-.1139198   .0351142   -3.24   0.001   -.1827424   -.0450971
edlevel3  | .2699573   .0316197    8.54   0.000    .2079838    .3319307
edlevel4  |  -.45572   .0417061  -10.93   0.000   -.5374626   -.3739775
  female  | 1.070587   .0173346   61.76   0.000    1.036612    1.104563
  hhkids  |  .141438   .0205879    6.87   0.000    .1010864    .1817895
```

```
   _cons   |-2.399443  .0513285 -46.75  0.000  -2.500046  -2.298841
-----------+----------------------------------------------------------
   sigma   | 1.375026  .0110233 124.74  0.000    1.35342   1.396631
     rho   |-.3620189  .0175305 -20.65  0.000   -.396378  -.3276597
----------------------------------------------------------------------
Likelihood ratio test for rho=0: chi2(1)= 0.00 Prob>=chi2 = 1.000
Robust Standard Errors presented.
```

Being out of work increases visits to the doctor by nearly three times greater than those who work. Since females with children are positively related to being out of work, they are likely to visit a doctor more frequently. The data are rather highly overdispersed, given sigma (σ) of 1.375. Rho (ρ) is a test for the endogeny of being out of work. It is negatively related, but significant. Note the boundary likelihood ratio test results for ρ.

Greene (1994) and Terza (1998) have recently developed maximum likelihood and two-step algorithms for count response sample selection models. We follow Greene's example using credit card reports to show an alternative method of how the model works and how it is to be interpreted. A negative binomial selection model is employed for the data.

The data contain records of major derogatory reports about credit card holders, with the goal of predicting the probability of a default on a credit card loan. Selection bias is inherent in the data since the reports are only gathered on those who already have credit cards. There is no information on individuals who have not yet been issued cards, but who would default if they had them. Since these individuals are excluded from the sample data, the sample is not completely random and exhibits selection bias. In order to remedy the bias it is necessary to model both the manner in which credit cards are issued as well as the actual counts of derogatory reports. The process of issuing cards may be modeled using a binary response model, with 1 indicating that a card has been issued and 0 that it has not. All potential applicants are thereby made part of the model, although individuals classified by the binary process as 0 are latent members of the data.

The sample selection model is therefore a two-part process, somewhat in the tradition of hurdle and zero-inflated models. Unfortunately, maximization of the two-part likelihood is much more complex than it is with any of the models we have thus far discussed. It must be maximized using either numerical integration or by using by simulation. LIMDEP, authored by Greene, uses simulation.

The model is structured so that the binary part, here a probit model, provides estimates of the probability of being issued a credit card. The count part is then estimated, but is adjusted by the probability values from the probit model. The

Table 13.9 *Sample selection models*

```
PROBIT
 Response
   cardhldr = 1: card been issued; 0: card not been issued
 Predictors
   agec     = age in years and twelfths of a year when applied
   income   = income in $10,000s
   ownrent  = own or rent home (1/0)
   curr_add = months residing at same address when applied for
              card
POISSON/NEGATIVE BINOMIAL
 Response
   majordrg = number of derogatory reports
 Predictors
   avgexp   = Average monthly expenditure
   inc_per  = income per dependent, in $10,000 units
   major    = 1/0 if applicant had another credit card at
              application
```

count model is said to be a selection-corrected Poisson or negative binomial. Predictors of each model are rarely identical since they are predicting different processes.

For our example, the models are specified as shown in Table 13.9. The probit model is entered into LIMEP as

```
probit;lhs=cardhldr;rhs=one,agec,income,ownrent,cur_add;hold
```

followed by the selection corrected Poisson

```
pois;lhs=majordrg;rhs=one,avgexp,inc_per,major;sel;mle$
```

The relevant models are displayed as

SAMPLE CORRECTED POISSON

```
------------------------------------------------------------------
 Poisson Model with Sample Selection.
------------------------------------------------------------------
Variable | Coefficient   St Error b/St.Er. P[|Z|>z]  Mean of X
---------+--------------------------------------------------------
         | Parameters of Poisson/Neg. Binomial Probability
 Constant | -3.65979959   .59816464 -6.118   .0000
 AVGEXP   |   .00078020   .00031097  2.509   .0121   238.602421
 INC_PER  |   .16237072   .07091514  2.290   .0220   2.21873662
  MAJOR   |   .22733512   .30623298   .742   .4579    .83968719
---------+--------------------------------------------------------
```

```
        | Parameters of Probit Selection Model
Constant|  .74148859   .15225187  4.870   .0000
AGEC    |  -.01027791  .00485007 -2.119   .0341    33.3853297
INCOME  |   .06174082  .02416541  2.555   .0106    3.36537604
OWNRENT |   .45569930  .08791258  5.184   .0000    .44048522
CUR_ADD |  -.00046311  .00063456  -.730   .4655    55.2676270
--------+-------------------------------------------------------
        | Standard Deviation of Heterogeneity
Sigma   |  1.16180092  .22171151  5.240   .0000
--------+-------------------------------------------------------
        | Correlation of Heterogeneity & Selection
Rho     |   .39658662 1.06023875   .374   .7084
----------------------------------------------------------------
```

Major does not contribute to the probit model and *cur_add* is not contributory to the selected corrected Poisson.

The sample corrected negative binomial model is displayed following the command:

SAMPLE CORRECTED NEGATIVE BINOMIAL

```
negb;lhs=majordrg;rhs=one,avgexp,inc_per,major;sel;mle$
```

```
-------------------------------------------------------------------
 Neg.Bin.Model with Sample Selection.
 Maximum Likelihood Estimates
-------------------------------------------------------------------
-------------------------------------------------------------------
Variable | Coefficient  St Error b/St.Er. P[|Z|>z] Mean of X
---------+---------------------------------------------------------
         | Parameters of Poisson/Neg. Binomial Probability
Constant |-3.07204954  .81142913 -3.786  .0002
AVGEXP   | .00082084  .00035875  2.288  .0221   238.602421
INC_PER  | .15419880  .07648030  2.016  .0438   2.21873662
MAJOR    | .23419460  .30789939   .761  .4469    .83968719
---------+---------------------------------------------------------
         | Parameters of Probit Selection Model
Constant |  .73676537 1.88621080   .391  .6961
AGEC     | -.01022903  .02646802  -.386  .6992   33.3853297
INCOME   |  .06194595  .15953633   .388  .6978   3.36537604
OWNRENT  |  .45529765 1.16379189   .391  .6956    .44048522
CUR_ADD  |-.00041768  .00125190  -.334  .7387   55.2676270
---------+---------------------------------------------------------
         |Overdispersion Parameter for Negative Binomial
Theta    | 1.60803441 1.31003489  1.227  .2196
---------+---------------------------------------------------------
         |Standard Deviation of Heterogeneity
Sigma    |  .49395748 1.15588222   .427  .6691
---------+---------------------------------------------------------
         |Correlation of Heterogeneity & Selection
Rho      |  .61978139 2.31339106  .268  .7888
-------------------------------------------------------------------
```

The parameters of the probit and the Poisson/negative binomial are fit at the same time. The *sigma* statistic that appears in both model outputs is the standard deviation of ν in $\lambda = \exp(\beta x + n)$. *Rho* is the correlation between ν and u in `Prob[d=1]==Prob(d'z + u>0)` in the probit model. The variance of u is 1. According to the model results, *rho* is not significant in the sample selection Poisson model, which is interpreted as meaning that selection is not an issue in these data. The interpretation of the negative binomial model is the same.

The negative binomial selection model here is apparently only weakly identified, which is no surprise given the Poisson results (*rho* approx $= 0$). What's going on is something like collinearity, but with the derivatives. It is likely that this model is overspecified. The reason is that the negative binomial model as initially constructed is the Poisson model with an additional term for heterogeneity. The selection model adds yet another source of latent heterogeneity to what is already intrinsic to the negative binomial. That is, the negative binomial is used to accommodate overdispersed Poisson models. But then selection adds an additional layer of accommodation to overdispersion. In this case I suspect it's too much, i.e. it is overspecified. The model may in fact not be adequate to pick up all this latent activity. Other modeling situations may require the extra accommodation. But, in this case, the probit selection model predictors are not significant, even though they were so when modeled alone. Thus the probit selection criteria with the Poisson model provide no support to the selection process.

13.3.3 Endogenous switching models

Endogenous switching models are also important to econometricians. Unlike sample selection models, the count response of endogenous switching (ES) models is always observed. The key characteristic of an ES model is that the response is a function of a binary switch variable. Like sample selection (SS) models though, and SS count models in particular, bias results if there are unobserved latent factors influencing the response that are correlated with unobserved latent factors influencing the switch.

The term *switch* appears to indicate a selection, as in *sample selection*. However, it merely indicates an influencing factor which changes the likelihood of the response based on its value. The key, however, to the switch is its endogenous nature. The general structure of the ES model can be expressed as:

$$\ln(\mu_i) = x_i'\beta + \theta S + \xi_i \tag{13.8}$$

where θ is the switch, ξ is the error term, and $x'\beta$ is defined as described for the sample selection model. The logic of the two methods is much the same, except that all of y is observed for ES models, but y is observed only when $S = 1$ for SS models.

The following representation of the ES Poisson probability function can be given as:

$$f(y_i|\xi_i) = \frac{\exp\left\{-\exp\left(x'_i\beta + \theta S + \xi_i\right)\right\}\left\{\exp\left(x'_i\beta + \theta S + \xi_i\right)\right\}^{y_i}}{y_i!} \tag{13.9}$$

where y is the count response, S is the endogenous binary variable, θ is an unobserved random coefficient characterizing S, and ξ is a random unobserved heterogeneity term.

Miranda (2004) designed a Stata command to specifically estimate ES Poisson models. Using it to estimate the number of doctor visits per year given the same exogenous and endogenous predictors used for the first sample selection example, the estimate appears as

```
. espoisson docvis married age edlevel2-edlevel4,
  ed(outwork) s(female hhkids)

Endogenous-Switch Poisson Regression
(6 quadrature points)
                                            Number of obs   =    27326
                                            Wald chi2(5)    =  1324.03
Log likelihood = -77335.134                 Prob > chi2     =   0.0000
------------------------------------------------------------------------------
  docvis |      Coef.  Std. Err.      z    P>|z|   [95% Conf. Interval]
---------+--------------------------------------------------------------------
docvis   |
 married | -.1031098    .017752   -5.81   0.000   -.1379031   -.0683166
     age |  .0202152   .0006453   31.33   0.000    .0189504      .02148
edlevel2 |  .0316484   .0429703    0.74   0.461   -.0525718    .1158687
edlevel3 | -.1180251   .0266013   -4.44   0.000   -.1701626   -.0658876
edlevel4 | -.4720279   .0315007  -14.98   0.000   -.5337681   -.4102878
   _cons | -.3212869   .0337193   -9.53   0.000   -.3873755   -.2551984
---------+--------------------------------------------------------------------
switch   |
  female |  1.049971   .0170466   61.59   0.000     1.01656    1.083382
  hhkids | -.1198469   .0172245   -6.96   0.000   -.1536064   -.0860875
   _cons | -.9854903   .0143149  -68.84   0.000   -1.013547   -.9574337
---------+--------------------------------------------------------------------
   sigma |  1.058598   .0045623  232.03   0.000    1.049694    1.067578
     rho |  .0904719   .0093325    9.69   0.000     .072152    .1087306
```

Being female and not having children are indicators of being out of work. Predictors of doctor visits, given the endogenous or indirect influence of being out of work, are being married, age, and the highest two educational levels, which entail being a university/college graduate. The overdispersion parameter is sigma (σ), which demonstrates substantial overdispersion and associated heterogeneity. If $\sigma = 0$, there is no endogeneity in the model, and the model is in fact purely exogenous. The test statistic for σ is a boundary likelihood test with an equal mixture in the likelihoods of the exogenous and endogenous switching models.

The test of the *rho*(ρ) coefficient is a test of endogenous switching. The test is performed using a likelihood ratio test of the two model likelihoods: $-2(L_{\text{exo}} - L_{\text{end}})$ with a *chi*2 degree of freedom of 1. The coefficient ρ is not highly influential, but is significant.

The data are modeled by the same command as used for sample selection, using quadrature. Note that *outwork* is given a coefficient. The two methods result in slightly differing coefficient values, but the differences are not significant. The values of σ and ρ, however, are some 30% greater and are of opposite signs respectively. Care must be taken, then, in interpreting these statistics. It is likely a good tactic to use several methods of analysis before providing a conclusion concerning the model.

```
. global exovars married age edlevel2-edlevel4

. ssm docvis outwork $exovars, s(outwork= $exovars female hhkids)
    adapt q(16) fam(poiss) link(log)

Endogenous Switch Poisson Regression
(Adaptive quadrature -- 16 points)
                                              Number of obs   =     27326
                                              Wald chi2(13)   =   8041.75
Log likelihood = -74397.715                   Prob > chi2     =    0.0000
-------------------------------------------------------------------------
  docvis |      Coef.   Std. Err.     z   P>|z|   [95% Conf. Interval]
---------+---------------------------------------------------------------
docvis   |
 outwork |   1.080154   .0535785  20.16  0.000    .9751418    1.185166
 married |  -.0841867   .0239364  -3.52  0.000   -.1311012   -.0372722
     age |   .0157214   .0010293  15.27  0.000     .013704    .0177388
edlevel2 |   .0552977   .0411851   1.34  0.179   -.0254236     .136019
edlevel3 |  -.1746433   .0356089  -4.90  0.000   -.2444354   -.1048513
edlevel4 |  -.2725468   .0430635  -6.33  0.000   -.3569497   -.1881438
   _cons |  -.6556359   .0441684 -14.84  0.000   -.7422044   -.5690674
---------+---------------------------------------------------------------
```

```
switch |
 married | -.0523007  .0217016  -2.41  0.016   -.094835  -.0097664
     age |  .0305701  .0008839  34.58  0.000   .0288376   .0323026
edlevel2 | -.1139198  .0352471  -3.23  0.001  -.1830028  -.0448367
edlevel3 |  .2699573  .0293061   9.21  0.000   .2125185   .3273961
edlevel4 |   -.45572  .0399806 -11.40  0.000  -.5340805  -.3773595
  female |  1.070587  .0174152  61.47  0.000   1.036454   1.104721
  hhkids |   .141438  .0203802   6.94  0.000   .1014934   .1813825
   _cons | -2.399443  .0454577 -52.78  0.000  -2.488539  -2.310348
---------+--------------------------------------------------------
   sigma |  1.375026  .0114819 119.76  0.000   1.352521    1.39753
     rho | -.3620189  .0173483 -20.87  0.000  -.3960208  -.3280169
------------------------------------------------------------------
Likelihood ratio test for rho=0: chi2(1)= 0.00 Prob>=chi2 =  1.000
```

13.4 Quantile count models

Quantile count models are quite unlike other models we have discussed thus far in the text, which are based on underlying probability distributions, namely the Poisson and negative binomial probability functions. For a given pattern of covariates, standard regression methods aim to model the mean parameter of the distribution, or μ. Quantile models on the other hand model the median, or other quantiles of the distribution. Median regression is a quantile model at a 50%, or 0.5, quantile. Quantile models assume a continuous or Gaussian distribution since this is the only manner in which quantiles make sense.

Researchers have attempted to design quantile count models according to two different methods. The first method is based on the research of Gourieroux *et al.* (1984). They estimated quantiles based on a semi-parametric modeling of the conditional mean of the count response, using a pseudo-likelihood algorithm. The problem with this approach is that the full range of the count distribution cannot be understood based on the predictors. Mullahy (1986), who also created hurdle models, proposed a fully parametric method that did describe all parts of the distribution, but the model he offered came with strong assumptions which were not truly applicable for discrete quantiles. Efron (1992) and others tried other approaches, but model interpretation became a stumbling block to their acceptance.

A second general approach was offered by Machado and Santos Silva (2005), who devised a method by which discrete counts could be modeled using traditional quantile regression techniques. They added a jitter to the Poisson or negative binomial counts structured in such a way that the resultant distribution appears as a continuous variable. The count response, y, is restructured as:

$$z = y + \text{uniform}(0, 1) \tag{13.10}$$

The new response, z, therefore has a density appearing as

$$f(z) = \begin{array}{l} \text{Prob}(1): 0 \le z < 1 \\ \text{Prob}(2): 1 \le z < 2 \\ \quad\vdots \end{array} \tag{13.11}$$

Given that the counts are non-negative, z is linearized at the conditional mean of each quantile as $\exp(x' \beta_q)$. The quantile count model algorithm makes multiple draws of the endpoints of the quantiles, attempting to obtain a smoother transition between them with the goal of attempting to have the distribution appear more continuous. In fact, there is a one-to-one relationship of quantile ranges of y and z. Machado and Santos Silva referred to these ranges as $Q_y(a)$ and $Q_z(a)$ respectively, with a as the quantile. Those desiring greater details of estimation are referred to the original journal article, which is posted on the web.

In order to illustrate how this type of model works, we use the **mdvis** data that we previously employed with hurdle models. First, it may be instructive to graph how the jitter affects the distribution of an otherwise count model. Figure 13.1 displays two graphs: on the left side are the quantiles of *numvisit*, the count variable which we shall employ as the model response. We have defined this as y in equation 13.1. On the right side is z, defined as y + uniform(). Note the continuous appearance of the distribution. It is the shape that allows the Machado–Santos Silva approach to work.

The response we use in the quantile model is *numvisit*, the number of visits to the doctor over a period of a year. Predictors include:

- *reform*: a binary variable indicating if the data were collected before (0) or after (1) health reform legislation took place
- *badh*: a binary variable indicating that the patient reports themselves as being in relatively good health (0) or bad (1)
- *educ*3: a binary variable indicating that the patient has post-baccalaureate education (1), compared with the reference, not high-school graduate

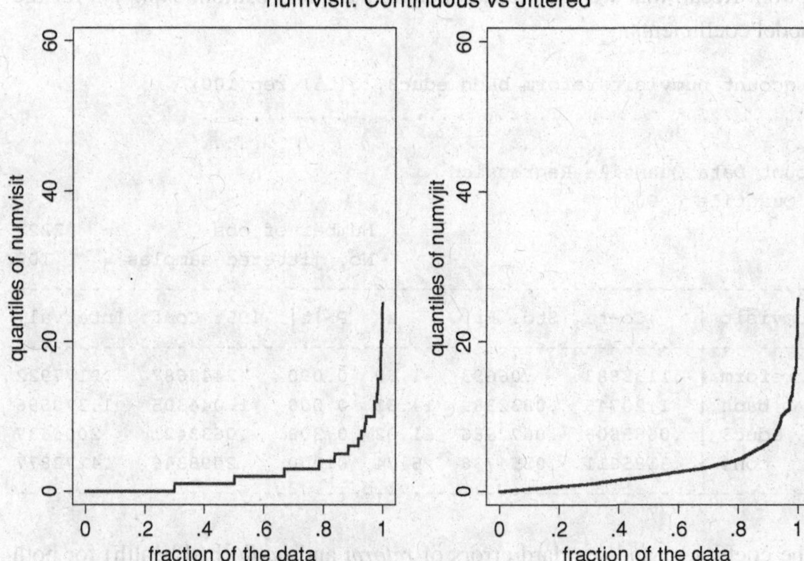

Figure 13.1 Quantiles of *y* (left) and *z* (right)

We begin by estimating a standard NB2 negative binomial model:

```
. glm numvisit reform badh educ3, fam(nb ml)
Generalized linear models              No. of obs      =       2227
Optimization     : ML                  Residual df     =       2223
                                       Scale parameter =          1
Deviance         =  2412.727878        (1/df) Deviance =   1.085348
Pearson          =  2702.803073        (1/df) Pearson  =   1.215836
Variance function: V(u) = u+(1.008)u^2 [Neg. Binomial]
Link function    : g(u) = ln(u)        [Log]
                                       AIC             =   4.105382
Log likelihood   = -4567.343292        BIC             =  -14723.07
-------------------------------------------------------------------
         |               OIM
numvisit |     Coef. Std. Err.     z    P>|z|  [95% Conf. Interval]
---------+---------------------------------------------------------
  reform |-.1264314 .0510852  -2.47  0.013  -.2265566   -.0263061
    badh |  1.15474 .0733899  15.73  0.000   1.010898    1.298581
   educ3 |-.0883131 .0545376  -1.62  0.105  -.1952048    .0185786
   _cons | .8179386 .0416287  19.65  0.000   .7363479    .8995293
-------------------------------------------------------------------
```

Next estimate the data using a median quantile count model written by Alfonso Miranda of Keele University. Other quantiles may be modeled using the *q()*

option. Recall that we use the exponential parameterization, $\exp(x'\beta)$, for the model coefficients.

```
. qcount numvisit reform badh educ3, q(.5) rep(100)
..................................................
..................................................
Count Data Quantile Regression
( Quantile 0.50 )
                                    Number of obs          =    2227
                                    No. jittered samples =      100
--------------------------------------------------------------------
numvisit |     Coef.  Std. Err.      z   P>|z|  [95% Conf. Interval]
---------+----------------------------------------------------------
  reform |-.1133883     .06693   -1.69  0.090  -.2445687   .0177922
    badh |  1.20745   .0832389   14.51  0.000   1.044305   1.370596
   educ3 | .0685608   .0672986    1.02  0.308  -.0633421   .2004637
   _cons | .3185611   .0554738    5.74  0.000   .2098346   .4272877
--------------------------------------------------------------------
```

The coefficient and standard errors of *reform* and *badh* (bad health) for both the negative binomial and quantile count models are similar.

Marginal effects of the coefficients are calculated at the means of the predictors. The initial output employs the jitter predictor, whereas the second table is in terms of unadjusted binary predictors.

```
. qcount_mfx
 Marginal effects after qcount
        y = Qz(0.50|X)
          =  2.02425 (0.0526)
--------------------------------------------------------------------
       |      ME       Std. Err.     z     P>|z|   [   95% C.I  ]     X
-------+------------------------------------------------------------
reform | -.17304384   .10197007   -1.7   0.0897 -0.3729  0.0268  0.51
badh   |  3.1161332     .336666   9.26   0.0000  2.4563  3.7760  0.11
educ3  |  .10569426   .10436291   1.01   0.3112 -0.0989  0.3102  0.34
--------------------------------------------------------------------
 Marginal effects after qcount
        y = Qy(0.50|X)
          = 2
--------------------------------------------------------------------
          |      ME   [95% C. Set]     X
----------+---------------------------------------------------------
reform    |      -1      -1   0       0.51
badh      |       3       2   3       0.11
educ3     |       0      -1   0       0.34
--------------------------------------------------------------------
```

If we compare the marginal effects of the above quantile model with the marginal effects of the tradititional negative binomial model, we obtain

```
. qui nbreg numvisit i.reform i.badh i.educ3,
. margins, dydx(*) atmeans
------------------------------------------------------------------------
         |            Delta-method
         |      dy/dx   Std. Err.      z    P>|z|     [95% Conf. Interval]
---------+--------------------------------------------------------------
1.reform |  -.297809   .1206791   -2.47   0.014   -.5343357   -.0612823
  1.badh |   4.48319   .4488606    9.99   0.000    3.603439    5.362941
 1.educ3 |-.2048335    .1248077   -1.64   0.101   -.4494522    .0397851
------------------------------------------------------------------------
Note: dy/dx for factor levels is the discrete change from the
      base level.
```

Actually, the comparisons are in terms of discrete change rather than marginal effects. But the relationship is such that we may make comparisons between the quantile and standard models. Note the differences between the two models. It is interesting that the effect is opposite for higher education. It is usually preferred to interpret quantile models using marginal effects or discrete change. Standard interpretations hold. There is not an easy interpretation for the coefficients in themselves.

Finally, of the two previous tables of statistical output, the bottom table represents the marginal effects for the conditional quantiles of the actual non-jittered count, $Q_y(a)$. The discrete values shown in the table are typical for this result.

An important consequence of estimating parameters using quantile count models is their robustness to outliers in the data. If data are highly skewed, or have several influential outliers, quantile regression may be a useful alternative. In addition, quantile regression models all of the quantiles of a distribution, covering the entire range of counts. Standard methods, as we mentioned earlier, concentrate on the mean value, or central moment of the distribution.

It is preferable to include at least one continuous predictor in the model, which we did not do with our example. Doing so assures that the quantile function is differentiable throughout the distribution. Aside from the original Machado and Santos-Silva article (2005), additional information may be obtained about this model in Winkelmann (2008) and Cameron and Trivedi (2009).

Summary

Models with endogenous predictors have an important role in econometrics, and have equal applicability to models in different domains (e.g. ecology, astronomy). We have only provided an overview of this area, discussing the major types of count models for which endogeny plays an essential role. As a means to adjust for heterogeneity, the methods discussed in this chapter have proved to be powerful. Refer to the sources listed in the chapter for additional information.

Stata, SAS, and LIMDEP are the primary statistical applications for use with models having endogenous predictors. Very little R code is available at present for these types of models.

Finite mixture models will prove to be used with increasing frequency in research. They provide a means to more realistically model a count consisting of two or three separate components. Quantile count models are valuable in this respect as well, and can prove valuable when the count data are not reflective of a known underlying PDF. Quantile count models are essentially designed for continuous responses, but innovation (jittering) has allowed them to be used for counts as well. The methodology is new, but can serve researchers well when a count response consists of severe outliers, or multiple distributional components that do not reflect an underlying probability distribution.

14

Count panel models

14.1 Overview of count panel models

A basic assumption in the construction of likelihood-based models is that constituent observations are independent. This is a reasonable assumption for perhaps the majority of studies. However, for longitudinal studies this assumption is not feasible, nor does it hold when data are clustered. For example, observations from a study on student drop-outs can be clustered by the type of schools sampled. If the study is related to intervention strategies, schools in affluent suburban, middle-class suburban, middle-class urban, and below-poverty-level schools have more highly correlated strategies within the school type than between types or groups. Likewise, if we have study data taken on a group of individual patients over time (e.g. treatment results obtained once per month for a year), the data related to individuals in the various time periods are likely to be more highly correlated than are treatment results between patients. Any time the data can be grouped into clusters, or panels, of correlated groups, we must adjust the likelihood-based model (based on independent observations) to account for the extra correlation.

We have previously employed robust variance estimators and bootstrapped standard errors when faced with overdispersed count data. Overdispersed Poisson models were replaced by negative binomial models, by adjusting the variance function of the basic Poisson model, or by designing a new log-likelihood function to account for the specific source of the overdispersion. Zero-inflated models as well as censored, truncated, and bivariate models are among the varieties of extended Poisson and negative binomial models discussed in the text. Negative binomial models can be overdispersed as well, leading to a variety of even more enhanced models.

In this chapter we begin by describing a group of models that add at least one extra parameter to the linear predictor, specifying how observations within panels are to be construed. New log-likelihoods are derived based on panels of correlated observations. The type of parameters that are added to the linear predictor, and the manner in which panels are treated will determine the type of panel model described. We shall first discuss generalized estimating equations (GEEs), or population averaged (PA) models. Thereafter we shall address a class of panel models commonly referred to as *subject-specific models*. Included in this class are fixed-, random-, and mixed-effects models. Fixed-effects models can be estimated as unconditional (all parameters estimated) or as conditional (the estimation is conditional on the value of a sufficient statistic for one or more parameters). Random-effects models are sometimes distinguished as random-intercept or as random-coefficient models. When a panel model consists of both fixed and random components, they can be estimated as mixed models, i.e. a mixture of fixed and random effects. Panel models can additionally be nested into a variety of levels of panels – which are commonly termed hierarchical or multilevel models.

Each of these types of panel model has software support in the major commercial packages. Both Stata and R provide support as well. We will find that some of these models, e.g. negative binomial multilevel models, are still being developed.

It should be noted at the outset that at this time no negative binomial GEE function is available using R software. The closest GEE model to the negative binomial is the quasi-Poisson family. As discussed earlier in the text, the quasi-Poisson GLM model is nothing more than a Poisson regression with standard errors scaled by the Pearson–dispersion. It is not a legitimate likelihood, nor is it a family member of GLM software outside of R.

It may be helpful at this point to clarify exactly what is meant by a panel structure. A panel structure relates to the manner in which data are collected, specifically to data collected on observations over time, or on which data are clustered on some identifying characteristic.

Data taken on observations or individuals longitudinally need to have a variable that indicates which rows of values are to be associated with a specific observation. Clustering takes the same form, but does not include a time component. An example of longitudinal panel data is given in Table 14.1. The data in Table 14.1 are structured into panels defined by ID number. Each set of observations (rows) associated with a given ID is a separate panel. Hence the term *panel data*.

ID is typically referred to as an *id* variable. In fact, many software applications specifically refer to it as simply *id=*. An ID value is assigned for all observations associated with a single individual or item. Fixed-effects terms can be understood in several ways. With respect to panel models, *fixed* in

Table 14.1 *Panel data*

ID	year	sex	age	married	score
001	2011	1	21	0	45
001	2012	1	22	0	47
001	2013	1	23	1	38
001	2013	1	24	1	45
001	2015	1	25	0	48
002	2011	0	18	0	49
002	2012	0	19	1	37
002	2013	0	20	1	35
002	2014	0	21	1	33
003	2013	1	27	1	28
003	2014	1	28	0	25
003	2015	1	29	0	22
003	2016	1	30	1	30
004	2011	1	19	0	22
004	2012	1	20	1	18
004	2013	1	21	0	23

general refers to a variable that does not change in value over panels. For example, *sex* is a fixed-effect variable. Once assigned, it does not (usually) change across observations within each *id*. *Year* and *age* are also fixed. Once *year* or *age* is assigned, subsequent yearly values are fully determined. Temporal variables like *year* or *age*, however, are not usually referred to as fixed-effects, but rather as *time* variables. Even though the variables are temporally fixed, they do not enter estimating algorithms as traditional fixed effects. The relationships between these types of variables will be more clear as we apply them to real data situations.

Married and *score* are generally considered to be random variables. An individual may change marital status within any given year. Likewise, *score*, which we use here in a generic sense, can take on any value within a predetermined range of values. Here we may be interested in determining if *score* is influenced by marital status or gender, and if there is perhaps some relationship related to *age,* or even *year*.

We shall begin our overview of count-based panel models by addressing the class of models known as GEE models. This class of models is perhaps one of the most popular collection of models dealing with longitudinal and clustered events.

14.2 Generalized estimating equations: negative binomial

14.2.1 The GEE algorithm

Generalized estimating equation (GEE) refers to a method of estimation first proposed in Liang and Zeger (1986). This class of models is an extension for panel data to the standard generalized linear models for individual data. Unlike the random-effects model, which is subject-specific, the most commomly employed GEEs constitute a class of population-averaged models for which marginal effects are averaged across individuals. Essentially, GEEs model the average response of individuals sharing the same predictors across all of the panels.

GEEs address data that are structured such that observations are grouped in panels, in a similar manner to fixed-effects and random-effects models. At the heart of the model specification, the variance function (identical to the GLM specification) is factored to include a parameterized within-panel correlation structure. This variance function is written:

$$\mathbf{V}(\mu_{ik}) = [\mathbf{D}(\mathbf{V}(\mu_{ik}))^{1/2}\mathbf{R}_{n_i \times n_i}\mathbf{D}(\mathbf{V}(\mu_{ik}))^{1/2}]_{n_i \times n_i} \tag{14.1}$$

where $\mathbf{V}(\mu_{ik})$ is the GLM variance function defined in terms of the mean. For example, the Poisson variance function is equal to the mean μ; the NB2 variance is $\mu + \alpha\mu^2$. Assuming independence within panels $\mathbf{R}_{n\times n} = \mathbf{I}_{n\times n}$, is called the independence correlation structure.

The benefit of the GEE approach is that the correlation matrix, which is sandwiched between the GLM variance functions, can be arbitrarily parameterized. The structural constraints of values that are substituted into this alternative matrix define the various types of correlation that are commonly supported in software. These include:

Foremost GEE correlation structures

1: Independence	2: Exchangeable	3: Unstructured
4: Autoregressive	5: Stationary	6: Non-stationary

The most commonly used correlation structure is the exchangeable, which we shall later describe in more detail. All of the structures define constraints on the values to be estimated. Those values are estimated from Pearson residuals obtained using the regression parameters. Pearson residuals are defined (Section 5.1), as

$$r_{ik} = \sum_{k=1}^{K}\sum_{i=1}^{n_i} (y_{ik} - \mu_{ik}) / \sqrt{V(\mu_{ik})} \tag{14.2}$$

The Poisson Pearson residual is defined as $\Sigma_i \Sigma_k (y_{ik} - \mu_{ik})^2 / \mu_{ik}$. The exchangeable correlation is defined as the average correlation between two arbitrary within-panel observations where that correlation is averaged over all possible pairs in all possible panels,

$$\hat{\alpha} = \frac{1}{\hat{\phi}} \sum_{i=1}^{n}\sum_{k=1}^{n_i} \left\{ \frac{\sum_{j=1}\sum_{k=1} r_{ij} r_{ik} - \sum_{j=1} r_{ij}^2}{n_i(n_i - 1)} \right\} \tag{14.3}$$

where the dispersion parameter, ϕ, is calculated as

$$\hat{\phi}^{-1} = \frac{\sum_i \sum_j r_{ij}^2}{N - p} \tag{14.4}$$

with N the number of model observations and p the number of model predictors. The exchangeable correlation structure has also been referred to as the compound symmetry matrix and the equal correlation structure. All off-diagonal values are equal to a single scalar constant.

Other correlation matrices are defined in different manners, depending on the purpose of the structure. However, all are inserted into the variance function as

$$\mathbf{V}(\boldsymbol{\mu}_{ik}) = [\mathbf{D}(\mathbf{V}(\boldsymbol{\mu}_{ik}))^{1/2} \mathbf{R}(|\mathbf{a}|)_{n_i \times n_i} \, \mathbf{D}(\mathbf{V}(\boldsymbol{\mu}_{ik}))^{1/2}]_{n_i \times n_i} \tag{14.5}$$

The GEE algorithm begins by estimating a model from a GLM member family, e.g. Poisson. After the initial iteration, Pearson residuals are calculated (equation 14.2) and put into the formula for calculating $\mathbf{R(a)}$ (equation 14.3). $\mathbf{R(a)}$ is then inserted into equation 14.5 in place of the identity matrix. The updated variance function is then used as such in the second iteration. Again, another updated variance function is calculated, and so on until the algorithm converges as does any GLM model. The form of the score equation, which serves as the estimating equation used to estimate parameters, is given by:

$$\sum_{i=1}^{n}\sum_{k=1}^{n_i} \left(\frac{d\mu_{ik}(\beta)}{d\beta} \right) V_{ik}^{-1} (Y_{ik} - \mu_{ik}(\beta)) = 0 \tag{14.6}$$

Since the resulting GEE model is not based on a probability function, the method is called a quasi-likelihood model. Recall that we used a similar appellation when an otherwise GLM variance function was multiplied by either a constant, or by another non-constant variable. In either case the working likelihood function is not based on a probability function. We therefore use a robust or sandwich variance estimator to adjust standard errors. Models using the independence correlation structure may find that such an adjustment is unnecessary, and if there is indeed no extra correlation in the data, model based standard errors are preferable. For our purposes, though, I'll use empirical standard errors as the default, allowing for a more consistent measure across models.

Variance estimators

Empirical (aka *sandwich* or *robust/semi-robust*). Consistent when the mean model is correctly specified (and if no missing data)

Model-based (aka *naive*). Consistent when both the mean model and the covariance model are correctly specified.

14.2.2 GEE correlation structures

Although GEE models are fairly robust to the use of incorrect correlation structures, it is nevertheless preferable to select the structure most appropriate to the data or to the goal of the study. Problems with convergence may occur as a result of selecting an inappropriate correlation structure. One may check the observed correlation matrix to determine if there is no known reason to select a specific matrix based on previous clinical studies. This might not provide a definitive solution as to which is the best correlation structure for the data, but it can nevertheless inform you about which type of structure is not appropriate.

The QIC statistic has recently been used to quantitatively decide on the preferred GEE correlation structure . The statistic, created by Pan (2001a), is called the *quasi-likelihood under the independence model information criterion*. It is similar to the AIC statistic, but tests correlation structures within the scope of generalized estimating equations. The QICu statistic, also developed by Pan (2001b), is aimed to assist the user in deciding on the best subset of model predictors for a particular correlation structure. Later research has shown, however, that it is preferable to use the QIC for evaluating both situations – deciding between families and links, and deciding the best model subset. New work by Hin and Wang (2008), Hilbe (2009), and Barnett *et al.* (2010) has indicated, though, that the QIC statistic is in fact itself a poor predictor of the "true" or appropriate model correlation structure. We discuss this situation at more length in Section 14.2.4.

Notwithstanding any bias with the QIC statistic, if two or more correlation structures result in nearly the same value of QIC, or other fit statistic (AIC), and no other factors can be used to assist in deciding which structure to use in a given modeling situation, the generally preferred choice is to employ the simplest structure i.e. the one with the least parameters.

There are a few summary guidelines that may be helpful in deciding which correlation structure to use when modeling a GEE.

1 Independence – if number of panels are small.
2 Exchangeable or Unstructured CS – if data relate to first-level clustered data
3 Autoregressive, Non-Stationary, and Stationary or m-dependent CS – when panel data relate to measurements over time periods.
4 Autoregressive CS – usually associated with longitudinal time-series data, where the correlations diminish over distance. This is a key mark of autoregressive models. Distances between time events are assumed to be equal, e.g. correlation between observations is assumed constant for observations having equal lags.
5 Unstructured – if panels are small and data are balanced and complete. Correlation of each pair of reponses differ.

An examination of the major correlation structures follows in this and the next subsection. 5×5 matrix schematics of the respective correlation structures are displayed, together with representative Poisson and negative binomial GEE models. Only the lower half of the symmetric matrix is completed. I also provide additional guidelines on when each structure should be used. Note that we shall not discuss the stationary or non-stationary structures. They are in fact rarely used in research, particularly with count response GEE models. I shall, however, provide a brief summary as to what type of data may lead one to consider one of these structures.

Independence correlation structure

SCHEMATIC

1				
0	1			
0	0	1		
0	0	0	1	
0	0	0	0	1

The independence correlation structure imposes the same structure on the GEE model as the standard GLM variance–covariance matrix. Observations are considered to be independent of one another. The use of this model is to set a base for evaluation of other GEE correlation structures. The structure assumes a zero correlation between subsequent measures of a subject within panels. Use this structure if the size of panels is small and if there is evidently no panel

Table 14.2 R: *GEE independence structure*

```
rm(list=ls())
library(geepack)
data(rwm5yr)
attach(rwm5yr)
geeind <- geeglm(docvis ~ female + edlevel2 + edlevel3 +
         edlevel4, family=poisson, corstr=independence,
         id=rwm$id, std.err = "san.se", data=rwm5yr)
library(haplo.ccs)
sandcov(geeind, rwm5yr$id)
sqrt(diag(sandcov(geeind, rwm5yr$id)))
summary(geeind)
exp(coef(geeind))
cor.mat
```

```
. xtgee docvis female edlevel2-edlevel4 , eform fam(poi) corr(indep)
   i(id) robust

GEE population-averaged model                Number of obs      =     19609
Group variable:                      id      Number of groups   =      6127
Link:                               log      Obs per group: min =         1
Family:                         Poisson                     avg =       3.2
Correlation:                independent                     max =         5
                                             Wald chi2(4)       =    173.84
Scale parameter:                      1      Prob > chi2        =    0.0000
Pearson chi2(19609):          204367.06      Deviance           = 119319.59
Dispersion (Pearson):          10.42211      Dispersion         =   6.08494
                                (Std. Err. adjusted for clustering on id)
---------------------------------------------------------------------------
         |             Semi-robust
  docvis |      IRR     Std. Err.      z    P>|z|    [95% Conf. Interval]
---------+-----------------------------------------------------------------
  female | 1.381491   .0519136     8.60   0.000    1.283399     1.48708
edlevel2 | .8399478   .0548827    -2.67   0.008    .7389827    .9547075
edlevel3 | .7348061   .0422671    -5.36   0.000    .6564633    .8224984
edlevel4 | .6093514   .0460152    -6.56   0.000      .52552    .7065555
---------------------------------------------------------------------------

. estat vce, cov

Estimated within-id correlation matrix R:

         c1      c2      c3      c4      c5
r1   1.0000
r2   0.0000  1.0000
r3   0.0000  0.0000  1.0000
r4   0.0000  0.0000  0.0000  1.0000
r5   0.0000  0.0000  0.0000  0.0000  1.0000
```

effect in the data. Again, model-based standard errors are generally employed with independence models unless the model is being compared with a model having another structure with empirical standard errors.

For the above example, we see that females have an approximate mean 38% greater increase in risk of going to the doctor than do males. Those patients with less education visit the doctor more often.

14.2.3 Negative binomial GEE models

Recall that, strictly speaking, a negative binomial is a member of the GLM family of distributions only if its heterogeneity or ancillary parameter (α) is known and then specified to the GLM algorithm as a constant. This results in a problem for the statistician, who must determine the appropriate value of α to be given the command or function. We discussed several search mechanisms in Chapter 8. However, the best solution is to have the GLM software call an external maximum likelihood function to estimate α, inserting the result into the GLM estimation algorithm. We have seen that this has been implemented using R's **glm.nb** function, Stata's **glm** command, and SAS's **GENMOD** procedure. This method is not, however, available in Stata's primary GEE command, **xtgee,** nor in **GENMOD**, which with the REPEATED option provides for a GEE model. The SAS algorithm employs a value of α for all structures based on the value obtained using a standard or independence model. At this time R's **gee**, **geeglm** and **yags** functions, all of which provide GEE modeling capability, fail to accommodate a negative binomial family.

When modeling a negative binomial GEE using Stata, we first must model a maximum likelihood negative binomial using **nbreg** or **glm, fam(nb ml)**, then insert the estimated value of α into the **xtgee** family option, i.e. *fam(nb 2.1414)* Of course, the standard NB2 model assumes an independence correlation structure i.e. the standard NB2 model is identical to a negative binomial GEE model with an independence correlation structure. When using other correlation structures, though, the independence-based ancillary parameter may not be quite as appropriate; but there is no alternative, unless we use a method similar to that employed in Table 8.4. It does not appear that using the value of α obtained from a pooled independence-structured NB2 model with GEEs having other structures results in substantial bias, but such bias does as a consequence exist. How much depends on the data and how close the selected structure is to the true structure for the given data. No studies exist in the literature addressing this question.

When modeling a negative binomial GEE, Stata's **xtgee** command works in a rather counterintuitive manner. The value for α, when entered into the family option as a constant, is inverted by the algorithm, producing a model with a displayed α of $1/\alpha$. The log-likelihood function is also parameterized with an inverted α. Recall the discussion of the parameterization of the NB2 PDF with $\alpha = 1/k$. In any case, the inversion in effect cancels the second inversion, maintaining the PDF that is shown as equation 8.8. The confusing part of this rests with the displayed α in the GEE output. The examples below should clarify how to handle this situation. Unfortunately there is no documentation in the Stata manuals or help files regarding this inversion. The above apparent inconsistency does not appear in other major commercial GEE implementations except S-Plus, upon which Stata's command was based.

We model the same data as above with a Poisson GEE, using the negative binomial family with independence correlation structure. A standard NB2 model is used to initially obtain a maximum likelihood value for α.

ML NB2 ESTIMATION – TO OBTAIN VALUE OF α

R

```
nbg <- glm.nb(docvis ~ female + factor(edlevel), data=rwm5yr)
library(haplo.ccs)
sandcov(nbg, rwm5yr$id)
sqrt(diag(sandcov(nbg, rwm5yr$id)))
summary(nbg)
exp(coef(nbg))
# no GEE NB in R
```

```
. nbreg docvis female edlevel2-edlevel4, irr robust cluster(id)
  robust

Negative binomial regression                Number of obs     =      19609
Dispersion             = mean               Wald chi2(4)      =     186.16
Log pseudolikelihood = -43110.7             Prob > chi2       =     0.0000
                         (Std. Err. adjusted for 6127 clusters in id)
------------------------------------------------------------------------------
             |              Robust
     docvis  |       IRR   Std. Err.      z    P>|z|     [95% Conf. Interval]
-------------+----------------------------------------------------------------
     female  | 1.391372   .0513088     8.96    0.000     1.294357    1.495659
   edlevel2  | .8331946   .0546531    -2.78    0.005     .7326763    .9475032
   edlevel3  | .7276139   .0420571    -5.50    0.000     .6496813    .8148949
   edlevel4  | .6001595   .0435391    -7.04    0.000     .5206136    .6918595
-------------+----------------------------------------------------------------
   /lnalpha  | .7558711   .0200549                      .7165642    .7951779
-------------+----------------------------------------------------------------
      alpha  | 2.129466   .0427062                      2.047387    2.214835
```

Now use the estimated value of α, 2.129466, as a constant in the GEE model.

GEE NB2 ESTIMATION – WITH α ENTERED AS A CONSTANT

```
. xtgee docvis female edlevel2-edlevel4 , i(id) corr(indep)
   fam(nb 2.129466)  eform robust

GEE population-averaged model            Number of obs      =     19609
Group variable:                    id    Number of groups   =      6127
Link:                             log    bs per group: min  =         1
Family: negative binomial(k=.4696) ←               avg  =       3.2
Correlation:              independent                 max  =         5
                                         Wald chi2(4)       =    196.15
Scale parameter:                    1    Prob > chi2        =    0.0000

Pearson chi2(19609):         82189.19    Deviance           =  48468.79
Dispersion (Pearson):        4.191401    Dispersion         =  2.471763
                               (Std. Err. adjusted for clustering on id)
-------------------------------------------------------------------------
         |              Semi-robust
  docvis |      IRR   Std. Err.      z    P>|z|   [95% Conf. Interval]
---------+---------------------------------------------------------------
  female | 1.391372  .0517178   8.89   0.000   1.293612    1.496521
edlevel2 | .8331946  .0546379  -2.78   0.005   .7327026    .9474693
edlevel3 | .7276139  .0422568  -5.48   0.000   .6493319    .8153334
edlevel4 | .6001595  .0439389  -6.97   0.000   .5199344    .6927633
-------------------------------------------------------------------------
```

The two models are identical, given rounding error. Notice the inverted value of α that is produced in the output:

```
. di 1/2.129466
.4696013                   /*  INVERTED VALUE OF ALPHA */
```

If we used the inverted value as a constant instead, the parameter estimates would be mistaken.

The correlation structure is the same as for any independence model.

```
. estat vce, cov
Estimated within-id correlation matrix R:
        c1       c2       c3       c4       c5
r1  1.0000
r2  0.0000   1.0000
r3  0.0000   0.0000   1.0000
r4  0.0000   0.0000   0.0000   1.0000
r5  0.0000   0.0000   0.0000   0.0000   1.0000
```

The coefficients and related standard error statistics for the above independence negative binomial model are displayed as:

NB GEE, Independence correlation structure, coefficients

```
                         (Std. Err. adjusted for clustering on id)
------------------------------------------------------------------------------
             |             Semi-robust
      docvis |      Coef.   Std. Err.      z    P>|z|     [95% Conf. Interval]
-------------+----------------------------------------------------------------
      female |   .3302906   .0371703     8.89   0.000     .2574381    .4031431
    edlevel2 |  -.1824881   .0655764    -2.78   0.005    -.3110154   -.0539607
    edlevel3 |  -.3179847   .0580758    -5.48   0.000    -.4318113   -.2041582
    edlevel4 |  -.5105598    .073212    -6.97   0.000    -.6540527   -.3670669
       _cons |   1.043704   .0303465    34.39   0.000     .9842261    1.103182
------------------------------------------------------------------------------
```

The characteristics of the independence structure have previously been mentioned. The model is rarely used in its own right, but is rather employed as a basis for models with other structures.

Exchangeable correlation structure

SCHEMATIC

```
1
A  1
A  a  1
A  a  a  1
A  a  a  a  1
```

EXAMPLE

```
1
.29  1
.29  .29  1
.29  .29  .29  1
.29  .29  .29  .29  1
```

```
. xtgee docvis female edlevel2-edlevel4 , c(exch) fam(nb 2.129466)
  i(id) eform robust

GEE population-averaged model               Number of obs      =     19609
Group variable:                         id  Number of groups   =      6127
Link:                                  log  Obs per group: min =         1
Family:  negative binomial(k=.4696)                        avg =       3.2
Correlation:                  exchangeable                 max =         5
                                            Wald chi2(4)       =    187.17
Scale parameter:                         1  Prob > chi2        =    0.0000
                         (Std. Err. adjusted for clustering on id)
```

```
------------------------------------------------------------------------------
             |             Semi-robust
     docvis  |       IRR   Std. Err.      z    P>|z|     [95% Conf. Interval]
-------------+----------------------------------------------------------------
     female  |    1.3822    .0505856    8.84   0.000     1.286527    1.484989
   edlevel2  |  .8513858    .0549046   -2.49   0.013     .7502978    .9660934
   edlevel3  |  .7263763    .0403438   -5.76   0.000     .6514556    .8099131
   edlevel4  |  .6341802    .0459056   -6.29   0.000     .5502978    .7308488
------------------------------------------------------------------------------

. estat vce, cov

Estimated within-id correlation matrix R:

          c1       c2       c3       c4       c5
r1    1.0000
r2    0.3324   1.0000
r3    0.3324   0.3324   1.0000
r4    0.3324   0.3324   0.3324   1.0000
r5    0.3324   0.3324   0.3324   0.3324   1.0000
```

SAS: *Exchangeable structure*

```
PROC GENMOD DATA=WORK.rwm5yr
PLOTS(ONLY)=ALL;
CLASS  ID;
 MODEL DOCVIS= FEMALE EDLEVEL2 EDLEVEL3 EDLEVEL4 / LINK=log
DIST=NEGBIN TYPE3 dscale pscale;
REPEATED subject=ID / TYPE=EXCH covb corrw modelse;
RUN;
```

The exchangeable correlation structure is the most commonly used structure among GEE models. It is the default for several of the major commercial software implementations and it is generally the appropriate model to use with clustered or first-level nested data. It is not to be used with longitudinal data.

The exchangeable correlation structure, defined in equation 14.3, assumes that the correlations between measurements within a panel are the same, irrespective of any time interval. Any correlation value within the structure may be exchanged with any other – hence the name *exchangeable*. The value of a, displayed in the schematic matrix above, is a scalar. It does not vary between panels. This is why it is not appropriate for assessing longitudinal effects. Its primary use is for use with clustered data.

The code in the box above uses the SAS/STAT GENMOD procedure with the REPEATED option to estimate a GEE. Unlike the GENMOD output for negative binomial models, which displays a value for α, which SAS terms k, no such statistic is displayed for GEE models. Stata requires a value to be given the algorithm, with a default of 1, but SAS requests no such value, nor does it display an estimate. As previously mentioned, it silently estimates a standard NB2 model, captures the value of *k*, and enters that into the GEE algorithm, regardless of the correlation structure used. Comparing Stata and SAS output for the models used here indicates a close approximation to estimates and SEs. SAS and Stata, however, employ different default divisors for ϕ. Use of the *nmp* option in Stata provides the appropriate adjustment so that SAS and Stata output are nearly identical. Stata has a default value of N, whereas SAS uses N-*p*, where n=model observations and *p*=model are predictors. The SAS formula was used in equation 14.4, which I believe is preferred.

SAS and Stata output will differ when there are single observation panels. It is not possible to draw a correlation from a single instance. SAS and Stata handle this situation in different manners and, as a result, when the data consist of a considerable number of single-observation panels, the model results will differ. I should also mention that use of R's **geeglm, gee**, and **yags** functions on the same data typically result in different model values. Researchers have discovered inconsistencies at times with R GEE results when compared with Stata and SAS. Care must be taken therefore when using these GEE functions in research, particularly for models involving a time component.

Unstructured correlation structure

SCHEMATIC

```
1
C1  1
C2  C5  1
C3  C6  C8  1
C4  C7  C9  C10  1
```

EXAMPLE

```
1
.34  1
.29  .28  1
.33  .14  .24  1
.21  .07  .11  .23  1
```

```
. xtgee docvis female edlevel2-edlevel4 , c(unst) fam(nb 2.129466)
i(id) t(year) eform robust

GEE population-averaged model              Number of obs      =     19609
Group and time vars:             id year   Number of groups   =      6127
Link:                                log   Obs per group: min =         1
Family:  negative binomial(k=.4696)                       avg =       3.2
Correlation:              unstructured                    max =         5
```

```
                                          Wald chi2(4)      =     184.02
Scale parameter:                      1   Prob > chi2       =     0.0000
                             (Std. Err. adjusted for clustering on id)
------------------------------------------------------------------------
         |             Semi-robust
  docvis |        IRR   Std. Err.       z    P>|z|    [95% Conf. Interval]
---------+--------------------------------------------------------------
  female |   1.374117   .0501441     8.71   0.000    1.279269    1.475998
edlevel2 |   .8558149    .055595    -2.40   0.017    .7535022    .9720199
edlevel3 |   .7259285   .0400149    -5.81   0.000    .6515887    .8087498
edlevel4 |   .6360158   .0460806    -6.25   0.000     .551819    .7330592
------------------------------------------------------------------------

. estat vce, cov
```

Estimated within-id correlation matrix, R:

```
        c1       c2       c3       c4       c5
r1  1.0000
r2  0.3683   1.0000
r3  0.3390   0.3576   1.0000
r4  0.3001   0.3218   0.5045   1.0000
r5  0.3245   0.2490   0.2655   0.2988   1.0000
```

Compare the correlation structure with that of the Poisson model on the same data. Notice the similarity of the two matrices.

Unstructured correlation structure with Poisson model

```
. estat vce, cov
```

Estimated within-id correlation matrix, R:

```
        c1       c2       c3       c4       c5
r1  1.0000
r2  0.3716   1.0000
r3  0.3474   0.3583   1.0000
r4  0.3071   0.3193   0.5092   1.0000
r5  0.3249   0.2519   0.2689   0.3082   1.0000
```

We would suspect that the negative binomial model itself differs little from the Poisson.

R: *Poisson GEE*

```
library(gee)
pgee <-
gee(docvis ~ female + factor(edlevel), id=id, family=poisson,
   data=rwm5yr, corstr="unstructured", scale.fix=TRUE
summary(pgee)
```

POISSON ESTIMATES

```
------------------------------------------------------------------------------
         |             Semi-robust
  docvis |        IRR   Std. Err.      z    P>|z|   [95% Conf. Interval]
---------+--------------------------------------------------------------------
  female | 1.364511   .0502951     8.43   0.000    1.269411    1.466736
edlevel2 | .8607318   .0563454    -2.29   0.022    .7570879    .9785642
edlevel3 | .7327953   .0401485    -5.67   0.000    .6581834    .8158652
edlevel4 | .6463942   .0481763    -5.85   0.000    .5585431    .7480631
------------------------------------------------------------------------------
```

We do find that the two tables of parameter estimates are similar.

In the unstructured correlation structure all correlations are assumed to be different; correlations are freely estimated from the data. This can result in the calculation of a great many correlation coefficients for large matrices. Because it (can have) has a different coefficient for each cell, the structure optimally fits the data. However, it loses efficiency, and hence interpretability, when models have more than about three predictors. The number of coefficients to be estimated is based on the size of the largest panel of observations.

$$\text{Number of coefficients} = p(p-1)/2$$

where p is the number of observations in the largest panel.

Use this correlation structure when the size of the panels is small, there are relatively few predictors, and there are no missing values.

Autoregressive correlation structure

SCHEMATIC

```
1
C^1  1
C^2  C^1  1
C^3  C^2  C^1  1
C^4  C^3  C^2  C^1  1
```

EXAMPLE

```
1
.48  1
.23  .48  1
.11  .23  .48  1
.05  .11  .23  .48  1
```

```
. xtgee docvis female edlevel2-edlevel4, i(id) t(year) corr(ar 1)
  force eform fam(nb 2.129466) robust
note:  some groups have fewer than 2 observations
       not possible to estimate correlations for those groups
       1150 groups omitted from estimation
```

```
GEE population-averaged model          Number of obs      =     18459
Group and time vars:         id year   Number of groups   =      4977
Link:                            log   Obs per group: min =         2
Family:  negative binomial(k=.4696)                   avg =       3.7
Correlation:                   AR(1)                  max =         5
                                       Wald chi2(4)       =    187.86
Scale parameter:                   1   Prob > chi2        =    0.0000
                                (Std. Err. adjusted for clustering on id)
------------------------------------------------------------------------
           |              Semi-robust
    docvis |       IRR    Std. Err.      z    P>|z|    [95% Conf. Interval]
-----------+------------------------------------------------------------
    female | 1.402319    .0539399    8.79   0.000    1.300486    1.512127
  edlevel2 | .8248012    .0561665   -2.83   0.005    .7217471    .9425699
  edlevel3 | .7200313    .0439616   -5.38   0.000     .638824    .8115618
  edlevel4 | .5970729      .04618   -6.67   0.000    .5130881    .6948046
------------------------------------------------------------------------

. estat vce, cov
```

Estimated within-id correlation matrix, R:

```
        c1       c2       c3       c4       c5
r1  1.0000
r2  0.3755   1.0000
r3  0.1410   0.3755   1.0000
r4  0.0529   0.1410   0.3755   1.0000
r5  0.0199   0.0529   0.1410   0.3755   1.0000
```

SAS: *Autoregressive 1 Correlation*

```
PROC GENMOD DATA=WORK.rwm5yr
PLOTS(ONLY)=ALL;
CLASS  ID year;
MODEL DOCVIS= FEMALE EDLEVEL2 EDLEVEL3 EDLEVEL4 / LINK=log
DIST=NEGBIN TYPE3;
REPEATED subject=ID / sorted withinsubject=year TYPE=AR covb
 corrw modelse;
```

The autoregressive (AR) correlation structure assumes that there is a marked decrease in correlation coefficient values with the corresponding increase in measurements within panel time intervals. Each off-diagonal from the main diagonal decreases by the square of the previous diagonal. One might consider the decrease in values to be increasing powers of the first off-diagonal.

```
. di (.3755)^2
.14100025

. di ((.3755)^2)^2
.01988107
```

Large matrices produce very small coefficient values. The depiction here is the case for AR(1) models, as it is for this example, but not for all AR levels. Note also that the use of the *force* option is required when time intervals are not equal. The intervals are generally equal in this model, except when missing values exclude a particular time interval. In any case, it is best to use this option when time-based GEE models are being estimated, in case there are non-equal intervals in the data. Use this correlation structure when each panel is a collection of data over time for the same person.

Stationary or *m*-dependent correlation and non-stationary structures

The stationary correlation structure specifies a constant correlation for each off-diagonal. The diagonals are then interpreted as lags or measurements. Correlations c lags apart are equal in value to one another, $c + 1$ lags apart are also equal to one another, and so forth until a defined stop, m, is reached. Correlations greater than m are defined as zero, hence the meaning of m-dependent. In larger matrices the correlation structure appears as a band. Since $m = 1$ in the above stationary model, correlations greater than the first off-diagonal have values of zero. The stationary correlation structure is primarily used when the off-diagonals, or lags, are thought of as time intervals.

The non-stationary correlation structure is the same as the stationary except that the values of each lag or off-diagonal are not constant. Of course, correlation values beyond m are all 0. Some statisticians use the non-stationary correlation structure when they have ruled out the others, but still have a limit to the range of measurement error or lags in the data.

14.2.4 GEE goodness-of-fit

As mentioned earlier in the chapter, for the past decade the QIC statistic has been regarded as a standard test for determining the “best” correlation structure to accommodate excess Poisson heterogeneity. Before reviewing some of the problems that have been identified with the statistic, it will be instructive to understand its logic.

The QIC was first developed by Pan (2001a), with an amendment made by Hardin and Hilbe (2003), as alternative to the AIC statistic. The problem with using the AIC with clustered and longitudinal data rests in the distributional assumption underlying the AIC that observations in the model are independent of one another. Clustering and longitudinal effects preclude independence. The QIC statistic, expressed as QIC(R) to represent its correlation structure form, is defined as:

$$\text{QIC}(R) = -2Q(g^{-1}(x\beta_R)) + 2\text{trace}(A_{\text{I}}^{-1}V_{\text{MS,R}}) \tag{14.6}$$

where

$2Q(g^{-1}(x\beta_R))$ is the value of the computed quasi-likelihood using the model coefficients with hypothesized correlation structure R. When evaluating the quasi-likelihood, $\mu = g^{-1}(x\beta_R)$ is substituted in place of $\mu = g^{-1}()$, where $g^{-1}()$ is the model inverse link function.

A_{I} is the independence structure variance matrix for the model.

$V(_{\text{MS,R}})$ is the modified robust or sandwich variance estimator of the model, with hypothesized structure $R(a)$. Some formulations use Σ_{R} as the symbol.

The AIC penalty term, $2p$, is expressed as $2\text{trace}(A_{\text{I}}{}^{-1}V_{\text{MS,R}})$ for the QIC statistic. The difference between the parameterization employed by Pan (2001a) and that given by Hardin and Hilbe (2003) rests in the second term of equation 14.6. Essentially, the term is the ratio of the robust covariance and the model-based covariance matrices. The Hardin–Hilbe statistic, referred to as QIC_{hh} by Hin and Wang (2008) and Barnett *et al.* (2010), specifies that the model-based coefficients derive from the estimation of an independence correlation structure. Pan's model-based coefficients, QIC_{p} derive from a non-independence structure. The ability of either version to select true correlation structures is rather low compared with the AIC statistic, as well as the Bayesian-based DIC statistic. However, work by Hin and Wang (2008) and Hilbe (2009) indicate that the second term of the QIC statistic, $2\text{trace}\left(A_{\text{I}}^{-1}V_{\text{R}}\right)$, is a significantly superior selector of the true correlation structure than QIC, and even for AIC. Hin and Wang termed the statistic the CIC, or correlation information criterion.

Barnett *et al.* (2010) constructed a number of simulations testing six structures on a model with a single fixed covariate and a model with a single random covariate. The exchangeable and autoregressive structures were employed, with a weak and a moderate correlation value. Summary results were given as shown in Table 14.3.

The AIC does a much better job selecting the true correlation structure, except with the unstructured correlation. Moreover, the QIC tends to select the unstructured when in fact the true structure is another. Therefore QIC it is somewhat unreliable when selecting the unstructured as well. Following Hilbe (2009), we refer to the 2∗trace statistic as CIC, as

Table 14.3 *Simulation: percentage true correlation structure selected*

	AIC		QIC	
	Fixed	Random	Fixed	Random
Independent:	70	76	3	2
Exchangeable (0.2)	97	98	3	0
Exchangeable (0.5)	100	100	25	30
Autoregressive (0.3)	97	89	14	10
Autoregressive (0.7)	100	100	81	89
Unstructured	27	13	56	40

discussed above. If both the CIC and AIC statistics are lower for a given GEE model than for alternatives, it is highly likely that these indicate the correct structure. If the two statistics do not cohere, then we must decide on the basis of considerations that take us beyond the present discussion. Refer to Hilbe (2009). See also Hardin and Hilbe (2003) for a text completely devoted to GEE and related models.

14.2.5 GEE marginal effects

The value that is reported by the **margins** command (without the -atmeans- option) is an average marginal effect that is the simple average of the marginal effects that are calculated for all observations used in the estimation. For example,

```
. qui xtgee docvis female edlevel2-edlevel4, fam(nb 2.13) i(id)
  robust
. margins, dydx(*)

Average marginal effects                        Number of obs =  19609
Model VCE    : Semirobust

Expression   : Exponentiated linear prediction, predict()
dy/dx w.r.t. : female edlevel2 edlevel3 edlevel4

------------------------------------------------------------------------------
             |            Delta-method
             |      dy/dx   Std. Err.      z    P>|z|     [95% Conf. Interval]
-------------+----------------------------------------------------------------
      female |   1.029924   .1174731     8.77   0.000     .7996807    1.260167
    edlevel2 |  -.5119443   .2065768    -2.48   0.013    -.9168274   -.1070612
    edlevel3 |   -1.01723   .1810877    -5.62   0.000    -1.372155   -.6623046
    edlevel4 |  -1.449133   .2349737    -6.17   0.000    -1.909673   -.9885933
------------------------------------------------------------------------------
```

The marginal effects may be calculated by hand using the following logic. First we demonstrate that the predicted fit is the same as the result of using the *predict* command for this type of model. Duplicating the Stata code in R is simple, and uses the same logic as used in previous margial effects coding.

```
. gen expxb = exp(_b[female]*female + _b[edlevel2]*edlevel2 +
    _b[edlevel3]*edlevel3 + _b[edlevel4]*edlevel4 +  _b[_cons])
. predict mu
. su expxb mu

Variable |  Obs       Mean  Std. Dev.       Min       Max
---------+-----------------------------------------------
   expxb | 19609  3.181951  .6730023  1.803849  3.931506
      mu | 19609  3.181951  .6730023  1.803849  3.931506
```

They are indeed the same. The marginal effect is obtained by multiplying the coefficient of the term by the fitted value. For *female*, we have

```
. gen dydfemale = _b[female] * mu
. sum dydfemale

 Variable |    Obs       Mean  Std. Dev.        Min        Max
----------+----------------------------------------------------
dydfemale |  19609  1.029924  .2178353   .5838642   1.272537
```

which is the same value displayed in the marginal effects output.

Likewise, if the *atmeans* option is specified, the marginal effects are computed at the simple mean of the variables. For the variable *female* we have

```
. sum female
. local mfemale = r(mean)
. sum edlevel2
. local med2 = r(mean)
. sum edlevel3
. local med3 = r(mean)
. sum edlevel4
. local med4 = r(mean)
. local mu2 = exp(_b[female]*`mfemale' + _b[edlevel2]*`med2' +
     _b[edlevel3]*`med3' + _b[edlevel4]*`med4' +  _b[_cons])
. di "dydfemale at means = " _b[female] * `mu2'
dydfemale at means = 1.0052001

. di "dydfemale at edlevel2 = " _b[edlevel2] * `mu2'
dydfemale at edlevel2 = -.4996549

. di "dydfemale at edlevel3 = " _b[edlevel3] * `mu2'
dydfemale at edlevel3 = -.99281098
```

```
. di "dydfemale at edlevel4 = " _b[edlevel4] * `mu2'
dydfemale at edlevel4 = -1.4143462

. margins, dydx(*) atmeans

Conditional marginal effects                    Number of obs   = 19609
Model VCE    : Semirobust

------------------------------------------------------------------------
         |            Delta-method
         |      dy/dx   Std. Err.      z    P>|z|   [95% Conf. Interval]
---------+--------------------------------------------------------------
  female |     1.0052   .1119555     8.98   0.000    .7857713   1.224629
edlevel2 | -.4996549    .2013257    -2.48   0.013    -.894246  -.1050638
edlevel3 |  -.992811    .1756741    -5.65   0.000   -1.337126  -.6484961
edlevel4 | -1.414346    .2275454    -6.22   0.000   -1.860327  -.9683654
------------------------------------------------------------------------
```

The values are identical. The predictors have been assumed to be continuous so that calculations for marginal effects could be demonstrated. The same logic that we used earlier for discrete change is applicable for binary and categorical variables.

Calculations for marginal effects are the same for other panel models as well, given that each is taken for the mean value of the predictor values per panel. The difference is based on the predicted values employed after the other panel commands. Again, R-users do not have GEE negative binomial modeling capability. If the Poisson model is used instead, marginal effects and discrete change may be calculated as demonstrated in Chapters 6 and 9, with the amendments given in this section.

14.3 Unconditional fixed-effects negative binomial model

Fixed-effects count models may be estimated in two ways – unconditionally and conditionally. We begin with a consideration of the unconditional fixed-effects Poisson model since it is the basis on which we can understand the negative binomial parameterizations.

Unconditional estimation of the fixed-effects Poisson model can be obtained using standard GLM software as well as the traditional maximum likelihood Poisson procedure. The model is specified by including a separate fixed effect for each defined panel in the data. The fixed effects are specified by indicator variables, as with estimating factor or categorical predictors. We represent this relationship as:

$$\ln(\mu_{ik}) = \exp(\beta x_{ik} + \delta_i) \tag{14.7}$$

where δ is the fixed effect associated with individual i, and subscript k indexes the observations associated with individual i. When a panel relates observations collected over a time period, it is customary to use the subscript t instead of k. We shall use k throughout our discussion, but with the knowledge that t is commonly used for longitudinal models. The log-likelihood for the unconditional fixed effects Poisson takes the form of

$$\mathcal{L}\left(x_i'\beta; y_i\right) = \sum_{i=1}^{n}\sum_{k=1}^{K}\left[y_{ik}\left(x_{ik}'\beta + \delta_i\right) - \exp\left(x_{ik}'\beta + \delta_i\right) - \ln\Gamma\left(y_{ik}+1\right)\right] \tag{14.8}$$

Note the similarity to that of the standard Poisson log-likelihood function defined in Chapter 6 as:

$$\mathcal{L}\left(x_i'\beta; y_i\right) = \sum_{i=1}^{n}\left\{y_i\left(x_i'\beta\right) - \exp\left(x_i'\beta\right) - \ln\Gamma\left(y_i+1\right)\right\} \tag{14.9}$$

I shall use the well-known *ships* data set that was used in McCullagh and Nelder (1989), Hardin and Hilbe (2003), and other sources. The data set contains values on the number of reported accidents for ships belonging to a company over a given time period. The variables are defined as:

```
accident   : number of accidents  (response)
ship       : ship identification (1-5)
op         : ship operated between the years 1975 and 1979 (1/0)
             op=0 ship operating between 1965-1974
co65_69    : ship was in construction between 1965 and 1969 (1/0)
co70_74    : ship was in construction between 1970 and 1974 (1/0)
co75_79    : ship was in construction between 1975 and 1979 (1/0)
service    : months in service
```

With *co*_65-69 as the reference for year of construction, and the natural log of the months of service specified as the offset, a basic Poisson model of the data is given as shown below.

Table 14.4 R: *Ships data – Poisson with offset*

```
rm(list=ls())
data(ships)
attach(ships)
poff <- glm(accident ~ op + co_70_74 + co_75_79 +
        offset(log(service)), family=poisson, data=ships)
summary(poff)
modelfit(poff)
```

```
. glm accident op co_70_74 co_75_79, fam(poi) lnoffset(service)

Generalized linear models                   No. of obs      =        34
Optimization      : ML                      Residual df     =        30
                                            Scale parameter =         1
Deviance          =  89.51040004            (1/df) Deviance = 2.98368
Pearson           =  127.7051909            (1/df) Pearson  = 4.25684
Variance function: V(u) = u                 [Poisson]
Link function     : g(u) = ln(u)            [Log]
                                            AIC             =  5.746379
Log likelihood    = -93.68844568            BIC             = -16.28042
--------------------------------------------------------------------
         |               OIM
accident |      Coef.   Std. Err.      z    P>|z|   [95% Conf. Interval]
---------+----------------------------------------------------------
      op |   .4667002   .1180644    3.95   0.000    .2352982   .6981022
co_70_74 |   .6001222   .1213314    4.95   0.000    .3623171   .8379273
co_75_79 |   .2338377   .1940506    1.21   0.228   -.1464944   .6141698
   _cons |  -6.556674   .0883009  -74.25   0.000    -6.72974  -6.383607
 service |  (exposure)
--------------------------------------------------------------------

. abic
AIC Statistic  =    5.746379           AIC*n        = 195.37689
BIC Statistic  =    5.837272           BIC(Stata) = 201.48233
```

Except for the upper year of construction group, predictors appear to be significant; however, the model is clearly overdispersed. We have purposefully ignored the correlation of values within each panel of ship in the above model. A negative binomial model can be used to generically account for the overdispersion.

Table 14.5 R: *Ships data – NB2 with offset*

```
nboff <- glm.nb(accident ~ op + co_70_74 + co_75_79 +
         offset(log(service)), data=ships)
summary(nboff)
modelfit(nboff)
```

```
. glm accident op co_70_74 co_75_79, fam(nb ml) lnoffset(service)

Generalized linear models                  No. of obs      =         34
Optimization      : ML                     Residual df     =         30
                                           Scale parameter =          1
Deviance          =  37.19091483           (1/df) Deviance =   1.239697
Pearson           =  51.13687296           (1/df) Pearson  =   1.704562
Variance function: V(u) = u+(.2256)u^2 [Neg. Binomial]
```

```
Link function     : g(u) = ln(u)           [Log]
                                            AIC            =  5.052949
Log likelihood    = -81.90013205            BIC            = -68.5999
------------------------------------------------------------------------
         |               OIM
accident |     Coef.  Std. Err.      z   P>|z|   [95% Conf. Interval]
---------+--------------------------------------------------------------
      op |   .3338351  .2786052   1.20   0.231   -.212221    .8798911
co_70_74 |   .5847677   .277535   2.11   0.035   .0408091    1.128726
co_75_79 |   .0580994  .3869406   0.15   0.881  -.7002903     .816489
   _cons |  -6.209554  .2392562 -25.95   0.000  -6.678487    -5.74062
 service | (exposure)
------------------------------------------------------------------------
Note: Negative binomial parameter estimated via ML and treated as
      fixed once estimated.

. abic
AIC Statistic    =    5.052949          AIC*n       = 171.80026
BIC Statistic    =    5.143842          BIC(Stata)  = 177.9057
```

Much of the overdispersion has been accommodated by the negative binomial model, but there is still evidence of extra correlation in the data. The AIC and BIC statistics are significantly lower in the negative binomial model. We also know from the data what may be causing overdispersion – the panel-specific effect of the individual ships. We assign a specific indicator to each panel. Each ship will have a separate intercept. This type of model is called an unconditional fixed-effects model. As a Poisson model we have the output shown below.

Table 14.6 R: *Ships data – fixed-effects Poisson*

```
fep <- glm(accident ~ op + co_70_74 + co_75_79 + factor(ship) +
        offset(log(service)), data=ships)
summary(fep)
modelfit(fep)
```

```
. tab ship, gen(ship)
. glm accident op co_70_74 co_75_79 ship2-ship5, fam(poi)
  lnoffset(service)

Generalized linear models                   No. of obs      =        34
Optimization      : ML                      Residual df     =        26
                                            Scale parameter =         1
Deviance          =  61.51666866            (1/df) Deviance =  2.366026
Pearson           =  68.35293608            (1/df) Pearson  =  2.628959
Variance function: V(u) = u                 [Poisson]
```

```
Link function     : g(u) = ln(u)             [Log]
                                             AIC              =  5.158328
Log likelihood    = -79.69157999             BIC              =  -30.1687
--------------------------------------------------------------------------
         |                OIM
accident |      Coef.   Std. Err.      z    P>|z|    [95% Conf. Interval]
---------+----------------------------------------------------------------
      op |   .4560211   .1181882    3.86   0.000     .2243764    .6876657
co_70_74 |   .3665811    .134349    2.73   0.006     .1032618    .6299004
co_75_79 |  -.0145843   .2069169   -0.07   0.944     -.420134    .3909653
   ship2 |  -.6237779   .1793695   -3.48   0.001    -.9753357   -.2722201
   ship3 |  -.7496662   .3292493   -2.28   0.023    -1.394983   -.1043494
   ship4 |  -.1183089   .2903265   -0.41   0.684    -.6873384    .4507206
   ship5 |   .3302558   .2359759    1.40   0.162    -.1322486    .7927601
   _cons |  -5.964478   .1933293  -30.85   0.000    -6.343396   -5.585559
 service | (exposure)
--------------------------------------------------------------------------

. abic
AIC Statistic    =    5.158328         AIC*n       = 175.38316
BIC Statistic    =    5.666301         BIC(Stata) = 187.59404
```

A substantial amount of the overdispersion present in the original Poisson model has been accounted for. However, the negative binomial handled the overdispersion better than the unconditional fixed-effects Poisson. Note the AIC and BIC statistics. They are significantly lower than the original Poisson model, but not as low as the negative binomial.

We next attempt to model an unconditional fixed-effects negative binomial. We suspect that it will better fit the data than any of the above models. Some statisticians will argue that robust variance estimators should be used to adjust standard errors, but we will maintain use of the model standard errors for the example. Neither parameters, fit, nor AIC/BIC statistics are affected.

Table 14.7 R: *Ships data – fixed-effects NB2*

```
fenb <- glm.nb(accident ~ op + co_70_74 + co_75_79 + factor(ship) +
        offset(log(service)), data=ships)
summary(fenb)
modelfit(fenb)
```

```
. glm accident op co_70_74 co_75_79 ship2-
ship5, fam(nb ml) lnoffset(service)

Generalized linear models                    No. of obs       =        34
Optimization     : ML                        Residual df      =        26
                                             Scale parameter  =         1
```

```
Deviance        =  36.44214112     (1/df) Deviance =  1.401621
Pearson         =   38.8973235     (1/df) Pearson  =  1.496051
Variance function: V(u) = u+(.0725)u^2  [Neg. Binomial]
Link function   : g(u) = ln(u)     [Log]
                                   AIC             =   4.84811
Log likelihood  = -74.41787634     BIC             = -55.24323
------------------------------------------------------------------
             |              OIM
    accident |     Coef.  Std. Err.     z   P>|z|  [95% Conf. Interval]
-------------+----------------------------------------------------
          op |  .3589723   .193599    1.85  0.064  -.0204748  .7384194
    co_70_74 |  .3600056  .2023644    1.78  0.075  -.0366214  .7566325
    co_75_79 | -.0512326   .282032   -0.18  0.856  -.6040051  .5015399
       ship2 | -.6233191  .2468374   -2.53  0.012  -1.107111 -.1395268
       ship3 |  -.706535  .3788082   -1.87  0.062  -1.448985  .0359154
       ship4 | -.1353214  .3490007   -0.39  0.698  -.8193501  .5487073
       ship5 |  .4383718  .3079319    1.42  0.155  -.1651635  1.041907
       _cons | -5.878417  .2652524  -22.16  0.000  -6.398302 -5.358532
     service | (exposure)
------------------------------------------------------------------
Note: Negative binomial parameter estimated via ML and treated as
      fixed once estimated.
Likelihood-ratio test of alpha=0: chibar2(01)= 10.55
                                  Prob>=chibar2 = 0.001
. abic
AIC Statistic   =   4.84811         AIC*n       = 164.83575
BIC Statistic   =  5.356083         BIC(Stata)  = 177.04663
```

The AIC and BIC statistics are substantially lower than the previous three models. Note that the value of α is 0.0725, approaching the Poisson. Given the value of the boundary likelihood-ratio test, however, we find that the data are better modeled as negative binomial than Poisson. (Note: the likelihood test is based on the **nbreg** model of identical data.)

A caveat on using this form of fixed-effects regression: use it only if there are a relatively few number of panels in the data. If there are more than 20 panels, it is preferrable to use the conditional fixed-effects model. The greater the number of panels, the greater the possible bias in parameter estimates for the levels or panels of the effect variable. This is called the 'incidental parameters problem', first defined by Neyman and Scott (1948). It is interesting that a number of econometricians have thought that the incidental parameters problem, which we shall refer to as the IP problem, affects the unconditional fixed-effects Poisson model. Woutersen (2002) attempted to ameliorate the IP problem with Poisson models by employing an integrated moment estimator. Other attempts include Lancaster (2002) and Vadeby (2002). Most of these attempted solutions are based on separating the main model parameters from the array of fixed-effects parameters. However, it has been demonstrated by Greene (2006a) and others that the IP problem is not real when applied to the Poisson model. This conclusion is based on the observation that the Poisson conditional fixed-effects estimator is numerically equal to the unconditional estimator, which means that there is no IP problem. On the other hand, the IP problem does affect the unconditional fixed-effects negative binomial. The fixed-effects negative binomial model has an additional problem. It is intrinsically different from the Poisson. Recall that the Poisson fixed-effects has a mean, μ_{ik}, value of $\exp(\beta x_{ik} + \delta_i)$. This means that the fixed effect is built into the Poisson mean parameter. The negative binomial fixed-effects model, though, builds the fixed effects into the distribution of the gamma heterogeneity, α, not the mean. This makes it rather difficult to interpret the IP problem with the negative binomial. One result is that the estimator is inconsistent in the presence of a large number of fixed effects. But exactly how it is inconsistent is still a matter of debate.

There is good evidence that in the presence of a large number of fixed effects the unconditional negative binomial will underestimate standard errors, resulting in insufficient coverage of the confidence intervals. That is, negative binomial predictors appear to enter the model as significant when in fact they do not. Simulation studies (Hilbe, 2011) have demonstrated that scaling the unconditional fixed-effects negative binomial model standard errors by the Pearson χ^2-based dispersion statistic produces standard errors that are closer to their nominal values. This is not the case when using deviance-based dispersion as the basis for scaling standard errors. This finding is consistent with what we have found when scaling other GLM-based count and binomial models. Simulation code and results are found on the book's website, *Negative Binomial Regression Extensions*. These facts need to be kept in mind when modeling unconditional fixed-effects count models.

14.4 Conditional fixed-effects negative binomial model

Panel-data models are constructed to control for all of the stable predictors in the model and to account for the correlation resulting from observations being associated within groups or panels. The value of conditional fixed-effects models is that a near infinite number of panels may be adjusted, while at the same time being conditioned out of the actual model itself. We do not have to deal with a host of dummy intercepts.

A conditional fixed-effects model is derived by conditioning out the fixed effects from the model estimation. Like unconditional fixed-effects models, there is a separate fixed effect, δ, specified in the linear predictor. Hence, $\eta = x\beta + \delta$. However, unlike the unconditional version, a revised log-likelihood function is derived to affect the conditioning out of the panel effects through a sufficient statistic, $\Sigma\, y_{it}$.

We first give the conditional fixed-effects Poisson probability function, upon which the log-likelihood function can be derived.

CONDITIONAL FIXED-EFFECTS POISSON PDF

$$f(y;\beta) = \left(\sum_{i=t}^{n_i} y_{it}\right) \prod_{t=1}^{n_i} \frac{\exp(x'_{it}\beta)^{y_{it}}}{y_{it}!\,\sum_s \exp(x'_{is}\beta)^{y_{it}}} \tag{14.10}$$

The conditional log-likelihood is then given as:

CONDITIONAL FIXED-EFFECTS POISSON LOG-LIKELIHOOD

$$\mathcal{L}(\beta;y) = \sum_{i=1}^{n}\left[\ln\Gamma\left(\sum_{t=1}^{n_i} y_{it}+1\right) - \sum_{t=1}^{n_i}\ln\Gamma\left(y_{it}+1\right) + \sum_{t=1}^{n_i}\left\{y_{it}\left(x'_{it}\beta\right) - y_{it}\,\ln\sum_{l=1}^{n_i}\exp(x'_{il}\beta)\right\}\right] \tag{14.11}$$

Following the derivation of the model as proposed by Hausman *et al.* (1984), the conditional fixed-effects negative binomial log-likelihood is shown as:

CONDITIONAL FIXED-EFFECTS NEGATIVE BINOMIAL LOG-LIKELIHOOD

$$\begin{aligned}\mathcal{L}(\beta;y) = \sum_{i=1}^{n}\Bigg[&\ln\Gamma\left(\sum_{t=1}^{n_i}\exp(x'_{it}\beta)\right) + \ln\Gamma\left(\sum_{t=1}^{n_i} y_{it}+1\right)\\ &-\ln\Gamma\left(\sum_{t=1}^{n_i}\exp(x'_{it}\beta)+\sum_{t=1}^{n_i} y_{it}\right) + \sum_{t=1}^{n_i}\left\{\ln\Gamma\left(\exp(x'_{it}\beta)+y_{it}\right)\right.\\ &\left.-\ln\Gamma(\exp(x'_{it}\beta)) - \ln\Gamma\left(y_{it}+1\right)\right\}\Bigg]\end{aligned} \tag{14.12}$$

or

$$\mathcal{L}(\lambda;y) = \sum_{i=1}^{n}\left[\ln\Gamma\left(\sum_{t=1}^{n_i}\lambda_{it}\right) + \ln\Gamma\left(\sum_{t=1}^{n_i} y_{it}+1\right) - \ln\Gamma\left(\sum_{t=1}^{n_i}\lambda_{it}+\sum_{t=1}^{n_i} y_{it}\right) + \sum_{t=1}^{n_i}\left\{\ln\Gamma\left(\lambda_{it}+y_{it}\right) - \ln\Gamma\left(\lambda_{it}\right) - \ln\Gamma\left(y_{it}+1\right)\right\}\right] \tag{14.13}$$

The heterogeneity parameter, δ, does not appear in the log-likelihood. It does not, as a result, appear in the model output. Complete derivations of both the Poisson and negative binomial log-likelihood functions can be found in Hardin and Hilbe (2003). I have placed a basic conditional fixed-effects negative binomial Stata command on the book's website, in *Negative Binomial Regression Extensions*.

We next model the same data using conditional fixed effects as we did with unconditional fixed effects (see Table 14.8).

Table 14.8 R: *Conditional fixed-effects Poisson*

```
library(lme4)
cefp <- lme4(accident ~ op + co_70_74 + co_75_79 + offset
       (Log(service)), data=ships, fixed=~1|id, family=poisson)
summary(cfep)
modelfit(cfep)
```

```
. xtpoisson accident op co_70_74 co_75_79, fe exposure(service)
  i(ship)

Conditional fixed-effects Poisson regression
                                       Number of obs        =    34
Group variable: ship                   Number of groups     =     5
                                       Obs per group: min   =     6
                                       avg =          6.8
                                       max =          7
                                       Wald chi2(3)         =  29.73
Log likelihood   = -66.052668          Prob > chi2          = 0.0000
------------------------------------------------------------------
accident |      Coef. Std. Err.    z    P>|z| 95% Conf. Interval]
---------+--------------------------------------------------------
      op |   .4560211 .1181882  3.86  0.000  .2243764    .6876657
co_70_74 |   .3665811  .134349  2.73  0.006  .1032618    .6299004
co_75_79 |  -.0145843 .2069169 -0.07  0.944  -.420134    .3909654
 service | (exposure)
------------------------------------------------------------------

. abic
AIC Statistic    =    4.061922             AIC*n        = 138.10533
BIC Statistic    =    4.079324             BIC(Stata) = 142.68442
```

The associated AIC statistic is 4.062, which is over a full one unit lower in value than the statistic for the unconditional Poisson model (5.15). Compare the above list of parameter estimates and standard errors output with that of the unconditional results in the previous section. The first thing that can be noticed is that the parameter estimates and standard errors for the unconditional and conditional fixed-effects Poisson models are identical, though this equality is not an equivalence of the two approaches. The conditional fixed-effects Poisson model does not include a constant, whereas the unconditional does. Of interest to note as well is the fact that the respective log-likelihoods differ (−54.64 to −68.28). We previously pointed out that the AIC statistics differ as well (4.06 to 5.15). We may conclude from this that, although the estimates and standard errors are the same, the two models intrinsically differ, with the preferred fit being that of the conditional fixed-effects Poisson model. No R functions are available for the negative binomial family.

We now turn to modeling the same data using the conditional fixed-effects negative binomial model. It can be displayed as:

```
. xtnbreg accident op co_70_74 co_75_79 , fe exposure(service)
  i(ship)

Conditional FE negative binomial regression
Group variable: ship                    Number of obs      =        34
                                        Number of groups   =         5
                                        Obs per group: min =         6
                                        avg =          6.8
                                        max =                        7
                                        Wald chi2(3)       =     17.03
Log likelihood  = -59.380917            Prob > chi2        =    0.0007
------------------------------------------------------------------------
accident |      Coef. Std. Err.      z  P>|z|    [95% Conf. Interval]
---------+--------------------------------------------------------------
      op |   .4713604 .1734299    2.72  0.007    .1314441    .8112767
co_70_74 |   .4357849 .1946556    2.24  0.025    .0542669    .8173029
co_75_79 |   .1075527 .2950438    0.36  0.715   -.4707226    .6858279
   _cons |  -6.657823 .5263946  -12.65  0.000   -7.689538   -5.626109
 service | (exposure)
------------------------------------------------------------------------

. abic
AIC Statistic   =    3.728289           AIC*n         = 126.76183
BIC Statistic   =    3.819182           BIC(Stata)    = 132.86728
```

The AIC and BIC statistics of the conditional fixed-effects negative binomial are substantially lower than the other unconditional or conditional Poisson models. Compare the above with the unconditional model in the previous section. The conditional and unconditional fixed-effects negative binomial models do not normally have the same parameter estimates. Also, note that, unlike the Poisson conditional fixed-effects, the conditional fixed-effects negative binomial has an intercept in the model. For a discussion on unconditional fixed effects intercepts, see Greene (2007). And, as previously indicated, the conditional negative binomial does not estimate or display a value for α; α is portioned across the panels, and no longer has a single value.

Unfortunately it has been discovered that the conditional fixed-effects negative binomial model is not a true fixed-effects model since it fails to control for all of its predictors. In addition, the α parameter that is conditioned out of the log-likelihood does not correspond to the different intercepts in the decomposition of μ. Allison and Waterman (2002) provide a full discussion, together with alternative models. The negative multinomial model has been suggested as an alternative for the conditional negative binomial. However, the negative multinomial produces the same estimators as a conditional Poisson, so does not provide any additional capability for handling overdispersion over what is available with Poisson options. The other foremost alternative is to revert to the unconditional negative binomial model. In fact, Allison and Waterman recommend that the unconditional negative binomial be used rather than the conditional. But, as previously discussed, it should also be accompanied by scaling the standard errors by the Pearson *chi*2 dispersion. If this strategy is unsatisfactory, then one should consider using other panel models, e.g. random-effects models or GEE models.

Finally, many statisticians prefer to calculate fixed-effects model standard errors using a bootstrap mechanism. Code to generate bootstrapped parameters and accelerated confidence intervals – those which are adjusted by the previous values – with clustering on ship panels, type:

```
. bootstrap _b, bca reps(100) cluster(ship) : xtnbreg accident
     op co_70_74 co_75_79 , fe exposure(service) i(ship)
```

Additionally, it should be noted that, with version 11.1, Stata's **xtpoisson** command with the *fe* option allows estimation of a fixed-effects Poisson with cluster-robust standard errors. This capability is not provided for the corresponding fixed-effects negative binomial, **xtnbreg**, *fe*.

I have placed a Stata conditional fixed-effects negative binomial program in *Negative Binomial Regression Extensions* on the book's website. It does not have all of the complexities of standard Stata commands, but it better enables one to visualize exactly how the model is estimated.

14.5 Random-effects negative binomial

Random-effects models begin with the same notation as fixed-effects models in that a heterogeneity parameter is added to the linear predictor. Moreover, the fixed-effects parameter, *d*, is now considered to be an *iid* random parameter rather than a fixed parameter. It is derived from a known probability distribution. In the case of Poisson, the random parameter can follow the usual Gaussian distribution, the gamma distribution, or the inverse Gaussian distribution. Gamma is the preferred random distribution to use since it is conjugate to the Poisson. The gamma distribution also allows an analytic solution of the integral in the likelihood. Other random distributions do not have these favorable features. However, as we shall discover when discussing random intercept models, the traditional random-effects model with Gaussian or normal distributed effects is what is typically referred to as a random-intercept model. For now, though, we shall emphasize models that have gamma distributed effects.

We shall use the term ν rather than δ for depicting the random parameter for random-effects count models. In so doing we shall be consistent with common terminology. We shall also use the standard GLM term μ rather than λ for the Poisson and negative binomial fitted value. λ is commonly found in the literature on count response models. But as with our choice of using ν, we shall use the term μ to maintain consistency for all count models that in some respect emanate from a GLM background.

It should be noted that random-effects models do not directly estimate random effects. Rather, the point of the algorithms is to estimate the parameters that describe the distribution of the random effects. The framework for the random-effects Poisson is

$$\ln(\mu_{it}) = x'_{it}\beta + \nu_i \tag{14.16}$$

with $\nu_i = \nu + \varepsilon_i$.

Following the derivation of the random gamma effects Poisson model by Hausman *et al.* (1984), we assume a random multiplicative effect on μ specified as:

$$\begin{aligned}
\Pr(y_{it}; \nu_i, x) &= \left\{\prod_{t=1}^{n_i} (\mu_{it}\nu_i)^{y_{it}} / y_{it}!\right\} \exp\left(-\sum_{t=1}^{n_i} \mu_{it}\nu_i\right) \\
&= \left(\prod_{t=1}^{n_i} \mu_{it}^{y_{it}} / y_{it}!\right) \exp\left(-\nu_i \sum_{t=1}^{n_i} \mu_{it}\right) \nu_i^{\sum_{t=1}^{n_i} y_{it}}
\end{aligned} \tag{14.17}$$

Summing of subject-specific observations are over panels with a mean given for each panel of

$$\mu_{it} = \exp(x'_{it}\beta) \tag{14.18}$$

where panels have separately defined means given as

$$\nu_i \mu_{it} = \exp(x'_{it}\beta + \eta_{it}) \tag{14.19}$$

With ν following a gamma distribution with a mean of 1 and a variance of θ, we have the mixture

$$f(\nu;\mu) = \frac{\theta^\theta}{\Gamma(\theta)} \nu_i^{\theta-1} \exp(-\theta \nu_i) \prod_{t=1}^{n_i} \frac{\exp(-\nu_i \mu_{it})(\nu_i \mu_{it})^{y_{it}}}{y_{it}!} \tag{14.20}$$

where the terms before the product sign specify the gamma-distributed random component and the terms from the product sign to the right provide the Poisson probability function. This mixture is of the same structural form as we derived for the NB1 probability function in Chapter 10.

Each panel is independent of one another, with their joint density combined as the product of the individual panels. The log-likelihood for the gamma-distributed Poisson random-effects model can be calculated by integrating over ν_i. The result is:

RANDOM-EFFECTS POISSON WITH GAMMA EFFECT

$$\begin{aligned} \mathcal{L}(\beta; y) = \sum_{t=i}^{n} \Bigg\{ & \ln\Gamma\left(\theta + \sum_{t=i}^{n_i} y_{it}\right) - \ln\Gamma(\theta) - \ln\Gamma(y_{it}+1) \\ & + \theta \ln(u_i) + \left(\sum_{t=i}^{n_i} y_{it}\right) \ln(1-u_i) \\ & - \left(\sum_{t=i}^{n_i} y_{it}\right) \ln\left(\sum_{t=i}^{n_i} \exp(x'_{it}\beta)\right) + \sum_{t=i}^{n_i} y_{it}(x'_{it}\beta) \Bigg\} \end{aligned} \tag{14.21}$$

where $\theta = 1/\nu$ with ν being the heterogeneity term and $u_i = \theta / \left(\theta + \sum_{t=1}^{n_i} \left(\exp\left(x'_{it}\beta\right)\right)\right)$.

We shall use the same data that were used for examining fixed-effects models for the examples of random-effects Poisson and negative binomial. The random-effects Poisson model, with a gamma effect, is shown on the next page.

The likelihood-ratio tests whether the data are better modeled using a panel structure or whether a pooled structure is preferred. Here we find that the random-effects (panel) parameterization is preferred over the pooled, or standard, Poisson model.

Table 14.9 R: *Random-effects Poisson*

```
library(lme4)
re1 <- lme4(accident ~ op+co+co_70_74+co_75_79+offset
    (log(service)), random=~1|ship, data=ships, family=poisson)
summary(re1)
modelfit(re1)  # see Table 5.3
```

```
. xtpoisson accident op co_70_74 co_75_79, exposure(service)
  i(ship) re

Random-effects Poisson regression          Number of obs      =       34
Group variable: ship                       Number of groups   =        5
Random effects u_i ~ Gamma                 Obs per group: min =        6
                                                          avg =      6.8
                                                          max =        7
                                           Wald chi2(3)       =    31.75
Log likelihood  =  -86.63664               Prob > chi2        = 0.0000
------------------------------------------------------------------------
accident |       Coef. Std. Err.     z    P>|z|  [95% Conf. Interval]
---------+--------------------------------------------------------------
      op |   .4557812    .11819   3.86  0.000  .2241331     .6874294
co_70_74 |      .3992 .1343059    2.97  0.003  .1359653     .6624347
co_75_79 |   .0191767 .2057828    0.09  0.926 -.3841502     .4225035
   _cons |  -6.167706 .2070758 -29.78  0.000 -6.573567    -5.761845
 service | (exposure)
---------+--------------------------------------------------------------
/lnalpha |  -2.150826 .8020862                -3.722886    -.5787661
---------+--------------------------------------------------------------
   alpha |    .116388 .0933532                 .0241641    .5605896
------------------------------------------------------------------------
Likelihood-ratio test of alpha=0: chibar2(01) =  14.10
                                  Prob>=chibar2 = 0.000
. abic
AIC Statistic    =      5.39039          AIC*n      = 183.27328
BIC Statistic    =    5.569637           BIC(Stata) = 190.90509
```

It is interesting to compare this output with that of an NB1 model on the same data. Recall that the NB1 PDF, like the NB2, can be derived by mixing the gamma random parameter with the Poisson probability function. Of course the NB1 PDF does not account for the panel structure of the data as the gamma-distributed random-effects Poisson does. However, because of the base similarity of the two models, we should expect that the outputs of the respective models will be similar, but not identical. This suspicion is indeed confirmed. Note also that the output above specifies "alpha" as the heterogeneity parameter. It is the same as what we have referred to as ν. Interpret δ in the same manner.

LINEAR NEGATIVE BINOMIAL (NB1)

Table 14.10 R: *Random-effects (intercept) NB1*

```
library(gamlss.mx)
re2 <- gamlssNP(accident+op+co_70_74+co_75_79+offset(log(service)),
    family=NBII, random=~1|id, data=ships, mixture="gq", K=20)
summary(re2)
modelfit(re2)
```

```
. nbreg accident op co_70_74 co_75_79, exposure(service)
     cluster(ship) disp(constant)

Negative binomial regression                    Number of obs    =  34
Dispersion             = constant               Wald chi2(2)     =   .
Log pseudolikelihood = -80.36411                Prob > chi2      =   .
                        (Std. Err. adjusted for 5 clusters in ship)
------------------------------------------------------------------------
         |              Robust
accident |      Coef.  Std. Err.     z   P>|z|   [95% Conf. Interval]
---------+--------------------------------------------------------------
      op |   .4762692  .1014479   4.69  0.000    .277435    .6751034
co_70_74 |   .6327833   .159793   3.96  0.000   .3195947    .9459719
co_75_79 |    .337207   .143428   2.35  0.019   .0560933    .6183208
   _cons |  -6.584051  .1075198 -61.24  0.000  -6.794786   -6.373316
 service | (exposure)
---------+--------------------------------------------------------------
/lndelta |   .4897027  .2644971                 -.0287021    1.008107
---------+--------------------------------------------------------------
   delta |   1.631831  .4316146                  .9717059     2.74041
------------------------------------------------------------------------

. abic
AIC Statistic    =    5.021418           AIC*n       = 170.72823
BIC Statistic    =    5.200665           BIC(Stata)  = 178.36002
```

We next compare the above with the standard NB2.

QUADRATIC NEGATIVE BINOMIAL (NB2)

R

```
nb <- glm.nb(los ~ hmo+white+factor(type), data=ships)
library(haplo.ccs)
sandcov(nb, ships$ship)
sqrt(diag(sandcov(nb, ships$ship)))
```

```
. nbreg accident op co_70_74 co_75_79, exposure(service)
cluster(ship)

Negative binomial regression                    Number of obs   = 34
Dispersion              = mean                  Wald chi2(2)    =   .
Log pseudolikelihood = -81.900132               Prob > chi2     =   .
                          (Std. Err. adjusted for 5 clusters in ship)
-------------------------------------------------------------------
         |              Robust
accident |      Coef.  Std. Err.      z   P>|z|  [95% Conf. Interval]
---------+---------------------------------------------------------
      op |   .3338351  .1719241   1.94   0.052    -.00313   .6708001
co_70_74 |   .5847677  .4053929   1.44   0.149  -.2097878   1.379323
co_75_79 |   .0580994  .5552667   0.10   0.917  -1.030203   1.146402
   _cons |  -6.209554  .5032703 -12.34   0.000  -7.195945  -5.223162
 service | (exposure)
---------+---------------------------------------------------------
/lnalpha |   -1.48894    .86228                 -3.178978   .2010978
---------+---------------------------------------------------------
   alpha |   .2256117  .1945405                  .0416282   1.222744
-------------------------------------------------------------------

. abic
AIC Statistic    =    5.111773          AIC*n        = 173.80026
BIC Statistic    =    5.291019          BIC(Stata) = 181.43207
```

When deriving the random-effects negative binomial, we begin with the same Poisson–gamma mixture as equation 14.22. By rearranging terms and not integrating out ν as we did for the Poisson, we have:

$$f(y;\mu,\nu) = \frac{\Gamma(\mu_{it}+y_{it})}{\Gamma(\mu_{it})\Gamma(y_{it}+1)}\left(\frac{1}{1+\nu_i}\right)^{\mu_{it}}\left(\frac{\nu_i}{1+\nu_i}\right)^{y_{it}} \tag{14.22}$$

which is the panel structure form of the NB1 model. For the random effect we select the beta distribution, which is the conjugate prior of the negative binomial, as gamma was the conjugate prior of the Poisson. With the dispersion defined as the variance divided by the mean, or $1+\nu$, it is stipulated that the inverse dispersion is distributed following a beta distribution. We have, therefore,

$$1/(1+\nu_{\mathrm{i}}) \sim \mathrm{Beta(a, b)} \tag{14.23}$$

which layers the random panel effect onto the negative binomial model. Deriving the log-likelihood function results in the following form of the function:

RANDOM-EFFECTS NEGATIVE BINOMIAL WITH BETA EFFECT PDF

$$f\left(y_{it};\beta,a,b\right) = \frac{\Gamma\left(a+b\right)+\Gamma\left(a+\sum_{t=1}^{n_i}(\exp\left(x_{it}'\beta\right))\right)+\Gamma\left(b+\sum_{t=1}^{n_i}y_{it}\right)}{\Gamma\left(a\right)\Gamma\left(b\right)\Gamma\left(a+b+\sum_{t=1}^{n_i}\left(\exp\left(x_{it}'\beta\right)\right)+\sum_{t=1}^{n_i}y_{it}\right)} \times \prod_{t=1}^{n_i}\frac{\Gamma\left(\exp\left(x_{it}'\beta\right)+y_{it}\right)}{\Gamma\left(\exp\left(x_{it}'\beta\right)\right)\Gamma\left(y_{it}+1\right)} \tag{14.24}$$

RANDOM-EFFECTS NEGATIVE BINOMIAL WITH BETA EFFECT LOG-LIKELIHOOD

$$\begin{aligned}\mathcal{L}\left(\beta;y,a,b\right) &= \sum_{i=1}^{n}\left\{\ln\Gamma\left(a+b\right)+\ln\Gamma\left(a+\sum_{t=1}^{n_i}\exp\left(x_{it}'\beta\right)\right)+\ln\Gamma\left(b+\sum_{t=1}^{n_i}y_{it}\right)\right.\\ &\quad -\ln\Gamma(a)-\ln\Gamma(b)-\ln\Gamma\left(a+b+\sum_{t=1}^{n_i}\exp\left(x_{it}'\beta\right)+\sum_{t=1}^{n_i}y_{it}\right)\\ &\quad \left.+\sum_{s=1}^{n_i}\left[\ln\Gamma\left(\exp\left(x_{is}'\beta\right)+y_{is}\right)-\ln\Gamma\left(y_{is}+1\right)-\ln\Gamma\left(\exp\left(x_{is}'\beta\right)\right)\right]\right\}\end{aligned} \tag{14.25}$$

Derivatives of the conditional fixed-effects and random-effects Poisson and negative binomial models are given in Greene (2006a). Output of the beta distributed random effect negative binomial is shown below for the data we have used in this chapter.

Table 14.11 R: *Random-effects (intercept) NB1*

```
library(gamlss.mx)
re3 <- gamlssNP(accident+op+co_70_74+co_75_79+offset(log(service)),
    family=NBI, random=~1|id, data=ships, mixture="gq")
summary(re3)
modelfit(re3)
```

```
. xtnbreg accident op co_70_74 co_75_79, exposure(service)
  i(ship) re

Random-effects negative binomial regression
                                        Number of obs      =        34
Group variable: ship                    Number of groups   =         5
Random effects u_i ~ Beta               Obs per group: min =         6
                                                       avg =       6.8
                                                       max =         7
                                        Wald chi2(3)       =     19.15
Log likelihood  = -79.778198            Prob > chi2        =    0.0003
------------------------------------------------------------------
accident |      Coef.  Std. Err.      z   P>|z|  [95% Conf. Interval]
---------+--------------------------------------------------------
      op |   .4723828  .1747434    2.70  0.007   .1298921   .8148735
co_70_74 |   .5146567  .1998413    2.58  0.010   .1229749   .9063384
co_75_79 |   .1980978  .2967118    0.67  0.504  -.3834466   .7796421
   _cons |  -6.713342  .5329034  -12.60  0.000  -7.757814  -5.668871
 service |  (exposure)
---------+--------------------------------------------------------
   /ln_r |   3.335776   1.34133                  .7068163   5.964735
   /ln_s |    3.68022   1.41294                  .9109095   6.449531
---------+--------------------------------------------------------
       r |   28.10017  37.69161                  2.027526   389.4498
       s |   39.65513   56.0303                  2.486583   632.4057
------------------------------------------------------------------
Likelihood-ratio test vs. pooled: chibar2(01) = 1.17
                                  Prob>=chibar2 = 0.140
. abic
AIC Statistic    =    5.045776          AIC*n        = 171.5564
BIC Statistic    =    5.325221          BIC(Stata) = 180.71455
```

A likelihood ratio test accompanies the output, testing the random-effects panel estimator with the pooled NB1, or constant dispersion, estimator. Here the random-effects model is preferred; *r* and *s* refer to the beta distribution values for the *a* and *b* parameters respectively. Note the extremely wide confidence intervals.

We shall use another example of a random-effects model. The data come from Thall and Vail (1990) and are used in Hardin and Hilbe (2003). Called the *Progabide* data set, the data are from a panel study of seizures in patients with epilepsy. Four successive two-week counts of seizures were taken for each patient. The response is *seizure*, with explanatory predictors consisting of the *progabide* treatment (1/0), a follow-up indicator called *time* (1/0), and an interaction of the two, called *timeXprog*. An offset, called *Period*, is given for weeks in the study, which are either 2 or 8. Since *Period* is converted by a natural log, the two values of *lnPeriod* are 2.079442 and 0.6931472. There are 295 observations on 59 epileptic patients (panels), with 5 observations, *t*, each. Results of modeling the data are given as shown below.

GAMMA-DISTRIBUTED RANDOM-EFFECTS POISSON

Table 14.12 R: *Random-effects (intercept) Poisson*

```
rm(list=ls())
library(gamlss.mx)
data(seizure)
attach(seizure)
re4 <- gamlssNP(seizures ~ time+progabide+timeXprog
 + offset(lnPeriod), family=PO, random=~1|id, data=seizure,
 mixture="gq")
summary(re4)
modelfit(re4)
```

```
. xtpoisson seizures time progabide timeXprog, offset(lnPeriod)
  re i(id)

Random-effects Poisson regression        Number of obs      =       295
Group variable: id                       Number of groups   =        59
Random effects u_i ~ Gamma               Obs per group: min =         5
                                                        avg =       5.0
                                                        max =         5
                                         Wald chi2(3)       =      5.73
Log likelihood  = -1017.3826             Prob > chi2        =    0.1253
------------------------------------------------------------------------
  seizures |      Coef.  Std. Err.     z    P>|z|   [95% Conf. Interval]
-----------+------------------------------------------------------------
      time |    .111836  .0468768    2.39   0.017    .0199591   .2037129
 progabide |   .0275345  .2108952    0.13   0.896   -.3858125   .4408815
 timeXprog |  -.1047258  .0650304   -1.61   0.107    -.232183   .0227314
     _cons |   1.347609  .1529187    8.81   0.000    1.047894   1.647324
  lnPeriod |   (offset)
-----------+------------------------------------------------------------
  /lnalpha |   -.474377  .1731544                    -.8137534  -.1350007
-----------+------------------------------------------------------------
     alpha |   .6222726  .1077492                     .4431915   .8737153
------------------------------------------------------------------------
Likelihood-ratio test of alpha=0: chibar2(01) =  2602.24
Prob>=chibar2 = 0.000
```

The model appears to favor the panel specification of the data (Prob>=chibar2=0.000). Note that alpha is 0.6222.

BETA-DISTRIBUTED RANDOM-EFFECTS NEGATIVE BINOMIAL

```
. xtnbreg seizures time progabide timeXprog, offset(lnPeriod)
  re i(id)
```

```
Random-effects negative binomial regression  Number of obs     =      295
Group variable: id                           Number of groups   =       59
Random effects u_i ~ Beta                    Obs per group: min =        5
                                                            avg =      5.0
                                                            max =        5
                                             Wald chi2(3)       =     6.88
Log likelihood  = -891.81046                 Prob > chi2        =   0.0759
--------------------------------------------------------------------------
  seizures |      Coef.  Std. Err.      z    P>|z|  [95% Conf. Interval]
-----------+--------------------------------------------------------------
      time |   .0637907   .0821828    0.78   0.438  -.0972846    .224866
 progabide |    .025827   .1822767    0.14   0.887  -.3314287   .3830827
 timeXprog |  -.2619935   .1158202   -2.26   0.024  -.4889969  -.0349901
     _cons |   .7961454   .1908792    4.17   0.000   .4220289   1.170262
  lnPeriod |   (offset)
-----------+--------------------------------------------------------------
     /ln_r |   1.261781   .1973829                   .8749176   1.648644
     /ln_s |    1.51837    .220681                   1.085843   1.950897
-----------+--------------------------------------------------------------
         r |   3.531705   .6970981                   2.398678   5.199925
         s |   4.564778    1.00736                   2.961936   7.034993
--------------------------------------------------------------------------
Likelihood-ratio test vs. pooled: chibar2(01) =   228.19
Prob>=chibar2 = 0.000
```

Again, the panel structure of the data is confirmed by the model (`Prob>=chibar2=0.000`). It is important to remember that this conclusion holds with respect to the subject specific parameterization of the data. A margin or population averaged model may yield contrary results, however, I believe that this would occur in very few instances. A thorough discussion of the derivation of the random-effects Poisson and negative binomial models can be found in Frees (2004) and Greene (2006a).

A disadvantage of a random-effects model is that it assumes that the subject-specific effects are uncorrelated with other predictors. The Hausman test is commonly used to evaluate whether data should be modeled using a fixed- or a random-effects model. The test is based on work done by Mundlak (1978) who argued that the fixed-effects model is more robust than the random-effects model to important predictors left out of the model. That the subject-specific effects are not correlated highly with model predictors is specified as the null hypothesis. See Freese (2004, p 247), for additional discussion.

Random-effects estimators are more efficient than fixed-effects estimators when the data come from within a larger population of observations, as well as when there are more panels in the data. Data coming from a smaller complete data set, with relatively few panels, prefer the fixed-effects estimator.

Random-effects models are subject-specific models in that the log-likelihood models individual observations rather than the average of panels, or marginal distribution. GEEs are population-averaging models, and care must be taken when interpreting GEE against random-effects model output.

14.6 Mixed-effects negative binomial models

Mixed-effects models are combinations of fixed and random effects. The most basic mixed-effects model is the random-intercept model, in which the model intercept is allowed to vary. The fixed effects are the standard predictors that are ordinarily in the model. In fact, in this sense, a standard Poisson or negative binomial model is a purely fixed-effects model. The addition of an extra parameter to indicate a randomly distributed intercept classifies the model as a random intercept. It is understood that random effects are the same within each panel of cluster, but they differ between clusters. The fixed coefficients are the same within and across clusters.

The effect of adding a random component to the linear predictor is to add extra correlation to the model, which in turn induces overdispersion. The Gaussian or normal distribution is typically used to describe or characterize the intercept randomness. Therefore the random intercept model is identical to the random-effects models described in the previous section, given that the random-effects model employs a Gaussian distribution.

Another type of mixed-effects model is the random-coefficients or slopes model. We have a random-coefficient model when the model coefficients are also allowed to vary. More complex mixed-effects models may be constructed by nesting levels within one another. For example, pupils may be nested within classrooms, which are themselves nested within schools, which may be nested further under school districts, and states. Each level can have random intercepts and coefficients, leading to rather complex models. Partitioning the variability within and between clusters, and between levels, is the task of the statistician dealing wih mixed-effects models. We shall only summarize the area, referring the reader to texts that are specifically devoted to their analysis.

14.6.1 Random-intercept negative binomial models

The most basic random-coefficient model is one in which only the regression intercept is allowed to vary. Such a model is called a *random-intercept model*. It is a subset of mixed-effects models. But it also is nearly identical to a GEE model with the exchangeable correlation structure and is the same as the standard random-effects model where the effects are normally distributed. The random-intercept model may be expressed as an equation:

$$y_{ik} = \beta_{0i} + \beta_1 X + \epsilon_{ik} \tag{14.26}$$

y_{ik} is the response for individual i in group k, or at time k (which we would change to t); β_{0i} refers to the regression intercept, varying over the individual i; $\beta_1 X$ is the coefficient for predictor x, and ϵ_{ik} is the error term, varying over both individuals and groups.

Using the **medpar** data set, we see that the random-effects Poisson is the same as the random-intercept Poisson. We consider that the data are more highly correlated within hospitals than between them, so assign *center* or *hospital* as the intercept. The random intercept was assumed to be gamma distributed in the previous section, thus relating the variability to the negative binomial. This is not the case when the distributon is Gaussian.

```
. use medpar
. gen center = real(provnum)
. tab type, gen(type)
```

RANDOM-EFECTS POISSON, WITH NORMALLY DISTRIBUTED INTERCEPT

R: *Random-intercept Poisson*

```
rm(list=ls())
data(medpar)
attach(medpar)
library(gamlss.mx)
rip <- gamlssNP(los~hmo+white+type2+type3, family=PO,
  random=~1|provnum,
    data=ships, mixture="gq")
summary(rip)
modelfit(rip)
```

```
. xtpoisson los hmo white type2 type3, re normal i(center)

Random-effects Poisson regression          Number of obs      =      1495
Group variable: center                     Number of groups   =        54
Random effects u_i ~ Gaussian              Obs per group: min =         1
                                                          avg =      27.7
                                                          max =        92
                                           Wald chi2(4)       =    109.81
Log likelihood  = -6528.3543               Prob > chi2        =    0.0000
```

```
-----------------------------------------------------------------------
los       |      Coef.  Std. Err.     z    P>|z|   [95% Conf. Interval]
----------+------------------------------------------------------------
hmo       | -.0907326  .0259212  -3.50  0.000  -.1415373  -.0399279
white     | -.0275539  .0302745  -0.91  0.363  -.0868908    .031783
type2     |  .2319145  .0247704   9.36  0.000   .1833655   .2804635
type3     |  .1226031  .0488912   2.51  0.012   .0267781    .218428
_cons     |  2.191951  .0649112  33.77  0.000   2.064728   2.319175
----------+------------------------------------------------------------
/lnsig2u  | -1.775147  .2222603  -7.99  0.000  -2.210769  -1.339525
----------+------------------------------------------------------------
sigma_u   |  .4116535  .0457471                  .3310836   .5118302
-----------------------------------------------------------------------
Likelihood-ratio test of sigma_u=0: chibar2(01) =   801.11
 Pr>=chibar2 = 0.000
```

RANDOM-INTERCEPT POISSON

```
. xtmepoisson los hmo white type2 type3 || center:

Mixed-effects Poisson regression         Number of obs       =    1495
Group variable: center                   Number of groups    =      54
                                         Obs per group: min  =       1
                                                        avg  =    27.7
                                                        max  =      92
Integration points =   7                 Wald chi2(4)        =   09.81
Log likelihood = -6528.6768              Prob > chi2         =  0.0000
-----------------------------------------------------------------------
los    |      Coef.  Std. Err.      z    P>|z|   [95% Conf. Interval]
-------+---------------------------------------------------------------
hmo    |  -.0907326  .0259212    -3.50  0.000  -.1415373  -.0399279
white  |  -.0275538  .0302745    -0.91  0.363  -.0868907   .0317831
type2  |   .2319144  .0247704     9.36  0.000   .1833654   .2804634
ype3   |   .1226027  .0488912     2.51  0.012   .0267778   .2184277
_cons  |   2.191951  .0649114    33.77  0.000   2.064727   2.319175
-----------------------------------------------------------------------

-----------------------------------------------------------------------
Random-effects Parameters  | Estimate Std. Err.[95% Conf. Interval]
---------------------------+-------------------------------------------
center: Identity           |
                 sd(_cons) | .4116553 .0457472 .3310851   .5118323
-----------------------------------------------------------------------
LR test vs. Poisson regression:  chibar2(01) =   800.46
  Prob>=chibar2 = 0.0000
```

It should perhaps be mentioned for clarity that a random-effects model has a random effect that is distributed *somehow.* There is no requirement that the distribution be anything in particular. However, there is often an analytic solution when a particular distribution is chosen for the random effect, but it is not a requirement. When the random effect has a distribution that is conjugate to the model distribution, e.g. beta is conjugate to the negative binomial, there is a closed-form solution and the model parameters may generally be more easily estimated – by maximum likelihood. A negative binomial model with Gaussian random effects generally requires estimation using quadrature or simulation, a more difficult method of estimation. Stata only provides for a beta negative binomial random-effects model, but R can be used for Gaussian effects.

We model the data using R's **gamlss** function, which employs a Gaussian quadrature method of estimation.

RANDOM-INTERCEPT NEGATIVE BINOMIAL, WITH GAUSSIAN-DISTRIBUTED EFFECTS

R: *Random-intercept negative binomial*

```
library(gamlss.mx)
rinb <- gamlssNP(los~ hmo +white+ type2 +type3, random=~1|provnum,
     data=medpar, family=NBI, mixture="gq", K=20)
summary(rinb)
```

```
Mu link function: log
Mu Coefficients:
             Estimate  Std. Error   t value  Pr(>|t|)
(Intercept)   2.23651    0.06568   34.0503 2.355e-249
hmo     -0.08994      0.05160    -1.7431  8.133e-02
white -0.04998        0.06662    -0.7502  4.532e-01
type2  0.22096        0.04890     4.5183  6.258e-06
type3  0.31361        0.07983     3.9287  8.560e-05
z      0.25033        0.02414    10.3716  3.683e-25
-------------------------------------------------------------------
Sigma link function: log
Sigma Coefficients:
              Estimate  Std. Error   t value   Pr(>|t|)
(Intercept)    -0.8935     0.0448    -19.94  6.371e-88
-------------------------------------------------------------------
No. of observations in the fit: 29900
Degrees of Freedom for the fit: 7
  Residual Deg. of Freedom: 1488
  at cycle: 1

Global Deviance: 9560.749
  AIC: 9574.749
  SBC: 9611.918
```

The *mixture* option relates to the manner of estimation – Gaussian quadrature (*gq*). The default number of quadrature points is 4. This needed to be raised to 20, which should provide more accurate estimates ($K = 20$). The dispersion parameter is given as the exponentiated intercept coefficient, `exp(-0.8935) = .40922097`. This parameter is stored in the saved `rinb$sigma.fv`:

```
summary(rinb$sigma.fv)
  Min. 1st Qu.  Median    Mean 3rd Qu.    Max.
 0.4092  0.4092  0.4092  0.4092  0.4092  0.4092
```

Although it is not given in gamlssNP output, the value of ρ (rho) may be calculated as

$$\rho = \frac{\sigma^2}{1+\sigma^2} = \frac{(0.4092)^2}{1+(0.4092)^2} = \frac{0.1674}{1.1674} = 0.1434$$

ρ informs us of the proportion of the total variance that is explained by the random effects, and is an important statistic to use when comparing models. The intracluster correlation is 0.16620, and is calculated by performing a one-way ANOVA of the response on the random-intercept variable. See Hosmer and Lemeshow (2000, p. 320) for details. The values of ρ and the intracluster correlation coefficient are typically close.

Predicted values, μ, are saved in `rinb$mu.fv`, and can easily be abstracted for use. In addition, the predictors appear to fit based on Wald values, which are *t*-squared and normally distributed. These may easily be converted from the output above.

The residuals employed in **gamlss** are normalized randomized quantile residuals (Dunn and Smyth, 1996). Pearson residuals, for example, are based on the two above described statistics and may be calculated as:

```
m<-rinb$mu.fv          # fitted values for extended model
s<-rinb$sigma.fv       # sigma
presid <- (medpar$los-m)/sqrt(m+s*m*m)
summary(presid)

  Min. 1st Qu.  Median    Mean 3rd Qu.    Max.
-1.5220 -1.0440 -0.3146  0.7409  1.3670 64.3500

hist(presid)
```

The graph of Pearson residuals (Figure 14.1) is highly skewed, which is typical with count data. Preferred model assessment graphs entail, among others, standardized Pearson or deviance residuals versus fitted value, `rinb$mu.fv`.

LIMDEP can also be used to estimate random-intercept negative binomial models. However, estimation may not be accomplished using the menu system,

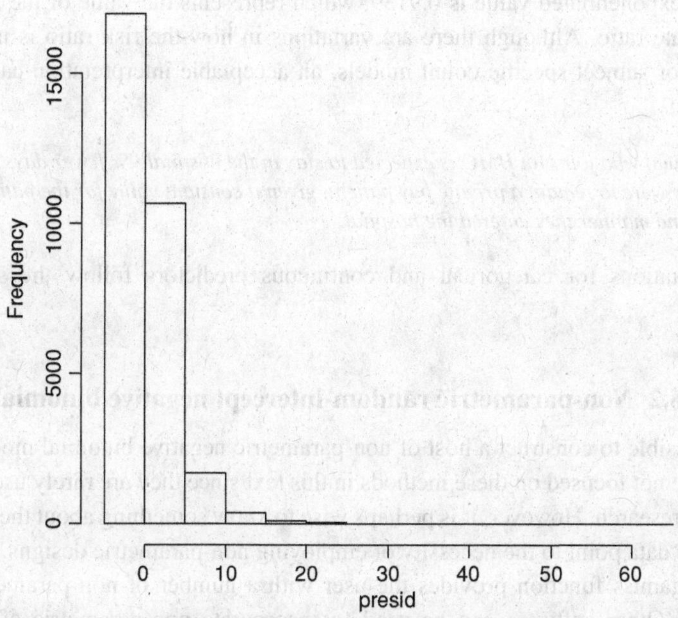

Figure 14.1 Histogram of random intercept Pearson residuals

but only by employing the command system. The following code may be used to construct the same model.

```
--
> Negb;lhs=los;rhs=one,white,hmo,type2,type3; pds= provnum; rpm;
    fcn=one(n); pts=20; halton $
```

ScalParm, given in LIMDEP output, is the negative binomial heterogeneity parameter, sigma, also calculated as 0.4112. Parameter estimates are also very closely similar to those of the R **gamlss** output.

It should be noted that convergence, and estimation, is optimal when the size of the panels, or clusters, are less than 10. However, if the value of the intracluster correlation is small, larger-sized panels appear to work as well. Very small values of the intracluster correlation indicate a problem with the model, and the estimation process usually fails. A GEE model may then be preferred.

Interpretation of random-intercept Poisson and negative binomial parameter estimates is more meaningful when done on exponentiated coefficients. The coefficient on *hmo*, whether the patient is a member of a health maintenance organization or has private insurance, is -0.09.

The exponentiated value is 0.9139, which represents the value of the incidence rate ratio. Although there are variations in how the risk ratio is interpreted for subject-specific count models, an acceptable interpretation can be given as:

> *A patient who joins an HMO is expected to stay in the hospital 9% fewer days than if they were to remain a private pay patient, given a constant value for the patient age and manner they entered the hospital.*

Interpretations for categorical and continuous predictors follow the same logic.

14.6.2 Non-parametric random-intercept negative binomial

It is possible to construct a host of non-parametric negative binomial models. We have not focused on these methods in this text since they are rarely used in current research. However, it is perhaps wise to know something about them in case the data point to the necessity of employing non-parametric designs.

R's gamlss function provides the user with a number of non-parametrtic options. Other software can be used to incorporate non-parametric effects into a model, but negative binomial models are rather conspicuously ignored. Since a simple amendment to the above **gamlssNP** code for random-intercept negative binomial models converts the model to a non-parametric model, we shall use it as an example. Note that the distribution of the random intercepts is a non-parametric discrete distribution, i.e. K intercept match points with their probabilities. The model attempts to estimate parameters for these K distributions.

In order to identify the model as non-parametric, we change the method of mixture estimation in above the random intercept model to *np*, for "non-parametric."

```
vcnb <- gamlssNP(los~ hmo +white+ type2 +type3, random=~1|provnum,
  data=medpar, family=NBI, mixture="np")
```

This algorithm uses the default $K = 4$ to provide four non-parametric match points, i.e. four different intercepts for *log*(*mu*) are fitted, obtained from `summary(vcnb)`: *intercept*, *intercept+MASS2*, *intercept+MASS3*, and *intercept+MASS4*. The values are: 1.699, 1.977, 2.306, and 3.321.

The summary table of parameter estimates and non-parametric match points is displayed below. Recall that **gamlss** refers to what we have termed NB2 negative binomial as NBI, or type I negative binomial model. NB1 is referred to as NBII, or type II.

```
summary(vcnb)
Family: c("NBI Mixture with NP", "Negative Binomial type I
 Mixture with NP" )

Mu link function: log
Mu Coefficients:

              Estimate  Std. Error  t value    Pr(>|t|)
(Intercept)   1.69933    0.30313    5.6060   2.163e-08
hmo          -0.05621    0.05146   -1.0923   2.747e-01
white        -0.07458    0.06629   -1.1250   2.606e-01
type2         0.21616    0.04887    4.4232   9.897e-06
type3         0.04860    0.10276    0.4729   6.363e-01
MASS2         0.27757    0.29932    0.9274   3.538e-01
MASS3         0.60719    0.29632    2.0491   4.050e-02
MASS4         1.62169    0.32357    5.0119   5.544e-07
-----------------------------------------------------------------
Sigma link function: log
Sigma Coefficients:
              Estimate  Std. Error  t value    Pr(>|t|)
(Intercept)  -0.8977     0.04535    -19.80   1.540e-84
-----------------------------------------------------------------
No. of observations in the fit: 5980
Degrees of Freedom for the fit: 12
  Residual Deg. of Freedom: 1483
  at cycle: 1

Global Deviance: 9539.474
  AIC: 9563.474
  SBC: 9627.193

> vcnb$prob
  MASS1       MASS2       MASS3       MASS4
0.004568576 0.165602171 0.795761515 0.034067738
```

The dispersion parameter may be calculated as before: exponentiating *sigma*. In this case the dispersion is 0.4075. *Rho* (ρ) is equal to 0.3895. Compare these values with the random intercept model of the same data in the previous section.

When modeling with non-parametric effects, one should attempt modeling with a different number of match points. The model with the lowest AIC statistic is preferred. Testing the above model using $K = 2$ (*vcnb*2) through 5 (*vcnb*5) results in,

```
GAIC(vcnb2,vcnb3,vcnb4,vcnb5)

      df      AIC
vcnb2  8 9591.208
vcnb3 10 9559.652 ←
vcnb4 12 9563.474
vcnb5 14 9567.456
```

Three match points appears to be preferred. Since the AIC value for a model with three points rests at a minimum, with either side growing larger, we can feel confident that this model is indeed the best fitted among these alternatives. See Rigby and Stasinopoulos (2008) for guidance on using **gamlss** and **gamlssNP** for non-parametric modeling.

14.6.3 Random-coefficient negative binomial models

We previously provided the formula for a random-intercept model, where the intercept varies over periods. The random-coefficient model expands this analysis to allow the regression coefficients to vary. The equation for a model in which the coefficient, β_{1i}, varies, but not the intercept, can be expressed as:

$$y_{ik} = \beta_0 + \beta_{1i}x + \epsilon_{ik} \tag{14.27}$$

We may allow both the intercept and coefficient to vary, giving us:

$$y_{ik} = \beta_{0i} + \beta_{1i}x + \epsilon_{ik} \tag{14.28}$$

Both of the above models are random coefficient models. More complex models can exist depending on the number of nested levels in the data.

We now use a random-coefficient model on a version of the German Health data set. We have employed several in this volume. This is from the year 1996, prior to the later Reform data, and after the **rwm5yr** data, that we have used in previous discussions,. Model variables include the following:

```
Response:
docvis    The number of visits to the doctor by a patient
           recorded over seven time periods.
Predictors include:
 age      Age  (25-64)
female    1=Female; 0=Male
educ      Years of schooling (7-18)
married   1=Married; 0 = Not married
hhninc    Net monthly house income in Marks/10000 (0-30.67)
hsat      Health satisfaction evaluation (0-10)
id        periods in which data were recorded for patients (1-7)
```

We let the coefficient on *hsat* vary in addition to the intercept; *hsat* has eleven levels. A random-coefficient Poisson model may be estimated in Stata as follows. A random intercept is also automatically estimated as well. The random coefficient must also be entered into the model as a fixed effect.

RANDOM-COEFFICIENT POISSON

```
. xtmepoisson docvis age female educ married hhninc hsat || id:
  hsat, nolog

Mixed-effects Poisson regression          Number of obs      =     27326
Group variable: id                        Number of groups   =      7293
                                          Obs per group: min =         1
                                                         avg =       3.7
                                                         max =         7

Integration points =    7                 Wald chi2(6)       =   4991.06
Log likelihood = -66630.606               Prob > chi2        =    0.0000
------------------------------------------------------------------------
docvis |      Coef.  Std. Err.      z   P>|z|     [95% Conf. Interval]
-------+----------------------------------------------------------------
   age |  .0147408  .0009972  14.78   0.000     .0127863    .0166954
female |  .3219193  .0309711  10.39   0.000      .261217    .3826216
  educ | -.0049411  .0065284  -0.76   0.449    -.0177366    .0078544
married|  .0025107  .0195583   0.13   0.898    -.0358227    .0408442
 hhninc| -.0129152  .0039179  -3.30   0.001    -.0205942   -.0052362
  hsat | -.2192573  .0034669 -63.24   0.000    -.2260523   -.2124622
 _cons |  1.253753  .0948593  13.22   0.000     1.067832    1.439674
------------------------------------------------------------------------

------------------------------------------------------------------------
Random-effects Parameters | Estimate Std. Err.  [95% Conf. Interval]
--------------------------+---------------------------------------------
id: Independent           |
                 sd(hsat) |.1423387  .0024482   .1376203    .1472188
                sd(_cons) |.8979792  .0151393   .8687915    .9281474
------------------------------------------------------------------------
LR test vs. Poisson regression:        chi2(2) = 47250.70
  Prob > chi2 = 0.0000
```

Recall that I earlier indicated that a random-effect model estimates the distributional parameters of the random effects rather than the random effects themselves. The variance of the coefficient and intercept of the normally distributed effects are given as 0.18517^2 (0.0343) and 1.2982^2 (1.6853). The estimates of the random effects may be obtained by,

```
. predict  re_hsat re_interc, reffects
. gen beta1 = _b[hsat] + re_hsat
. gen beta0 = _b[_cons] + re_interc
. by id, sort: gen repanel = _n==1
. list id beta1 beta0 if repanel & (id <=4 | id >= 51)
```

```
                              Random
                        Coef  Intercept
--------+-----------------------------+
        |   id        beta1      beta0|
--------+-----------------------------|
     1. |     1   -.2902111   .8386426|
     4. |     2   -.2535112   1.045504|
     8. |     3   -.1317606   1.320989|
    12. |     4   -.1094025   1.690978|
 27323. |  7290   -.0538211   1.912194|
        |-----------------------------|
 27324. |  7291   -.1545169   1.621851|
 27325. |  7292   -.0658055   2.780611|
 27326. |  7293   -.1837293   1.430506|
--------+-----------------------------+
```

A boundary likelihood-ratio test can be given to determine if random coefficients accommodate more of the correlation in the data than a more simple random-intercept model. In order to use the test one must make certain to use the identical fixed-effect terms, or predictors, in both models.

A random-coefficient negative binomial may be estimated using LIMDEP, as follows. Recall that the negative binomial random-coefficients model is adding more adjustment for overdispersion into the model than the random-coefficient Poisson, or even random-coefficient negative binomial. Care must be taken to check if the model is explaining too much. That is, we must determine if the model provides more adjustment for correlation than is warranted by the data.

RANDOM-COEFFICIENT NEGATIVE BINOMIAL

```
--> Negb;lhs=docvis;rhs=one,age,female,educ,married,hhninc,hsat;
    pds=_groupti;rpm;fcn=one(n),hsat(n);cor;pts=20;halton $

----------------------------------------------------------
Random Coefficients   NegBnReg Model
Info. Criterion: AIC =            4.09253
  Finite Sample: AIC =            4.09254
Info. Criterion: BIC =            4.10446
Info. Criterion:HQIC =            4.09666
Restricted log likelihood      -46669.83
Chi squared                     67951.15
Degrees of freedom                     3
Prob[ChiSqd > value] =          .0000000
----------------------------------------------------------

----------------------------------------------------------------------
Variable  Coefficient Stand Error b/St.Er.  P[|Z|>z] Mean of X
----------------------------------------------------------------------
          Nonrandom parameters
 AGE        .01680197   .00159059   10.563   .0000   44.3351586
 FEMALE     .37039825   .02990835   12.384   .0000    .42277339
```

```
EDUC      -.03147513  .00813808  -3.868  .0001  10.9408707
MARRIED    .09847528  .04141046   2.378  .0174   .84538573
HHNINC     .01194585  .08915694    .134  .8934   .34929630
          Means for random parameters
Constant 1.65242649   .12955696  12.754  .0000
HSAT      -.24630006  .00653340 -37.699  .0000  6.69640844
          Dispersion parameter for NegBin distribution
ScalParm 1.24095398   .02905699  42.708  .0000

Implied covariance matrix of random parameters
Matrix Var_Beta has  2 rows and  2 columns.
              1             2
        +----------------------------
      1|    .61380       -.06887
      2|   -.06887        .02121

Implied standard deviations of random parameters
Matrix S.D_Beta has  2 rows and  1 columns.
              1
        +--------------
      1|    .78346
      2|    .14562
Implied correlation matrix of random parameters
Matrix Cor_Beta has  2 rows and  2 columns.
              1             2
        +----------------------------
      1|   1.00000       -.60364
      2|   -.60364       1.00000
```

Recall from the previous section that LIMDEP output *ScalParm* is the negative binomial heterogeneity parameter, sigma. We have generally symbolized it as α.

The negative binomial by construction already picks up some heterogeneity that manifests itself in the overdispersion. The random coefficients formulation is an extension that gathers together other time-invariant heterogeneity across individuals. Random-coefficient models allow the randomness of the coefficients to explain the heterogeneity across individuals as well as the heterogeneity across groups. This heterogeneity, in sum, results in the differences found in the responses, y_{ik}, due to changes in the predictors. The fact that levels of nesting, or of additional unexplained heterogeneity, can be explained by these models, makes them attractive to those who wish to model count data having such a structure.

The only caveat to keep in mind when using negative binomial random coefficient models is to be careful of overspecification, i.e. multiple adjustments being given to the otherwise Poisson counts. Care must be taken to assure that our model does not make too much adjustment.

14.7 Multilevel models

Multilevel models are sometimes called hierarchical models, particularly in educational and social science research. However, the majority of statisticians now tend to draw a distinction between multilevel and hierarchical models, primarily because of the manner in which the methods define order of levels, or nesting. Regardless, the idea behind multilevel models is to model the dependence that exists between nested levels in the data. For instance, we may model visits to the doctor within groups of different hospitals. Unlike GEE models, multilevel models are not based on the framework of generalized linear models. Rather, they are an extension to the random-effects models we discussed in the previous section.

Until recently, nearly all discussion, and application, of multilevel models has been of continuous response models. Binary response models, especially logistic models, were introduced around 2000. Only in the last few years have Poisson models been discussed within the domain of multilevel regression. Negative binomial models have been largely ignored. As of this writing, only LIMDEP provides the capability of modeling multilevel negative binomial random-coefficient models. Matlab and other similar software that can easily be programmed are capabable of modeling multilevel models of considerable complexity.

The subject of multilevel or hierarchical models is beyond the scope of this text. However, with respect to the negative binomial, there are currently few implementations of multilevel models beyond the simple two-level models we considered in the previous section. The logic of estimating deeper nested models is generally the same as that which is used for multilevel Poisson models. Stata, R, LIMDEP, and SAS are capable of estimating these models. Discussions of them can be found in Skrondal and Rabe-Hesketh (2004), Lee, Nelder and Pawitan (2006), Zuur *et al.* (2009), and Rabe-Hesketh and Skrondal (2008). Only Zuur *et al.* discuss the negative binomial deeper than the random-intercept model.

Summary

The advantage of GEE over random-effects models relates to the ability of GEE to allow specific correlation structures to be assumed within panels. Parameter estimates are calculated without having to specify the joint distribution of the repeated observations.

In random-effects models a between-subject effect represents the difference between subjects conditional on them having the same random effect. Such models are thus termed conditional or subject-specific models. GEE parameter estimates represent the average difference between subjects, thus are known as marginal, or population-averaged models. Which to use depends on the context, i.e. on the goals of the study.

In this chapter I presented overviews of the foremost panel models: unconditional and conditional fixed-effects models, random-effects models, and generalized estimating equations. Multilevel mixed models are a comparatively new area of research, with multilevel count models being the most recent. LIMDEP is the only commercial software supporting negative binomial linear mixed models, with its initial application in 2006.

Hierarchical GLMs, called HGLMs, and double HGLMs have recently been developed, primarily by John Nelder and Yougjo Lee. However, they have not employed HGLM theory to the negative binomial, and we do not discuss them here. HGLMs are supported only by GenStat software.

15
Bayesian negative binomial models

15.1 Bayesian versus frequentist methodology

Bayesian statistics has continually been growing in popularity since the 1980s. Currently there are societies devoted to Bayesian methodology, and the major statistical associations typically have committees devoted to the promotion of Bayesian techniques. In the discipline of astrostatistics, for example, the majority of articles related to the statistical analysis of astronomical data employ some type of Bayesian approach. Not all disciplines are so committed to Bayesian methodology of course, but it is the case that there are a growing number of researchers who are turning to Bayesian modeling. In this chapter we shall present a brief overview of this approach, and give an example of a Bayesian negative binomial model.

Bayesian statistics is named after Thomas Bayes (1702–1761), a British Presbyterian minister and amateur mathematician who was interested in the notion of inverse probability, now referred to as *posterior probability*. The notion of posterior probability can perhaps best be understood with a simple example.

Suppose that 90% of the students in a statistics course pass. Students who passed are given the symbol P; students who failed are given an F. Therefore, 10% of the students failed the course. Everyone who passed purchased the course text, while only half of the students who failed did so. Let's suppose we observe a student with the course text (assume they purchased it). What is the probability that this student failed? This is an example of posterior probability. Bayes developed a formula that can be used to determine this value, which has become known as *Bayes' Theorem*. With "∼" indicating "not", we have:

P(F) = 0.1 probability that the student failed notwithstanding other information

P(F~) = 0.9 probability that the student passed notwithstanding other information

P(T|F) = 0.5 probability that a student purchased the text given that they failed

P(T|F~) = 1 probability that a student purchased the text given that the passed

P(T) = 0.95 probability of randomly selecting a student who purchased the text

$$P(F)^*P(T|F) + P(F\sim)^*P(T|F\sim) = 0.1^*0.5 + 0.9^*1 = 0.05 + 0.9 = 0.95$$

$$P(F|T) = \frac{P(T|F)\,P(F)}{P(T)} = \frac{0.5^*0.1}{0.95} = 0.0526$$

The posterior probability of that particular student having failed, given that they purchased a text, is approximately 5%. This statistic makes sense given the specified conditions. But it is not an outcome that is addressed by traditional methods, as discussed later in this section.

The above formula is an instance from the generic expression of Bayes' Theorem, which can be given as:

BAYES' THEOREM

$$P(B|D) = \frac{P(D|B)\,P(B)}{P(D)} \tag{15.1}$$

or, in terms of Bayesian modeling,

$$p(\theta|y) = \frac{p(y|\theta)\,\pi(\theta)}{p(y)} = \frac{L(\theta)\,\pi(\theta)}{\int L(\theta)\,\pi(\theta)\,d\theta} \tag{15.2}$$

In Bayesian terminology P(B) is termed the *prior probability*, which is the probability that a belief or hypothesis is true, prior to our having any data or information to evaluate it. If we were testing a coin for bias, we would be likely to give the prior probability of the coin coming up heads on a single throw as 0.5. P(D|B) can be termed the likelihood, i.e. the conditional probability of the data being the case given the truth of our belief. Finally, P(D) is probability of the data being the case. P(D) has also been referred to as the marginal probability, and as a normalizing constant (which we explained when discussing the binomial normalization term). In the context of a statistical model, it is the term that is required to assure that the distribution sums to 1. This notion is displayed more clearly in equation 15.2, where the prior probability of the data is expressed as an integral. Since $p(y)$ has no inferential value for the parameter θ, and only seems to insure that the posterior distribution is standardized, i.e. sums or integrates to 1, it is typically not considered in the calculation of the posterior probability of a model parameter. As a result, the Bayesian posterior is construed to be proportional to the product of the likelihood and the prior, but not necessarily equal to it.

$$P(B|D) \propto P(D|B)\,P(B) \tag{15.3}$$

or

$$p(\theta|y) \propto L(\theta)\,\pi(\theta) \tag{15.4}$$

$p(\theta|y)$ is therefore the posterior probability of a model parameter, given the data. It is a fundamental statistic of interest in Bayesian methodology, and differs from the approach taken to parameter estimation using traditional statistical procedures.

The important point to remember with Bayesian probability is that when new data or facts become available, a revised posterior probability may be constructed. It is not fixed, and is accepted or rejected with a degree of belief. We shall explore how Bayesians use these concepts in the following section.

The above approach and manner of viewing probability is not the most commonly accepted view among statisticians. The majority of statisticians have adopted the *frequentist* view of probability after the work of Jerzy Neyman (1894–1981) and Egon Pearson (1895–1980), who developed the approach in the early middle part of the last century. Pearson in turn was influenced by the work of his father, Karl Pearson (1857–1936), who created, among other important statistical methods, the concept of employing a pre-determined probability level, or α, for accepting or rejecting a hypothesis. In fact, a foremost criterion of the frequentist approach is the use of hypothesis testing when assessing the statistical worth of a model parameter.

A second foremost criterion of the frequentist position relates to how probability is defined. In clear contrast to the method outlined above that is used to determine the posterior probability of a parameter, the frequentist defines probability in terms of the long-run expected frequency of events. Typically the view is expressed as a ratio, i.e. the number of times event A occurs in N trials or opportunities. To the Bayesian, probability is a measure of the plausibility, or credibility, of an event occuring, given our incomplete knowledge of the data.

Bayesian modeling methods are based on the notion that unknown parameters (e.g. means) have probability distributions. The distribution for a population mean is based on our prior belief about a mean – prior to the subsequent information we gain from a given data situation.

A Bayesian, therefore, interprets probabilities as the degree of belief we have in a given hypothesis or posterior quantity of interest.

To the frequentist, the mean of a large population has a true value, but, based on a sample of the data, its value can only be estimated together with a certain confidence interval. The aim is to arrive at an unbiased estimate of the mean, which is generally understood as having a certain degree of probability. The Bayesians, on the other hand, argue that the mean is a random parameter having a certain distribution, e.g. Gaussian or normal. A *credible interval* is believed to surround the sample mean, but it is adjusted based on the particular data on hand, and as new information becomes available. The Bayesian can then assert that there is a 95% probability that the confidence interval contains the true mean. Technically, the frequentist cannot make such a statement. Frequentists must conclude that, given a very large number of samples from the data, each with a calculated confidence interval, 95% of them contain the true or population mean. Admittedly this is a rather convoluted way of expressing a confidence interval, and in reality most frequentists tend to use the Bayesian meaning of confidence interval when reporting results. Transferring the interpretation of a Bayesian credible interval to that of a frequentist confidence interval may appear to make more sense, but it is mistaken.

Again, in Bayesian philosophy, uncertainty is attributed to the model parameters, while the sampled data are construed to be given. Owing to this, parameters are modeled by distributions. Any prior knowledge or beliefs that are entertained about the parameters before data are considered are referred to as prior knowledge, which is expressed as a distribution, i.e. a prior distribution. The data are brought into consideration by conjoining them with a prior distribution, resulting in a parameter posterior distribution. Each parameter in a model has its own distribution, and therefore its own posterior distribution. The mean – or mode or median – of the posterior distributions for each model parameter is then determined, which are the Bayesian (posterior) parameter point estimates of the model. The interval estimates of the parameters are referred to as *credible intervals*.

The logic of Bayesian model estimation process, as outlined above, will be given in more detail in the next section. The key to differentiating Bayesian models from frequentist-based models is in how parameters are regarded. In frequentist methodology, an aim is to estimate the coefficients or parameter estimates of the predictors using IRLS, maximum likelihood, quadrature, or some similar method. In Bayesian methodology, parameters are considered to have a particular distribution, which is estimated based on our prior

Table 15.1 *Differences in Bayesian vs frequentist approaches to modeling*

Item	Bayesian	Frequentist
Starting point	prior distribution	presumed ignorance
Fixed quantity	data	unknown parameters
Parameter	randomly distributed	fixed constant
Inference	posterior distribution	(log)likelihood; setting α in advance
Parameter estimate	Mean (or other) of posterior	ML estimate
Fit test	DIC, HPD intervals, etc.	AIC, BIC, LR, etc.

beliefs about it, as well as the predictor data. The mean or mode of the resultant posterior distribution of a parameter is referred to as the *parameter of interest.*

When dealing with complex and multidimensional data, it is difficult to determine the mean or other parameter of interest of the posterior distribution. A Markov Chain Monte Carlo (MCMC) simulation method has been developed to have the software continually sample from the posterior until the posterior quantity of interest is available, with an acceptable degree of precision. A Markov chain, named after Russian mathematician Andrey Markov (1856–1922), steps through a series of random variables in which a given value is determined on the basis of the value of the previous element in the chain. A homogeneous chain is employed for the Bayesian modeling of members of the family of generalized linear models. The most basic type of homogeneous Markov chain has a state-space of [0,1], and is regulated by a transition matrix which ideally runs until it reaches its limiting distribution. A state-space is merely a term that identifies the range of possible parameter (θ) values, whether discrete or continuous. We shall discuss the transition matrix in more detail later in this Chapter. In any case, MCMC diagnostics are provided by standard Bayesian software advising users if convergence is achieved. The details are beyond the scope of this discussion, but can be found in most of the standard texts on the subject.

15.2 The logic of Bayesian regression estimation

The goal in Bayesian modeling is to determine the parameter of interest of the posterior distribution of a predictor together with a credible interval. The credible interval is somewhat similar to a confidence interval, informing us that the true parameter of interest can be found within the range of the interval 95% of the time. I shall assume that the parameter of interest is the mean of the posterior distribution, but the median and mode are also commonly used.

We begin a Bayesian modeling procedure by first selecting a prior distribution for the predictor of interest. Each predictor will have a separate, but possibly similar, prior. If we use a uniform distribution for the prior, producing a *non-informative prior*, the result of the Bayesian modeling effort will be a mean value that is close in value to the traditional maximum likelihood parameter estimate. An example of the use of a non-informative prior for a continuous predictor in a Bayesian negative binomial model will be shown in the following section. You will notice that the mean of the posterior distribution is extremely close to that of the maximum likelihood estimate. This result is typical when using a non-informative prior. Such a prior is used when we do not have any reason to give additional information about the predictor than is already contained in the data. Recall that each predictor in a model can have a different prior. When engaged in estimating a Bayesian model, one should use what is called an *informative prior* when information is available that can amend the results of the estimate if used.

Before MCMC methods were used for sampling a posterior distribution, it was common, if not necessary, to employ a prior that was *conjugate* to the distribution of the model. In the context of Bayesian modeling, a conjugate distribution is one where the posterior distribution for a model parameter has the same form as the prior. For example, the gamma distribution is conjugate for the Poisson mean; the beta distribution is conjugate to both the binomial and negative binomial event probability.

It is important to recognize that conjugate priors (e.g. gamma, beta) for negative binomial models are really only applicable for intercept-only models, and where the parameter of interest is the simple probability of an event, p. Model interpretation in this type of case is that the response is the number of y failures before the rth success in a series of Bernoulli trials (where the probability of an event p is the parameter of interest). In this case, the beta prior for p is conjugate. Informative priors for the negative binomial regression coefficients or for the dispersion, α, are useful in that they allow more information to be brought into the model than exists in the predictor data; however, they are not conjugate priors. Given the sophistication of current sampling algorithms, conjugate priors are more of a mathematical convenience than they are essential components in informative priors. It is therefore more useful to focus on whether non-informative or informative priors are conjoined with the model likelihood when constructing posterior distributions and their parameters.

MCMC sampling therefore allows non-conjugate informative priors to be used to multiply with the likelihood. The likelihood, of course, reflects the type of model being used on the data. A Bayesian negative binomial model determines the posterior distribution by multiplying the negative binomial likelihood function by the selected prior and normalizing constant. Strictly speaking, however, one can only use this method of calculating the posterior distribution in a limited number of cases, and for simple models. For example, such a method can be used for a normal likelihood with conjugate prior, or an intercept-only Poisson model with a gamma prior. More complex models must resort to sampling algorithms like MCMC and Gibbs' Sampling that do not require the normalizing constant. For almost all Bayesian models, and Bayesian negative binomial models in general, MCMC, or alternative sampling methods, must be used to construct a posterior distribution, and posterior parameters of interest. It should be mentioned that when modeling a Bayesian negative binomial the dispersion parameter is also provided a posterior distribution, for which a mean must be determined. Given that the dispersion parameter, α, has a range from 0 to ∞, it is reasonable to select either the normal or gamma as the prior – preferably the latter. The beta distribution would restrict the range of the resultant posterior to (0,1), which is not acceptable. It should be understood that the beta is only useful for parameters in (0,1), such as event probabilities. Binary predictors in a regression model have coefficients that are generally unrestricted in range, and it is the coefficients that are the parameters being modeled. Therefore, in a Bayesian model, a prior such as the uniform or normal is typically used with binary predictors, although beta priors are also used in practice. Table 15.2 provides an overview of the major steps in the process of modeling.

Prior to examining examples of Bayesian negative binomial modeling, it will be helpful to discuss the logic of MCMC sampling in a bit more depth. Sampling from the posterior distribution has been a cornerstone of Bayesian modeling, so it is therefore important to understand the general nature of its algorithm

As previously mentioned, MCMC methods construct chains in which values at each link are dependent on only the values of the previous link. The idea is that if the chain is allowed to run long enough it will find a solution for the desired posterior parameter of interest, e.g. the posterior mean.

The Markov chain is a type of stochastic process, for which any given state in the chain, θ^k, is only dependent on the previous value in the series, θ^{k-1}. Recall that a stochastic process is a consecutive set of random values that are defined on the basis of a known state-space. The chain therefore moves about

Table 15.2 *Outline: Logic of Bayesian modeling – negative binomial model*

Select a probability distribution, and therefore likelihood function, which relates our understanding of the data.

Select a probability distribution as the prior, which represents our beliefs about the parameters independent of related data. Choose either a non-informative or informative prior for each predictor in the model. The NB2 dispersion parameter, α, is included as well. A conjugate distribution is often not feasible for the negative binomial.

Using MCMC, or a related simulation method (e.g. Metropolis–Hastings algorithm; Gibbs sampler; adaptive rejection sampling algorithm), repeatedly sample from the posterior until a posterior distribution is constructed, and a credible point estimate of the parameter of interest is achieved, e.g. posterior mean or mode.

Compute Monte Carlo standard error (MCSE) and credible interval or highest posterior density (HPD) interval for parameter of interest.

Assess Markov chain convergence and test the reliability of sampling using software diagnostics (e.g. Gelman–Rubin, Geweke, Raftery–Lewis tests). Make certain all parameters have converged, not only those of primary interest. Using MCSE mentioned before, finding point estimates usually involves finding the mode or the mean of the parameter's distribution.

Assess comparative model fit using deviance information criterion (DIC) or similar test.

---

the state-space, with only the previous value at any point in the chain having a bearing on its value. However, there must be a *transition matrix* which defines the probabilities of moving from one to another link in the chain. That is, a transition matrix must be established that maps transitions to their probability of occurrence. A simple example of a Markov chain should help clarify the process.

Suppose a simple 2×2 transition matrix consists of two selections, θ_1 and θ_2. We may state that those who select θ_1 have a 60% probability of continuing to select θ_1. Those who select θ_2 have a 80% probability of continuing to select it. The transition matrix appears as:

$$\begin{array}{cc} & \textit{stage } i+1 \\ & \begin{array}{cc}\theta_1 & \theta_2\end{array} \\ \textit{stage } i \begin{array}{c}\theta_1 \\ \theta_2\end{array} & \begin{bmatrix} 0.6 & 0.4 \\ 0.2 & 0.8 \end{bmatrix} \end{array}$$

Initial values must be provided the selection process. We assume that, independent of our knowledge of the above transition probabilities, half of those engaged in the selection process select θ_1 and half select θ_2. This is a reasonable assumption without prior information. The first stage is then calculated as

$$\begin{bmatrix} 0.5 & 0.5 \end{bmatrix} \begin{bmatrix} 0.6 & 0.4 \\ 0.2 & 0.8 \end{bmatrix} = \begin{bmatrix} 0.4 & 0.6 \end{bmatrix}$$

then,

$$\begin{bmatrix} 0.4 & 0.6 \end{bmatrix} \begin{bmatrix} 0.6 & 0.4 \\ 0.2 & 0.8 \end{bmatrix} = \begin{bmatrix} 0.36 & 0.64 \end{bmatrix}$$

$$\begin{bmatrix} 0.36 & 0.64 \end{bmatrix} \begin{bmatrix} 0.6 & 0.4 \\ 0.2 & 0.8 \end{bmatrix} = \begin{bmatrix} 0.344 & 0.656 \end{bmatrix}$$

$$\begin{bmatrix} 0.344 & 0.656 \end{bmatrix} \begin{bmatrix} 0.6 & 0.4 \\ 0.2 & 0.8 \end{bmatrix} = \begin{bmatrix} 0.3376 & 0.6624 \end{bmatrix}$$

$$\begin{bmatrix} 0.33334 & 0.66665 \end{bmatrix} \begin{bmatrix} 0.6 & 0.4 \\ 0.2 & 0.8 \end{bmatrix} = \begin{bmatrix} 0.33333 & 0.66666 \end{bmatrix}$$

The transition matrix converges to 0.33 and 0.67, or one-third and two-thirds. The transition matrices guide through the state-space, converging to stationary values. This example is overly simple compared with the complex transition matrices and sampling mechanism that takes place in MCMC methodology. But the basic logic is the same.

15.3 Applications

Bayesian techniques are available for a wide range of statistical concerns. We are focusing on how Bayesian methodology can be applied to negative binomial regression, which is a subset of Bayesian modeling in general. Bayesian methods can in general be implemented more easily for members of the family of generalized linear models (GLM) than for models not based on an exponential distribution. A reason for this is the fact that GLMs have clearly identified likelihood and deviance functions, and except for Gaussian models have variances that are based on the means. Moreover, many conjugate distributions are also members of the GLM family. This compatibility of GLM and Bayesian methodology in general has allowed SAS software, for example, to incorporate Bayesian modeling options into its GLM procedure. The SAS/STAT GENMOD Procedure not only provides a framework for estimating a wide range of generalized linear models, but also provides options within the procedure for GEE and Bayesian modeling.

It should be noted that R also has substantial Bayesian modeling capabilities. However, most R users employ R with Winbugs for the actual sampling process. Winbugs is freeware produced by the Imperial College of Science, Technology and Medicine in the UK, and can be downloaded from http://www.mrc-bsu.cam.ac.uk/bugs/. The R package **R2WinBUGS**, posted to CRAN, allows the user to write a data and input file in R, and then to run an associated script in WinBugs, with output returned back into the R session. WinBugs performs MCMC sampling as well as runs diagnostics.

WinBugs may also be accessed from within a Stata session using the **winbugs** command, authored by John Thompson, Department of Health Sciences, University of Leicester. A variety of supporting **winbugs** commands provide the capability of entering, running and testing Bayesian models in Stata using Winbugs software. Whether one uses Winbugs directly as a standalone package, or from within R or Stata, parameters of a Bayesian negative binomial model may be estimated using the Winbugs *dnegbin(p, r)* function. It should be recognized that the parameterization of the negative binomial function in Winbugs is identical to equation 8.21a, and is therefore consistent with Stata and R parameterizations. It is also the same as the distribution used in SAS.

When modeling count data most statisticians tend to use a Poisson distribution, and hence likelihood, together with a gamma or normal prior. A gamma prior makes sense if prior information is added to the Poisson likelihood. The gamma shape allows for a rather wide range of data, and is similar to a standard negative binomial (NB2) maximum likelihood model. If the data being modelled are highly dispersed, however, a Bayesian negative binomial may more closely fit the data.

For our examples we shall use the **badhealth** data, which is a subset of the German health data used earlier in the text. The data have only three variables: *numvisit*, or the number of visits a patient makes to the doctor during the year 1998; *badh*, a binary predictor for which a patient identifies self as in poor health (1), or not in poor health (0); and *age*, a positive-valued continuous variable ranging from 20 to 60. We begin by giving a non-informative prior to the two predictors, assuming that the information contained in the predictors is sufficient for explaining the response.

SAS's GENMOD Procedure is particularly easy to work with for GLM-based Bayesian models. We therefore use it to determine the posterior mean values of the two predictors. A non-informative prior is used. The first output is a straightforward maximum likelihood model. Header output is not displayed, but is given on the book's website. Note that SAS's MCMC Procedure can be used for fitting Bayesian models with arbitrary priors and likelihood functions.

Table 15.3 *SAS code for examples*

```
libname data 'sasdata';
ods graphics on / imagefmt=wmf;
proc genmod data=data.badhealth;
   class BADH;
   model numvisit=age badh/d=nb;
   bayes sampling=gam
         diagnostics=none;
run;
data NormalPrior;
input _type_ $ Intercept Age Badh0 Badh1;
datalines;
Var 1e6  1e6   .01631 1e6
Mean 0.0 0.0 -1.3540  0.0
;
proc genmod data=data.badhealth;
  class BADH;
  model numvisit=age badh/d=nb;
  bayes sampling=gam
        nmc=100000
        thin=10
        diagnostics=none
        cprior=normal(input=NormalPrior);
run;
ods graphics off;
```

MAXIMUM LIKELIHOOD

Analysis of maximum likelihood parameter estimates

Parameter		DF	Estimate	Standard Error	Wald 95% Confidence Limits	
Intercept		1	1.5115	0.1755	1.1675	1.8554
AGE		1	0.0070	0.0033	0.0004	0.0135
BADH	0	1	-1.1073	0.1109	-1.3247	-0.8900
BADH	1	0	0.0000	0.0000	0.0000	0.0000
Dispersion		1	1.0025	0.0696	0.8749	1.1488

Note: The negative binomial dispersion parameter was estimated by maximum likelihood.

Seed values are shown that initiate the Markov chain, followed by the DIC goodness-of-fit statistic and the mean of the posterior estimates. Also included are the credible intervals and a correlation matrix. Similar to the AIC and BIC statistics, when comparing fit, the model with the lowest DIC value is considered the better fit. If two models have nearly the same DIC, neither is preferred over the other, as in case of the non-informative and informative models below.

BAYESIAN NEGATIVE BINOMIAL – NON-INFORMATIVE UNIFORM

```
                      Bayesian Analysis
           Model Information
  Data Set                   DATA.BADHEALTH Written by SAS
  Burn-In Size                          2000
  MC Sample Size                      100000
  Thinning                                10
  Sampling Algorithm                Gamerman
  Distribution             Negative Binomial
  Link Function                          Log
  Dependent Variable              NUMVISIT visits to doctor in 1998

  Uniform Prior for Regression Coefficients
           Parameter     Prior
           Intercept     Constant
           AGE           Constant
           BADH0         Constant
Algorithm converged.

Independent Prior Distributions for Model Parameters
                                     Hyperparameters
              Prior                            Inverse
Parameter     Distribution           Shape       Scale
Dispersion    Gamma                  0.001       0.001

                    Initial Values of the Chain
Chain      Seed  Intercept       AGE      BADH0    BADH1   Dispersion
    1 245153001  1.511457 0.006952   -1.10734     0      1.002524

                        Fit Statistics
DIC (smaller is better)                        4475.355
pD (effective number of parameters)               4.033

                      Posterior Summaries
                            Standard              Percentiles
Parameter     N       Mean Deviation       25%        50%        75%
Intercept  10000    1.5124    0.1766    1.3924     1.5116     1.6318
AGE        10000   0.00697   0.00337   0.00472    0.00697    0.00927
```

```
BADHO      10000 -1.1078    0.1119  -1.1835  -1.1068 -1.0312
Dispersion 10000  1.0096    0.0706   0.9612   1.0061  1.0552

                        Posterior Intervals
Parameter    Alpha  Equal-Tail Interval       HPD Interval
Intercept    0.050     1.1692   1.8545      1.1615   1.8457
AGE          0.050  0.000424    0.0135   0.000510    0.0135
BADHO        0.050    -1.3289  -0.8932     -1.3275  -0.8928
Dispersion   0.050     0.8787   1.1553      0.8774   1.1531

                  Posterior Correlation Matrix
Parameter    Intercept        AGE      BADHO  Dispersion
Intercept        1.000     -0.813     -0.672       0.002
AGE             -0.813      1.000      0.166       0.007
BADHO           -0.672      0.166      1.000      -0.007
Dispersion       0.002      0.007     -0.007       1.000
```

A diagnostic plot for *age* is shown in Figure 15.1 (but is also generated for other parameters including the intercept and α). The top frame shows that MCMC (Gamerman sampling algorithm) convergence is steadily reaching the appropriate mean value of the posterior. The lower left frame displays autocorrelation, and indicates the approximate independence of the samples from the posterior distribution. The lower right frame displays a kernel density estimate of the marginal posterior distribution of parameter *age* using the sampling mechanism. The posterior mean is at the top of the distribution, showing where the majority of mass rests. Estimating from the graph it appears that the posterior mean of *age* is approximately 0.007. The "true" Bayesian value is 0.00699, which is 0.007; 0.00467 to 0.00923 are the 25th and 75th percentiles of the posterior distribution of *age*. Each parameter has a similar diagnostic table displayed.

A wide variety of diagnostic tests have been designed for Bayesian models. Alternative methods of sampling are also available, but all take us beyond the overview we are giving. You should be aware of them, however, and should read the chapter on Bayesian Analysis (Chapter 7) in the SAS-STAT manual, or in the Winbugs manual.

BAYESIAN NEGATIVE BINOMIAL – INFORMATIVE PRIOR

We next add prior information to the model. For simplification purposes, and to emulate how Bayesian models are often used to incorporate outside knowledge into a model, specific additional information will be supplied to the binary *badh* variable.

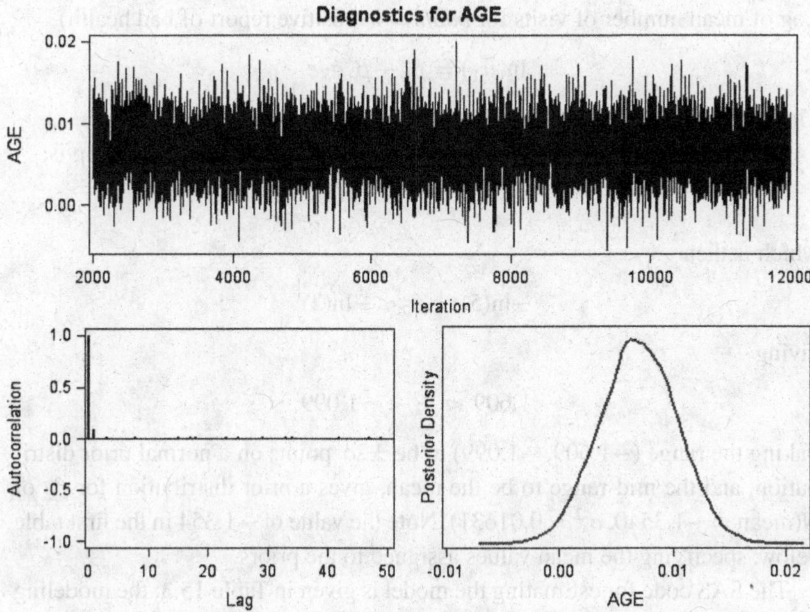

Figure 15.1 Diagnostics of posterior distribution of age – non-informative

The formula for the negative binomial model previously developed may be expressed as:

$$\ln(\mu) = \beta_0 + \beta_1^* age + \beta_2^* badh0$$

Recall that, in SAS, the reference level for a binary predictor is 1, not 0 as in Stata and R. To non-SAS users this relationship may seem strange, but the modeling results are identical if the appropriate reference is kept in mind.

In order to bring prior information into the model that we previously estimated using maximum likelihood, and then as a Bayesian negative binomial model with non-informative priors, we may suppose that researchers have access to a baseline report related to attidudes regarding the self-reporting of health status. The report shows that self-reports of bad health leads to 3–5 times the number of visits to a physician throughout the year compared with those who report that their health is satisfactory or better. This information may be incorporated into the model as a Bayesian informative prior.

In the GENMOD parameterization, the prior may be related to the original model using the following schema.
Log of mean number of visits for *badh* = 0 (a negative report of bad health)

$$\ln(\mu_0) = \beta_0 + \beta_1^* age + \beta_2$$

Log of mean number of visits for *badh* = 1 (positive report of bad health)

$$\ln(\mu_1) = \beta_0 + \beta_1^* age$$

Therefore, the coefficient of *badh*0 is $\ln(\mu_1/\mu_0) = -\beta_2$.

Bringing in the information obtained from the report, the relationship is:

$$3 < \mu_1/\mu_0 < 5$$

which is then

$$-\ln(5) < \beta_2 < -\ln(3)$$

giving

$$1.609 < \beta_2 < -1.099$$

Taking the range (−1.609, −1.099) to be $\pm 2\sigma$ points on a normal prior distribution, and the mid-range to be the mean, gives a prior distribution for β_2 of N(mean $= -1.3540$, $\sigma^2 = 0.01631$). Note the value of −1.354 in the first table below, specifying the mean values assigned to the priors.

The SAS code for estimating the model is given in Table 15.3; the modeling results are displayed below. The results show a stronger effect of the *badh* indicator than the noninformative analysis. Note, for the informative analysis, the predicted increase in the mean is exp(1.2170) = 3.377, which shows that the data pulled the predicted increase to the low side of the prior. I should also mention that Bayesians tend to prefer, and to report, the *precision* rather than the variance. The precision is simply the inverse of the variance, or V^{-1}, and is reported in the first output frame below.

```
Independent Normal Prior for Regression Coefficients
          Parameter          Mean      Precision
          Intercept             0           1E-6
          AGE                   0           1E-6
          BADH0            -1.354       61.31208
Algorithm converged.

Independent Prior Distributions for Model Parameters

                                   Hyperparameters
                 Prior                       Inverse
Parameter     Distribution     Shape           Scale
Dispersion       Gamma         0.001           0.001

                      Initial Values of the Chain
Chain       Seed     Intercept      AGE    BADH0   BADH1 Dispersion
   1   422853000      1.627209 0.006393 -1.21518     0     1.004406
```

```
                  Fit Statistics
DIC (smaller is better)                        4475.517
pD (effective number of parameters)               3.635

                  Posterior Summaries
                               Standard                 Percentiles
Parameter         N       Mean Deviation        25%         50%        75%
Intercept     10000     1.6288    0.1602     1.5215      1.6286     1.7379
AGE           10000    0.00641   0.00339    0.00417     0.00639    0.00865
BADHO         10000    -1.2170    0.0862    -1.2736     -1.2174    -1.1598
Dispersion    10000     1.0110    0.0702     0.9619      1.0085     1.0582

                  Posterior Intervals
Parameter     Alpha   Equal-Tail Interval            HPD Interval
Intercept     0.050      1.3114     1.9426       1.3192        1.9469
AGE           0.050    -0.00012     0.0132     -0.00016        0.0131
BADHO         0.050     -1.3895    -1.0503      -1.3812       -1.0435
Dispersion    0.050      0.8795     1.1538       0.8818        1.1552

              Posterior Correlation Matrix
Parameter    Intercept        AGE      BADHO Dispersion
Intercept        1.000     -0.852     -0.573      0.025
AGE             -0.852      1.000      0.126     -0.001
BADHO           -0.573      0.126      1.000     -0.044
Dispersion       0.025     -0.001     -0.044      1.000
```

Figure 15.2 provides the diagnostic plot for *badh*, which has an informative prior defined by the specified conditions. From the *Posterior Summaries* table above we find that the posterior mean is -1.217, and from the *Posterior Intervals* table the credible interval values are given as -1.38 to -1.0435. The value halfway between the extremes of the credible interval is -1.1689. The fact that the estimated posterior mean is calculated at -1.217 indicates that more mass exists to the more negative side, which is clearly displayed in the Kernal density plot below.

Figure 15.3 is a set of diagnostic plots for the dispersion. The estimated value of α of 1.011 (from the *Posterior Summaries* table above) is well represented by the kernel density plot. Note that the prior for the dispersion in both cases is nearly the same (1.0096 to 1.011). It is a gamma prior, but the mean and variance are such that it is flat relative to the likelihood, and so it is in fact non-informative.

The plots for both non-informative and informative cases appear to show convergence and good "mixing," i.e. independence of samples. The remainder of the plots, both non-informative and informative – for the intercept, predictors, and dispersion parameters – are available on the book's website.

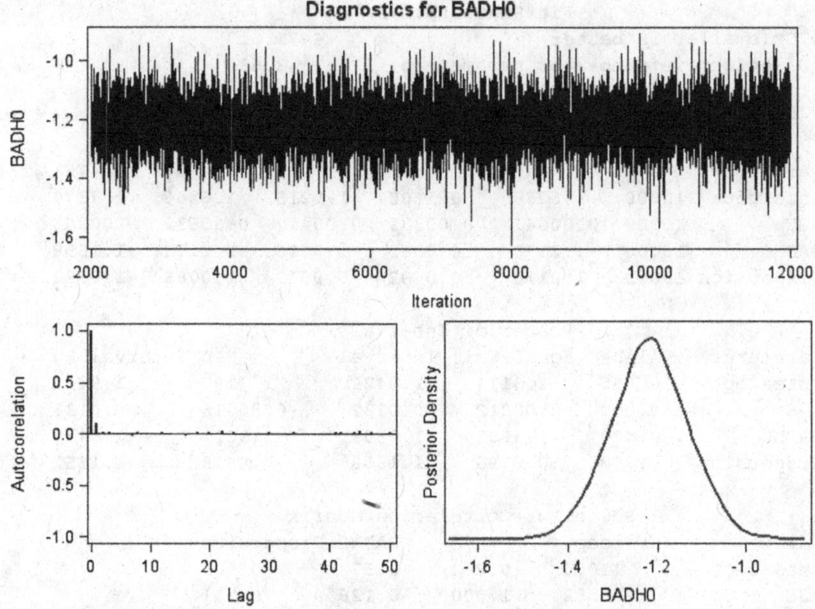

Figure 15.2 Diagnostics for posterior distribution of badh – informative

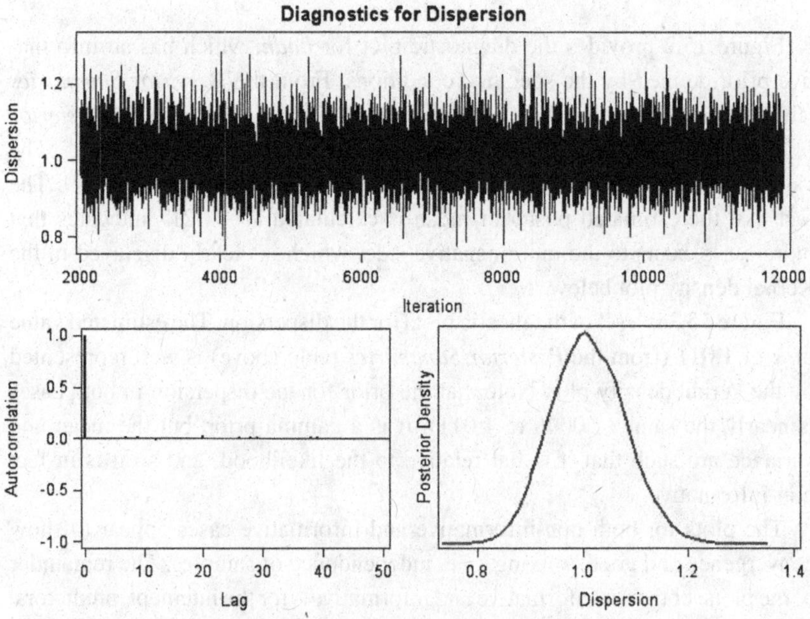

Figure 15.3 Diagnostics for posterior distribution of dispersion, α

More complicated models can be developed, e.g. specifying prior information about *age*. For instance it may be that there is prior information regarding the effect of *age* on reporting health status – information that is in addition to the simple reporting of *age*. In that case perhaps percentage values can be used as prior information as they were for *badh*, or the prior information can be expressed as a prior distribution, e.g. gamma. The logic of Bayesian modeling, however, is the same as displayed above.

Excellent resources exist to assist in learning how to develop Bayesian models, from the simple to the extremely complex. As mentioned, the SAS/STAT manual has a chapter on Bayesian models that is clearly written and informative. The Winbugs manual is also a fine resource. Other excellent resources that have discussions of Bayesian count models include Gelman *et al* (2004), Gelman and Hill (2007), Gill (2002), and Gill (2010).

Appendix A
Constructing and interpreting interaction terms

Interaction terms play an important role in statistical modeling. However, it is a topic that relates to regression models in general and not specifically to count models. The subject was, therefore, not addressed in the text. On the other hand, there are few sources that provide specifics on how interactions are to be constructed and interpreted for non-linear models, and in particular for count models. This appendix provides a brief overview of interactions and their construction.

Recall that there are a variety of types of predictors. The three most common are binary (1/0), categorical (1, 2, . . . , *k*), and continuous. Continuous predictors are often constrained to have positive real numbers as allowable values. In addition, continuous predictors that do not begin at zero or 1 are usually centered, as discussed in Section 6.3.1 the text.

Many researchers believe that the coefficients of interaction terms are interpreted in the same manner as other predictors in a model. However, this is not the case, and in particular it is not the case for interactions in discrete response regression models.

I shall provide a brief overview of how to construct and interpret Binary × Binary, Binary × Continuous, Categorical × Continuous, and Continuous × Continuous interactions. These appear to be the most commonly employed interactions. Understanding these allows the construction and interpretation of any of the other options.

A comprehensive analysis of interaction terms for binary response logistic regression models can be found in Hilbe (2009), where an entire chapter is devoted to their analysis. The same logic used for the construction of logistic model interaction terms also applies to the construction of count model interactions.

Binary × Binary interactions

The German health survey, **rwm5yr**, data used in the text will be employed here for the purpose of constructing model interaction terms. We begin by using a Poisson model of *docvis* on *outwork* and *female,* with patients who are out of work being *outwork* == 1 and females being *female* == 1. The model appears as

```
. glm docvis outwork female, fam(poi) eform nohead

------------------------------------------------------------------
         |                 OIM
  docvis |       IRR  Std. Err.      z    P>|z|  [95% Conf. Interval]
---------+--------------------------------------------------------
 outwork | 1.423918  .0127564  39.45  0.000   1.399134    1.449141
  female | 1.223422  .0110059  22.42  0.000    1.20204    1.245184
------------------------------------------------------------------
```

We may suspect an interactive effect between the predictors if we exclude one of them and find that the risk ratio of the remaining predictor has substantially changed. Excluding *female* results in the following model:

```
. glm docvis outwork, fam(poi) eform nohead

------------------------------------------------------------------
         |                 OIM
  docvis |       IRR  Std. Err.      z     P>|z|  [95% Conf. Interval]
---------+--------------------------------------------------------
 outwork | 1.556847    .012541  54.95    0.000    1.53246    1.581622
------------------------------------------------------------------
```

The risk ratio of *outwork* has increased by some 13%, indicating a possible interactive effect.

An interaction term is created by taking the product of the two binary predictors. The resulting product term is added to the model, including the two main effects predictors. You *must* have the main effects in the model, just as we keep lower terms in an ANOVA when there are interactions. The same logic obtains.

```
. gen byte oxf = outwork * female
. glm docvis outwork female oxf, fam(poi) eform nohead

------------------------------------------------------------------
         |                 OIM
  docvis |       IRR  Std. Err.      z    P>|z|  [95% Conf. Interval]
---------+--------------------------------------------------------
 outwork | 1.950039  .0272785  47.74  0.000   1.897301    2.004244
  female | 1.466181  .0160671  34.92  0.000   1.435026    1.498012
     oxf | .6114415  .0108215 -27.80  0.000   .5905955    .6330234
------------------------------------------------------------------
```

oxf indeed appears to be significant. The main effects terms in the interaction model have the following interpretation:

> *The risk ratio of **outwork** reflects the effect of outwork on the number of visits to the doctor when female == 0. Likewise, the risk ratio of **female** reflects the effect of female on the number of visits to the doctor when outwork ==0.*

To determine the actual interaction of *outwork* and *female* we use the following two formulae:

$$\mathrm{A}: \exp(\beta_1 + \beta_3^* \text{female})$$

and

$$\mathrm{B}: \exp(\beta_2 + \beta_3^* \text{outwork})$$

We therefore need the coefficients:

```
. glm docvis outwork female oxf , fam(poi) nohead

------------------------------------------------------------------------------
             |                 OIM
      docvis |      Coef.   Std. Err.      z    P>|z|     [95% Conf. Interval]
-------------+----------------------------------------------------------------
     outwork |   .6678495   .0139887    47.74   0.000     .6404322    .6952668
      female |   .3826608   .0109585    34.92   0.000     .3611827     .404139
         oxf |  -.4919359   .0176983   -27.80   0.000    -.5266239    -.457248
       _cons |   .8388561   .0070617   118.79   0.000     .8250153    .8526968
------------------------------------------------------------------------------
```

1 The incidence rate ratio of unemployed females to unemployed males is $\exp(0.6678495 + (-0.4919359 * 1)) = 1.192335$
Unemployed females have 19% more visits to the doctor during the year than do unemployed males.

2 The incidence rate ratio of unemployed males to employed males is $\exp(0.6678495 + (-0.4919359 * 0))) = 1.9500392$
Unemployed males have 95% more visits to the doctor during the year than do employed males.

3 The incidence rate ratio of unemployed females to employed females is $\exp(0.3826608 + (-0.4919359 * 1)) = .89648376$
Unemployed females have 10% fewer visits to the doctor than do employed females.

4 The incidence rate ratio of employed females to employed males is $\exp(0.3826608 + (-0.4919359 * 0)) = 1.4661806$
Employed females have 47% more visits to the doctor than do employed males.

Binary × Continuous interactions

Using the same data, it is possible to construct an interaction between the binary *married* and the continuous variable, *age*. For the purpose of exposition we shall not center *age* for this example, but in genuine studies we may wish to do so. The difference with what we show here exists only when we interpret the interaction, as explained in Section 6.3.1.

We determine the base interaction coefficient in the same manner as we did for Binary × Binary interactions. Again, we find the product of the two predictors, referring to it as *mxa*.

```
. gen mxa = married * age
. glm docvis married age mxa, fam(poi) nohead

------------------------------------------------------------------------------
             |                 OIM
      docvis |      Coef.   Std. Err.      z    P>|z|     [95% Conf. Interval]
-------------+----------------------------------------------------------------
     married |  -.218668   .0366924    -5.96   0.000    -.2905837   -.1467523
         age |  .0225065   .0006485    34.71   0.000     .0212355    .0237775
         mxa |  .0026277   .0007823     3.36   0.001     .0010945     .004161
       _cons |   .211727   .0298829     7.09   0.000     .1531576    .2702963
------------------------------------------------------------------------------
```

The following formula can be used to determine the interactions. There are as many interactions as there are values of the continuous predictor – for each level of the binary predictor. Typically, however, the continuous predictor will be categorized into five to ten levels. We have,

$$\text{IRR}_{\text{B}\times\text{C}} = \exp[\beta_1 + \beta_3 x]$$

where $\text{IRR}_{\text{B}\times\text{C}}$ is the interaction of Binary × Continuous predictors, β_1 is the coefficient of the binary term, β_3 is the coefficient of the interaction, and x is the value of the continuous predictor.

For our model we have:

$$\text{IRR}_{\text{m}\times\text{a}} = \exp[\beta_1 + \beta_3 age]$$

$$\text{IRR}_{\text{m}\times\text{a}} = \exp[-0.218668 + 0.0026277^{*} age]$$

In order to calculate the Binary × Continuous interactions, we construct a table with specified values of *age*. There will therefore be a separate interaction IRR value for each age we specify. Ages 30, 40, 50 and 60 are defined.

$\exp(-0.218668 + 0.0026277^{*}30) = 0.86949995$
$\exp(-0.218668 + 0.0026277^{*}40) = 0.89265064$
$\exp(-0.218668 + 0.0026277^{*}50) = 0.91641771$
$\exp(-0.218668 + 0.0026277^{*}60) = 0.94081759$

The next problem is in determining the standard errors (SEs) and confidence intervals (CIs) for the various instances of the interaction. The SEs and CIs will differ for each value of *age*. The IRR standard errors for the interactions are determined by first calculating the variance. The formula is:

$$V_{\mathrm{B\times C}} = V(\beta_1) + x^2 V(\beta_3) + 2 \times \mathrm{Cov}(\beta_1, \beta_3)$$

In order to obtain the values necessary to calculate the variance, and therefore the standard error, of the interaction, we need the model variance–covariance matrix. Given the above model, it is:

```
. estat vce , cov

Covariance matrix of coefficients of glm model

---------------------------------------------------------------------
             | docvis
        e(V) |    married          age          mxa        _cons
-------------+-------------------------------------------------------
docvis       |
     married |  .00134633
         age |   .0000186    4.205e-07
         mxa |  -.0000277   -4.205e-07    6.119e-07
       _cons | -.00089299    -.0000186     .0000186    .00089299
---------------------------------------------------------------------
```

Inserting values into the variance formula above, we have,

$$V_{\mathrm{B\times C}} = 0.00134633 + \mathrm{age}^2 * 0.0000006119 + 2 * \mathrm{age} * -0.0000277$$

Taking the square root of the variance produces the standard error. Combining the operations:

STANDARD ERRORS

Age 30:

```
. di sqrt(.00134633 + (30*30* 0.0000006119) + (2*30*-.0000277))
.01533101
```

Age 40:

```
. di sqrt(.00134633 + (40*40* 0.0000006119) + (2*40*-.0000277))
.01045801
```

Age 50:

```
. di sqrt(.00134633 + (50*50* 0.0000006119) + (2*50*-.0000277))
.01029951
```

Age 60:

```
. di sqrt(.00134633 + (60*60* 0.0000006119) + (2*60*-.0000277))
.01500567
```

Finally the confidence intervals must be calculated: The formula is

$$[\beta_1+\beta_3 x] +/- z_{1-a/2} {}^*\text{SE}$$

I shall show the confidence intervals for the *mxa* interaction when age = 30. Other confidence intervals can be calculated using the same logic. I use the 95% CI, or $p = 0.05$. This statistic is calculated as 1.96. Since we exponentiated the coefficients to express incidence rate ratios, we exponentiated the coefficient confidence intervals to generate IRR confidence intervals.

Confidence intervals

LOW CI FOR AGE = 30

```
. di exp((-.218668  + .0026277*30) - 1.96*.01533101)
.84376119
```

HIGH CI FOR AGE = 30

```
. di exp((-.218668  + .0026277*30) + 1.96*.01533101)
 .89602387
```

The interaction IRR of *married* and *age* at *age* = 30 is reported as: IRR = 0.8695 (0.8438, 0.8960).

The remaining confidence intervals are likewise calculated. The manner of calculating standard errors and confidence intervals is the same for all types of two-term interactions.

Interpretation

Married individuals at age 30 have some 13% (95% CI: 10%–16%) fewer visits to the doctor than unmarried individuals.

Significance of interaction

The significance of the exponentiated interaction can be determined if the confidence interval includes 1.0. If it does, the predictor is not statistically significant. Likewise, if the interaction is left un-exponentiated and the confidence interval includes 0, it is not significant. Since the confidence interval for the IRR of *mxa* at age 30 does not include 1.0, it is statistically significant at the $p = 0.05$ level.

Binary × Categorical predictors are calculated in the same way as Binary × Continuous. The difference is that with continuous predictors we select the levels we wish to calculate, whereas with categorical predictors the values are already determined. By categorizing *age* into categories of 30, 40, 50, and 60, we have artificially made a categorical predictor. Of course, we can calculate as many values of the continuous predictor as we wish. Interpretation of the IRR is the same.

Categorical × Continuous interactions

A Categorical × Continuous interaction is constructed in the same manner as a Binary × Continuous interaction, except that there are multiple Binary–Continuous equations. Given a three-level categorical predictor with level 1 as the reference, a Categorical × Continuous interaction is constructed using two equations. One equation relates the continuous predictor with a binary predictor composed of categorical levels 2 and 1, and a second equation relates the continuous predictor with categorical levels 3 and 1. For the first equation, the binary component has a value of 1 for categorical level 2 and 0 for the reference (level 1). For the second equation, the binary component has a value of 1 for categorical level 3, and 0 for the reference.

Using the same **rwm5yr** data, we shall develop an interaction between 4-level *edlevel* and *age*. Again, we shall not center *age*, but only for pedagogical purposes. We can let the software generate the appropriate coefficients by multiplying *age* with each non-reference level of *edlevel*.

*ed*1: not a high-school graduate
*ed*2: high-school graduate
*ed*3: college/university
*ed*4: graduate work

```
. tab edlevel, gen(ed)         /* categorize edlevel: 4 binary
                                  predictors */
. gen byte axed2 = age * ed2
. gen byte axed3 = age * ed3
. gen byte axed4 = age * ed4
. glm docvis age ed2 ed3 ed4 axed2 axed3 axed4, fam(poi) nohead

------------------------------------------------------------------
        |             OIM
 docvis |    Coef.  Std. Err.      z    P>|z|  [95% Conf.  Interval]
--------+---------------------------------------------------------
    age | .0232975  .0004022  57.93   0.000   .0225092   .0240858
    ed2 |-.0538255    .07402  -0.73   0.467   -.198902   .0912511
    ed3 | .1275363   .058218   2.19   0.028   .0134311   .2416415
```

```
    ed4 |-.4711621   .091667  -5.14  0.000   -.650826   -.2914981
  axed2 |-.0001867  .0016307  -0.11  0.909  -.0033829    .0030095
  axed3 |-.0086629  .0013273  -6.53  0.000  -.0112643   -.0060615
  axed4 | .0001118  .0020647   0.05  0.957  -.0039349    .0041585
  _cons | .1454206  .0196208   7.41  0.000   .1069645    .1838767
-----------------------------------------------------------------
```

The coefficients for the above model are specified, for the purpose of interpreting interaction effects, as:

$\beta_0 = \text{intercept}$
$\beta_1 = age$
$\beta_2 = ed2$, $\beta_3 = ed3$, $\beta_4 = ed4$
$\beta_5 = axed2$, $\beta_6 = axed3$, $\beta_7 = axed4$

The actual interactions are developed using the same formula we employed earlier. Each interaction group is therefore expressed as:

axed2: $\beta_{\text{Cat2}\times\text{Con}} = \beta_2 + \beta_5 x = -0.0538255 + (-0.0001867*\text{age})$
axed3: $\beta_{\text{Cat3}\times\text{Con}} = \beta_3 + \beta_6 x = 0.1275363 + (-0.0086629*\text{age})$
axed4: $\beta_{\text{Cat4}\times\text{Con}} = \beta_4 + \beta_7 x = -0.4711621 + (0.0001118*\text{age})$

To obtain the incidence rate ratios (IRR), simply exponentiate each of the above formulae, or the resultant coefficient. For high-school graduates now 30 and 60 years of age, the IRRs are

HIGH SCHOOL GRADUATE

age = 30: $\exp(-0.0538255 + (-0.0001867*30)) = 0.94230479$
age = 60: $\exp(-0.0538255 + (-0.0001867*60)) = 0.93704169$

GRADUATE SCHOOL

age = 30: $\exp(-0.4711621 + (0.0001118*30)) = 0.62637371$
age = 60: $\exp(-0.4711621 + (0.0001118*60)) = 0.6284781$

Remember, though, that the above IRRs are in comparison to the reference level. We may interpret the 30-year-old individual with a graduate school education as follows:

30-year-olds with graduate-school education visit the doctor some 37% fewer times during the year than 30-year-olds without a high-school diploma.

Other interaction IRR values may be determined and interpreted in the same manner. Moreover, standard errors and confidence intervals may be calculated as described for Binary × Continuous interactions.

Continuous × Continuous interactions

Adding a Continuous × Continuous interaction to a model can result in a host of coefficients, or slopes. For examples, if *age* and *los* are both continuous variables, having n and m distinct values respectively, their interaction will result in $n \times m$ separate coefficients. It is easy to get lost in the morass of slopes.

The solution is to select one of the continuous predictors, factoring it into five to ten categories. More may be specified if deemed reasonable for the purposes of the study. Suppose that *age* ranges from 20 to 90. For a continuous main effect, this is typically called the *moderator* variable. We could then select values for the moderator at 20, 30, 40, 50, 60, 70, 80, and 90, treating it as a categorical predictor. The moderator can be named *agegrp*, so that slopes are calculated after estimating the above model including the interaction, as

$$\text{IRR}_{\text{losXagegrp}} = \exp[\beta_2(los) + \beta_3 * agegrp]$$

where β_1 is the coefficient of *age*, β_2 is coefficient of *los*, and β_3 is the coefficient of the interaction of *age* and *los*. If $\beta_2 = 1.056$ and the interaction coefficient $\beta_3 = -0.0598$, the Continuous × Quasi-continuous coefficients can be determined as:

Slope($agegrp = 20$): $1.056 - 0.0598{*}20 = -0.140$ IRR $= \exp(-0.140) = 0.869$
Slope($agegrp = 30$): $1.056 - 0.0598{*}30 = -0.738$ IRR $= \exp(-0.738) = 0.478$

⋮

Slope($agegrp = 90$): $1.056 - 0.0598{*}90 = -4.326$ IRR $= \exp(-4.326) = 0.013$

Interpretation is the same as described for Categorical × Continuous interactions. Of course, to find the full complement of slopes, one must calculate the coefficients for various levels of *los*. A moderator termed *losgrp* can then be used to obtain coefficients and IRRs at various levels of the corresponding side of the interaction.

$$\text{IRR}_{\text{ageXlosgrp}} = \exp[\beta_1(age) + \beta_3(losgrp)]$$

The Continuous × Continuous interaction, when actually used in practice, is in effect, for each side of the interaction, a Categorical × Continuous predictor, and is interpreted as such. There are two components to the interaction, however, which is not the case for standard Categorical × Continuous interactions.

When factoring a continuous variable into discrete categories, information is lost. If we think it to be substantial, then the new categorical variable should have more categories or levels. Remember, though, that the values we depict along the range of a continuous variable are in fact discrete, and therefore in a sense are categorical. The question becomes how many levels we are willing to deal with in an analysis.

Brief look at software

Beginning with Stata version 11, a binary or categorical variable can be specifically declared as such by proceeding it with an "i;" "c" declares a continuous variable, e.g. *i.married* and *c.age*. By placing a "##" operator between i and c declared predictors, the appropriate interaction coefficients will be displayed, from which subsequent calculations may be developed. In R, the corresponding operator is "*". These operators display both main effects as well as the appropriate interaction coefficients.

```
Stata:   . glm docvis i.married##i.female, fam(poi)
R    :   > glm(docvis ~ married*female, family=poisson)
```

Appendix B
Data sets, commands, functions

Data files (Stata:.dta; R:.RData) used in text: Chapter, section first used

titanic	2.1	azpro	5.2.1.1	medpar	6.3.1	rwm5yr	6.3.1
fasttrakg	6.7.3	lbw	9.4	lbwgrp	9.4	affairs	9.5;1
titanicgrp	9.5;3	mdvis	9.5;4	rwm	10.5	azdrg112	13.1.1
loomis	13.3.1	ships	14.4	seizure	14.5	badhealth	15.3

SAS Code : WORK.rwm5yr - Ch 14.2.3 badhealthA.sas - Ch 15.3
ASCII text: smokedrink.raw - Table 10.18
R data and functions in COUNT package

R *script files* *<Additional scripts and Stata programs on book's website>*

modelfit.r	Table 5.3	AIC/BIC statistics	function
poisson_syn.r	Table 6.4	syn Poisson	function
p.reg.ml.r	§6.2.1	Poisson optimization	
mysim.r	Table 6.5	Monte Carlo simulation Poisson	
nbr2_6_2_2.r	Table 6.6	Changes to univariable model	
nbr2_6_2_3.r	Table 6.7	Changes to multivariable model	
syn.poissono.r	Table 6.12	syn Poisson-offset	
poi.obs.pred.r	Table 6.15	Poisson: observed vs predicted probs	function
nbr2_7_1.r	§7.1	syn poisson	
ml.nb2.r	Table 8.7	max like NB2	function
nb2_syn.r	Table 9.3	syn NB2	function
abic.r	§8.4	AIC, BIC tests	
nbsim.r	Table 9.5	syn Monte Carlo NB2	

(*cont.*)

(cont.)

syn.nb2o.r	Table 9.9	syn NB2 offset	
nb.reg.ml.r	Table 9.11	max like NB2 optimiz	
logit_syn.r	Table 9.20	syn logit	function
probit_syn.r	Table 9.21	syn probit	function
syn.bin_logit.r	Table 9.22	syn grp logistic regression	
nb2.obs.pred.r	Table 9.27	NB2 observed vs predicted prob	function
myTable.r	Table 9.40	frequency table	function
geo_rng.r	Table 10.2	syn geometric via NB2	
syn.cgeo.r	Table 10.3	syn canonical geometric	
syn.geo.r	Table 10.4	syn geometric	
nb1syn.r	Table 10.6	syn NB1	
ml.nb1.r	Table 10.8	max like NB1	function
nbcsyn.r	Table 10.9	syn NB-C	function
nbc.reg.ml.r	Table 10.11	max like NB-C	
ml.nbc.r	Table 10.12	max like NB-C	function
binegbin.glm.r	Table 10.16	bivariate NB	
syn.hurdle_lnb2.r	Table 11.4	syn logit-NB hurdle	
ZAP.r	§11.3	zero adjusted Poisson	
ZINBI	§11.3	zero-inflated NB2	
ZANBI.r	§11.3.4	zero-altered NB2	
syn.ppfm.r	Table 13.2	syn finite mixture Poisson-Poisson	
syn.nb2nb2fm.r	Table 13.3	syn finite mixture NB2-NB2	

STATA do files				Selected Specialized Author Commands			
poi_rng.do	Table 6.2	poi_sim.ado	Table 6.3	abic.ado	5.2.1.1	hnblogit.ado	11.2
poio_rng.do	Table 6.19	nb2ml_sim.ado	Table 9.6	negbin	8.4.1	hnblogit_p	11.2
nb2_rng.do	Table 9.4	nb2o_rng.do	Table 9.10	cnbreg.ado	10.1.2	cpoissone.ado	12.1.2
cgeo_rng.do	§10.1.2	geo_rng.do	§10.1.3	chnbreg.ado	10.4	cpoisson.ado	12.2
nb1_rng.do	§10.2	nbc_rng.do	§10.3.2	hnbregl.ado	10.4	censornb.ado	12.2
nb2logit_hurdle.do	Table 11.5	nbr2_f11.11.do	Table 11.13	ztcnb	11.1		

References and further reading

Akaike, H. (1973). Information theory and extension of the maximum likelihood principle, in *Second international Symposium on Information Theory*, ed. B. N. Petrov and F. Csaki, Budapest: Akademiai Kiado, pp. 267–281.

Allison, P. D. and R. Waterman (2002). Fixed-effects negative binomial regression models, unpublished manuscript.

Amemiya, T. (1984). Tobit models: A survey, *Journal of Econometrics* **24**: 3–61.

Anscombe, F. J. (1948). The transformations of Poisson, binomial, and negative binomial data, *Biometrika* **35**: 246–254.

Anscombe, F. J. (1949). The statistical analysis of insect counts based on the negative binomial distribution, *Biometrics* **5**: 165–173.

Anscombe, F. J. (1950), Sampling theory for the Negative Binomial and Logarithmic Series Distributions, *Biometrika* **37** (3/4): 368–382.

Anscombe, F. J. (1972). Contribution to the discussion of H. Hotelling's paper, *Journal of the Royal Statistical Society – Series B* **15**(1): 229–230.

Bartlett, M. S. (1947). The use of transformations, *Biometrics* **3**: 39–52.

Beall, G. (1942). The Transformation of data from entomological field experiments so that that analysis of variance becomes applicable, *Biometrika* **29**: 243–262.

Bliss, C. I. (1958). The analysis of insect counts as negative binomial distributions. *Proceedings of the Tenth International Congress on. Entomology* **2**: 1015–1032.

Bliss, C. I., and R. A. Fisher (1953). Fitting the negative binomial distribution to biological data and note on the efficient fitting of the negative binomial, *Biometrics* **9**: 176–200.

Bliss, C. I. and A. R. G. Owen (1958). Negative binomial distributions with a common *k*, *Biometricka* **45**: 37–58.

Blom, G. (1954). Transformations of the binomial, negative binomial, Poisson, and χ^2 distributions, *Biometrika* **41**: 302–316.

Boswell, M. T. and G. P. Patil (1970). Chance mechanisms generating the negative binomial distribution, in *Random Counts in Models and Structures*, Volume 1, ed. G. P. Patil, University Park, PA: Pennsylvania State University Press, pp. 1–22.

Bozdogan, H. (1987). Model selection and Akaike's Information Criterion (AIC): The general theory and its analytical extensions, Special Section, *Psychometrika* **52**(3): 345–370.

Breslow, N. E. (1984). Extra-Poisson variation in log-linear models, *Applied Statistics* **33** (1): 38–44.

Bulmer, M. G. (1974). On Fitting The Poisson Lognormal Distribution to Species-Abundance Data, *Biometrics* **30**:101–110.

Cameron, A. C. and P. K. Trivedi (1986). Econometric models based on count data: Comparisons and applications of some estimators, *Journal of Applied Econometrics* **1**: 29–53.

Cameron, A. C. and P. K. Trivedi (1990). Regression-based tests for overdispersion in the Poisson model, *Journal of Econometrics* **46**: 347–364.

Cameron, A. C. and P. K. Trivedi (1998). *Regression Analysis of Count Data*, New York: Cambridge University Press.

Cameron, A. C. and P. K. Trivedi (2010). *Microeconometrics Using Stata*, revised edition, College Station, TX: Stata Press.

Collett, D. (1989). *Modelling Binary Data*, London: Chapman & Hall.

Consul, P. and G. Jain (1973). A generalization of the Poisson distribution, *Technometrics* **15**: 791–799.

Consul, P. C. and R. C. Gupta (1980). The generalized binomial distribution and its characterization by zero regression, *SIAM Journal of Applied Mathematics* **39**(2): 231–237.

Consul, P. and F. Famoye (1992). Generalized Poisson regression model, *Communications in statistics – Theory and Method* **21**: 89–109.

Cui, J. (2007). QIC program and model selection in GEE analysis, *Stata Journal* **7**(2): 209–220.

Dean, C. and Lawless, J. F. (1989). Tests for detecting overdispersion in Poisson regression models, *Journal of the American Statistical Association* **84**: 467–472.

Deb, P. and P. K. Trivedi (2002), The structure of demand for medical care: latent class versus two-part models, *Journal of Health Economics* **21**: 601–625.

Deb, P. and P. K. Trivedi (2006), Maximum simulated likelihood estimation of a negative binomial regression model with multinomial endogeneous treatment, *Stata Journal* **6**: 246–255.

Dohoo, I., W. Martin and H. Stryhn (2010). *Veterinary Epidemiologic Research*, Charlottetown, Prince Edward island, CA: VER, Inc.

Drescher, D. (2005). Alternative distributions for observation driven count series models, Economics Working Paper No 2005–11, Christian-Albrechts-Universitat, Kiel, Germany.

Dunn, P.K. and G.K. Smyth (1996). Randomized quantile residuals, *Journal of Computational and Graphical Statistics* **5**(3): 236–244.

Edwards, A. W. F. (1972). *Likelihood*, Baltimore, MD: Johns Hopkins University Press.

Eggenberger F. and G. Polya (1923). Uber die Statistik Verketteter Vorgange, *Journal of Applied Mathematics and Mechanics* **1**: 279–289.

Englin, J. and J. Shonkwiler (1995). Estimating social welfare using count data models: An application under conditions of endogenous stratification and truncation, *Review of Economics and Statistics* **77**: 104–112.

Evans, D. A. (1953). Experimental evidence concerning contagious distributions in ecology, *Biometrika* **40**: 186–211.

Fair, R. (1978). A theory of extramarital affairs, *Journal of Political Economy* **86**: 45–61.

Famoye, F. (1995). Generalized binomial regression model, *Biometrical Journal* **37**(5): 581–594.

Famoye, F. and K. Singh (2006). Zero-truncated generalized Poisson regression model with an application to domestic violence, *Journal of Data Science* **4**: 117–130.

Faraway, J. (2006). *Extending the Linear Model with R*, Boca Raton, FL: Chapman & Hall/CRC Press.

Fisher, R. A. (1941). The negative binomial distribution, *Annals of Eugenics*, 11, 182–187.

Frees, E. (2004). *Longitudinal and Panel Data*, Cambridge: Cambridge University Press.

Fridstrøm, L., J. Ifver, S. Ingebrigsten, R. Kulmala and L. K. Thomsen (1995). Measuring the contribution of randomness, exposure, weather, and daylight to the variation in the road accident counts, *Accident Analysis and Prevention* **27**(1): 1–20.

Gelman, A., J. B. Carlin, H. S. Stern and D. B. Rubin (2004). *Bayesian Data Analysis*, second edition, Boca Raton, FL: Chapman & Hall/CRC.

Gelman, A. and J. Hill (2007). *Data Analysis Using Regression and Multilevel/Hierarchical Models,* Cambridge: Cambridge University Press.

Gerdtham, U. G. (1997). Equity in health care utilization: Further tests based on hurdle models and Swedish micro data, *Health Economics* **6**: 303–319.

Geweke. J.F. (2005). *Contemporary Bayesian Econometrics and Statistics*, New York: Wiley.

Gill, J. (2002). *Bayesian Methods*, Boca Raton, FL: Chapman & Hall/CRC.

Gill, J. (2010). Critical differences in Bayesian and non-Bayesian inference and why the former is better, in *Statistics in the Social Sciences*, ed. S. Kolenikov, D. Steinley and L. Thombs, New York: Wiley.

Goldberger, A. S. (1983). Abnormal selection bias, in *Studies in Econometrics, Time Series, and Multivariate Statistics*, ed. S. Karlin, T. Amemiya and L. A. Goodman, New York: Academic Press, pp. 67–85.

Gould, W., J. Pitblado and W. Scribney (2006). *Maximum Likelihood Estimation with Stata*, third edition, College Station, TX: Stata Press.

Gourieroux, C., A. Monfort and A. Trognon (1984). Pseudo maximum likelihood methods: theory, *Econometrica* **52**: 681–700.

Greene, W. H. (1992). Statistical Models for Credit Scoring, Working Paper, Department of Economics, Stern School of Business, New York University.

Greene, W. H. (1994). Accounting for excess zeros and sample selection in Poisson and negative binomial regression models, **EC-94–10**, Department of Economics, Stern School of Business, New York University.

Greene, W. H. (2006a). *LIMDEP Econometric Modeling Guide*, Version 9, Plainview, NY: Econometric Software Inc.

Greene, W. H. (2006b). A general approach to incorporating 'selectivity' in a model, Working Paper, Department of Economics, Stern School of Business, New York University.

Greene, W. H. (2007). *Econometric Analysis*, fifth edition, New York: Macmillan.

Greenwood, M. and H. M. Woods (1919). *On the Incidence of the Industrial Accidents upon Individuals with special Reference to Multiple Accidents.* Report of the Industrial Fatigue Research Board. 4, 1–28. London: His Majesty's Stationery Office.

Greenwood, M. and G. U. Yule (1920). An inquiry into the nature of frequency distributions of multiple happenings, with particular reference to the occurrence of multiple attacks of disease or repeated accidents, *Journal of the Royal Statistical Society A*, **83**: 255–279.

Gurmu, S. and P. K. Trivedi (1992). Overdispersion tests for truncated Poisson regression models, *Journal of Econometrics* **54**: 347–370.

Hannan, E. J. and B. G. Quinn (1979) The determination of the order of an autoregression, *Journal of the Royal Statistical Society*, B **41**: 190–195.

Hardin, J. W. (2003). The sandwich estimate of variance, in *Maximum Likelihood of Misspecified Models: Twenty Years Later*, ed. T. Fomby and C. Hill, Advances in Econometrics volume 17, Oxford: Elsevier, pp. 45–73.

Hardin, J. W. and J. M. Hilbe (2001). *Generalized Linear Models & Extensions*, College Station, TX: Stata Press.

Hardin, J. W. and J. M. Hilbe (2002). *Generalized Estimating Equations*, Boca Raton, FL: Chapman & Hall/CRC Press.

Hardin, J. W. and J. M. Hilbe (2007). *Generalized Linear Models & Extensions, second edition*, College Station, TX: Stata Press. Third edition (April, 2012).

Hausman, J., B. Hall and Z. Griliches (1984). Econometric models for count data with an application to the patents – R&D Relationship, *Econometrica* **52**: 909–938.

Heckman, J. (1979). Sample selection bias as a specification error, *Econometrica* **47**: 153–161.

Heilbron, D. (1989). Generalized linear models for altered zero probabilities and overdispersion in count data, Technical Report, Department of Epidemiology and Biostatistics, University of California, San Francisco.

Hilbe, J. M. (1993a). Log negative binomial regression as a generalized linear model, Technical Report COS **93/94–5-26**, Department of Sociology, Arizona State University.

Hilbe, J. M. (1993b). Generalized linear models, *Stata Technical Bulletin*, STB-11, sg16.

Hilbe, J. M. (1993c). Generalized linear models using power links, *Stata Technical Bulletin*, STB-12, sg16.1.

Hilbe, J. M. (1994a). Negative binomial regression, *Stata Technical Bulletin*, STB-18, sg16.5.

Hilbe, J. M. (1994b). Generalized linear models, *The American Statistician*, **48**(3): 255–265.

Hilbe, J. M. (2000). Two-parameter log-gamma and log-invese Gaussian models, in *Stata Technical Bulletin Reprints*, College Station,TX: Stata Press, pp. 118–121.

Hilbe, J. M. (2005a). CPOISSON: Stata module to estimate censored Poisson regression, Boston College of Economics, Statistical Software Components, http://ideas.repec.org/c/boc/bocode/s456411.html

Hilbe, J. M. (2005b), CENSORNB: Stata module to estimate censored negative binomial regression as survival model, Boston College of Economics, Statistical Software Components, http://ideas.repec.org/c/boc/bocode/s456508.html

Hilbe, J. M. (2005c), Censored negative binomial regression, EconPapers, RePec, Research Papers in Economics, Boston School of Economics, Nov 30, 2005. http://fmwww.bc.edu/repec/bocode/c/censornb.ado

Hilbe, J. M. (2009). *Logistic Regression Models*, Boca Raton, FL: Chapman & Hall/CRC.

Hilbe, J. M. (2010a), Modeling count data, in *International Encyclopedia of Statistical Science*, ed. M. Lovric, New York: Springer.

Hilbe, J. M. (2010b), Generalized linear models, in *International Encyclopedia of Statistical Science*, ed. M. Lovric, New York: Springer.

Hilbe, J. M. (2010c), Creating synthetic discrete-response regression models, *Stata Journal* **10**(1): 104–124.

Hilbe, J. M. (2011). *Negative Binomial Regression Extensions*. Cambridge University Press website for the text: www.cambridge.org/9780521857727.

Hilbe, J. and B. Turlach (1995). Generalized linear models, in *XploRe: An Interactive Statistical Computing Environment*, ed. W. Hardle, S. Klinke and B.Tulach, New York: Springer-Verlag, pp. 195–222.

Hilbe, J. and D. Judson (1998). Right, left, and uncensored Poisson regression, in *Stata Technical Bulletin Reprints*, College Station, TX: Stata Press, pp. 186–189.

Hilbe, J. and W. Greene (2007). Count response regression models, in *Epidemiology and Medical Statistics*, ed. C. R. Rao, J. P. Miller and D. C. Rao, Elsevier Handbook of Statistics Series, London: Elsevier.

Hilbe, J.M and A.P. Robinson (2011), *Methods of Statistical Model Estimation*, Boca Raton, FL: Chapman & Hall/CRC.

Hin, L.-Y. and Y.-G. Wang (2008). Working-correlation-structure identification in generalized estimating equations, *Statistics in Medicine* **28**(4): 642–658.

Hinde, J. and C. G. B Demétrio (1998). Overdispersion: models and estimation, *Computational Statistics & Data Analysis* **27**(2): 151–170

Hoffman, J. (2004). *Generalized Linear Models*, Boston, MA: Allyn and Bacon.

Hosmer, D. and S. Lemeshow (2000), *Applied Logistic Regression*, second edition, New York: Wiley.

Huber, P. J. (1967). The behavior of maximum likelihood estimates under nonstandard conditions, in *Proceedings of the Fifth Berkeley Symposium on Mathematical Statistics and Probability*, Berkeley, CA: University of California Press, pp. 221–233.

Hurvich, C. M. and C. L. Tsai (1989), Regression and time series model selection in small samples, *Biometrika*, **76**: 297–307.

Irwin, J. O. (1968). The generalized Waring distribution applied to accident theory, *Journal of the Royal Statistical Society*, A **131**(2): 205–225.

Iwasaki, M. and H. Tsubaki (2005a). A bivariate generalized linear model with an application to meteorological data analysis, *Statistical Methodology* **2**: 175–190.

Iwasaki, M. and H. Tsubaki (2005b). A new bivariate distribution in natural exponential family, *Metrika* **61**: 323–336

Iwasaki, M. and H. Tsubaki (2006). Bivariate negative binomial generalized linear models for environmental count data, *Journal of Applied Statistics* **33**(9): 909–923.

Jain, S. K. (1959). Fitting the negative binomial distribution to some data on asynaptic behaviour of chromosomes, *Genetica* **30**: 108–122.

Jain, G. C. and P. C. Consul (1971). A generalized negative binomial distibution, *SIAM Journal of Applied Mathematics*, **21**(4): 501–513.

Johnson, N. L., A. W. Kemp and S. Kotz (2005). *Univariate Discrete Distributions*, third edition, New York: Wiley.

Jones, A. M., N. Rice, T.B. d'Uva and S. Balia (2007). *Applied Health Economics*, New York & Oxford: Routledge, Taylor and Francis.

Jones, O., R. Maillardet and A. Robinson (2009). *Scientific Programming and Simulation Using R*, Boca Raton, FL: Chapman & Hall/CRC

Karim, M. R. and S. Zeger (1989). A SAS macro for longitudinal data analysis, Department of Biostatistics, the Johns Hopkins University: Technical Report **674**.

Katz, E. (2001). Bias in conditional and unconditional fixed effects logit estimation, *Political Analysis* **9**(4): 379–384.

King, G. (1988). Statistical models for political science event counts: Bias in conventional procedures and evidence for the exponential Poisson regression model. *American Journal of Political Science* **32**: 838–863.

King, G. (1989). Event count models for international relations: generalizations and applications, *International Studies Quarterly* **33**: 123–147.

Lambert, D. (1992). Zero-inflated Poisson regression with an application to defects in manufacturing, *Technometrics*, **34**: 1–14.

Lancaster, T. (2002). Orthogonal parameter and panel data, *Review of Economic Studies* **69**: 647–666.

Lawless, J. F. (1987). Negative binomial and mixed Poisson regression, *Canadian Journal of Statistics* **15**(3): 209–225.

Lee, Y., J. Nelder and Y. Pawitan (2006). *Generalized Linear Models with Random Effects*, Boca Raton, FL: Chapman & Hall/CRC Press.

Leisch, F. and B. Gruen (2010). *Flexmix: Flexible Mixture Modeling*, CRAN

Liang, K.-Y. and S. Zeger (1986). Longitudinal data analysis using generalized linear models, *Biometrika* **73**: 13–22.

Long, J. S. (1997). *Regression Models for Categorical and Limited Dependent Variables*, Thousand Oaks, CA: Sage.

Long, J. S. and J. Freese (2006). *Regression Models for Categorical Dependent Variables using Stata*, second edition, College Station, TX: Stata Press.

Loomis, J. B. (2003). Travel cost demand model based river recreation benefit estimates with on-site and household surveys: Comparative results and a correction procedure, *Water Resources Research* **39**(4): 1105.

Machado, J. A. F. and J. M. C. Santos Silva (2005), Quantiles for counts, *Journal of the American Statistical Association* **100**: 1226–1237.

Marquardt, D. W. (1963). An algorithm for least-squares estimation of non-linear parameters, *Journal of the Society for Industrial and Applied Mathematics* **11**: 431–441.

Martinez-Espiñeira, R., J. Amaoko-Tuffor, and J. M. Hilbe (2006). Travel cost demand model-based river recreation benefit estimates with on-site and household surveys: comparative results and a correction procedure – revaluation, *Water Resource Research* **42**.

McCullagh, P. (1983). Quasi-likelihood functions, *Annals of Statistics* **11**: 59–67.

McCullagh, P. and J. A. Nelder (1989). *Generalized Linear Models*, second edition, New York: Chapman & Hall.

Melkersson, M. and D. Roth (2000). Modeling of household fertility using inflated count data models, *Journal of Population Economics* **13**: 189–204.

Miaou, S.-P. (1996). Measuring the goodness-of-fit of accident prediction models, FHWA-RD-96–040, Federal Highway Administration, Washington, DC.

Muenchen, R. A. and J. M. Hilbe (2010). *R for Stata Users*, New York: Springer.

Mullahy, J. (1986). Specification and testing of some modified count data models, *Journal of Econometrics* **33**: 341–365.

Mundlak, Y. (1978). On the pooling of time series and cross section data, *Econometrica* **46**: 69–85.

Murphy, K. and R. Topel (1985). Estimation and inference in two step econometric models, *Journal of Business and Economic Statistics* **3**:370–379.

Mwalili, S., E. Lesaffre and D. DeClerk (2005). The zero-inflated negative binomial regresson model with correction for misclassification: an example in Caries Research, Technical Report 0462, LAP Statistics Network Interuniversity Attraction Pole. Catholic University of Louvain la Neuve, Belgium. www.stat.ucl.ac.be/IAP.

Nelder, J. A. (1994). Generalized linear models with negative binomial or beta-binomial errors, unpublished manuscript.

Nelder, J. A. and R. W. M. Wedderburn (1972). Generalized linear models, *Journal of the Royal Statistical Society*, A **135**(3): 370–384.

Nelder, J. A. and D. Pregibon (1987). An extended quasi-likelihood function, *Biometrika* **74**: 221–232.

Nelder, J. A. and Y. Lee (1992). Likelihood, quasi-likelihood, and pseudo-likelihood: some comparisons, *Journal of the Royal Statistical Society*, B **54**: 273–284.

Nelson, D. L. (1975). Some remarks on generalized of negative binomial and Poisson distributions, *Technometrics* **17**: 135–136.

Newbold, E. M. (1927). Practical applications of the statistics of repeated events, particularly to industrial accidents. *Journal of the Royal Statistical Society* **90**: 487–547.

Neyman, J. and E. L. Scott (1948). Consistent estimation from partially consistent observations, *Econometrica*, **16**(1): 1–32

Nylund, K. L, T. Asparouhov, and B. O. Muthén (2007). Deciding on the number of classes in latent class analysis and growth mixture modeling: A Monte Carlo simulation study, *Structural Equation Modeling* **14**(4): 535–569.

Pan, W. (2001a). Akaike's information criterion in generalized estimating equations, *Biometrics* **57**: 120–125.

Pan, W. (2001b). On the robust variance estimator in generalized estimating equations, *Biometrika* **88**(3): 901–906.

Piegorsch, W. W. (1990) Maximum likelihood estimation for the negative binomial dispersion parameter, *Biometrics* **46**: 863–867.

Pierce, D. A. and D. W. Schafer (1986). Residuals in generalized linear models, *Journal of the American Statistical Association* **81**: 977–986.

Rabe-Hesketh, S. and A. Skrondal (2004). *Generalized Latent Variable Modeling*, Boca Raton, FL: Chapman & Hall/CRC Press.

Rabe-Hesketh, S. and A. Skrondal (2005). *Multilevel and Longitudinal Modeling Using Stata*, College Station, TX: Stata Press.

Rigby, R. and M. Stasinopoulos (2008). A fexible regression approach using GAMLSS in R, Handout for a short course in GAMLSS given at International Workshop of Statistical Modelling, University of Utrecht.

Rodríguez-Avi, J., A. Conde-Sánchez, A. J. Sáez-Castillo, M. J. Olmo-Jiménez, and A. M. Martínez-Rodríguez (2009), A generalized Waring regression model for count data, *Computational Statistics & Data Analysis* **53**(10): 3717–3725.

Rouse, D. M. (2005). Estimation of finite mixture models, Masters thesis, North Carolina State University.

SAS/STAT 9.22 *User's Guide* (2010), Cary, NC: SAS Institute

Schwarz, G. E. (1978), Estimating the dimension of a model, *Annals of Statistics* **6**(2): 461–464.

Shaw, D. (1988). On-site samples' regression, *Journal of Econometrics* **37**: 211–223.

Shults, J., W. Sun, X. Tu, H. Kim, J. Amsterdam, J. M. Hilbe and T. Ten-Have (2009). Comparison of several approaches for choosing between working correlation structures in generalized estimating equation analysis of longitudinal binary data, *Statistics in Medicine* **28**: 2338–2355.

Simon, L. J. (1960). The negative binomial and Poisson distributions compared, *Proceedings of the Casuality and Actuarial Society* **47**: 20–24.

Simon, L. (1961). Fitting negative binomial distributions by the method of maximum likelihood, *Proceedings of the Casuality and Actuarial Society* **48**: 45–53.

Stata Reference Manual, version 11 (2009), College Station, TX: Stata Press.

Student (1907). On the error of counting with a haemacytometer, *Biometrika* **5**: 351–360.

Terza, J. V. (1998). A tobit-type estimator for the censored Poisson regression model, *Econometric Letters* **18**: 361–365.

Thall, P. and S. Vail (1990). Some covariance models for longitudinal count data and overdispersion, *Biometrika* **46**: 657–671.

Twist, J. W. R. (2003). *Applied Longitudinal Data Analysis for Epidemiology*, Cambridge: Cambridge University Press.

Vadeby, A. (2002). Estimation in a model with incidental parameters, LiTH-MAT-R-2002–02 working paper.

Venables, W. and B. Ripley (2002). *Modern Applied Statistics with S*, fourth edition, New York: Springer-Verlag.

Vogt, A. and J. G. Bared (1998). Accident models for two-lane rural roads: segments and intersections, publication No. FHWA-RD-98–133, Federal Highway Administration, Washington, DC, http://www.tfhrc.gov/safety/98133/ch05/ch05_01.html

Vuong, Q. H. (1989). Likelihood ratio tests for model selection and non-nested hypotheses, *Econometrica* **57**: 307–333.

Wedderburn, R. W. M. (1974). Quasi-likelihood functions, generalized linear models and the Gauss–Newton method, *Biometrika* **61**: 439–47.

White, H. (1980). A heteroskestasticity-consistent covariance matrix estimator and a direct test for heteroskedasticity, *Econometrica* **48**(4): 817–838.

Williamson, E. and M. H. Bretherton (1963). *Tables of the Negative Binomial Distribution*, New York: Wiley.

Winkelmann, R. (1995). Duration dependence and dispersion in count-data models, *Journal of Business and Economic Statistics* **13**: 467–474.

Winkelmann, R. (2008). *Econometric Analysis of Count Data*, Fifth Edition, New York: Springer.

Winklemann, R. and K. F. Zimmermann (1995). Recent developments in count data modelling, theory and application, *Journal of Economic Surveys* **9**: 1–24.

Woutersen, T. (2002). Robustness against incidental parameters, University of Western Ontario, Department of Economics Working papers, 20028.

Xekalaki, E. (1983). The univariate generalized Waring distribution in relation to accident theory: Proneness, spells or contagion?, *Biometrics* **39**(3): 39–47.

Xekalaki, E. (1984). The bivariate generalized Waring distribution and its application to accident theory, *Journal of the Royal Statistical Society* A **147**(3): 488–498.

Yang, Z., J. Hardin and C. Abby (2006). A score test for overdispersion based on generalized poisson model, Unpublished manuscript.

Yule, G. U. (1910). On the distribution of deaths with age when the causes of death act cumulatively, and similar frequency distributions, *Journal of the Royal Statistical Society* **73**: 26–35.

Zeger, S. L., K.-Y. Liang, and P. S. Albert (1988). Models for longitudinal data: A generalized estimating equation approach, *Biometrics* **44**: 1049–1060.

Zuur, A. F., E. N. Ieno, N. J. Walker, A. A. Saveliev, and G. M. Smith (2009). *Mixed Effects Models and Extensions in Ecology with R*, New York: Springer.

Zuur, A. F., A. A. Saveliev, and E. N. Ieno (2012). *Zero-Inflated Models and Generalized Linear Mixed Models*, To be published.

Index

Printed in the United States
By Bookmasters